PORTLAND COMMUNITY COLLEGE
LEARNING RESOURCE CENTER
WITHDRAWN

D0222334

DICTIONARY OF CERAMICS

DICTIONARY OF CERAMICS

Arthur Dodd

THIRD EDITION

Revised and updated by
David Murfin

THE INSTITUTE OF MATERIALS

Book 588
This edition first published in 1994 by
The Institute of Materials
1 Carlton House Terrace
London SW1Y 5DB

© The Institute of Materials 1994
All rights reserved

British Library Cataloguing-in-Publication Data

Dodd, Arthur Edward
Dictionary of Ceramics. – 3Rev.ed
I. Title II. Murfin, David
666.03

ISBN 0-901716-56-1

Typeset by
Dorwyn Ltd, Rowlands Castle
Printed and bound at
The University Press, Cambridge

Dedication

The first and second editions carried no dedications, but A.E. Dodd might like to be remembered for other reasons than this dictionary. Some lines from his *Weaver Hills and Other Poems* seem appropriate:

Rock, crag, Earth's skull
Known by lens and X-ray beam;
Soil, silt, Earth's skin,
Analysed to micron grain;
Grass, tree, Earth's hair,
Numbered, classified and named;
Atom tagged, star docketed,
Computed by electron brain.

These myriad facts all known
And yet
Another's mind unknown
Except
By Mind, by Mind alone.

Words are as leaves,
They shrivel, trees stand bare.
Words are as leaves,
Leaf-mould, seeds root there.
We once more tread the pack-horse way
And smile to think how small the lasting worth
May be of so much toil.

A.E. Dodd, 1967

For my own part I might hope to fulfil:

The Preacher sought to find out acceptable words, and that which was written was upright, even words of truth.

Ecclesiastes 12:10

and be allowed to put my dedication to the third edition in the form of a silica tetrahedron:

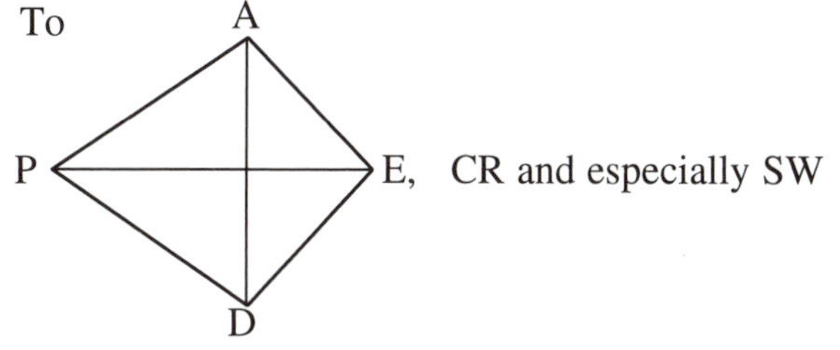

Preface to the Third Edition

When I joined BCRA (as it was then known) to work for Dr A.E. Dodd, the second edition of this dictionary was well in hand. As a newly-qualified physicist, I contributed a few specialised entries. Nearly 30 years later, many more are required. The dictionary has been considerably expanded to take account of new developments in engineering ceramics, electroceramics; of new processes in whitewares; of new machinery and new test methods; of the impact of environmental legislation. The same broad scope has been adopted and coverage of glass, vitreous enamel and cement industries widened. It has seemed wise, in an industry which has moved much further to being science-based rather than craft-based, to include some terminology in relevant areas of basic science: in particular crystal structure, fracture mechanics and sintering. Here, and for some advanced ceramics produced in a wide variety of forms, some longer entries have been introduced to avoid extensive use of cross-references and to provide a coherent understanding of closely-related materials and processes.

Abbreviations have become something of a disease. Those in general use have been listed in their appropriate alphabetical positions, but where authors substitute (unrecognized and sometimes unrecognizable) initials for well-known processes as a shorthand, the obscurity resulting from carrying them over unexplained to another context cannot be clarified even by a multitude of dictionary entries. Such have been omitted.

About 2000 new terms have been added. Though some existing terms have been modified to take account of changed emphases (roller hearth kilns are not the rarities of 30 years ago), very few have been removed. This is a twofold tribute to Dodd's original work. His definitions, concentrating on the essence of the meaning, have stood the test of time, even when the words now have much broader application in details. Where words have become obsolete, it has been thought worthwhile to preserve his accurate and authoritative definitions for historical record.

My thanks are due to The Institute of Materials for giving me the opportunity to revive Dodd's Dictionary; to Mrs Dodd for permission to use the original copyright material; to Dr R.C.P. Cubbon, then Deputy Chief Executive of British Ceramic Research Ltd, for allowing me to use the Company's library – without which this revision would have been impossible; to the staff of the Information Department there, past and present, who have all shared in gathering the information on which this book is based; and finally, to all my other ex-colleagues and friends in the industry, whose continued penchant for asking awkward questions has been the driving force and reason for that collection.

D. Murfin
Madeley, 1994

7.6 Temperature. See SOFTENING POINT.
12.0 Temperature. See ANNEALING POINT.
'A' Glass. A glass composition first used for making glass fibre, it was a container-glass type containing 10–15% alkali, and was soon superseded by special compositions.
Abbe Number or Abbe Value. A measure of the optical dispersion of a glass proposed by the German physicist E. Abbe (1840–1905). Its usual symbol is ν and is defined by $\nu = (n_D - l)\ (n_F - n_C)^{-1}$ where n_C n_D and n_F are the refractive indices at wavelengths equivalent to the spectral lines *C*, *D* and *F*. Also known as CONSTRINGENCE and as the ν-VALUE.
Aberson Machine. See SOFT MUD PROCESS.
Abbreviations. Abbreviations have become a disease, in that authors substitute (unrecognized and sometimes unrecognizable) initial letters for well-known general processes whose names are more than a few letters long, e.g. LPS for liquid-phase sintering, CTE for coefficient of thermal expansion. What is explained and used as a shorthand in one technical paper gets carried over to others, and becomes a source of obscurity. This dictionary lists abbreviations in general use in their appropriate alphabetical positions.
Ablation. The removal of a surface skin; the term has been applied to the volatilization or burning away of a special protective coating, of ceramic or other material, on space vehicles. The heat absorbed during ablation effectively dissipates the heat generated by re-entry of the space vehicle into the Earth's atmosphere.
Abradability Index. See ABRASION RESISTANCE and BS 1902 Pt 4.6
Abrams' Law. States that the strength, *S*, of a fully compacted concrete is related to the ratio, *R*, of water to cement by the equation $S = A/B^R$ where *A* and *B* are constants. (D. A. Abrams, *Struct. Mat. Res. Lab. Lewis Inst.*, *Bull.* 1, Chicago, 1918.)
Abrasion. Wear caused by the mechanical action of a solid, e.g. wear of the stack lining of a blast furnace by the descending burden (cf. CORROSION and EROSION).
Abrasion Resistance. The ability of a material to resist mechanical rubbing. Some ceramic products are specially made to have this property, for example blue engineering bricks for the paving of coke chutes. There are several types of apparatus for the evaluation of this property; see BENDER ABRASION METHOD, KESSLER ABRASION TESTER, MIDLAND IMPACT TEST, MORGAN-MARSHALL TEST, RATTLER TEST, SCC TEST, TABER ABRASER. BS 1902 Pts 4.6 and 4.7 specify its determination for refractories at ambient and elevated temperatures, the first using air-blown alumina abrasive, the second alumina/zirconia abrasive fed by gravity. BS 6431 Pt 20 specifies a wet test (PEI TEST) and a dry test (MCC TEST) for the surface abrasion of glazed tiles. Pt 14 specifies a deep abrasion test for unglazed tiles, in which alumina abrasive, grain size 80, is fed under a vertical rotating wheel to cut a groove in the tile, whose depth is measured after a standard number of revolutions of the wheel. BS 1344 Pt 4 is an obsolescent test for vitreous enamels. ASTM C.704 specifies a test for refractory bricks and castables, using an air blast with 1kg of graded SiC abrasive.
Abrasive. A hard material used in the form of a wheel or disk for cutting, or as a powder for polishing. The usual abrasives are silicon carbide and fused alumina. Abrasive wheels are commonly

made by bonding abrasive grains with clay and feldspar (a proportion of FRIT (q.v.) is also often added) and firing the shaped wheel; organic bonds e.g. synthetic resins, rubber or shellac, are also used, in which case the wheels are not fired. For the standard marking system see under GRINDING WHEEL.

Abrasive Jet Machining. Abrasive particles are entrained in a high-velocity fluid jet which acts as the cutting tool. If the fluid is a gas, up to 50g/min of abrasive powder is carried through the jet nozzle at supersonic speed by pressures of 200 to 850 kPa. In WATERJET CUTTING, discharge pressures up to 400MPa provide jet velocites up to 700 m/s, carrying 1.5 to 2.0 kg/min of abrasive in a jet which can be confined to a smaller diameter without spreading.

Absolute Temperature (°Abs). See KELVIN TEMPERATURE SCALE.

Absorption. See POROSITY; WATER ABSORPTION.

Absorption Ratio. See SATURATION COEFFICIENT.

Accelerated Tests. Test methods which attempt to determine the effect of slow changes on materials by adopting more severe or more rapid changes, and correlating the test result with actual behaviour in service. Examples are tests for weathering, wear, moisture expansion. ASTM-E632 gives criteria for designing such accelerated tests.

Accelerator. A material that, when added to hydraulic cement or to plaster, accelerates the setting. With cement, care must be taken to distinguish between the effect on the initial set and the effect on the longer-term hydration – some additions accelerate setting but inhibit the final development of strength; also, a compound that acts as an accelerator at one level of concentration may become a retarder at some other concentration. In general, alkali carbonates are strong accelerators for portland cement. For plaster, the best accelerator is finely-ground gypsum, which provides nuclei for crystallization; potash alum has also been used.

Acceptance Angle of an optical fibre, see NUMERICAL APERTURE.

Acceptance Testing. With increased emphasis on quality control and quality assurance, raw materials or intermediates are subjected to a series of tests of relevant properties to ensure that they match the producers' specification. The *acceptance quality level*, which is the percentage which is acceptable to the buyer, is usually specified, and compliance checked by testing and statistical analysis of the results. BS 3921 lays down a detailed sampling scheme for building bricks. See also ACTION LIMIT, QUALITY CONTROL; ATTRIBUTES.

Accessory Mineral. A mineral that is present in a rock or clay in relatively small quantity, usually as an impurity; e.g. mica is usually found as an accessory mineral in china clay.

Acheson Furnace. An enclosed electric resistance furnace, of the type first used by Acheson in 1891 to make silicon carbides from carbon and sand, which can attain temperatures of 2500 °C.

Acicular. see MORPHOLOGY.

Acid Annealing. Dipping of the metal base for vitreous enamelling into dilute acid (generally HCl) followed by annealing; this promotes scaling before the pickling process.

Acid Embossing. The etching of glass with HF or a fluoride.

Acid Frosting. The etching of glass hollow-ware with HF or a fluoride.

Acid Gold. A form of gold decoration for pottery introduced in 1863 by J. L. Hughes at Mintons Ltd., Stoke-on-Trent.

The glazed surface is etched with dilute HF prior to application of the gold; the process demands great skill and is used for the decoration only of ware of the highest class. A somewhat similar effect can be obtained by applying a pattern in low-melting flux on the glaze and goldbanding on the fluxed area; this is known as MOCK ACID-GOLD, c.f. BRIGHT GOLD and BURNISH GOLD.

Acid Open-hearth Furnace. An OPEN-HEARTH FURNACE (q.v.) used in the refining of hematite iron; little such iron is now made. The particular feature is that the hearth is made of acid refractories – silica bricks covered with a fritted layer of silica sand.

Acid Refractory Material. General term for those types of refractory material that contain a high proportion of silica, e.g. silica refractories (> 92% SiO_2) and siliceous refractories (78–93% SiO_2). The name derives from the fact that silica behaves chemically as an acid and at high temperatures reacts with bases such as lime or alkalis.

Acid Resistance. In USA, the acid resistance of ceramic decorations on glass is specified in ASTM C724 and C735. The acid resistance of vitreous enamelled ware at (nominal) room temperature is determined by exposing the enamelled surface to 10% citric acid for 15 min at 26 ± 1°C. (ASTM C282) Five classes of enamelware are distinguished according to their subsequent appearance:

- AA: no visible stain and passes dry-rubbing test
- A: passes blurring-highlight test and wet-rubbing test
- B: passes blurring-highlight test: fails wet-rubbing test
- C: fails blurring-highlight test; passes disappearing-highlight test
- D: fails disappearing-highlight test.

A similar classification has been adopted in B.S. 1344 (Part 2) but the time of exposure to the 10% citric acid is 20 min and the temperature is 20 °C. BS 1344 Pt 3 tests resistance to H_2SO_4; other parts test resistance to hot acids, alkalis and detergents.

Acid-resisting Brick or Tile. A fired clay brick or tile of low porosity and high resistance to a variety of chemicals. B.S. 1902 Pt 3.12 specifies an acid resistance test for refractories, Pt 3.15 a test for bulk material. ASTM C-279 specifies three grades according to resistance to H_2SO_4. ASTM C-410 specifies industrial floor bricks.

Acid-resisting Cement. The principal types are as follows: 1. *Silicate*: an inert filler bonded with silica gel that has been precipitated *in situ* from Na- or K- silicate in the presence of Na_2SiF_6, or from silicon ester. 2. *Rubber Latex*: essentially cement sand mixes impregnated with rubber. 3. *Synthetic Resin*: with an inert filler. 4. *Sulphur Cements*: usually with sand as filler. 5. *Bituminous Cements*. ASTM specifications for these materials are listed in Vol 4-05 of the Annual Book of ASTM Standards, and number some 25, including specifications for mortars and grouts.

Acid Scaling. Dipping or spraying metal with acid, then annealing at red heat, to remove oil, rust and dirt before vitreous enamelling. cf. ACID ANNEALING.

Ackerman Blocks. Hollow extruded clay blocks, with thin walls and large voids, for ceiling and floor construction, available in sizes from 150 to 220 mm. (Polish Stand. PN B-12005, 1969).

ACL Kiln or Lepol Kiln. ACL is a trade-mark of the Allis Chalmers Mfg Co., USA: Lepol is a trade-mark of Polysius Co., Germany. Both terms refer to a travelling-grate preheater for portland cement batch prior to its being fed to a rotary cement kiln; with this attachment, the length of a rotary cement kiln can be halved.

ACOP. Approved Code of Practice (in HEALTH and SAFETY, q.v.).

Acoustic Emission. Internal structural changes in materials (crack growth, phase transitions, plastic deformation) may release strain energy manifested as ultrasonic waves, whose detection and analysis gives an insight into the type of changes taking place. In particular, crack growth may be continuously monitored and warning of impending failure obtained.

Acoustophoresis. Dense suspensions are subjected to ultrasonic vibrations. (~ 1 MHz) The amplitude of the resulting particle vibration is considered to be proportional to the ZETA POTENTIAL.

ACS. American Ceramic Society.

Actinic Green. An emerald green glass of the type used for poison bottles.

Action Limit. In statistical QUALITY CONTROL (q.v), the number of defective items per sample which will only be exceeded by chance in a given number of tests, in practice usually taken to be 1 in 1000. The *warning limit* is usually taken to be 1 test in 40 revealing a defective item.

Activated Alumina. A synthetic high-surface area alumina, produced by appropriate heat-treatment, and usually a transition alumina containing a small proportion of OH^- ions.

Activated Carbon. According to ASTM D-2652, activated carbons are a family of carbonaceous substances manufactured by processes that develop absorptive properties. This standard defines over 100 terms relating to the use of granular and porous carbons as absorbant materials, and the ASTM Annual Book of Standards Vol 15–01 contains some 20 test methods for relevant properties.

Activated Clay. Acid treatment of bentonite and other clays to improve bleaching and absorption properties.

Activated Sintering. Sintering in the presence of a sintering aid – a material such as a foreign oxide which produces supersaturated surface phases providing higher surface energy.

Activation Energy. A threshold energy value which must be acquired by a system before some types of reaction or change can occur. See ARRHENIUS' LAW.

Active Brazing. Joining ceramics with metal brazes which wet well the surface of the ceramic, obviating the need for prior metallising.

Active Filler-Controlled Pyrolysis. (AFCOP) In this technique for making multiphase ceramic matrix composites, an active filler material (a transition metal or compound thereof which will yield a carbide or other ceramic) is mixed with an organometallic polymer and pyrolysed. The kind, content and structure of the filler control the kinetics of the polymer pyrolysis and the resulting microstructure, which may contain disordered or glassy areas. (M. Seibold and P. Greil, *1st European Conf on Adv. Mater & Processes.*)

ACZS. Alumina-Chrome-Zirconia-Silica refractory, used in glass tanks.

A-D Goodness-of-Fit Test. See K-S TEST.

Adams Chromatic Value System. A method for the quantitative designation of colour in terms of (1) lightness, (2) amount of red or green, and (3) amount of yellow or blue. The system has been used in the examination of ceramic colours (E. Q. Adams, *J. Opt. Soc. Am.,* **32**, 168, 1942).

Adams Process. A method for the removal of iron compounds from glass-making sands by washing with a warm solution of acid (Na oxalate containing a small quantity of $FeSO_4$; see *J. Soc. Glass Tech.,* **19**, 118, 1935).

Adams-Williamson Annealing Schedule. A procedure, derived from first principles, for determining the optimum annealing conditions for a particular glass; (L. H. Adams and E. D. Williamson, *J. Franklin Inst.*, **190**, 597, 835, 1920).

Adapter. A connecting-piece, usually made of fireclay, between a horizontal zinc-retort and the condenser in which the molten zinc collects.

Adherence. The strength of the bond between two materials, especially that between a coating and the substrate. ASTM C313 specifies a test for the 'Adherence of Porcelain Enamel and Ceramic Coatings to Metal.' *Adherence failure* is measured as the ratio of bright metal to remaining adherent enamel after indentation by a specified plunger. ASTM C633 specifies a text for the 'Adhesion or Cohesive Strength of Flame-sprayed Coatings'.

Adiabatic Diesel Engines. Ceramic components have been made to withstand the high operating temperatures in uncooled (waterless) diesel engines designed for minimum heat loss, usually incorporating exhaust turbochargers. (The term 'adiabatic' is not used with its strict meaning of no heat gained or lost during a thermodynamic process.)

Adobe. Mud used in tropical countries for making sun-dried (unfired) bricks; the term is also applied to the bricks themselves.

Adsorption. Process of taking up a molecular or multi-molecular film of gas or vapour on a solid surface. This film is not readily removed except by strongly heating the body. The adsorption of moisture on the internal surface of a porous body may cause the body to expand slightly; this is known as MOISTURE EXPANSION (see also GAS ADSORPTION METHOD). Langmuir's adsorption isotherm, with the equation $f = ap/(1 - ap)$, is based on the assumption that adsorption is limited to a monomolecular layer (f is the fraction of the surface covered; p is the gas pressure; a is a constant).

Advanced Ceramics. A general term for ceramics, usually of high purity or carefully controlled composition and microstructure, used in technical applications where their mechanical, thermal, electrical and/or optical properties are important. cf. SPECIAL CERAMICS, FINE CERAMICS, TECHNICAL CERAMICS, ENGINEERING CERAMICS, ELECTROCERAMICS, OPTICAL APPLICATIONS.

Aebi Kiln. A car tunnel kiln built to the design of Robert Aebi, Zurich. One such design is top fired by the CRYPTO SYSTEM (q.v.).

Aerated Concrete. Concrete with a high proportion of air spaces resulting from a foaming process; the bulk density may vary from about 560–1440 kg/m^3 (35–90 lb/ft^3). Aerated concrete is chiefly used for making pre-cast building units. It is also known as GAS CONCRETE, CELLULAR CONCRETE or FOAMED CONCRETE.

Aeration of Cement. The effect of the atmosphere on portland cement during storage. Dry air has no effect, but if it is exposed to moist air both moisture and carbon dioxide are absorbed with erratic effects on the setting behaviour (cf. AIR ENTRAINING).

Aeroclay. Clay, particularly china clay, that has been dried and air-separated to remove any coarse particles.

Aerogel. A highly porous solid formed by replacing the liquid in a GEL by a gas.

Aerograph. A device for spraying powdered glaze or colour on the surface of pottery by means of compressed air.
Aerox. Alumina filter ceramic with uniform micron-sized pores, made by Aerox Ltd, which company was acquired by Doulton Industrial Products, Stone, now part of Fairey Engineering.
A.F.A. Rammer. Apparatus designed by the American Foundrymen's Association for the preparation of test-pieces of foundry sand; in 1948 the Association changed its name to the American Foundrymen's Society. See A.F.S. and SAND RAMMER.
AFNOR. Abbreviation for the French Standards Association; Association Française de Normalisation, 23 Rue Notre-Dame-des-Victoires, Paris 2.
AFRSI. Advanced Flexible Reusable Surface Insulation comprises an inner layer of leached silica glass fabric; a central layer of high-purity fused silica fibre; an outer layer of quality fabric. The material was developed as in insulating blanket for the Space Shuttle by NASA, Ames Research Centre and Johns-Manville, USA.
After Contraction. The permanent contraction (usually expressed as a linear percentage) that may occur if a fired or chemically-bonded refractory product is re-fired under specified conditions of test. Fireclay refractories are liable to show after contraction if exposed to a temperature above that at which they were originally fired (cf. FIRING SHRINKAGE).
After-glow. See JUMP.
After Expansion. The permanent expansion (usually expressed as a linear percentage) that may occur when a refractory product that has been previously shaped and fired, or chemically bonded, is re-fired under specified conditions of test. Such expansion may take place, for example, if the product contains quartz or kyanitc, or if bloating occurs during the test (cf. FIRING EXPANSION).
A.F.S. American Foundrymen's Society, Des Plaines, Illinois, USA. This Society has issued standard methods for testing foundry sands.
Afwillite. A hydrated calcium silicate, $3CaO.\ 2SiO_2\ .\ 3H_2O$; it is formed when portland cement is hydrated under special conditions and when calcium silicate is autoclaved (as in sand-lime brick manufacture).
Agalmatolite. An aluminosilicate mineral that is very similar to PYROPHYLLITE (q.v.).
Agate. A cryptocrystalline variety of silica that has the appearance of a hard, opaque glass; it usually contains coloured LIESEGANG RINGS. It is used in the ceramic industry for burnishing gold decoration on pottery and for making small mortars and pestles for preparing samples for chemical analysis.
Agate Glass. Decorative glassware, simulating AGATE, made by blending molten glasses of two or more colours, or rolling glasses of several colours together in the plastic state.
Agate Ware. Decorative earthenware made, chiefly in the 18th century in Staffordshire, by placing thin layers of differently coloured clays one on another, pressing them together, slicing across the layers and pressing the slab of variegated coloured clay into a mould. When fired, the ware had the appearance of natural agate.
Ageing. (1) A process, also known as SOURING (q.v.), in which moistened clay, or prepared body, is stored for a period to permit the water to become more uniformly dispersed, thus improving the plasticity. A similar effect can be achieved more quickly by TEMPERING (q.v.).

(2) The process of allowing vitreous-enamel slip to stand, after it has been milled, to improve its rheological properties.

Agglomerate. To form assemblages of smaller particles, rigidly joined together electrostatically or by intergrowth. (Or, as a noun, such an assemblage.) Agglomeration defeats mixing on the scale of the finest particles, and can produce variations in packing density which impair strength and other properties.

Aggregate. See CONCRETE AGGREGATE; LIGHTWEIGHT EXPANDED CLAY AGGREGATE.

Aglite. Trade-name: a lightweight expanded clay aggregate made by the Butterley Co. Ltd, Derby, England, from colliery shale by the sinter-hearth process. The bulk density is: (12–18 mm) 0.5 kg/l; (5–12 mm) 0.55 kg/l; (<5 mm) 0.8 kg/l.

Agricultural Pipes. See FIELD-DRAIN PIPES.

Agricultural Tile. See DRAIN TILE.

Air Bell. A fault, in the form of an irregularly shaped bubble, in pressed or moulded optical glass.

Air-borne Sealing. A process for the general, as opposed to local (cf. SPRAY WELDING), repair of a gas retort by blowing refractory powder into the sealed retort, while it is hot; the powder builds up within any cracks in the refractory brickwork and effectively seals them against gas leakage. (West's Gas Improvement Co. Ltd., Brit. Pat. 568159; 21/3/45.)

Air Brick. A fired clay brick, of the same size as a standard building brick, but having lateral holes for the purpose of ventilation, e.g. below floors. BS 493 specifies units for use externally (Class 1) and internally (Class 2).

Air Chain. A line of air bubbles, forming a defect in glass or vitreous enamel.

Air Content. The volume of air voids in cement paste, morter or concrete, excluding the pore space in the aggregate. Air entrained by mixing or gases generated chemically may be included. (ASTM C-125).

Air Entraining. The addition of a material to portland cement clinker during grinding, or to concrete during mixing, for the purpose of reducing the surface tension of the water so that 4–5 vol.% of minute air bubbles become trapped in the concrete. This improves workability and frost resistance and decreases segregation and bleeding. The agents used as additions include: 0.025–0.1% of alkali salts of wood resins, sulphonate detergents, alkali naphthenate, or triethanolamine salts; or 0.25–0.5% of the Ca salts of glues (from hides); or 0.25–1.0% of Ca lignosulphonate (from paper-making).

Air-floated. Material, e.g. china clay, that has been size-classified by means of an AIR-SEPARATOR (q.v.).

Air-hardening Refractory Cement or Mortar. See CHEMICALLY BONDED REFRACTORY CEMENT.

Airless Drying. The drying chamber is surrounded by an airtight inner shell and an insulated outer shell. The original air atmosphere is recirculated and heated, so that it is progressively replaced as heat transfer medium by superheated steam from the drying product. No air is allowed to enter the dryer during the drying process.

Air Line. A fault, in the form of an elongated bubble, in glass tubing- also known as HAIR LINE.

Air Permeability. See PERMEABILITY.

Air Pollution. BS 1747 specifies methods of test for the determination of fine suspended particles (smoke); SO_2, NO_x. See also TA-LUFT and HEALTH AND SAFETY.

Air Ramming. US term for the process of shaping a refractory brick or furnace hearth with a pneumatic rammer.
Air Seal. A method for the prevention of the escape of warm gases from the entrance or exit of a continuous furnace, or tunnel kiln, by blowing air across the opening.
Air Separator. A machine for the size classification of fine ceramic powders, e.g. china clay; the velocity of an air current controls the size of particles classified.
Air-setting Refractory Cement or Mortar. See CHEMICALLY BONDED REFRACTORY CEMENT.
Air-swept Ball Mill. See BALL MILL.
Air Twist. Twisted capillaries as a form of decoration within the stem of a wine glass.
Aired Ware. Pottery-ware that has a poor glaze as a result of volatilization of some of the glaze constituents. The term was used more particularly when ware was fired in saggars in coalfired kilns, 'air' escaping from a faulty saggar into the kiln while kiln gases at the same time penetrated into the saggar. The term is also sometimes applied to a glaze that has partially devitrified as a result of cooling too slowly between 900 and 700 °C; the most common crystallization is that of calcium silicate or zinc silicate.
Akermanite. $2CaO.MgO.2SiO_2$; m.p. 1454°C; sp.gr. 3.15; thermal expansion (~1200°C) 10.6×10^{-6} K^{-1}. This mineral is found in some slags.
Alabaster Glass. A milky-white glass resembling alabaster; transmitted light is diffused without any fiery colour.
Albany Clay. A clay found in the neighbourhood of Albany, New York; a quoted analysis is (%) SiO_2 57.6; Al_2O_3 14 .5; Fe_2O_3 5.2; CaO 5.7; MgO 2.7; Alkalis 3.1. A somewhat similar clay occurs in Ontonagon County, Michigan. Because of its fine particle size and high flux content, this clay fuses at a comparatively low temperature to form a greenish-brown glaze suitable for use on stoneware; it is also used as a vitrifying bonding material.
Alberene Stone. A US stone having properties similar to those of POLYPHANT STONE (q.v.).
Albite. See FELDSPAR.
Albolite. A silica and magnesia plastic cement.
Alcove. A narrow covered extension to the working end of a glass-tank furnace; it conveys molten glass to a forehearth or revolving pot.
Alginates. Organic binders and suspending agents derived from sea plants. Ammonium alginate and sodium alginate have found some use as additions to suspensions of glazes and enamels.
Ali Baba. Popular name for a large chemical stoneware jar of the type used for the bulk storage of acids; these jars are made in sizes up to 5000 litres capacity.
Alite. The name given to one of the crystalline constituents of portland cement clinker by A. E. Törnebohm (*Tonindustr. Ztg.*, **21**, 1148, 1897). Alite has since been identified as the mineral $3CaO. SiO_2$
Alkali–Aggregate Reaction. The siliceous components of aggregates may react detrimentally with the alkali componenets of Portland cement, after the concrete has hardened.
Alkali Attack. See DURABILITY OF ON-GLAZE DECORATION.
Alkali Neutraliser. See under NEUTRALISER.
Alkoxide. Compounds formed from the reaction of an alcohol with an alkali metal. They are used in SOL-GEL processes (q.v) for preparing ceramics.

All-basic Furnace. Abbreviation for All-Basic Open Hearth Steel Furnace. The whole of the superstructure of such a furnace-hearth, walls, roof, ports, ends, is built of basic refractories These furnaces were introduced in Europe in about 1935, the object being to make it possible to operate at a higher temperature than that possible with basic O.H. furnaces having a silica roof.

Alligator Hide. (1) A vitreous-enamelware fault in the form of severe ORANGE PEEL (q.v) and TEARING (q.v.); causes include too-rapid drying and too-heavy application of enamel.
(2) A decorative glaze effect, having the appearance indicated by its name, sometimes used by studio potters; it is achieved by loading the glaze with refractory oxides to increase the viscosity so that during firing, the glaze does not flow uniformly over the surface of the ware.

Alon. Aluminium Oxy-Nitride (q.v). (Alon has previously been used as a tradename for fused alumina, by the Cabot Corp, USA.)

Allophane. An amorphous gel consisting, usually, of 35–40% Al_2O_3, 20–25% SiO_2 and 35% H_2O. It is often found in association with halloysite.

Allport Oven. A pottery BOTTLE OVEN (q.v.) in which the hot gases from the firemouths entered the oven nearer to its centre than the usual points of entry around the oven walls; another feature was preheating of the secondary air. (S. J. Allport, Brit. Pat. 570 335; 3/7/45; 606 084; 5/8/48.)

Alluvial Clay. A brickmaking clay of river valleys typified in England by the clays of the Humber estuary, the Thames valley, and the Bridgwater district. The composition is variable.

Aloxite. Trade-name: Fused alumina abrasives and abrasive products made by the Carborundum Co., Ltd.

Alsing Cylinder. An early type of BALL MILL (q.v.) invented by J. R. Alsing in 1867.

Alumina. Aluminium oxide, Al_2O_3. This oxide exists in several forms: principally γ-Al_2O_3, stable up to about 1000°C and containing traces of water or of hydroxyl ions, and α-Al_2O_3, the pure form obtained by calcination at high temperature. The so-called β-Al_2O_3 is the compound $Na_2O.11Al_2O_3$. BS 4140 specifies powder properties and methods of determining chemical contaminants of aluminium oxide powders, which are also specified in ASTM F.7. See also ALUMINA (FUSED); ALUMINA (SINGLE CRYSTAL); ALUMINA (SINTERED); ALUMINA (TABULAR); CORUNDUM.

Alumina Cement. See HIGH-ALUMINA CEMENT.

Alumina Clogging. Nozzles for secondary steelmaking ladles can become blocked by build-up of dense products of the reaction between the nozzle refractories and the steel and slag, in particular by attack on the alumina phase of the refractory.

Alumina (Fused). In spite of its high m.p. (2050 °C), alumina can be fused in an oxy-hydrogen flame or in an electric arc. By the former method, large single crystals (boules) can be produced; they are used as bearings, and as dies for wire-drawing, and for other purposes demanding high abrasion resistance. Fused alumina made in the electric arc furnace is usually crushed, and sieved into a range of grit sizes for use as an abrasive; when re-bonded and sintered it is used as a special refractory material.

Alumina Porcelain. Defined in ASTM-C242 as: A vitreous ceramic whiteware, for technical application, in which alumina is the essential crystalline phase.

Alumina (Single Crystal). Made by the flame-fusion of the powder; used in bearings, lasers, fibre-drawing dies, etc.

Alumina (Sintered). Alumina, sometimes containing a small amount of clay or of a mineralizer, and fired at a high temperature to form a dense ceramic. Sintered alumina has great mechanical strength: compressive strength 3 GN/m^2 (20 °C), 0.9 GM/m^2 (1000°C), 1 GN/m^2 (1500 °C); tensile strength 175 MN/m^2 (20°C), 140 MN/m^2 (1000 °C), 35 MN/m^2 (1400 °C); Young's Modulus, 350 GN/m^2 (20 °C), 140 MN/ m^2 (1000 °C), 140 GN/m^2 (1500°C). Sintered alumina also has high abrasion resistance, high dielectric strength, and low power factor. Thermal expansion (0–1000 °C) 8×10^{-6}. Because of these properties, sintered alumina is used in thread guides, tool tips, and grinding media; as the ceramic component of sparking plugs, electronic tubes, ceramic-to-metal seals, etc. Sintered alumina coatings can be applied to metals by flame-spraying. Alumina ceramics for electrical and electronic applications are specified in ASTM D2442.

Alumina (Tabular). A pure (99.5%) alumina calcined at 1650°C and containing very little (e.g. 0.02%) Na_2O; it consists of coarse tablet-like crystals of CORUNDUM (q.v.).

Alumina Whiteware. Defined in ASTM – C242 as: Any ceramic whiteware in which alumina is the essential crystalline phase.

Aluminalon. A composite, 90% alumina, 10% aluminium nitride, formed by hot-pressing. It has high strength and ductility to 1400 °C, and is acid resistant to 1200 °C.

Aluminium Borates. Two compounds have been reported: $9Al_2O_3.\ 2B_2O_3$, which melts incongruently at 1950 °C; $2Al_2O_3.\ B_2O_3$, which dissociates at 1035°C. The compound $9Al_2O_3.\ 2B_2O_3$, when made by electrofusion, has been proposed as a special refractory material: R.u.L., 1500°C; thermal expansion, 4.2×10^{-6} (20–1400°C); good thermal-shock resistance.

Aluminium Boride. The usual compound is AlB_2; sp.gr. 3.2. AlB_2 dissociates above 980°C to form AlB_{12} (sp.gr. 2.6) and Al.

Aluminium Carbide. Al_4C_3; the dissociation pressure at 1400°C is approx. 1×10^{-5} atm.; m.p. 2800 °C; sp.gr. 2.99. This carbide is slowly attacked by water at room temperature.

Aluminium Enamel. A vitreous enamel compounded for application to aluminium. There are three main types: containing lead, phosphate glass, or barium. Lithium compounds have also been used in these enamels.

Aluminium Nitride. AlN; m.p. 2200°C in 4 atm. N_2; dissociates at 1700–1800°C *in vacuo*. The theoretical density is 3.26. Thermal expansion (25–1000°C) 5.5×10^{-6}. Thermal conductivity 550 W/m.K. That of alumina is about 30 W/m.K, so AlN is tape cast (q.v) into substrates and chip carriers for printed circuits, using CaO or Y_2O_3 as densification additives. AlN is a raw material for SIALONS (q.v), forming a series of POLYTYPOIDS (q.v) with constant (Si+Al):(O+N) ratio. When shaped into crucibles AlN can be used as a container for molten aluminium. It is a good electrical insulator, the resistivity exceeding 10^{12} ohm. cm at 25 °C. Modulus of rupture, 250 MN/m^2 at 25°C and 120 MN/m^2 at 1400 °C. Hardness, 1230 (K100).

Aluminium Oxynitride. AlON is an infrared transmitter in the wavelength range 1 to 4µm.

Aluminium Phosphate. Generally refers to the orthophosphate, $AlPO_4$; sp.gr. 2.56, m.p. approx. 2000 °C but decomposes. It is supplied in various grades, particularly for use as a binder in the refractories industry. The bonding action has been attributed to reaction between the phosphate and basic oxides or Al_2O_3 in the refractory. This type of bond preserves its strength at intermediate temperatures (500–1000 °C), a range in which organic bonds are destroyed and normal ceramic bonding has not yet begun.

Aluminium Titanate. $Al_2O_3.TiO_2$ melts at 1860 °C, but slowly dissociates below 1150 °C to Al_2O_3 and TiO_2, producing a microstructure with a network of fine cracks. This gives good thermal shock resistance and a thermal expansion coefficient up to 700°C which is almost zero, but non-uniform. It may be stabilized with small amounts of Fe_2TiO_5 or $MgTiO_3$ to provide more uniform expansion.

Alumino-silicate Refractory. A general term that includes all refractories of the fireclay, sillimanite, mullite, diaspore and bauxite types. BS 1902 Pt 2 defines it as a refractory containing 8 to 45% Al_2O_3, the balance being predominantly silica.

Aluminous Cement. See CIMENT FONDU.

Aluminous Fireclay Refractory. This type of refractory material is defined in B.S. 1902 Pt 2 as a refractory containing in the fired state, > 45% Al_2O_3, the balance being predominantly silica.

Alundum. Trade-name: Fused alumina made by Norton Co.

Alunite. $KAl_3(SO_4)_2(OH)_6$ occurs in Australia, USA (Arizona, Nevada, Utah), Mexico, Italy, Egypt, and elsewhere. The residue obtained after calcination and leaching out of the alkali can be used as raw material for high-alumina refractories.

Amakusa. The Japanese equivalent of CHINA STONE (q.v.).

Amberg Kaolin. A white-firing micaceous kaolin from Hirschau, Oberpfalz, Germany. A quoted analysis is (per cent): SiO_2, 48.0; Al_2O_3 37.5; Fe_2O_3 0.5; TiO_2 0.2; CaO 0.15; alkalis 2.6; loss-on-ignition 12.2.

Amber Glass. Whilst other colours of CONTAINER GLASS are selected for aesthetic reasons (clear and green being those generally available) the amber or brown glass is designed to absorb ultraviolet light to protect some foods and pharmaceutical products sensitive to UV. The amber colour is imparted by iron and sulphur.

Ambetti or Ambitty. Decorative glass containing specks of opaque material; the effect is produced by allowing the glass to begin to crystallize.

Amblygonite. A lithium mineral of variable composition, $(Li,Na)AlPO_4(F,OH)$; sp. gr. 3 0; m.p. 1170°C. It occurs in various parts of southern Africa, but is relatively uncommon and therefore finds less use as a source of lithium than do PETALITE and LEPIDOLITE (q.v.).

Amborite. Hot-pressed fine-grained boron nitride with a ceramic bond, used as an abrasive (de Beers Co).

Amboy Clay. An American siliceous fireclay; it is plastic and has a P.C.E. above 32.

AMDCB. Applied Moment Double Cantilever Beam. See FRACTURE TOUGHNESS TESTS.

American Bond. Brickwork of stretcher courses with a header course every 5th, 6th or 7th course.

American Hotel China. A vitreous type of pottery-ware. A typical body composition is (per cent): kaolin, 35; ball clay, 7; feldspar, 22; quartz, 35; whiting, 1.

Ammonium Bifluoride. $NH_4F.HF$; used as a saturated solution in hydrofluoric acid (usually with other additions) for the etching of glass.

Ammonium Vanadate. NH_4VO_3; used as a source of vanadium in ceramic pigments, e.g. tin-vanadium yellow, zirconium vanadium yellow and turquoise, etc.
Ampelite. A carbonaceous schist containing alumina, silica and sulphur, sometimes used as a refractory.
Amperit. Tradename. Mixed alumina-titania powders supplied by Hoechst AG.
Amphiboles. See SILICATE STRUCTURES.
Amphoteric. Capable of reacting as acid or base (e.g. Al_2O_3).
Ampoule. A small glass container that is sealed (after it has been filled with liquid) by fusing the narrow neck. The glass must resist attack by drugs and chemicals, but must be readily shaped by mass-production methods; normally, a borosilicate glass of fairly low alkali and lime contents is used.
Amsler Volumeter. An apparatus for the measurement of the bulk density of powdered or granular materials (see BS 1902 Pt 3.6). It comprises a cylindrical steel container, whose lid is retained by a screwed collar to provide a mercury seal. A guarded capilary tube with a reference mark is fitted into the domed lid. A plunger with a graduated micrometer, fitted into the side of the cylinder, controls and measures the displacement of mercury within.
Anaconda Process. A method for the shaping of silica refractories formerly used at some refractories works in USA. The bricks were first 'slop-moulded', then partially dried, and finally re-pressed. The name derives from the town of Anaconda, Montana, USA, where the process was first used, early in the present century, by the Amalgamated Copper Co.
Analysis. See DIFFERENTIAL THERMAL ANALYSIS; PARTICLE-SIZE ANALYSIS; RATIONAL ANALYSIS; ULTIMATE ANALYSIS.
Anatase. A tetragonal form of titania, TiO_2; sp. gr. 3.84. The other forms are BROOKITE (q.v.) and RUTILE (q.v.). All three structures have similar Ti-O bond distance but differ in the linking of the TiO_6 octahedra. Anatase has been observed in some fireclays and kaolins. It is rapidly converted into rutile when heated above about 700 °C.
Anchors. Specially shaped metal or ceramic pins to hold ceramic fibre blankets or panels in place to form furnace linings. Other heavier unshaped linings (of refractory concrete) are also anchored to the furnace structure by notched or shaped rods and bars.
Andalusite. A mineral having the same composition (Al_2SiO_5) as sillimanite and kyanite but with different physical properties. Thc principal source is S. Africa; it also occurs in California, Nevada, and New England (USA). When fired, it breaks down at 1350°C to form mullite and cristobalite; the change takes place without significant change in volume (cf. KYANITE). Andalusite finds some use as a refractory raw material.
Andreasen Pipette. An instrument used in the determination of the particle size of clays, by the SEDIMENTATION METHOD (q.v.); it was introduced by A. H. M. Andreasen, of Copenhagen (*Kolloid Zts.*, **49**, 253, 1929).
Andrews' Elutriator. A device for particle-size analysis. It consists of: a feed vessel or tube; a large hydraulic classifier; an intermediate classifier; a graduated measuring vessel. (L. Andrews, Brit. Pat. 297 369; 14/6/27.)
Angle Bead. A special shape of wall tile (see Fig. 7, p350).
Angle of Drain. The angle at which vitreous enamel ware is placed on a rack

after dipping, to obtain the required thickness of coating.
Angle of Nip. The angle between the two tangents to a pair of crushing rolls at the point of contact with the particle to be crushed. This angle is important in the design of crushing rolls for clay; it should not exceed about 18°.
Angle Tile. A purpose-made clay or concrete tile for use in an angle in vertical exterior tiling.
Ångstrom Method. See THERMAL DIFFUSIVITY.
Ångstrom Unit, Å. 10nm, still met in X-ray crystallography.
Angular. See MORPHOLOGY.
Anhydrite. $CaSO_4$; there are soluble and insoluble forms. Anhydrite begins to form when gypsum is heated at temperatures above about 200°C. The presence of anhydrite in plaster used for making pottery moulds can cause inconsistent blending times; the moulds tend to be soft, and to give slow casting, slow release of casts from the mould, and flabby casts.
Anionic Exchange. See IONIC EXCHANGE.
Anisotropy. The state of having properties whose values differ according to the direction of measurement. This is quite frequently met in ceramics, and can arise from the basic crystal structure (e.g. sheet-like crystal structures with weaker, asymmetric forces between the layers, have much lower tensile strength perpendicular to the sheet). On the microstructural scale, directional forming processes such as uniaxial PRESSING (q.v) can make anisotropic such properties as strength and shrinkage.
Anneal. To release stresses from glass by controlled heat treatment; the process is usually carried out in a LEHR (q.v.). See also ADAMS–WILLIAMSON ANNEALING SCHEDULE; REDSTON–STANWORTH ANNEALING SCHEDULE; TREBUCHON–KIEFFER ANNEALING SCHEDULE.
Annealing Point or Annealing Temperature. The temperature at which glass has a viscosity of 10^{13} poises; also known as the 13.0 TEMPERATURE. When annealed at this temperature glass becomes unstressed in a few minutes. A method of test by fibre elongation is given in ASTM – C336; C598 specifies a beam bending test.
Annealing Range. The range of temperature in which stresses can be removed from a glass within a reasonable time, and below which rapid temperature changes do not cause permanent internal stress in the glass. When comparing glasses, the Annealing Range is taken as the temperature interval between the ANNEALING POINT (q.v.) and the STRAIN POINT (q.v.).
Annular Kiln. A large continuous kiln, rectangular in plan, of a type much used in the firing of building bricks. The bricks are set on the floor of the kiln and the zone of high temperature is made to travel round the kiln by progressively advancing the zone to which fuel is fed. There are two principal types: LONGITUDINAL-ARCH KILN (q.v.) and TRANSVERSE-ARCH KILN (q.v.)
Annulus Brick. A DOME BRICK with the large faces equally inclined to each other in breadth and length, so that one end face is smaller than the other.
Anode Pickling. See ELECTROLYTIC PICKLING.
Anorthite. See FELDSPAR.
Anorthosite. A coarse-grained rock consisting almost exclusively of plagioclase (see under FELDSPAR). It occurs in Canada, New York State, Norway and Russia; trials have been made with this material as a flux in ceramic bodies.

Antiferroelectric. A dielectric of high permittivity which undergoes a change in crystal structure on cooling through a CURIE TEMPERATURE (q.v.) but which possesses no spontaneous polarisation. The state is visualised as polarisation of adjacent lines of atoms in opposite directions.

Antiferromagnetic. A material in which the atomic magnetic moments are aligned antiparallel so that the total moment is zero and there is no observable spontaneous magnetization. The alignment disappears at the CURIE TEMPERATURE (q.v.). Antiparallel ordering gives rise to an antiferromagnetic or to a ferrimagnetic depending on the crystal structure.

Antimony Oxide. Sb_2O_3; m.p. 655°C; sp. gr. approx. 5.5. Used as an opacifier in enamels and as a decolorizing and fining agent in glass manufacture, particularly in pot melting; in pottery manufacture it is a constituent of Naples Yellow.

Antimony Yellow. See LEAD ANTIMONATE.

Antioch Process. Plaster moulds are produced by pouring an aqueous plaster of Paris slip over a mould, steam treating, allowing to set in the air and oven drying.

Anti-piping Compound. HOT-TOPPING COMPOUND, (q.v).

Antique Glass. Flat glass made by the CYLINDER PROCESS (q.v.) and with textured surfaces resembling old glass; it is used in the making of stained-glass windows (cf. CATHEDRAL GLASS).

Anti-static Tiles. Floor tiles of a type that will dissipate any electrostatic charge and so minimize the danger of sparking; such tiles are used in rooms, e.g. operating theatres, where there is flammable vapour. One such type of ceramic tile contains carbon. The National Fire Protection Association, USA, stipulates that the resistance of a conductive floor shall be less than 1 megohm as measured between two points 3 ft (1 m) apart; the resistance of the floor shall be over 25 000 ohm between a ground connection and any point on the surface of the floor or between two points 3 ft (1 m) apart on the surface of the floor.

A.P. US abbreviation for ANNEALING POINT (q.v.).

Apatite. Calcium phosphate, $CaPO_4$ though other negative ions (F, Cl, CO_3 or OH) may be present in natural rocks. Apatite is a chief constituent of bone. Apatite and hydroxyapatite (containing the OH– ion) are used in bioceramics, particularly if it is desired to achieve bone intergrowth with the artificial component or prosthesis.

APGS. Advanced Plasma Gun System for thermally sprayed coatings.

Aplite. A rock mined in Virginia for use in glass manufacture; it consists principally of albite, zoisite and sericite.

Apparent Initial Softening. When applied to the REFRACTORINESS-UNDER-LOAD TEST (q.v.), this term has the specific meaning of the temperature at which the tangent to the curve relating the expansion/contraction and the temperature departs from the horizontal and subsidence begins. (For typical curve see Fig. 5, p230.)

Apparent Porosity. See under POROSITY.

Apparent Solid Density. A term used when considering the density of a porous material, e.g. a fireclay or silica refractory. It is defined as the ratio of the mass of the material to its APPARENT SOLID VOLUME (q.v.) (cf. TRUE DENSITY).

Apparent Solid Volume. A term used when considering the density and volume of a porous solid, particularly a refractory brick. It is defined as the volume of the solid material plus the volume of any sealed pores and also of the open pores.

Apparent Specific Gravity. For a porous ceramic, the ratio of the mass to the mass of a quantity of water that, at 4°C, has a volume equal to the APPARENT SOLID VOLUME (q.v.) of the material at the temperature of measurement. ASTM C20 specifies a boiling water test; ASTM C830 a vacuum pressure test.

Application Weight. The weight of vitreous enamel coating applied, per unit area covered.

Applied Moment Double Cantilever Beam. See FRACTURE TOUGHNESS TESTS.

Apron. See under SAND SEAL.

APS. Air plasma spraying – see PLASMA SPRAYING.

AQL. Acceptance Quality Level. See ACCEPTANCE TESTING.

Arabian Lustre. The original type of on-glaze lustre used by the Moors from the 9th century onwards for the decoration of pottery; the sulphides or carbonates of copper and/or silver are used, the firing-on being in a reducing atmosphere so that an extremely thin layer of the metal is formed on the glaze.

Arbor. US term for the spindle of a grinding machine on which the abrasive wheel is mounted.

Arc Furnace. See ELECTRIC FURNACE.

Arc ion plating. Metal is evaporated and ionised, then a reactive gas of non-metallic elements is introduced and reacted with the metal vapour at the surface of a substrate, whereon a ceramic film is formed. Usually the substrate is given a negative bias voltage.

Arc-image Furnace. See IMAGE FURNACE.

Arc-spraying. See PLASMA SPRAYING.

Arch. In addition to its normal meaning this term is sometimes applied to a furnace roof. See also ARCH BRICK; ARCHING; POT ARCH.

Arch Brick. A building brick or refractory brick having the two large faces inclined towards each other. This may be done in two ways: in an END ARCH (q.v.) one of the end faces is smaller than the opposite end face; in a SIDE ARCH (q.v.) one of the side faces is smaller than the opposite side face. Arch bricks are used in the construction of culverts, furnace roofs, etc. (See Fig. 1, p 15.)

Arching. (1) The preheating of glass-pots before use. (2) The jamming of material in a hopper or shaft furnace, an empty space being formed beneath the 'arch' thus produced.

Architectural Glass. Glass used in the construction industry. ASTM C-724 specifies Test Methods for Acid Resistance of Ceramic Decorations on Architectural Glass. Four drops of each of 10% citric acid and nominal 3.7% HCl are placed on the glass surface for 15 min, and acid attack assessed visually against a 7-point scale.

Archless Kiln. Alternative name for SCOVE (q.v.).

Argillaceous. Containing clay; clayey. (From French *argile,* clay).

Argillite. A sedimentary rock rich in clay minerals; the term is now generally applied to thick-bedded clay rock that has no distinct cleavage.

Aridised Plaster. Plaster that has been treated, while being heated in the 'kettle', with a deliquescent salt, e.g. $CaCl_2$; it is claimed that this produces a strong plaster having more uniform properties.

Ark. A large vat used in the pottery industry for the mixing or storage of clay slip.

Arkose. A rock of the sandstone type but containing 10% or more of feldspar; it is formed by decomposition of granite.

Armour. Ceramic armour stops projectiles by absorbing their kinetic energy to generate cracks in the hard, brittle ceramic. Of light weight compared to metal armour, ceramics such as boron carbide and alumina are useful as body armour.

Armouring. Metal protection for the refractory brickwork at the top of the stack of a blast furnace; its purpose is to prevent abrasion of the refractories by the descending burden (i.e. the raw materials charged to the furnace).

Arrest Line. A RIB MARK (q.v) defining the crack front shape of an arrested crack, before the crack is spread again by a new stress.

Arrhenius' Law. A logarithmic relationship between reaction rate and temperature which applies to creep, diffusion and solid state phase transformations (among many other chemical and physical reactions).

$$\log v = A - q/kT$$

where v is the reaction rate, A is a constant, q is the activation energy, T is the temperature and k is Boltzmann's constant. The activation energy is a threshold energy value which the system must have before the reaction can occur.

Arris. The sharp edge of a building brick or ridge-tile; an ARRIS TILE is a specially shaped tile for use in the ridge or hip of a roof. The ARRIS EDGE on glass is a bevel up to 1/16 in. (1.5 mm) wide and at an angle of 45°.

Arsenic Oxide. As_2O_3; sublimes at 193°C; sp. gr. 3.87. Used as a fining and decolorizing agent (particularly in association with selenium) in glass manufacture.

ASA. Prefix to specifications of the American Standards Association, 70 East 45th Street, New York 17.

Artifical Porcelain. See SOFT PASTE PORCELAIN.

Asbestine. A fibrous mineral consisting of mixed hydrated silicates of Mg and Ca; it occurs in New York State.

Asbestos. A fibrous, heat-resistant mineral, which may be formed into boards or woven into fabric or paper. Formerly much used as a heat insulating material, the health hazard of ASBESTOSIS has led to its replacement by forms of calcium silicate board. There are two types of asbestos. *Chrysotile* is a hydrated magnesium silicate, stable up to 500 °C. It is used as a reinforcing fibre in *asbestos cement*, when mixed with water and Portland cement. Heat-resistant boards, sheets and pipes can be made, containing about 10% asbestos. Relevant British standards are BS486, 567, 690, 835, 3497, 3656, 4624 and 5247. *Amphiboles* (e.g. *tremolite*, a hydrated calcium magnesium silicate, or *crocidolite*, a hydrated sodium iron silicate) also known as *blue asbestos*, cause ASBESTOSIS. They are less strong than chrysotile, though with higher elastic moduli, and weaken markedly above 300 °C. The Annual Book of ASTM standards, Vol 4.05, lists some 50 standards for asbestos fibre and asbestos-cement products and relevant test methods.

Asbestosis. A severe inflammatory disease of the lungs, caused by inhaling amphibole (blue) asbestos.

Asbolite. Impure cobalt ore used by the old Chinese potters to produce underglaze blue.

Ascera Method. A tensile testing technique using test pieces of simple geometry and large test volume (cylinders 9 to 10 mm diameter, 120 mm long). Non-tensile forces are minimised by careful design of the gripping, alignment and force application devices. (ASEA Ceramica AB, Sweden).

Ashfield Clay. A fireclay associated with the Better Bed coal, Yorkshire, England.

The raw clay contains about 57% SiO_2, 27% Al_2O_3, 1.7% Fe_2O_3 and 1.5% alkalis.

ASTM. American Society for Testing and Materials. This Society is responsible for most of the American standard specifications for ceramics. Address: 1916 Race St., Philadelphia 3, Pa., USA.

Ashlar Brickwork. Clay brick or blockwork, made to resemble stonework by the size, arrangement and surface appearance of the units.

Aspirating Screen. A vibrating SCREEN (q.v) through which the fines are drawn by suction.

Astbury Model. A theoretical model to describe the plasticity of clays, and in particular their non-elliptical (except bentonite) hysteresis loops. The clay is represented as an arrangement of purely elastic elements (springs) and purely viscous elements (dashpots) with the proviso that as more energy is stored, the elastic elements break down and are converted into viscous elements. (N.F. Astbury. *Science of Ceramics* ,Vol. 2, 1965).

Atomiser. A device to produce a fine spray of small particles or liquid droplets. In a SPRAY DRYER, the atomiser may be a nozzle, or a spinning head which produces the spray centrifugally.

Atritor. Trade-name: a machine that simultaneously dries and pulverizes raw clay containing up to 18% moisture; it consists of a feeder, metal separator, pulverizer, and fan. (Alfred Herbert Ltd., Coventry, England.)

Attaclay. See ATTAPULGITE.

Attapulgite. One of the less common clay minerals; it was first found at Attapulgus, USA, and is characterized by a high (10%) MgO content. It has a large surface area and is used in oil refining, as a carrier for insecticides, etc. Also known as PALYGORSKITE (q.v.).

Atterberg Test. A method for determining the plasticity of clay in terms of the difference between the water content when the clay is just coherent (tested by rolling out the clay by hand) and when it begins to flow as a liquid; the difference is the Atterberg Number. The test was first proposed by A. Atterberg *(Tonindustr.* Ztg., **35**, 1460, 1911).

Attributes. Qualitative characteristics of products, whose presence or absence is subject to inspection to assess the acceptance quality level of a sample. (Inspection by *variables* involves numerical measurement) BS 1902 Pt 3.13 relates to the attributive properties of corner, edge and surface defects in refractories. See ACCEPTANCE TESTING.

Attrition. Particle-size reduction by a process depending mainly on impact and/or a rubbing action.

Attrition Mill. A disintegrator depending chiefly on impact to reduce the particle size of the charge. Attrition mills are sometimes used in the clay building materials industry to deal with the tailings from the edge-runner mill.

Attritor. Tradename. A bead mill, made by Torrance & Sons, Bristol.

Aubergine Purple. A ceramic colour, containing Mn, introduced in the 18th century, when it was used for underglaze decoration.

Auger. (1) An extruder for clay, or clay body, the column being forced through the die by rotation of a continuous screw on a central shaft. (2) A large 'corkscrew' operated by hand or by machine to take samples of soft material, e.g. clay, during prospecting.

Autoclave. A strong, sealed, metal container, fitted with a pressure gauge and safety valve; water introduced into

the container can be boiled under pressure, the gauge pressure developed at various temperatures being as follows:
Gauge pressure (p.s.i.): 0 25 50 100 150
(0 175 350 700 1050 kPa)
Temperature (°C): 100 130 148 170 186
A common method of testing glazed ceramic ware for crazing resistance is to expose pieces of ware to the steam in an autoclave, at 50 p.s.i. in B.S. 3402 and B.S. 4034, or at 5~250 p.s.i. (increasing the pressure in steps of 50 p.s.i.) in ASTM – C424. The method is unsatisfactory because it confuses the effects of moisture expansion (the prime cause of crazing) and thermal shock. An autoclave test is also sometimes used in the testing of vitreous enamelware. As applied to the testing of portland cement, the autoclave test reveals unsoundness resulting from the presence of any MgO or CaO.

Autocombustion System. An electronically controlled impulse system for the oil firing (from the top or side) of ceramic kilns.

Autogenous Grinding Mill. In this type of rotating cylinder mill, the only grinding medium is the incoming coarse feed. (Autogenous = self-produced).

Automation. The operation of processes under automatic control (nowadays usually by digital computer) without human intervention. Older U.S. literature, especially that relating to the brick industry, used the term when mechanisation was meant – power assistance without automatic control.

Autosieve. An automatic sieve fractionation system for particle size analysis. Four sieves are mounted on the outside of a hollow plastics cylinder, into which powder is fed through an axial part. Cylinders are interchangeable, with different sets of sieves. The finest sieve is at the bottom. Sieves are mechanically oscillated and fluid jets blown through the sieves to fluidize the powder. The fines are automatically weighed at each oscillation of the sieves.

Auxiliary Metal Bath Process. Molten metals act as solvents to promote reactions between inorganic materials to form powders of borides, carbides or nitrides. (*Powder Metallurgy Int.* **4**, (4), 191, 1972).

Autostic. A water-based, air-setting cement much used for small tasks such as mounting thermocouple sheaths. (Tradename of Carlton, Brown & Partners, Sheffield).

Aventurine. A decorative effect achieved on the surface of pottery or glass by the formation of small, bright, coloured crystals (generally hematite, Fe_2O_3, on pottery bodies, but chrome or copper on glassware).

Ayrshire Bauxitic Clay. A non-plastic fireclay formed by laterisation of basalt lava and occurring in the Millstone Grit of Ayrshire, Scotland; there are two types, the one formed *in situ,* the other being a sedimentary deposit. Chemical analysis (per cent) (raw): SiO_2 42; Al_2O_3 38; TiO_2, 3–4; Fe_2O_3, 0.5; alkalis, 0.2.

AZS Refractories. Alumina-zirconia-silica refractories, used in contact with the molten glass in glassmaking. These refractories are fusion-cast for high density, low porosity, and resistance to corrosion by ingress of molten glass. ASTM C1223 tests for glass exudation from AZS refractories, on heating to 1510 °C.

B25 Block. A hollow clay building block, 25×30×13.5 cm, introduced by the Swiss brick industry in 1953. It is economical to lay (22 blocks per m^2), giving a wall 25 cm thick with low thermal transmittance.

Babal Glass. Name for Barium Boro-Aluminate glass proposed by

M. Monneraye, J. Serindat and C. Jouwersma (*Glass Techn.*, **6**, (4), 132, 1965).

Babic's Mortar. A cement mortar made water-repellant by incorporating an aqueous siliconate. (Build Trades J. **183** (5439) 34 1982)

Babosil. Trade-name: a frit for pottery glazes, so-named because it contains barium, boron and silica. It comprises:

0.06 K_2O	0.125 Al_2O_3	0.43 BaO
0.50 Na_2O	0.68 B_2O_3	2.45 SiO_2

Backbone. A system of polystyrene backing for brick panels. (United Research & Development Corp, and BASF).

Backface Wear. Abrasion of the rear face of a refractory by differential movement between it and a rotating or flexing structure, e.g. in a rotary cement kiln.

Back Heat. In an annular kiln, the temperature in the furthest back of the chambers being fired; sometimes the term is extended to include the hottest of the chambers being cooled.

Back Ring. See HOLDING RING.

Back Stamp. The maker's name and/or trademark stamped on the back of pottery flatware or under the foot of hollow-ware.

Backing. The common brickwork behind facing bricks in a brick building, or low-duty refractories or heat insulation behind the working face of a furnace.

Backing Strip. A strip of metal welded to the back of a metal panel prior to its being enamelled; the purpose is to prevent warping.

Bacor; Bakor. A Russian corundum-zirconia refractory for use more particularly in the glass industry; the name is derived from BADDELEYITE (q.v.) and CORUNDUM (q.v.). There are various grades, e.g. BAKOR-20 (62% Al_2O_3, 18% ZrO_2, 16% SiO_2) and BAKOR-33 (50% Al_2O_3, 30% ZrO_2, 15% SiO_2).

Baddeleyite. Natural zirconia, ZrO_2; there are deposits of economic importance in Australia, Brazil and Russia.

Badging. The application, usually by transfer or silk-screen, of crests, trade marks, etc. to pottery or glass-ware (cf. BACK STAMP).

Baffles. Thin detachable batts, placed vertically around the edge of a kiln car as a shield, held in place by small wedges known as *baffle pegs*.

Baffle Mark. A fault that may occur on a glass bottle as a mark or seam caused by the joint between the blank mould and the baffle plate.

Baffle Wall. See SHADOW WALL.

Bag Wall. A wall of refractory brickwork built inside a downdraught kiln around a firebox, each firebox having its own bag wall. The purpose is to direct the hot gases towards the roof of the kiln and to prevent the flames from impinging directly on the setting (cf. FLASH WALL).

Bain Deformation. The conversion of one crystal lattice into another by a simple expansion or contraction along the orthogonal axes. A MARTENSITIC TRANSFORMATION (q.v.) can be described by a Bain deformation combined with a lattice-invariant deformation (one which does not change the crystal structure) to produce an undistorted plane; the transformed (product phase) lattice is then rotated to coincide with the present lattice along the *habit plane*.

Bait. (1) A tool used in the glass industry- it is dipped into the molten glass in a tank furnace to start the drawing process.

(2) To shovel solid fuel into the firebox of a kiln.

Bakor. See BACOR.

Balance; Balancing. A grinding wheel is said to be in STATIC BALANCE if it will remain at rest, in any position, when it is mounted on a frictionless, horizontal spindle; such a wheel is also in DYNAMIC BALANCE if, when rotated, there is no whip or vibration. The process of testing for balance, and the making of any necessary adjustment to the wheel, is known as BALANCING.

Baldosin Catalan. A Spanish type of extruded ceramic floor tile.

Ball Clay. A sedimentary kaolinitic clay that fires to a white colour and which, because of its very fine particle size, is highly plastic. The name is derived from the original method of winning in which the clay was cut into balls, each weighing about 15kg. As dug, ball clays are often blue or black owing to their high content of carbonaceous matter. In England, ball clays occur in Devon and Dorset; in USA, they are found in Kentucky and Tennessee. Ball clays are incorporated in ceramic bodies to give them plasticity during shaping and to induce vitrification during firing. In addition to their use in most pottery bodies and as suspending agents for vitreous enamels, ball clays are used to bond nonplastic refractories such as sillimanite. Current consumption is about 400 000 tons in USA and 100 000 tons in Britain.

Ball-in-hand Test. A ball of mixed castable refractory, shaped by hand, is allowed to fall from a specified height and caught in the open fingers of the hand. The consistency is judged with the aid of photographs. The test is specified in ASTM C-860 and in BS 1902 Pt. 7.3.

Ball Mill. A fine-grinding unit of a type much used in the pottery, vitreous enamel, and refractories industries. It consists of a steel cylinder or truncated cone that can be rotated about its horizontal axis (see CRITICAL SPEED); the material to be ground is charged to the mill together with alloy-steel balls, pebbles, or specially made ceramic grinding media (generally high-alumina). The mill may be operated dry or wet; wet ball-milling is usually a batch process but dry ball-milling may be continuous, the 'fines' being removed by an air current (AIR-SWEPT BALL MILL).

Ball Test. See KELLY BALL TEST.

Balling. (1) The tendency of some ceramic materials or batches, when moist, to aggregate into small balls when being mixed in a machine.

(2) One method of shaping pottery hollow-ware in a JOLLEY (q.v.); a ball of the prepared body is placed in the bottom of the plaster mould and is then 'run up' the sides of the mould as it rotates.

Ballotini. Transparent glass spheres less than about 1.5 mm diameter; presumably a derivative of the Italian *ballotta,* a small ball used for balloting.

Baltride. Tradename. A coating of TITANIUM NITRIDE (q.v.) applied, especially to twist drills, by plasma vapour deposition. (Balzer's High Vacuum Co).

Bamboo Tile. A term used in Africa and Asia for SPANISH TILE (q.v.) on account of their resemblance, when placed in position, to a roof of split bamboo.

Bamboo Ware. A bamboo-like type of CANE WARE (q.v.), somewhat dark in colour, first made by Josiah Wedgwood in 1770.

Banbury Mixer. A heavy-duty mixer for viscous pastes, with a pair of counter-rotating rotors.

Banding. The application, by hand or by machine, of a band of colour to the edge of a plate or cup.

Band Theory of Solids. The outer electrons of the atoms in a crystal lattice behave as though unbound, but subject to a periodic electrical potential corresponding to the lattice spacing. Their possible energies fall into *bands* of *allowed* and *forbidden* states. See also ELECTRICAL CONDUCTIVITY.
Bank Kiln. A primitive type of pottery kiln used in the Far East; it is built on a bank, or slope, which serves as a chimney.
Banks. The sloping parts between the hearth of an open-hearth steel furnace and the back and front walls. They were constructed of refractory bricks covered with fritted sand (Acid O.H. Furnace) or burned-in magnesite or dolomite (Basic O.H. Furnace). cf. BREASTS.
Bannering. Truing the rim of a saggar (before it is fired) by means of flat metal or a wooden board, to ensure that the rim lies in one horizontal plane and will in consequence carry the load of superimposed saggars uniformly.
Bansen. A unit of gas permeability expressed in $[(m^3/h.).cm]/[m^2(mm\ water\ pressure)]$ (H. Bansen, *Arch. Eisenhuttenw.* **1**, 667, 1927–28).
Baraboo Quartzite. A quartzite of the Devil's Lake region of Wisconsin, USA, used in silica brick manufacture. A quoted analysis is (per cent): SiO_2, 98.2; Al_2O_3, 1.1; Fe_2O_3, 0.2; Na_2O + K_2O, 0.1.
Barbotine. French word for SLIP (q.v.).
Barelattograph. A French instrument for the automatic recording of the contraction and loss in weight of a clay body during drying under controlled conditions; (for description see *Bull. Soc. Franc. Céram.* No. 40, 67, 1958).
Barium Boride. BaB_6; m.p., 2270°C; sp. gr., 4.32; thermal expansion, 6.5×10^{-6}; electrical resistivity (20°C), 306 μohm.cm.
Barium Carbonate. $BaCO_3$; decomposes at 1450°C; sp. gr. 4.4. Occurs naturally in Durham, England, as witherite. The principal use in the ceramic industry of the raw material is for the prevention of efflorescence on brickwork; for this purpose it is added to brick-clays containing soluble sulphates. The pure material is used in the manufacture of barium ferrite permanent magnets, and in some ceramic dielectrics to give lower dielectric loss. In the glass industry this compound is used in optical glass and television tubes; it is also used in some enamel batches.
Barium Chloride. $BaCl_2$ a soluble compound sometimes used to replace $CaCl_2$ as a mill-addition in the manufacture of acid-resisting vitreous enamels. It also helps to prevent scumming. Poisonous.
Barium Osumilite. A glass-ceramic formulation $BaMg_2Al_3(Si_9Al_3O_{30})$ patented by J. J. Brennan et al (U.S. Pat 4589900, 1986, used as the foundation material for composites.
Barium Oxide. BaO; m.p. 1920°C; sp. gr. 5.72.
Barium Stannate. $BaSnO_3$; it is sometimes added to barium titanate bodies to decrease the Curie temperature. An electroceramic containing approx. 90 mol% $BaTiO_3$ and 10 mol%, $BaSnO_3$ has a very high dielectric constant (8000–12000 at 20°C).
Barium Sulphate. $BaSO_4$; m.p. 1580°C; sp. gr. 4.45.
Barium Titanate. $BaTiO_3$; m.p. 1618°C. Made by heating a mixture of barium carbonate and titania at 1300–1350°C. Because of its high dielectric constant (1350–1600 at 1 MHz and 25°C) and its piezoelectric and ferroelectric properties, it finds use in electronic components; its Curie temperature is 120–140°C. The properties can be altered by the

formation of solid solutions with other ferroelectrics having the perovskite structure, or by varying the BaO:TiO_2 ratio on one or other side of the stoichiometric.

Barium Zirconate. $BaZrO_3$; m.p. 2620 C; sp. gr. 2.63. Synthesized from barium carbonate and zirconia and used as an addition (generally 8–10%) to barium titanate bodies to obtain high dielectric constant (3000–7000) and other special properties.

Barker-Truog Process. A process described by G. J. Barker and E. Truog (*J. Amer. Ceram. Soc.*, **21**, 324, 1938; **22**, 308, 1939; **24**, 317, 1941) for the treatment of brickmaking clays with alkali, this being claimed to facilitate shaping and to reduce the amount of water necessary to give optimum plasticity. According to their patent (US Pat. 2 247 467) the clay is mixed with alkali to give pH 7–9 if it was originally acid, or pH 8–10 if originally non-acid; it is also stipulated that the total amount of alkali added shall be limited by the slope of the curve relating the pH to the quantity of alkali added, this slope being reduced to half its original value. The alkali is usually added as soda ash (Na_2CO_3) and the amount commonly required is about 0.2%.

Bar Kiln, Bar Dryer. A pre-fabricated, steel-cased tunnel kiln for the rapid firing of heavy clay products is combined in a two-storey arrangement with a rapid dryer in which the ware is set on the same kiln cars as are used for firing. (Occidental Industries, Paris).

Barratt-Halsall Firemouth. A design for a stoker-fired firemouth for a pottery BOTTLE OVEN (q.v.); a subsidiary flue system links all the firemouths around the oven wall to assist in temperature equalization. The design was patented by W. G. Barratt and J. Y. M. Halsall (Brit. Pat. 566 838; 16/1/45). Also known as the GATER HALL DEVICE because it was first used at the factory of Gater, Hall & Co., Stoke-on-Trent, where J. Y. M. Halsall was General Manager and which was at the time associated with the Barratt pottery.

Barrelling. The removal of surface excrescences and the general cleaning of metal castings by placing them in a revolving drum, or barrel, together with coarsely crushed abrasive material such as broken biscuit-fired ceramic ware (cf. RUMBLING).

Barton Clay. A clay of the Eocene period used for brickmaking near the coast of Hampshire and in the Isle of Wight.

Barytes. Naturally occurring BARIUM SULPHATE (q.v.) It is the principal source of BARIUM CARBONATE (q.v.) but is also used to some extent in the natural state in glass manufacture.

Basalt Ware. A type of ceramic artware introduced in 1768 by Josiah Wedgwood and still made by the firm that he founded. The body is black and vitreous, iron oxide and manganese dioxide being added to achieve this; a quoted composition is 47% ball clay, 3% china clay, 40% ironstone and 10% MnO_2. The ware has something of the appearance of polished basalt rock (cf. FUSION-CAST BASALT).

Base Coat. A fired coating over which another coating is applied. See GROUND-COAT; ENGOBE.

Base Code. Alternative name for PUNT CODE (q.v.).

Base Exchange. See IONIC EXCHANGE.

Base Metal. (1) In the phrase 'base-metal thermocouple, this term signifies such metals and alloys as copper, constantan, nickel, tungsten, etc.

(2) In the vitreous-enamel industry, the term means the metal (steel or cast iron) to which the enamel is applied.

Basic Fibre. Glass fibres before they have been processed. A number of fibres may subsequently be bonded together to form a strand.

Basic Open-hearth Furnace. An OPEN-HEARTH FURNACE (q.v.) used in the refining of basic pig-iron. The hearth is built of basic refractory bricks covered with burned dolomite or magnesite.

Basic Oxygen Steelmaking. Various related processes in which oxygen is blown into the molten iron bath by a retractable refractory lance. The vessels are lined with basic refractories, and lime may also be injected through the lance. The BOS, BOF or LD/AC process is top-blown, in that the oxygen is introduced from above. The Q-BOP (quiet or quick BOP) or OBM (oxygen bottom Maxhuette) process is bottom-blown through a nozzle in the lower lining, which shrouds the oxygen with a concentric jet of hydrocarbons to protect the lining. See L-D PROCESS.

Basic Refractory. A general term for those types of refractory material that contain a high proportion of MgO and/or CaO, i.e. oxides that at high temperatures behave chemically as bases. The term includes refractories such as magnesite, chrome-magnesite, dolomite, etc.

Basket-weave Checkers. CHECKERS (q.v.) built to form continuous vertical flues.

Basse-taille. Vitreous enamelled artware in which a pattern is first cut in low relief on the metal backing, usually silver; the hollows are then filled with translucent enamel, which is subsequently fired on.

Basset Process. For the simultaneous production of hydraulic cement and pig-iron by the treatment, in a rotary kiln, of a mixture of limestone, coke, and iron ore. (L. P. Basset, French Pat. 766 970, 7/7/34; 814 902, 2/7/37.)

Bastard Ganister. A silica rock having many of the superficial characters of a true GANISTER (q.v.) such as colour and the impression of rootlets, but differing from it in essential details, e.g. an increased proportion of interstitial matter, variable texture, and incomplete secondary silicification.

Bat. (1) A refractory tile or slab as used, for example, to support pottery-ware while it is being fired.

(2) A slab of plaster used for various purposes in the pottery industry.

(3) A roughly shaped disk of pottery body – as prepared on a batting-out machine prior to the jiggering of flatware, for example.

(4) A short building brick, either made as such or cut from a whole brick.

(5) In USA, a piece of brick that has been broken off.

Bat Printing. A former method of decorating pottery; it was first used, in Stoke-on-Trent, by W. Baddeley in 1777. A bat of solid glue or gelatine was used to transfer the pattern, in oil, from an engraved copper plate to the glazed ware, colour then being dusted on. The process was still in use in 1890 and has now been developed into the MURRAY-CURVEX MACHINE (q.v.).

Batch. A proportioned mixture of materials. The term is used in most branches of the ceramic industry; in glass-making, the batch is the mixture of materials charged into the furnace.

Batch-type Mixer. See MIXER.

Ban. See BAT.

Batdorf Theory. A theory of brittle fracture which assumes random flaw orientation and a consistent crack geometry, proposing that reliability predictions should be based on linear elastic fracture mechanics combined with the weakest link theory. See FRACTURE MECHANICS. (S. B. Batdorf and J. G. Crose. J Appl.Mech **41** (1974) p459).

Batt Wash. See REFRACTORY COATING.

Batter. In brickwork, the name given to an inclined wall surface or to the angle of such a surface to the vertical.

Battery. Term applied to the large refractory structure in which coke is produced in a series of adjacent ovens (cf. BENCH).

Battery Casting. A system of mechanised handling, tilting and automatic filling of the heavy plaster moulds used to slip-cast sanitaryware. Several moulds are grouped together in a 'battery' and handled as one unit. The sytem was invented by Shanks Ltd of Glasgow (Br.Pat.), and developed by the Armitage-Shanks Group.

Battery Top Paving. Refractory brickwork, cast blockwork or cast refractory concrete laid or placed on the top of a coke-oven BATTERY.

Batting Out. The process of making a disk of prepared pottery body for subsequent shaping in a JIGGER (q.v.).

Battledore. A tool used in the hand-made glass industry for shaping the foot of a wine glass; also known as a PALLETTE.

Baudran Expansion Apparatus. A device for measuring the thermal expansion of ceramics up to 1500°C; expansion of the test-piece is directly transmitted by a lever system to an inductive displacement detector. This procedure is used for testing refractories (B.S. 1902, Pt. lA); it was devised by A. Baudran *(Bull. Soc. Franc. Ceram. (27), 13,* 1955).

Baumé Degrees (°Bé). A system, introduced by a Frenchman, A. Baumé, in 1768 for designating the specific gravity of liquids. There are two scales, one for liquids lighter than water and one for heavier liquids; only the latter is normally of interest in the ceramic industry, where it is sometimes used in describing the strength of sodium silicate solutions. The precise equivalence between the Baumé scale and the specific gravity is in some doubt but the following conversion factor has been standardized in USA and may be generally accepted:

sp. gr. = 145/(145–°Bé)

Bauxite. A mineral containing a high proportion of aluminium hydroxides, formed by weathering in tropical climates. Its principal sources are Australia, Brazil, France, French Guinea, Guyana, Hungary, Jamaica, Surinam, Russia and former Yugoslavia. It is purified in the BAYER PROCESS (q.v.) to produce calcined alumina used in the abrasives, electroceramics, refractories and other branches of the ceramics industry.

Bauxitland Cement. See KUHL CEMENT.

Bayer Process. A process for the extraction of alumina from bauxite, invented by K. J. Bayer in 1888. The bauxite is digested with hot NaOH and the Al_2O_3 is then extracted as the soluble aluminate. Because of this method of extraction, the calcined alumina used in the ceramic industry usually contains small quantities of Na_2O.

Bayerite. A trihydrate of alumina, $Al_2O_3.3H_2O$, named after K. J. Bayer, originator of the BAYER PROCESS. (q.v.); however, the trihydrate formed in this process is GIBBSITE (q.v).

BDKG. Berichte der Deutschen Keramische Gesellschaft.

Bead. (1) Any raised section round glass-ware.
(2) A small piece of glass or ceramic tubing surrounding a wire.

Beading. (1) The application of vitreous enamel slip, usually coloured, to the edge of enamelware.
(2) Removal of excess vitreous enamel slip from the edges of dipped ware.
Bead Test. The softening and flow characteristics of a glass are assessed by comparing the bahaviour of a bead of specified size and shape with beads of standard compositions.
Bear. See SALAMANDER
Bearer Arch. See RIDER BRICKS.
Bed Alternative name for a SETTING (q.v.) of gas retorts.
Bed Joint. A horizontal joint in brickwork.
Bedder. A plaster shape conforming to that of the ware to be fired and used to form the bed of powdered alumina on which bone china is fired.
Bedding. (1) The placing of pottery flatware on a bed of refractory powder, e.g. silica sand or calcined alumina, to provide uniform support during the firing process, thus preventing warpage.
(2) The compacted base (of gravel or other material) on which sewer pipes are laid and supported. ASTM C896. The Clay Pipe Development Association, UK has published tables relating to bedding and loads on vitrified clay pipes.
Beehive Kiln. See ROUND KILN.
Beer-Lambert Law. = LAMBERT'S LAW (q.v.).
Beidellite. A clay mineral first found at Beidell, Colorado, USA; it is an Al-rich member of the montmorillonite group.
Belfast Sink. A domestic or industrial sink with an overflow and with its top edge plain; cf. EDINBURGH SINK and LONDON SINK.
Belgian Kiln. A type of annular kiln patented by a Belgian, D. Enghiens (Brit.Pats.18281, 1891; 22008, 1894; 7914, 1895). It is a longitudinal-arch kiln with grates at regular intervals in the kiln bottom; it is side-fired on to the grates. Such kilns have been popular for the firing of fireclay refractories at 1200–1300°C.
Belite. The name given to one of the crystalline constituents of portland cement clinker by A. E. Törnebohm (*Tonindustr. Ztg.*, **21**, 1148, 1897). This mineral has since been identified as one form of $2CaO.SiO_2$.
Bell. The socket of a sewer pipe, into which the SPIGOT fits. See also TRUMPET.
Bellarmine. A fat, salt-glazed bottle or jug, usually decorated with a bearded face stamped on the narrow neck.
Bell Damper. A damper of the sand-seal type, and bell-shaped. Such dampers are used, for example, in annular kilns.
Belleek Ware. A distinctive type of pottery made at Belleek, County Fermanagh, Northern Ireland. The factory was established in 1857 and the ware is characterized by its thinness and slightly iridescent surface; the body contains a significant proportion of frit.
Bell Kiln. See TOP HAT KILN.
Bench Paving. Brickwork on the top surfaces of coke oven side benches, to resist coke spillage, heat and abrasion.
Belly. (1) The part of a blast furnace where the cross-section is greatest, between the STACK (q.v.) and the BOSH (q.v.); this part of a blast furnace is also known as the WAIST.
(2) The part of a converter in which the refined steel collects when the converter is tilted before the steel is poured.
Bender Abrasion Method. The Concrete Masonry Association of Australia test for ABRASION RESISTANCE (q.v.) based on the test developed by Perth City Council, derived from ASTM C779–82. The abrasion index Ia is given by
Ia = (number of thousands of revolutions)½/depth of penetration.

Steel ball bearings are driven at constant load across the surface and the depth of penetration measured.

Belshazzar. A 16-quart wine bottle.

Belt Kiln. A tunnel kiln through which ware is carried on an endless belt made of a wire-mesh woven from heat-resisting alloy. Such belts, if slightly overheated, can leave *belt marks* on the ware, a common fault with glassware from the lehr. In the pottery industry such kilns have found some use for glost and decorating-firing.

Bench. (1) A series of chamber ovens or gas retorts built of refractory bricks to form a continuous structure.
(2) The floor of a pot furnace for making glass.
(3) A clay pit, if beyond a certain depth. may be worked in a a series of horizontal layers, known as Benches: these may coincide with different strata in the working face.

Bend Test. See MODULUS OF RUPTURE.

Beneficiation. US term for any process, or combination of processes, for increasing the concentration of a mineral in an ore; the term is now replacing the older term 'mineral dressing' in the UK. An example of beneficiation in the ceramic industry is the treatment of certain pegmatites to produce a SPODUMENE (q.v.) concentrate.

Bent Glass. FLAT GLASS (q.v.) which has been reshaped when hot.

Bentonite. A natural clay, named from Fort Benton, USA, where it was first found. It consists chiefly of MONTMORILLONITE (q.v.). It occurs as two varieties – sodium bentonite and calcium bentonite; the former swells very considerably when it takes up water. The principal source is USA: Western bentonite (the Na variety from the Wyoming area) and Southern bentonite (the Ca variety from the Mississippi area). The particle size of bentonite is all less than 0.5μm; It has remarkable bonding properties and is used as a binder for non-plastics, particularly for foundry sands.

Bentonite Number. A measure of sedimentation resistance, expressed in terms of the heights of clear and turbid portions of the liquid column, and its overall density. A sharp boundary between the suspension and the clear liquid is formed by adding bentonite. (Glass Ceram. **39** (5) 224 1983).

Berg Method. See DIVER METHOD.

Berkeley Clay. A plastic, refractory kaolin from S. Carolina; P.C.E.34.

Berkovich Indenter. A 3-faced diamond pyramid for indentation hardness tests. A sharper tip can be maintained compared to the 4-faced Vickers pyramid indenter.

Berl Saddle. A chemical stoneware shape for the packing of absorption towers; it is saddle-shaped, size about 25mm, and approx. 60000 pieces are required per m^3. A packing of Berl Saddles provides a very large contact surface and is a most efficient type of ceramic packing. (Patented in Germany by E. Berl 1928).

Berlin Porcelain. Porcelain, particularly laboratory porcelain, made at the Berlin State Porcelain Factory. A quoted body composition is 77% purified Halle clay and 23% Norwegian feldspar, all finer than 10mm. The ware is fired at 1000°C and is then glazed and refired at 1400°C.

Bernitz Blocks. Perforated refractory blocks designed for the passage of air and steam in water-gas plant.

Berry Machine. See SOFT MUD PROCESS.

Beryl. $3BeO.Al_2O_3.6SiO_2$; this mineral is the only economic source of beryllium oxide and other beryllium compounds. It occurs in Brazil, South Africa, India, and

Madagascar. Beryl has a low, but anisotropic, thermal expansion.

Beryllia. See BERYLLIUM OXIDE.

Beryllides. A group of intermetallic compounds of potential interest as special ceramics. Cell dimensions and types of structure have been reported for the beryllides of Ti, V, Cr, Zr, Nb, Mo, Hf and Ta (A. Zalkin *el al.*, *Acta Cryst.*, **14**, 63, 1961).

Berylliosis. A serious lung disease acquired by exposure to very small quantities of the dust of beryllia or other beryllium compounds. It may take many years to reveal itself.

Beryllium Carbide. Be_2C; decomposes at 2100°C with volatilization of Be; sp. gr. approx. 2; Knoop hardness (25-g load) 2740; thermal expansion (25–800°C), 10.5×10^{-6}. This refractory carbide is slowly attacked by water at room temperature and by O_2 and N_2 at 1000°C. It has been used as a moderator in nuclear engineering.

Beryllium Nitride. Be_3N_2; m.p. 2200°C but oxidizes in air when heated to above 600°C. There are two forms: cubic and hexagonal.

Beryllium Oxide or **Beryllia.** BeO; m.p. approx. 2530°C; sp. gr. 3.02. A special refractory oxide notable for its abnormally high thermal conductivity (1.4 W/mK at 60°C) and mechanical strength (crushing strength 1.5 GN/m^2; transverse strength, 200 MN/m^2 tensile strength 100 MN/m^2-all at 20°C). Thermal expansion (20–1000°C), 9×10^{-6}. Thermal-shock resistance is good. Electrical properties: Volume resistivity (300°C) 10^{13}–10^{15} ohm-cm; dielectric constant (lMHz) 6–7; power factor (1 MHz) 6.5×10^{-4}. ASTM F356 relates to its electrical and electronic applications. It acts as a moderator for fast neutrons and is used for this purpose in nuclear reactors; it is also sometimes used as a constituent of special porcelains. Instruments have been mounted on BeO blocks in space-craft. Beryllium compounds are toxic.

Bessemer Converter. See CONVERTER.

Best Gold. See BURNISH GOLD.

Besto wing. The cover of fired bricks (usually three courses) for the setting of a CLAMP (q.v.).

Beta-Alumina. (β-alumina) The compound $Na_2O.11Al_2O_3$. It has applications as a solid electrolyte.

Bethel-Bagnall Nozzle. A composite refractory nozzle designed to give a constant teeming rate in the casting of steel. A nozzle having the diameter required for the latter part of the teeming is fitted in the ladle well, and a secondary nozzle of smaller diameter is then fitted below the primary nozzle; the nozzle of smaller diameter is removed when the ladle has been partly emptied. (H. A. Bethel and F. T. Bagnall, *Brit. Pat.*, 549 212, 11/11/42).

B.E.T. Method. See BRUNAUER, EMMETT and TELLER METHOD.

Better Bed Fireclay. A siliceous fireclay occurring under the Better Bed Coal of the Leeds area, England.

Bevel Brick. A brick having one edge replaced by a bevel (see Fig. 1, p39).

Bevelled. Cut at an angle. Flat glass is given *bevelled edges* for aesthetic effect, especially on mirrors. *Bevelled bricks* are so made to fit corners, or to enhance the appearance of the top of a wall. *Bevelled pipes* have angled ends to fit complementary ends or other components at an angle.

BG-sign. A sign that the slip resistance of a surface has been tested by the Berufsgenossenschaftliche Institut für Arbeitssicherheit. The sign is accompanied by a classification of the open voids and the slip resistance.

BHN. BRINELL HARDNESS (q.v.) NUMBER.

BI. Bridge Indentation. See FRACTURE TOUGHNESS TESTS.
Bicheroux Process. A method developed in 1918 by Max Bicheroux, of Aachen, for the production of plate glass; molten glass from a pot is cast between rollers.
Bidet. A low ceramic bowl specially designed for washing the private parts. BS5505 provides a specification.
Bierbaum Scratch Hardness. A diamond point is drawn across the surface to be measured, under a specified pressure, and the scratch width is measured with a travelling microscope.
Bigot Curve. Drying shrinkage curves for clays which correlate lineal shrinkage with water loss in air-drying to 110°C. Plasticity figures are derived by multiplying the shrinkage by the ratio of weights of water of plasticity to total water absorbed. (A. Bigot, *Comptes Rendues*, **172**, 755, 1921).
Binder. A substance added to a ceramic raw material of low plasticity to facilitate its shaping and to give the shaped ware sufficient strength to be handled; materials commonly used for this purpose include sulphite lye, sodium silicate, molasses, dextrin, starch, gelatine, etc. The term BOND (q.v.) is also sometimes used in this sense but is better reserved to denote the intergranular material that gives strength to the fired ware.
Bingham Body or **Bingham Material.** A material that behaves elastically up to a yield stress but at higher stresses deforms at a rate proportional to the stress in excess of the yield value. (Named after E. C. Bingham; see BINGHAM PLASTOMETER).
Bingham Plastometer. A device for the measurement of the rheological properties of clay slips by forcing the slip through a capillary under various pressures; a curve is drawn relating the rate of flow to the pressure. E. C. Bingham, *Proc. A.S.T.M.*, **19**, Pt.2, 1919; **20**, Pt. 2, 1920.)
Bioceramics. Ceramics used in biomedical applications. The chief applications are as DENTAL CERAMICS (q.v.) and PROSTHESES (q.v.). Alumina and synthetic apatite or hydroxyapatite are the most frequently used for prostheses, which must be compatible with body fluids.
Bioglass. Registered Tradename of University of Florida for a series of bioactive glasses.
Bioglass-Ceramics. Glass-ceramics (q.v.) usually based on calcium phosphates, used in biomedical applications
Biomimetic Process. The gradual incorporation into the body, by a natural process, of a synthetic material mimicking a natural material. For example, the gradual intergrowth of bone material into synthetic porous hydroxyapatite prostheses.
Biot Number. The heat-transfer ratio *hr/k*, where *h is* the heat transfer coefficient, *r* is the distance from the point or plane under consideration to the surface, and *k* is the thermal conductivity. The Biot Number is a useful criterion in assessing thermal-shock resistance. (M. A. Biot, *Phil. Mag.*, **19**, 540, 1935.)
Biotite. A mica containing appreciable amounts of iron and magnesium; it sometimes occurs as an impurity in such ceramic raw materials as feldspar and nepheline syenite.
Bird's Beak. A special type of wall tile (see Fig. 7, p350).
Birdcage. An imperfection occasionally occurring in bottle manufacture, a glass thread (or threads) spanning the inside of the bottle.
Birefringence. Double refraction, with the formation of two images through a

crystal such as calcite, which has different refractive indexes for the two planes of polarization of the incident light. The greater this difference, the greater the birefringence, or separation of the images. BS 7604 specifies two tests (Pt. 1a tensile test; Pt.2 a bending test) for determining the stress optical coefficient of glass, allowing stresses to be computed from birefringence measurements.

Biscuit. (1) Pottery that has been fired but not yet glazed. Biscuit earthenware is porous and readily absorbs water; vitreous ware and bone china are almost non-porous even in the biscuit state. (2) In the vitreous-enamel industry, the dried (but still unfired) coating of enamel slip.

Biscuit Firing. The process of kiln firing potteryware before it has been glazed. Earthenware is biscuit-fired at 1100–1150°C; bone china is biscuit-fired at 1200–1250°C.

Bismuth Oxide. Bi_2O_3; m.p. 825°C; sp. gr. approx. 8.5. Four crystalline forms exist.

Bismuth Stannate. $Bi_2(SnO_3)_3$: sometimes added to barium titanate bodies to modify their dielectric properties; particularly to produce bodies having an intermediate dielectric constant (1000–1250) with a negligible temperature coefficient.

Bismuth Telluride. Bi_2Te_3; m.p. 585°C. A semiconductor that has found some use as a thermoelectric material.

Bisque. The older form, still preferred in USA, of the term BISCUIT (q.v.).

Bit Gatherer. The man whose job is to gather small quantities of glass for use in decorating hand-blown glass-ware.

Bitstone. Broken pitchers or calcined flints (in either case crushed to about 25mm) formerly used in the bottom of a saggar for glost firing to prevent sticking.

Black Ash. This term has been applied to crude barium sulphide, which has been used as a substitute for barium carbonate as an additive to clay to prevent scumming.

Black Body. As applied to heat radiation, this term signifies that the surface in question emits radiant energy at each wavelength at the maximum rate possible for the temperature of the surface, and at the same time absorbs all incident radiation. Only when a surface is a Black Body can its temperature be measured accurately by means of an optical pyrometer.

Black Core or **Black Heart.** Most fireclays and brick-clays contain carbonaceous matter; if a brick shaped from such clays is fired too rapidly, this carbonaceous matter will not be burned out before vitrification begins. The presence of carbon and the consequent reduced state of any iron compounds in the centre of the fired brick result in a 'Black Core' or 'Black Heart'.

Black Edging. A black vitreous enamel applied over the GROUND-COAT (q.v.) and subsequently exposed, as an edging, by brushing away the COVER-COAT (q.v.) before the ware is fired.

Black Speck. A fault, in glass, particularly in the form of a small inclusion of chrome ore. On pottery-ware the fault is generally caused by small particles of iron or its compounds. In vitreous enamelware also, the fault is caused by contamination.

Black Silica. A carbonaceous silica ash produced by the controlled incineration of rice husks (q.v.)

Blackboard Enamel. A slightly rough, matt vitreous enamel which will accept writing in chalk.

Blade. A section of a setting in a batch kiln.

Blaes. A Scottish name for carbonaceous shales, of a blue-grey colour, associated in the Lothians with oil-shales but differing from these in having a much lower proportion of bituminous matter, in being brittle rather than tough, and in producing when weathered a crumbling mass which, when wetted, is plastic.

Blaine Test. A method for the evaluation of the fineness of a powder on the basis of the permeability to air of a compact prepared under specified conditions. The method was proposed by R. L. Blaine *(Bull. A.S.T.M.,* (123), 51, 1943) and is chiefly used in testing the fineness of portland cement. The equipment is essentially the same as RIGDEN'S APPARATUS (q.v.).

Blake's Profiles. Auxilary tools to ensure accurate brickwork.

Blake-type Jaw Crusher. A jaw crusher invented by E. W. Blake in 1858. One jaw is fixed, the other being pivoted at the top and oscillating at the bottom. Output is high but the product is not of uniform size (cf. DODGE-TYPE JAW CRUSHER).

Blakely Test. See TUNING FORK TEST.

Blank. (1) In the vitreous-enamel industry, a piece cut from metal sheet ready for the shaping of ware. (2) A piece of glass that has received preliminary shaping. See also CLOT; OPTICAL BLANK.

Blanc-de-Chine. White, glazed Chinese porcelain.

Blanc Fixe. BARYTES (q.v.)

Blank Mould. The metal mould in which the PARISON (q.v.) is shaped in glass hollow-ware production.

Blanket Feeding. The charging of batch (to a glass tank furnace) as a broad, thin layer; this gives a uniform distribution across the width of the furnace.

Blast Furnace. A shaft furnace, 30–36m high, for the extraction of iron from its ore; a smaller type of blast furnace is used for the extraction of lead. In iron-making, ore, limestone and coke are charged at the top and hot air is blasted (hence the name) in near the bottom. The whole furnace is lined with refractory material carbon/SiC in bosh and stack; aluminosilicate, silicon carbide, firebrick or gunned aluminosilicate in the upper stack; alumina-chrome, mag-chrome or SiC/ graphite castables may be used in the hearth.

Bleaching Clay. A clay, usually of the bentonite or fuller's earth type, used for decolorizing petroleum products etc.; the liquid is allowed to percolate through a layer of the clay, which adsorbs the colouring matter on its surface.

Bleb. A raised blister on the surface of faulty pottery-ware.

Bleeding. (1) The appearance of water at the surface of freshly placed concrete, hence the alternative name WATER GAIN; it is caused by sedimentation of the solid particles, and can to some extent be prevented by the addition of plasticizers and or by AIR ENTRAINING (q.v.). A test for the bleeding of cement pastes and mortars is provided in ASTM C243. See also LAITANCE.
(2) The movement of gold or platinum decoration on pottery ware during the early stages of the firing-on; the movement is caused by the passage of an excess of hot vaporized solvent over the ware as a result of overloading the kiln or inadequate draughting.

Blending Batch. A batch charged to a glass furnace during a changeover in composition of the finished glass. It has a composition intermediate between that of the glass already in the furnace and that finally required.

Bleu Persan. A form of pottery decoration in which a white pattern was painted over a dark blue background; the name derives from the fact that the pattern generally had a Persian flavour.

Blibe. A fault, in glass-ware, in the form of an elongated bubble intermediate in size between a seed and a blister. GREY BLIBE consists of undissolved sodium sulphate.

Blinding. (1) The clogging of a sieve or screen.
(2) US term for a glaze fault revealed by a reduction in gloss, and caused by surface devitrification.

Blister. A large bubble sometimes present as a fault in ceramic ware. In glass-ware, if near the surface it is a SKIN BLISTER and if on the inside surface of blown glass-ware it is a PIPE BLISTER (q.v.). The common causes of blisters in vitreous enamelware are flaws in the base-metal, surface contamination of the base-metal, and too high a moisture content in the atmosphere of the enamelling furnace.

Bloach (US). A blemish on plate glass resulting from stopping the grinding before all the hollows have been removed; a 'bloach' is an area of the original rough surface.

Bloating. The permanent expansion exhibited by some clays and bricks when heated within their vitrification range; it is caused by the formation, in the vitrifying clay, of gas bubbles resulting either from entrapped air or from the breakdown of sulphides or other impurities in the clay (cf. EXFOLIATION and INTUMESCENCE).

Block. (1) Defined in the UK (B.S. 3921) as: 'A walling unit exceeding in length, width or height the dimensions specified for a BRICK' (q.v.).
(2) A rectangular piece of KILN FURNITURE (q.v.), with cavities in its upper surface to hold small items such as spark plugs.

Block Handle. A cup handle of the type that is attached to the cup by a solid bar of clay (which is, of course, integral with the handle) (cf. OPEN HANDLE).

Block Mill. See PAN MILL.

Block Mould. A one-piece mould, especially a glassmaking mould.

Block Rake. A surface blemish, having the appearance of a chain, sometimes occurring on plate glass.

Block Reek. CULLET CUT (q.v.)

Blocking. (1) The shaping of glass in a wooden or metal mould.
(2) POLING (q.v.) with a block of wood.
(3) The removal of surface blemishes from glass-ware by reprocessing.
(4) The setting of optical glass blanks in a carrier prior to grinding and polishing.
(5) A US meaning is: running a glass-furnace idle at a reduced temperature.
(6) Synonym of THWACKING (q.v.).

Bloom. (1) Surface treatment of glass (e.g. lenses) by vapour deposition; this decreases reflection at the air/glass interface.
(2) A surface film on glass caused by weathering or by the formation of sulphur compounds during annealing.
(3) A fault on vitreous enamels, especially blacks, caused by their reaction with sulphur gases in the dryer or enamelling furnace to form a surface film of sodium sulphate; the fault may not become visible until the enamelware has been in a moist atmosphere for some time.

Blotter. US term for a disk of compressible material, e.g. blotting-paper stock, for use between a grinding wheel and its mounting flanges.

Blow-and-Blow. The process used to blow small-neck glass containers with the IS MACHINE (q.v.). The PARISON (q.v.) is blown and the second blowing process produces the final shape of the ware.

Blow-in Lining. A sacrifical refractory lining to protect the permanent lining of a blast furnace when recommissioned after re-lining. It normally covers BOSH to HEARTH.

Blow Mould. The metal mould in which blown glass-ware is given its final shape.

Blow pipe Spray Welding. See SPRAY WELDING.

Blowing. See LIME BLOWING, GLASS BLOWING.

Blowing Iron. An iron tube used in making hand-blown glassware

Blown Away. See HOLLOW NECK.

Blown Enamel. Vitreous enamel showing surface ridges. They form during wet-spraying with too thick or too fluid a coating, or too high an air pressure.

Blowpipe. See BLOWING IRON.

Blue Brick. See ENGINEERING BRICKS.

Blue Coring. A blue colouring at the centre of fired ware, caused by residual reduction of iron- or titanium-bearing materials in the body, after the organics have been burned out.

Blue Enamel. An area in dry-process vitreous enamelware where the coating is too thin and the ware appears blue.

Blueing. The production of blue engineering bricks, quarries or roofing tiles by controlled reduction during the later stages of firing; this alters the normal state of oxidation of the iron compounds in the material and results in a characteristic blue colour.

Blunger. A machine for mixing clay and/or other materials to form a slip; it usually consists of a large hexagonal vat with a slowly rotating vertical central shaft on which are mounted paddles. The process of producing a slip in such a machine is known as BLUNGING.

Blurring-highlight Test. A test to determine the degree of attack of a vitreous-enamelled surface after an acid-resistance test; (see ASTM – C282).

Blushing. A pink discoloration sometimes occurring during the glost-firing of pottery. It can be caused by traces of Cr in the kiln atmosphere arising, for example, from CHROME-TIN PINK (q.v.) fired in the same kiln, or by firing a tin-opacified glaze in a kiln containing traces of Cr vapour from the firing of Chrome Green.

Board. See WORK BOARD.

Boccaro Ware. Red, unglazed stoneware with relief decoration.

Bock Kiln. See BULL'S KILN.

Body. (1) A blend of raw materials awaiting shaping into pottery or refractory products.

(2) The interior part of pottery, as distinct from the glaze.

(3) The condition of molten glass conducive to ready working.

(4) The cylindrical part of a steel converter.

Body Mould. In the pressing of glass, that part of the mould which gives shape to the outer surface of the ware.

Body Stain. See STAIN (4).

Boehme Hammer. A device for the compaction of test-pieces of cement or mortar prior to the determination of mechanical strength, it consists of a hammer, pivoted so that the head falls through a definite are on the test-piece mould to cause compaction under standard conditions.

Boehmite. A monohydrate of alumina: often a constituent of bauxite and bauxitic clay.

Boetius Furnace. A semi-direct coal-fired pot furnace for melting glass; this was the first pot-furnace to use secondary air to increase the thermal efficiency.

Bogie Kiln or Truck Chamber Kiln. An intermittent kiln of the BOX KILN (q.v.)

type distinguished by the fact that the ware to be fired is set on a bogie which is then pushed into the kiln; the bogie has a deck made of refractory material (cf. SHUTTLE KILN).

Bohemian Glass. A general term for Czechoslovakian glass, particularly tableware and chemical ware; it is generally characterized by hardness and brilliance.

Boiling. A fault in vitreous enamelware (particularly in enamelled sheet-steel) visible as blisters, pinholes, specks, dimples or a spongy surface. The usual cause is undue activity of the GROUND-COAT (q.v.) during the firing of the first COVER-COAT (q.v.), but gases are also sometimes evolved during the firing of the cover-coat (cf. BOILING THROUGH).

Boiling Through. A fault in vitreous enamelware, small dark specks appearing on the surface of the ware, usually in consequence of gas evolution from the base metal.

Bole. A friable earthy clay highly coloured by iron oxide.

Bolley's Gold Purple. A colour that has been used on porcelain. A solution of stannic ammonium chloride is left for some days in contact with granulated tin and is then treated with dilute gold chloride solution. The gold purple is precipitated.

Bolt-hole Brush. A stiff brush with a straight metal guide, used to remove vitreous enamel from the edges of ware to prevent chipping and to neaten the fired product.

Bolus Alba. CHINA CLAY (q.v.)

Bomb. The large, approximately hemispherical pad of silicone rubber or, originally, gelatine which transfers the colour in the MURRAY-CURVEX printing process (q.v.).

Bond. (1) The arrangement of bricks in a wall; the bond is usually such that any crossjoint in a course is at least one quarter the length of a brick from joints in adjacent courses. For special types of bonds see AMERICAN BOND, DUTCH BOND, ENGLISH BOND, FLEMISH BOND, FLYING BOND, IN-AND-OUT BOND, MONK BOND, QUETTA BOND. To begin or end a course, a *bonder*, a brick of special size or shape, may be required.
(2) The placing of roofing tiles so that the joint between two tiles in one course is at or near the centre of the tile of the course below.
(3) The intergranular material, glassy or crystalline, that gives strength to fired ceramic ware.
(4) The bond in abrasive wheels may be ceramic, silicate of soda, resin, shellac, rubber, or magnesium oxychloride.

Bond Fireclay. A U.S. term for a fireclay of sufficient natural plasticity to bond nonplastic materials.

Bond's Law. A theory of grinding which postulates that the grinding rate of a solid is proportional to the crack propagation speed in the solid. (F.C.Bond, *Trans. Amer. Inst. Min. (metall) Engrs.* **193**, 1952, p484).

Bond Strength. In ASTM C952 for the bond strength of mortar to masonry, a crossed-brick couplet tensile test is used for clay bricks; a stacked bond flexural test for concrete blocks. ASTM E518 is a flexural test for the bond strength of masonry, in which a transverse force is applied to a beam of bricks, using an air bag.

Bond and Wang Theory. A theory of crushing and grinding: the energy (h) required for crushing varies inversely as the modulus of elasticity (E) and specific gravity (S), and directly as the square of the compressive strength (C) and as the approximate reduction ratio (n). The energy in hp.h required to crush a short ton of material is given by the following equation, in which all quantitics are in f.p.s. units:

$$h = \left[\frac{0.001748C^2}{SE} \right] \left[\frac{(n+2)(n-1)}{n} \right]$$

The theory is due to F. C. Bond and J. T. Wang *(Trans. Amer. Inst. Min(metall.) Engrs.,* **187**, 875, 1950).

Bondaroy's Yellow. An antimony yellow developed by Fourgeroux de Bondaroy in 1766: white lead, 12 parts; potassium antimonate, 3 parts; alum, 1 part; sal ammoniac, 1 part.

Bonded Roof. A term for the roof of a furnace when the transverse joints in the roof are staggered (cf. RIGGED ROOF).

Bonder. A brick that is half as wide again as a standard square (rectangular or arch); such bricks are sometimes used to begin or end a course of bonded brickwork. (See Fig. 1, p39).

Bondley Process. See under METALLIZING.

Bone Ash. Strongly calcined bone, approximating in composition to $Ca_3(PO_4)$. The m.p. of bone ash is about 1670°C, density 3.14. It is used in the making of CUPELS (q.v.). Less strongly calcined bone constitutes about 50% of the BONE CHINA body. It is occasionally used, to the extent of about 2%, in some vitreous enamels. See also CALCIUM PHOSPHATE.

Bone China. Vitreous, translucent pottery made from a body of the following approximate composition (per cent): calcined bone, 45–50; china clay, 20–25, china stone, 25–30. The COMBINED NOMENCLATURE (q.v.) defines this ware as: 'Completely vitrified, hard, impermeable (even before glazing), white or artificially coloured, translucent and resonant. Bone China contains calcium phosphate in the form of bone ash; a translucent body is thus obtained at a lower firing temperature than with hard porcelain. The glaze is normally applied by further firing at a lower temperature, thus permitting a greater range of underglaze decoration.' The British Pottery Manufacturers' Federation defined this material as: 'Ware with a translucent body containing a minimum of 30% of phosphate derived from animal bone and calculated as calcium phosphate. B.S.5416 'China Tableware' now specifies that bone china shall contain 'at least 35% by mass of the fired body of tricalcium phosphate'.In the USA, ASTM-C242 permits the term 'bone china' to be applied to any translucent whiteware made from a body containing as little as 25% bone ash. Bone china, though delicate in appearance, is very strong. It was first made by Josiah Spode, in Stoke-on-Trent, where by far the largest quantity of this type of high-class pottery is still made (cf. PORCELAIN).

Bonnet Hip. See HIP TILE.

Bonnybridge Fireclay. A fireclay occurring in the Millstone Grit in the Bonnybridge district of Scotland. A typical per cent analysis (fired) is: SiO_2, 56–57; Al_2O_3, 36; Fe_2O_3, 3–4; alkalis, 0.75. P.C.E. 32–33.

Bont. N. Staffordshire term for one of the iron hoops used to brace the outside brickwork of a BOTTLE OVEN (q.v.).

Boost Melting. The application of additional heat to molten glass in a fuel-fired tank furnace by the passage of an electric current through the glass.

Boot. Alternative name, preferred in USA, for POTETTE (q.v.).

Borax. $NaB_4O_7.10H_2O$; sp. gr. 2.36 (anhydrous), 1.7 (hydrated). Occurs in western USA and is an important constituent of vitreous enamel and glaze frits, and of some types of glass. On heating, the water of crystallization is lost by a series of steps:

$Na_2B_4O_7.10H_2O$ 62°C → $Na_2B_4O_7.5H_2O$ 130°C → $Na_2B_4O_7.3H_2O$ 150°C →

$Na_2B_4O_7.2H_2O$ 180°C → $Na_2B_4O_7.H_2O$ 318°C → $Na_2B_4O_7$.

The anhydrous borax, (sodium tetraborate) melts at 741°C.

Borax Glass. Vitreous, anhydrous, sodium tetraborate (NaB_4O_7).

Borazon. Trade-name (General Electric Co., USA); cubic form of BORON NITRIDE (q.v.).

Boric Oxide. B_2O_3; m.p. 450°C; sp. gr. 1.84. Solid B_2O_3 is commonly available only in the vitreous state; two crystalline forms exist, however, α-B_2O_3 (hexagonal) and β-B_2O_3 (a denser form).

Borides. A group of special ceramic materials. Typical properties are great hardness and mechanical strength, high melting point, low electrical resistivity and high thermal conductivity; impact resistance is low but the thermal-shock resistance is generally good. For the properties of specific borides see under the borides of the following elements: Al, Ba, Ca, Ce, Cr, Hf, La, Mo, Nb, Si, Sr, Ta, Th, Ti, U, V, W, Zr.

Boroaluminate. See ALUMINIUM BORATE.

Borocalcite. See COLEMANITE.

Boron Carbide. B_4C; synthesized by the reaction of C and B_2O_3 at high temperatures. Boron carbide has a Knoop Hardness (100g) of 2300–2800, which is second only to that of diamond; for this reason it is used for grinding and drilling. Its transverse strength is approx. 280 MN/m^2 and compressive strength over 3GN/m^2 Boron carbide is also refractory (m.p. 2450°C) although oxidation becomes severe above 1100°C; it is otherwise chemically resistant and abrasion resistant, and finds use in nozzles and other high-temperature locations. The sp. gr. is 2.52. Thermal expansion (25–800°C) 4.5×10^{-6}; (25–2000°C) 6.5×10^{-6}.

Boron Nitride. BN; oxidizes in air at 800°C. but its m.p. under a pressure of nitrogen is >3000°C. Sp. gr. 2.2. This ceramic exists in forms that correspond to the graphite (hexagonal) and the diamond (cubic) forms of carbon. The cubic form is as hard as diamond and was originally made (R. H. Wentorf, *J. Chem. Phys.,* **26**, 956, 1957) by the simultaneous application of very high pressure (85 000 atm.) and temperature (1800°C). It has since been found that, in the presence of a catalyst, both thc temperature and pressure can be reduced. It is made under the trade-name Borazon by General Electric Co., USA. The hexagonal form of BN is readily produced by high-temperature reaction between B_2O_3 or BCl_3 and ammonia. Hot-pressed hexagonal BN is machinable. Its electrical resistivity is 1.7×10^{13} ohm.cm at 20°C and 2.3×10^{13} ohm.cm at 500°C. Uses include dielectric valve spacers, crucibles and rocket nozzles; it is not wetted by some molten glasses.

Boron Oxide. See BORIC OXIDE.

Boron Phosphate. BPO_4; vaporizes at 1400°C; sp. gr. 2.81; related structurally to high-cristobalite. It has been used as a constituent of a ceramic body that fires to a translucent porcelain at 1000°C.

Boron Phosphide. There are two compounds: BP, Sp. gr. 2.97; Knoop Hardness (100g) 3200; decomposes at 1130°C.; $B_{13}P_2$, sp. gr. 2.76; Knoop Hardness (100 g) 3800; melts or sublimes at approx. 2000°C.

Boron Silicides. See SILICON BORIDES

Borosilicate Glass. A silicate glass containing at least 5% B_2O_3 (ASTM definition); a characteristic property of borosilicate glasses is heat resistance.

Bort. Industrial diamond of the type used as an abrasive for cutting and grinding.

Bosh. (1) The part of a blast furnace between the tuyere belt and the lintel; it is usually lined with high-grade fireclay refractory and is water-cooled.
(2) In the glass industry, a tank containing water for cooling glass-making tools.

Boson Box. Incorrect form of BOZSIN Box (q.v.).

Boss; Bossing. In the process of pottery decoration known as GROUND-LAYING (q.v.), brush marks are removed by BOSSING, i.e. striking, the ware with a pad, or Boss, made by stuffing cottonwool into a silk bag.

Böttger Ware. A dark red stoneware.

Botting Clay. Prepared plastic refractory material for use in the stopping of the tap-holes in cupolas. A typical composition would be 50–75% fireclay, up to 50% black sand, 10% coal dust and up to 5% sawdust.

Bottle Brick. A hollow clay building unit shaped like a bottomless bottle, 305mm long, 76mm o.d., 50mm. i.d. and weighing 1kg. The neck of one unit is placed in the end of another to build beams, arches or flat slabs; steel reinforcement can be used. Bottle bricks have been used in France (where they are known as 'Fusées Céramiques'), in Switzerland, the Netherlands, and in S. America.

Bottle Oven. A type of intermittent kiln, usually coal-fired, formerly used in the firing of pottery; such a kiln was surrounded by a tall brick hovel or cone, of typical bottle shape.

Bottom Pouring or Uphill Teeming. A method of teeming molten steel from a ladle into ingot moulds. The steel passes through a system of refractory fireclay tubes and enters the moulds at the bottom; the refractory tubes are of various shapes. See TRUMPET, GUIDE-TUBE, CENTRE BRICK and RUNNER BRICKS.

Bottom Stirring Elements. Refractory blocks including apertures or pipes for passing gas or powder into the steel charge, built into the bottom lining of a BASIC OXYGEN FURNACE (q.v.).

Bottom Tuyeres. Refractory blocks inserted separately to introduce gases into the charge of a BASIC OXYGEN FURNACE (q.v.). Cf. BOTTOM STIRRING ELEMENTS.

Boudouard Reaction. The reaction leading to dissociation of CO into CO_2; see CARBON MONOXIDE DISINTEGRATION. (O. Boudouard, *Compt.Rend.,* **128,** 98, 824, 1522, 1899).

Boulder Clay. A glacial clay used in making building bricks, particularly in the northern counties of England.

Boule. A fused mass of material, pear-shaped, particularly as produced by the VERNEUIL PROCESS (q.v.). Sapphire (99.9% Al_2O_3) boules, about 50 mm long, are produced in this way, and are used, for example, in making thread guides, bearings and gramaphone needles.

Bourry Diagram. A diagram relating the water-loss and shrinkage of a clay body to the drying time; the two curves are shown on the one diagram. (E. Bourry, the 19th century French ceramist.)

Bouyoucos Hydrometer. A variable-immersion hydrometer. The original instrument was graduated empirically to indicate the weight of solids per unit volume of suspension; it was subsequently developed for particle-size analysis. (G. J. Bouyoucos, Soil Sci.,**23,** 319, 1927, **25**, 365, 1928; **26**, 233, 1928.)

Bowl. That part of the feeder which delivers molten glass to the following unit.

Bowmaker Test. A method of forecasting the durability of refractory glass-tank blocks proposed by E. J. C. Bowmaker (*J. Soc. Glass Tech.*, **13,** 130, 1929). The loss in weight of a sample cut from the tank block is determined after the sample has been immersed for 3 h. in HF/H_2SO_4 at 100°C; the acid mixture is 3 parts by vol. HF (commercial 50–60% HF) and 2 parts by vol. pure conc. H_2SO_4. The test is no longer considered valid.
Box-car roof. Popular name for the KREUTZER ROOF (q.v.).
Box Feeder. A device for feeding clay to preparation machines. It consists of a large metal box, open topped, with the bottom usually formed by a steel-band conveyor or by a conveyor of overlapping steel slats; for plastic clay the feeding mechanism may be a number of revolving screw shafts.
Box Kiln. A relatively small industrial intermittent kiln of box-like shape
Boxing. The placing of biscuit hollow-ware, e.g. cups, rim to rim one on another; this helps to prevent distortion during firing.
Boxing-in. A method of setting in a kiln so that, for example, special refractory shapes can be fired without being stressed and deformed; also known as POCKET SETTING.
Boyd Press. A toggle-press in which pressure is exerted both on the top and bottom of the brick in the mould. It was introduced in 1888 and has since been widely used for the dry-pressing of building bricks and refractories. (Chisholm, Boyd and White Co., Chicago, USA).
Bozsin Box. A box, with heat-insulated walls, containing a temperature recorder; it was designed by M. Bozsin to travel with the ware through a vitreous-enamelling furnace.
Brabender Plastograph; Brabender Plasti-Corder. Trade-names: instruments designed in USA to assess the plasticity of clays and other materials on the basis of stress measurement during a continuous shearing process.
Brackelsberg Furnace. A rotary furnace, originally fired by pulverized coal, for the melting of cast iron; the lining of the first such furnaces was a rammed siliceous refractory, but silica brick linings have also been used successfully (C. Brackelsberg, Brit. Pat. 283 381; 13/4/27.)
Bracken Glass. Old English glass-ware made from a batch in which the ash from burnt bracken supplied the necessary alkali.
Bracklesham Beds. Pale-coloured clays intermingled with glauconite sand occurring in parts of Southern England and worked for brickmaking to the S.W. of London and near Southampton.
Bragg Angle, Bragg's Law. See X-RAY CRYSTALLOGRAPHY.
Brake Linings. See FRICTION ELEMENTS.
Brasqueing. A process sometimes used for the preparation of the interior of a fireclay crucible prior to its use as a container for molten metal. The crucible is lined with a carbonaceous mixture; it is then covered with a lid and heated to redness (From French word with the same meaning.)
Bravais Lattice. See CRYSTAL STRUCTURE; SPACE GROUP.
Bravaisite. A clay mineral containing Mg and K, and of doubtful structure; it has variously been stated to be a mixture of kaolinite and illite or of montmorillonite and illite.
Brazing. Ceramics may be joined to each other, or to metals, by providing a molten interlayer which wets the surfaces to be joined, then solidifies. Metal alloys are usually used, though

inorganic glasses have been used for silicon nitride. Brazing may be *direct*, when the braze material itself (possibly including a suitable *activator*) wets the surfaces; or *indirect*, used for the many ceramic surfaces which are difficult to wet with low-melting brazes. The ceramic is first metallized, and then brazed using an alloy which wets the metallization layer. The most common method is the MOLY-MANGANESE process.

Brazilian Test. A method for the determination of the tensile strength of concrete, ceramic, or other material by applying a load vertically at the highest point of a test cylinder or disk (the axis of which is horizontal), which is itself supported on a horizontal plane. The method was first used in Brazil for the testing of concrete rollers on which an old church was being moved to a new site (cf BRITTLE-RING TEST).

Break-out. Defined (ASTM – C286) as: 'In dry-process enamelling, a defect characterized by an area of blisters with well-defined boundaries.'

Break-up of Matt Glaze. The term BREAK-UP is applied more particularly to the glazes containing rutile used on wall tiles. Some of the added rutile dissolves in the glaze, the yellow or brown titanates thus formed subsequently collecting round the undissolved rutile crystals to give the marbled effect known as the RUTILE BREAK or BREAK-UP.

Breast Wall. (1) The side-wall of a glass-tank furnace above the tank blocks, also known as CASING WALL, CASEMENT WALL or JAMB WALL.
(2) The refractory wall between the pillars of a glass-making pot furnace and in front of the pot.

Breasts. The sloping parts joining the hearth of an openhearth furnace to the furnace ends below the ports and adjoining brickwork (cf. BANKS).

Brecem. High alumina cements blended with granulated blast furnace slag, developed by the Building Research Establishment, Watford.

Bredigite. The form of CALCIUM ORTHOSILICATE (q.v.) that is stable from about 800–1447°C on heating, persisting down to 670°C on cooling.

Breezing. A thin layer of crushed anthracite or of coarse sand spread on the siege of a pot furnace before setting the pots.

Brémond Porosimeter. Apparatus for the evaluation of pore size distribution by the expulsion of water from a saturated testpiece. (P. Brémond, *Bull. Soc. Franc. Ceram.* (37), 23, 1957).

Brenner Gauge. An instrument for the non-destructive determination of the thickness of a coating of vitreous enamel; it depends on the measurement of the force needed to pull a pin from contact with the enamel surface against a known magnetic force acting behind the base metal.

Breunnerite. Magnesite containing 5–30% ferrous carbonate; some of the Austrian 'magnesite' used as a raw material for basic refractories is, more strictly, breunnerite.

Brewster. Unit of photoelasticity: 1 brewster is equivalent to a relative retardation of 10^{-13} cm^2/dyne. Named after Sir D. Brewster who, in 1816, demonstrated that glass becomes birefringent when stressed.

Brianchon Lustre. A lustre (q.v.) for pottery-ware developed by M. Brianchon, in Paris, in 1856. The reducing agent necessary to form the thin deposit of metal is incorporated as bismuth resinate so that a reducing atmosphere in the kiln is not needed. Although the easiest lustre to apply, it is

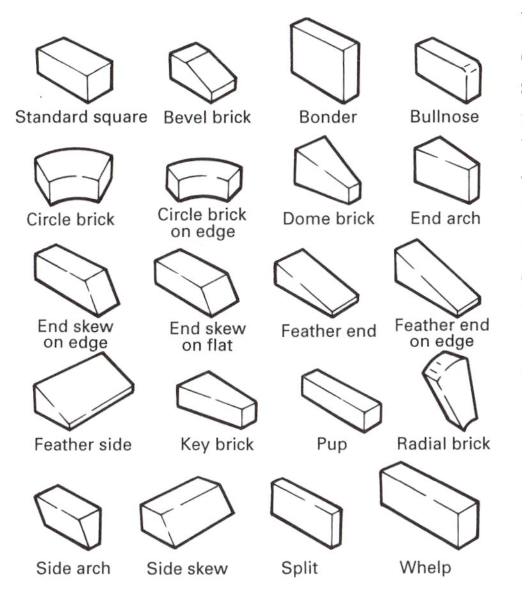

Fig. 1 Brick types

less durable than lustres produced in a reducing fire.

Brick. A brick is a clay or calcium silicate walling unit as big as can be conveniently handled in one hand. BS3921 for clay bricks, defines their *co-ordinating size* to be 225 × 112.5 × 75mm, with a *working size* (derived by subtracting a 10mm allowance for a mortar joint) of 215 × 102.5 × 65mm. BS3921 also specifies requirements for compressive strength, water absorption, soluble salt content, efflorescence and sampling. cf. BLOCK.

See following types: ACID-RESISTING; CLAY BUILDING; CONCRETE; FACING; FLOOR; PAVING; PERFORATED; SAND LIME; SEWER. See also CLADDING. For Brick Shapes, both for building and for furnace construction, see Fig. 1, and BS 3446 Pt.1. BS 3056 specifies dimensions of refractory bricks for various applications. BS 1902 Pt 3.11 specifies the measurement of dimensions and shapes of refractory bricks and blocks. ASTM C861 specifies metric dimensions for refractory bricks; C909 a series based on a 38-mm module for refractory rectangular and tapered bricks. Measurement of dimensions and warpage are specified in C134.

Brick Clays. Clays suitable for the manufacture of building bricks occur chiefly in the carboniferous and more recent geological systems. In the UK about 30% of the bricks are made from carboniferous clays, 30% from the Oxford clays, 10% Glacial clays, 6% Keuper Marl; the remaining 24% are made from Alluvial clays, the so-called Brick-earths, Tertiary, Cretaceous, Devonian, Silurian, and Ordovician deposits. Brick clays are impure and most of them vitrify to give bricks of adequate strength when fired at 900–1100°C.

Brick Earth. An impure loamy clay, particularly that of the Pleistocene of the Thames Valley, used for brickmaking.

Brick Shapes. See Fig. 1 and the further references in Appendix A. The German DIDIER CO publishes a detailed coding system for refractory bricks and shapes, which is a *de facto* standard in Europe.

Brick Slip. A fired ceramic architectural facing varying from about 13 to 38mm in thickness and produced either as a unit itself or cut from a larger unit.

Bricking Centre. A curved wooden former for the construction of arches.

Bricking Rig. A metal framework with compressed air plungers, for constructing arches or cylindrical refractory structures.

Bricklaying. The process of building a wall or other structure by bonding together individual BRICKS with MORTAR or other methods. BS 5628 Pt3 includes recommendations for bricklaying.

Brickwork. See MASONRY; BRICKLAYING.

Bridge; Bridge Wall. A refractory wall separating two parts of a furnace. The bridge wall (or firebridge) in a boiler furnace terminates the combustion chamber. In a glass-tank, the bridge wall separates the melting end from the working end of the furnace; the wall is in this case usually double, is pierced by the THROAT (q.v.) and is spanned at the top by refractory tiles known as *bridge covers.*

Bridge Crack. A defect sometimes found in the clay column from an extruder; such a crack is likely to be present if the bridge that supports the auger-shaft is too close to the mouthpiece.

Bridge Wall. See BRIDGE.

Bridge Indentation. A modified SINGLE-EDGE PRECRACKED BEAM. See FRACTURE TOUGHNESS TESTS.

Bridging. A phenomenon preventing the free flow of powder from a hopper. The particles stick together temporarily to form a bridge across the outlet nozzle, effectively reducing its area. See also CRACK BRIDGING; TILE BRIDGING.

Bridging Oxygen. An oxygen ion placed between two silicon ions, e.g. in the structure of a silicate glass.

Bright Annealing. Heating metal to be vitreous enamelled to red heat or above in a reducing atmosphere, to produce a clean, bright surface.

Bright Gold or Liquid Gold. A material for the decoration of pottery-ware; it consists essentially of a solution of gold sulphoresinate together with other metal resinates and a flux (e.g. a bismuth compound) to give adhesion to the ware. This form of gold decoration is already bright when drawn from the decorating kiln (cf. ACID GOLD and BURNISH GOLD).

Brights. Any portion of decorated glass forming part of a design, but which has not been acid treated.

Brilliant Cut. Process of decorating flat glass by cutting a pattern with abrasive wheels followed by polishing.

Brimsdown Frit. A lead bisilicate frit for glazes made from 1913 until 1928 by Brimsdown Lead Works, Brimsdown, Middlesex, England. The batch consisted of litharge, silica, and Cornish Stone. Although the frit contained 64% PbO its solubility, when tested by the Home Office method then used, was only 1–2%.

Brindled Brick. A building brick made from a ferruginous clay and partially reduced at the top firing temperature; it has a high crushing strength.

Brinell Hardness. An evaluation of the hardness of a material in terms of the size of the indentation made by a steel ball when pressed against the surface. Symbol – HB supplemented by numbers indicating the diameter of the ball used and the load applied; the abbreviation adopted by the American Ceramic Society is BHN-Brinell Hardness Number. The test is primarily for metals; it has been applied to clay products, e.g. building bricks, but without great success. (I. E. Brinell, *Comnm. Congr. Internat. Math Essai,* **2**, 83, 1901)

Bristol Glaze. A feldspathic type of glaze, generally opaque, maturing at 1200–1250°C and suitable for use on once-fired stoneware. Both the transparent and the opaque types are compounded from feldspar, whiting, ZnO, china clay and flint; the opaque type contains more feldspar and less whiting. Bristol glazes tend to be rather dull and often show pin-holes.

BRITE. Basic Research in Industrial Technologies in Europe/European Research on Advanced Materials – a European Community programme of research including advanced ceramics.

Britmag. Trade-name: dead-burned magnesia made by the seawater process in Britain. (Steetley Co. Ltd., Worksop, England).

Brittle Fracture, Brittleness. See FRACTURE. Brittle, or fast fracture occurs when the intensified stress at a crack tip exceeds the fracture stress of the material (i.e. the force is sufficient to break the interactive bonds). The crack then propagates catastrophically rapidly. Brittle fracture is characterized by a very small (sub-millimetre) microcrack zone at the crack tip, low (0.01 kJm^{-2}) fracture energy and a linear relationship between the load and the load-point displacement in fracture tests. This type of fracture is typical of low-toughness ceramics such as glass and fine-grained monolithic ceramics.

Brittle-ring Test. A test to determine the behaviour of a ceramic material under tensile stress; a test-piece in the form of an annulus is loaded along a diameter so that maximum tensile stresses develop on the inner periphery of the annulus in the plane of loading. For theory of this test see *Mechanical Properties of Engineering Ceramics,* p. 383, 1961. (cf. BRAZILIAN TEST.)

Broad Glass. Larger panes of flat glass were made by blowing a tall cylinder instead of a sphere, and cutting it lengthwise and flattening it by reheating. See CROWN GLASS.

Broken-joint Tile. A single-lap roofing tile of a size such that the edge of one tile, when laid, is over the centre of the head of a tile in the course next below.

Broken Seed. See SEED.

Brokes. Term used in the English ball-clay mines for clay that will not cut into balls; such clay is generally of low plasticity and poor fired colour.

Brongniart's Formula. A formula relating the weight (W, oz) of solid material in 1 pint of slip (or slop glaze), the weight (P, oz) of 1 pint of the slip, and the specific gravity (S) of the dry solid material:

$$W = (P - 20) \times S/(S-1)$$

The formula was established for slop glazes by A. Brongniart *(Traité des Arts Céramiques,* Vol.1, p.249, 1854).

Brookfield Viscometer. An electrically operated rotating cylinder viscometer in which the drag is recorded directly on a dial; it has been used in the testing of vitreous-enamel slips (*J. Amer. Ceram. Soc.*, **31**, 18, 1948).

Brookite. The orthorhombic form of titania, TiO_2; sp. gr. 4.17. The other forms are ANATASE (q.v.) and RUTILE (q.v.). It is comparatively rare and is rapidly transformed into rutile at temperatures above about 800°C.

Broseley Tile. An old name for a plain clay roofing tile; such tiles were made in Broseley, Shropshire, England.

Brownies. Term sometimes applied to brown spots in white vitreous-enamel ground-coats; more commonly known as COPPERHEADS.

Brown China. ROCKINGHAM WARE, q.v.

Brownmillerite. A calcium aluminoferrite that occurs in portland and high-alumina cements. It was originally thought to have the composition $4CaO.Al_2O_3.Fe_2O_3$; there is, however, a continuous series of solid solutions in the $CaO\text{-}Al_2O_3\text{-}Fe_2O_3$ system, the usually accepted Brownmillerite composition being only a particular point in the series. M.p. 1415°C; thermal expansion (0–1000°C), 10.1×10^{-6}.

Brucite. $Mg(OH)_2$; deposits in Ontario and Quebec (Canada) and in Nevada (USA), are used as a source of MgO.

Bruise. A concentration of cracks in the surface of glass-ware caused by localized impact.

Brulax System. An impulse system of oil firing, particularly for the top-firing of annular kilns, developed by A. A. Niesper, in Switzerland, in 1955.

Brunauer, Emmett and Teller Method. A procedure for the determination of the total surface area of a powder or of a porous solid by measurement of the volume of gas (usually N_2) adsorbed on the surface of a known weight of the sample. The mathematical basis of the method was developed by S. Brunauer, P. H. Emmett and E. Teller – hence the usual name B.E.T. METHOD (*J. Amer. Chem. Soc.*, **60**, 309, 1938). ASTM C1069 measures the specific surface area of alumina or quartz by nitrogen adsorption.

Brunner's Yellow. An antimony yellow recipe given by K. Brunner in 1837: 1 part tartar emetic, 2 parts lead nitrate, 4 parts NaCl. The mixture is calcined and then washed free from soluble salts prior to its use as a ceramic colour.

Brush Marks. A surface imperfection found on the exterior of some bottles; the marks resemble a series of fine vertical laps and are also known as SCRUB MARKS.

Brushing. The removal of bedding material from pottery-ware after the biscuit firing.

Brussels Nomenclature. Term given to the International Tariff Nomenclature agreed and issued in 1955. Chapter 69 related to ceramic products. The Brussels Nomenclature has been superseded for tariff purposes by the COMBINED NOMENCLATURE OF THE EUROPEAN COMMUNITIES (q.v.)

BS; BSI. Abbreviations for British Standard and British Standards Institution. The Institution is responsible for the preparation (through industry committees on which interested parties are represented) of national standards for Britain; copies of these standards, and of foreign standards, can be obtained from the Institution at Linford Wood, Milton Keynes, MK14 6LE

BSCCO. Bi – Sr – Ca ± Cu – O SUPERCONDUCTORS.

Bubble Alumina. Sintered bauxite is fused with carboneous materials and steel borings in an electric arc furnace. A slag containing nearly 90% Al_2O_3 is formed, and dispersed into bubbles by blowing air or steam into the molten slag. The alumina bubbles are leached in dilute sulphuric acid to remove remaining oxide impurities. Alumina over 99.7% pure is obtainable, either as bubbles for insulating applications, or as very fine, hard crystals for abrasive applications, obtained by crushing the bubbles. (B.T. Horsfield. US Pat 1682675, 1925).

Bubble Cap. A small, hollow, chemical stoneware hemisphere with serrations round the bottom edge; used on stoneware trays in de-acidifying towers in the chemical industry.

Bubble Glass. Glass-ware containing gas bubbles sized and arranged to produce a decorative effect (cf. FOAM GLASS).

Bubble-pressure Method. A technique for the determination of the maximum size of pore in a ceramic product; this size is calculated from the pressure needed to force the first bubble of air through the ceramic when it is wetted with a liquid of known surface tension. The method is used, for example, in the testing of ceramic filters (B.S. 1752).

Bubble Opal. OPAL GLASS (q.v.) containing many tiny pockets of gas, rendering it opaque.

Bubble Structure. The relative abundance, size and distribution of gas bubbles in a ceramic glaze or in a vitreous enamel. In the latter, the dull enamels contain most bubbles, glossy enamels being relatively bubble-free.

Bubbling. See GAS BUBBLING BRICK.
Bubbly Clay. A clay which, because it contains small amounts of organic matter, causes bubbles if used in vitreous enamels.
Buck. A support used in the firing of heavy vitreous enamel-ware.
Buckstave or Buckstay. A steel bracing designed to take the thrust of the brickwork of a furnace.
Buckyball. See FULLERENES.
Buffer Course. Refractory brickwork installed to separate two materials, to prevent chemical reaction between them.
Buffer Layer. The layer with composition and properties intermediate between body and glaze, found at the body/glaze interface.
Bugholes. Small holes or pits in the surface of formed concrete.
Bührer Kiln. The ZIG-ZAG KILN (q.v.) invented by J. Bührer (Brit. Pat. 562; 1867).
Building Brick. See CLAY BUILDING BRICK; CONCRETE BRICK; SAND-LIME BRICK.
Building Clays. See BRICK CLAYS.
Bulb Edge. The rounded edge of sheet-drawn glass.
Bulged Finish. See FINISH.
Bulk Density. A term used when considering the density of a porous solid, e.g. an insulating refractory. It is defined as the ratio of the mass of the material to its BULK VOLUME (q.v.). ASTM C20 specifies a boiling water test, ASTM C830 a vacuum pressure test for refractories; ASTM C914 a wax immersion test. ASTM C357 and C493 are boiling water and mercury displacement tests for granular refractory materials. ASTM C134 specifies a bulk density determination from measurements of the size of refractory bricks using steel straight edges to measure dimensions and warpage. BS 1902 Pt 3.6 specifies four tests for the GRAIN DENSITY (q.v.) of refractories.
Bulk Specific Gravity. For a porous ceramic, the ratio of the mass to that of a quantity of water which, at 4°C, has a volume equal to the BULK VOLUME (q.v.) of the material at the temperature of measurement.
Bulk Volume. A term used relative to the density and volume of a porous solid, e.g. a refractory brick. It is defined as the volume of the solid material plus the volume of the sealed and open pores present.
Bulkhead. A panel of thinner brickwork built into a furnace wall to permit early replacement.
Bulking. The tendency of fine particles, e.g. sand, to occupy a greater volume when slightly moist; the finer the particles the more pronounced is this effect of surface moisture.
Bull Float. A long-handled finishing tool for levelling large areas of concrete.
Bull's Eye. (1) A thick round piece of glass or lens.
(2) See ELECTRODE RING.
Bull's Kiln. A CLAMP (q.v.) of a type designed by W. Bull (Brit. Pat. 1977; 31/5/1875) in which the bricks to be fired are set in a trench below ground level; this type of kiln finds some use in India. (Also known as a BOCK KILN.)
Bullers' Rings. Annular rings (63.5mm dia. with a central hole 22mm dia.) made by pressing a blend of ceramic materials and fluxes; the rings are not fired. The constituents are so proportioned that the contraction of the rings during firing can be used as a measure of the temperature to which they have been exposed in a kiln; the

contraction is measured, after the firing process and when the rings are cold, by means of a gauge. The rings were introduced in about 1900 by Bullers Ltd., Stoke-on-Trent, England, from whom four types are now available for use in various temperature ranges within the overall range 960–1400°C.

Bullhead. See KEY BRICK.

Bullion. Flat glass of uneven thickness made by the handspinning of a gob of glass at the end of an iron rod.

Bullnose. A building brick or refractory brick having one end face rounded to join one side face. Such bricks built above one another can be used to form a rounded jamb, hence the alternative name JAMB BRICK (q.v.) (See Fig. 1, p39).

Bump Check. PERCUSSION CONE (q.v.)

Bung. (1) A vertical stack of saggars or of ware.
(2) A refractory brick shape used in the roof of a malleable iron furnace.

Bunsen's Extinction Coefficient. The reciprocal of the thickness that a layer of glass, or other transparent material, must have for the intensity of transmitted light to be decreased to one-tenth of its intensity as it falls on the layer.

Burgee. Contaminated sand resulting from the grinding of plate glass. (N. England dialect; the same word is used for poor quality coal).

Burger's Vector. See CRYSTAL STRUCTURE

Burgos Lustre. A red lustre for porcelain made by suitably diluting a gold lustre with a bismuth lustre; some tin may also be present. (From Burgos, Spain.)

Burley Clay. A refractory clay, intermediate in alumina content between a flint clay and a diaspore clay, that occurs in Missouri, USA; elsewhere it is known as Nodular Fireclay. The name derives from the diaspore oolites, known to the local miners as 'burls', found in these clays. The Al_2O_3 content of the raw clay is 55–63%.

Burning. An alternative (but less appropriate) term for FIRING (q.v.)

Burner Block. See QUARL BLOCK.

Burning Bar; Burning Point; Burning Tool. A support, usually made of heat-resisting alloy, for vitreous enamelware during firing. Defects in the ware caused by contact with these supports are known as BURNING TOOL MARKS.

Burning Off. See BURN OFF.

Burnish Gold or Best Gold. The best gold decoration on pottery is made bright by BURNISHING, i.e. rubbing, usually with a blood-stone or agate. It is applied to the ware as a suspension of gold powder in essential oils, with or without the addition of a proportion of BRIGHT GOLD (q.v.); other ingredients are a flux (e.g. lead borosilicate or a bismuth compound) to promote adhesion to the ware, and an extender such as a mercury salt. This type of gold decoration is dull as taken from the kiln, hence the need for subsequent burnishing. cf. ACID GOLD.

Burn Off. A fault in vitreous enamelling resulting from the apparent burning-away of the ground coat. The fault may be due to the enamel having become saturated with iron oxide; to prevent this, the fusion temperature of the ground-coat should be raised by altering its composition. Other causes are too thin a coat, wetness of the metal before dipping, or shaking the dipped metal too vigorously.

Burn-out. Heating a green ceramic shape to remove the (usually organic) binder. When this immediately precedes sintering, the rate of heating must be carefully controlled, to avoid trapping the resulting gases in pores as these close, and to prevent excess porosity.

Burnover. An underfired STOCK BRICK (q.v.) from the outside of a clamp; such bricks are usually re-fired.

Burr. (1) A partially fused waste brick from a kiln, or several such bricks fused together.
(2) A rough edge on the base-metal used in vitreous enamelling; it must be removed before the enamel is applied.

Burr Mill. A mill for pigment preparation, comprising two ribbed discs rotating against each other.

Bursting Expansion. In the refractories industry this term has the specific meaning of surface disintegration of basic refractories caused by the absorption of iron oxide. The expansion that leads to this form of failure results from solid solution of magnetite (Fe_2O_4) in the chrome spinel that forms a major constituent of chrome and chrome-magnesite refractories. A laboratory test submits a test-piece cut to the size of a 50mm cube, to the action of 40 g of mill-scale (crushed to pass a 30 B.S. sieve) for I h at 1600°C; the expansion is expressed as a linear percentage.

Bursting Off. Breaking blown glass-ware from the end of the blowing iron.

Bushing. (1) An electric glass-melting unit for the production of glass fibres, which are drawn through platinum orifices in the base.
(2) A liner fitted in the feeder that delivers glass to a forming machine; this liner is also known as an ORIFICE RING.

Bustle Pipe or Hot-blast Circulating Duct. A metal tube of large diameter which surrounds a blast furnace at a level a little above the tuyeres; it is lined with refractory material and distributes the hot air from the hot-blast-stoves to the pipes known as GOOSENECKS which in turn carry the air to the tuyeres.

Butterfly Bruise. PERCUSSION CONE (q.v.)

Button. Part of a piece of pressed glass designed to produce a hole when knocked out; also sometimes called a CAP or KNOCK

Button Test. A test for the fusibility of a vitreous enamel frit, or powder, first proposed by C. J. Kinzie (*J. Amer. Ceram.* **15**, 357, 1932) and subsequently standardized (ASTM C374); Also known as the FUSION-FLOW TEST and as the FLOW BUTTON TEST.

Bwlchgwyn Quartzite. A quartzite from Bwlchgwyn, N. Wales, used as a raw material for silica-brick manufacture. Chemical analyses are:

	SiO_2	Al_2O_3	Alkalis
As quarried	96.6	1.1	0–4
Washed	97.4	0.7	0.3

°C. Degrees Celsius (formerly, and still more commonly, known as Degrees Centigrade). For conversion to Fahrenheit $T°C = (\frac{9}{5}T + 32)°F$.

'C' Glass. (Chemical) An acid-resistant glass composition for glass-fibre, it contains 7–10% Na_2O.

C_{60}. See BUCKYBALL.

C_{80}. See CONTRAST RATIO.

Cabal Glass. A special glass consisting solely of **Ca**lcium oxide, **B**oric oxide and **Al**umina, hence its name. Cabal glass is known for its low dielectric loss. It was first made by H. Jackson for the UK Ministry of Munitions in 1917–19.

Cable Cover. A fired clay (in this context generally, but erroneously, known as 'earthenware') or concrete conduit for covering underground electric cables; these covers warn of the cable's presence and protect it from excavating tools. For specification see B.S. 2484.

CACT. Center for Advanced Ceramic Technology, at Alfred University, USA.
CAD/CAM. Computer Aided Design/ Computer Aided Manufacture
Cadmium Carbonate. $CdCO_3$. Small amounts (up to 2%) are sometimes added to CADMIUM SELENIDE (q.v.) red colours to improve their stability.
Cadmium Niobate. $Cd_2Nb_2O_7$; an anti-ferroelectric compound; the Curie temperature lies between – 85°C and – 100 °C.
Cadmium Nitrate. $Cd(NO_3)_2 .4H_2O$ m.pt. 59.5 °C used to add cadmium for colouring glass to vitreous enamel.
Cadmium Oxide. CdO; sp. gr. approx. 7.0. The vapour pressure is approx. 1 mm at 1000°C and 7 mm at 1100°C.; sublimes rapidly at 1500°C. It is occasionally used in ceramic colours. The oxide is poisonous.
Cadmium Selenide. This compound, or the polyselenide, is responsible for the colour in selenium ruby glass. With some CdS in solid solution it forms the basis of Cd-Se red ceramic (especially vitreous enamel) colours; the firing temperature must not exceed about 850°C. The presence of up to 3% V_2O_5 is claimed to stabilize the colour; blackening in the presence of a glaze or flux containing Pb is caused by the formation of PbSe.
Cadmium Silicate. $CdSiO_3$; sp. gr. 4.9; m.p. 1240°C.
Cadmium Sulphide. CdS; sp. gr. 4.8; m.p. 1750°C. A constituent of some ruby-coloured glass, of some yellow enamels, and of Cd-Se red ceramic colours which can be fired at temperatures up to about 850°C. Sintered CdS can be used as a photoconductive ceramic.
Cadmium Sulphoselenide. CdS.*x* CdSe. A ceramic pigment, the basis of a range of yellow, orange and red ceramic colours. See CADMIUM SELENIDE; ENCAPSULATED COLOURS.
Cadmium Titanate. $CdTiO_3$; a ferroelectric ceramic having the ilmenite structure at room temperature; the Curie temperature is approx. -220°C.
Cadmium Yellow. Cadmium sulphide, coprecipitated with barium sulphate.
Cadmium Zirconate. $CdZrO_3$; occasionally used as an addition to barium titanate bodies, the effect being to reduce the dielectric constant and Curie temperature.
Calcine: Calcination. Heat treatment intended to produce physical and/or chemical changes in a raw material. The calcination of bauxite, for example, eliminates water and yields a product that is relatively free from further firing shrinkage; when kyanite is calcined, on the other hand, the change is not in chemical but in mineralogical composition, mullite and cristobalite being formed.
Calcite. $CaCO_3$, the mineral constituent of limestone, chalk, and marble. Used as a constituent of soda-lime glass and, as WHITING (q.v.), in some pottery bodies; it is a major component of the batch used in portland cement manufacture.
Calcium Aluminates. The five compounds (q.v.) are: TRICALCIUM ALUMINATE($3CaO.Al_2O_3$); DODECACALCIUM HEPTALUMINATE ($12CaO.7Al_2O_3$); CALCIUM MONO-ALUMINATE ($CaO.Al_2O_3$); CALCIUM DIALUMINATE ($CaO. 2Al_2O_3$) and CALCIUM HEXALUMINATE ($CaO.6Al_2O_3$). TRICALCIUM PENTALUMINATE (q.v.) is a mixture of the mono- and di-aluminates.
Calcium Boride. CaB_6; m.p. 2235 °C: sp. gr. 2.45; thermal expansion (25–1000 °C.) 5.8×10^{-6}; electrical resistivity (20 °C), 124 μohm.cm.
Calcium Carbonate. See CALCITE; WHITING.
Calcium Chloride. $CaCl_2$; used for FLOCCULATION (q.v.) in the preparation of glazes; about 0.05% is normally

sufficient. It is also sometimes used as a mill-addition in the preparation of vitreous enamel slips. Calcium chloride solution accelerates the rate of setting of portland cement; it is also sometimes added to concrete mixes as an integral waterproofer.

Calcium Dialuminate. $CaO.\ 2Al_2O_3$; melts incongruently at 1762 ± 5 °C; thermal expansion, 5.0×10^{-6} · Present in high alumina cement but does not itself have cementing properties.

Calcium Ferrite. In the binary system, two ferrites are formed – $CaO.Fe_2O_3$ and $2CaO.Fe_2O_3$; the former may occur in some HIGH-ALUMINA CEMENT (q.v.). Thermal expansion (0–1000 °C): CaO. Fe_2O_3, 11.7×10^{-6}; $2CaO.Fe_2O_3$, 10.5×10^{-6}.

Calcium Fluoride. See FLUORSPAR.

Calcium Hafnate. $CaHfO_3$; m.p. 2470 ± 20°C; sp. gr. 5.73; thermal expansion (10–1300 °C), 7×10^{-6}.

Calcium Hexaluminate. $CaO.6Al_2O_3$; melts incongruently at 1830 ± 15 °C to form corundum and a liquid.

Calcium Metasilicate. See WOLLASTONITE.

Calcium Monoaluminate. $CaO.Al_2O_3$; melts incongruently at 1602 ± 5 °C; thermal expansion (0–1200 °C), 6.8×10^{-6}. A major constituent of HIGH-ALUMINA CEMENT (q.v.).

Calcium Orthosilicate or Dicalcium Silicate. $2CaO.SiO_2$; m.p. 2130 °C. Occurs in four crystalline forms: α, stable above 1447 °C; α', bredigite, stable from about 800–1447°C on heating, 1447 °C–670 °C on cooling; β, larnite, stable or metastable from 520-670 °C; γ stable below 780–830 °C. Material in which a considerable amount of $2CaO.SiO_2$ has been formed by high-temperature reaction, falls to a powder – 'dusts' – on cooling because of the inversion (accompanied by a 10% increase in volume) to the γ-form at 520 °C. The inversion can be prevented by the addition of a stabiliser, e.g. B_2O_3 or P_2O_5. Calcium orthosilicate is a constituent of portland cement and may be formed in dolomite refractories.

Calcium Oxide. See LIME.

Calcium Phosphate. The compound of most interest is $3CaO.P_2O_5$ or $Ca_3(PO_4)_2$; melts with some dissociation at 1755 °C. Crucibles made of this phosphate withstand FeO melts at 1600 °C and permit the refining of steel to a phosphate content of 0.025%. It is also used in the manufacture of opal glass. See also BONE ASH.

Calcium Silicate. The four compounds (q.v.) are: WOLLASTONITE ($CaO.SiO_2$); RANKINITE ($3CaO.2SiO_2$); CALCIUM ORTHOSILICATE ($2CaO.SiO_2$); and TRICALCIUM SILICATE ($3CaO.SiO_2$). Autoclaved calcium silicate board products, reinforced with cellulose fibres, have been developed as substitutes for asbestos. BS 3958 Pt 2 specifies the properties and dimensions of preformed calcium silicate insulation suitable for use up to 650°C and 950°C.

Calcium Silicate Bricks. The approved term (replacing sand-lime brick and flint-lime brick) for bricks made by autoclaving a mixture of sand (or crushed siliceous rock) and lime. Requirements and qualities are specified in B.S. 187, and in the USA, in ASTM C73 for calcium silicate facing bricks. Dimensions are specified in BS 4729.

Calcium Stannate. $CaSnO_3$; sometimes used as an additive to barium titanate bodies, one effect being to lower the Curie temperature.

Calcium Sulphate. See ANHYDRITE; GYPSUM; PLASTER OF PARIS.

Calcium Sulpho-Aluminates. Two compounds exist: the 'high' form, $3CaO.Al_2O_3.3CaSO_4.30–32H_2O$; and the 'low' form, $3CaO.Al_2O_3.CaSO_4.12H_2O$. Both forms may be produced by reaction between $3CaO.Al_2O_3$ and gypsum during the hydration of portland cement. The 'high' form is also produced when cement and concrete are attacked by sulphate solutions.

Calcium Titanates. Three compounds exist: $CaO.TiO_2$, m.p. 1915°C; $4CaO.3TiO_2$, melts incongruently at 1755 °C; $3CaO.2TiO_2$ melts incongruently at 1740 °C. Calcium titanate refractories slowly deform under load at high temperature; they resist attack by basic O.H. slag and portland cement clinker but react with materials containing SiO_2 and/or Al_2O_3. $CaO.TiO_2$ has a dielectric constant of 150–175; it finds use as a component of more complex titanate dielectrics.

Calcium Zirconate. $CaZrO_3$; a material having useful dielectric and refractory properties. As a dielectric it is normally used in minor amounts (3–10%) in titanate bodies; as a refractory it can be used, even under reducing conditions, up to 1700°C provided that it is not in contact with siliceous material. M.p. approx. 2350°C, sp. gr. 4.74; thermal expansion 8–11 × 10^{-6} (25–1300 °C).

Calcrete. Gravel and sand cemented with calcium carbonate.

Calcspar. Old name for CALCITE (q.v.).

Calculated Brickwork. Brickwork designed according to the engineering principles set out in BS 5628 'Code of Practice for the Use of Masonry'. The calculations take into account the load on the wall, its height, the crushing strength of the bricks, the type of mortar, the stiffening effect of the floors, and other factors. Walls of calculated brickwork can be thinner, and hence more economic, than traditional walls, yet still be fully adequate to withstand the stresses that will be imposed on them. See also LIMIT STATE DESIGN.

Calculated Mineralogy. Modern terminology for RATIONAL ANALYSIS (q.v.)

Calculon. A building-brick size equal to 6.75 in (17 cm).

Caledonia Body. A ceramic body for making coarse glazed pottery; typically, it contained 70% yellow-firing clay, 20% china clay and 10% flint.

Calgon. Proprietary name for a complex sodium phosphate (q.v.) sometimes used as a deflocculant for clay slips; (Albright and Wilson, Ltd., Oldbury, England).

Callow. A term, of localised use, for the overburden of a clay-pit.

Calorific Value (CV). The quantity of heat released when a unit weight (or volume) of fuel is completely burned. Units: 1 Btu/lb = 0.556 cal/g = 2326 J/kg. Typical values: coal, 25–35 MJ/kg; liquid fuels, 40–50 MJ/kg; natural gas 40 MJ/m^3.

Calorite. Trade-name: a pyroscope similar to a pyrometric cone but cylindrical; they are made for use between 500 and 1470 °C (Wengers Ltd, Stoke-on-Trent, England).

Cameo. (1) Glass jewellery, etc., made of two or more layers of glasses of different colours and carved in relief. (2) Similar ornamental pieces made of JASPER WARE (q.v.)

Cam Lining. A lining, usually of refractory brick, whose thickness varies regularly around a kiln to induce feed movement and improve heat exchange.

Campaign. The working life of an industrial furnace between major repairs; the term is particularly applied to blast furnaces and to glass-tank furnaces.

Canal. The section of a flat-glass tank-furnace through which molten glass flows to the drawing chamber.

Candle. See CERAMIC FILTER.

Candlot's Salt. CALCIUM SULPHO-ALUMINATE (q.v.).
Cane. Glass rods of small or medium diameter.
Cane Marl. Local name for one of the low-quality fireclays associated with the Bassey Mine, Littlerow and Peacock coal seams of N. Staffordshire, England.
Cane Ware. Eighteenth-century English stoneware of a light brown colour; it was a considerable advance on the coarse pottery that preceded it but, for use as tableware, cane ware was soon displaced by white earthenware. During the 19th and the earlier part of the 20th century, however, cane ware continued to be made in S. Derbyshire and the Burton-on-Trent area as kitchen-ware and sanitary-ware; it had a fine-textured cane-coloured body with a white engobe on the inner surface often referred to as CANE AND WHITE.
Cank. An indurated clay with cemented particles, or a useless mixture of clay and stone, the term is chiefly used in the Midlands and North of England.
Cannock. South Staffordshire term for a ferruginous nodule occurring in a fireclay; the name derives from the town of Cannock in that area.
Cannon. A small (15–60 ml) thick-walled glass bottle of the type used to contain flavouring essences.
Cannon Pot. A small POT (q.v.) for glass melting.
Cant. The bottom outside edge of a SAGGAR (q.v.); it is important that this should be rounded to help the saggar to resist thermal and mechanical shock.
Cantharides Lustre. A silver lustre for artware similar in appearance to the lustre on the wings of the cantharides beetle; it is yellow and has usually been applied to a blue lead-glaze.
Canton Blue. A violet-blue ceramic colour made by the addition of barium carbonate to cobalt blue. A quoted recipe is (per cent): cobalt oxide, 40; feldspar, 30; flint, 20; $BaCO_3$, 10.
Caolad Flint. A form of cryptocrystalline silica occurring at Cloyne, Co. Cork, Eire. It has a sp. gr. of 2.26 and is readily ground, without the need for pre-calcination, for use in pottery bodies.
Cap. (1) See BUTTON. (2) A COLLAR (q.v.) on the end of a POST (q.v.) to increase the area of contact.
Capacitor. See DIELECTRIC.
Capon Abrasion Test. A method for the determination of deep abrasion resistance. A flat test-piece is placed to touch tangentially a 200mm diameter steel disc rotating at 75 rpm, and white fused alumina of grain size 80 (32 GB 1971) is fed into the grinding zone at a rate of 100g/100 revolutions of the disc. The volume of material removed after 150 revolutions of the disc is measured by measuring the length of groove ground in the test-piece. The equipment is calibrated against Austrian standard granite. The test is the British and European standard test for unglazed ceramic floor tiles (BS 6431, Pt14, 1983 EN 102) and has also been adopted for clay pavers.
Capping. (1) The formation of a weak region at the end of a pressed compact. In the region just below the top punch, the compact is subjected to a compressive force only, with no shear component die to interaction with the die walls. This region is consequently weak. This fault is also known as END-CAPPING.
(2) capping is also the term used for the various methods (cardboard or cement spacers, grinding) used to ensure that the faces of crushing strength test specimens are plane, parallel and perpendicular to the applied force.
Carbides. A group of special ceramic materials; see under the carbides of the

following elements: Al, Be, Ce, Cr, Hf, La, Nb, Si, Ta, Th, Ti, U, V, W, Zr.

Carbon. See CARBON REFRACTORIES; GRAPHITE.

Carbon-Ceramic Refractory. See PLUMBAGO.

Carbon Dioxide Process. A method of bonding refractory grains by mixing them with a solution of sodium silicate, moulding to the required shape and then exposing the shape to CO_2. The process was first mentioned in Brit. Pat. 15 619 (1898) but did not come into general use until about 1955, when it began to be employed in the bonding of foundry sands and cores. The process has been tried for bonding rammed linings in small ladles.

Carbonisation. The removal of volatile components from pitch-bonded, resin-bonded or impregnated basic refractories, by heating to 980 to 1000 °C immersed in metallurgical coke in an airtight container, to provide a reducing atmosphere. The refractories retain RESIDUAL CARBON (q.v.) The *carbonisation mass loss* is the loss in mass during this process. ASTM C-607 specifies a coking or carbonization procedure for large shapes of caron-bearing materials. Cf. CARBON YIELD.

Carbon Monoxide Disintegration. The breakdown of refractory materials that sometimes occurs (particularly with fireclay refractories) when they are exposed, within the temperature range 400–600 °C, to an atmosphere rich in carbon monoxide. The disintegration is due to the deposition of carbon around 'iron spots' in the brick, following the well-known dissociation reaction:

$$2CO = CO_2 + C$$

BS 1902 Pt 3.10 and Pt 7.6 specify tests for resistance to CO, as does ASTM C288.

Carbon Refractories. These refractories, consisting almost entirely of carbon, are made from a mixture of graded coke, or anthracite, pitch and tar; the shaped blocks are fired (packed in coke). The fired product has an apparent porosity of 20–25% crushing strength 50–70 MNm^{-2}; R.u.L. (350 kPa), 1700°C; thermal expansion (0–1000 °C), 0.65%. The principal use is in the lining of blast furnaces, particularly in the hearth and bosh (cf. PLUMBAGO).

Carbon Yield. See RESIDUAL CARBON.

Carborundum. Trade-mark of The Carborundum Co., Niagara Falls, USA, and Trafford Park, Manchester, England. This firm pioneered the industrial synthesis of silicon carbide and their trade-mark is often used as a synonym of SILICON CARBIDE (q.v.), particulary in its use as an abrasive.

Carbothermal Reduction. High temperature reactions in which carbon acts as a reducing agent so that one ceramic may be prepared from another. e.g. AlN may be prepared by heating C + Al_2O_3 mixtures in N_2. The carbon reduces the alumina, to Al metal which reacts with the nitrogen gas.

Carboxymethylcellulose (CMC). An organic compound that, in the form of its sodium salt, finds use in the ceramic industry as an additive to glazes and engobes to prevent friability before the coating is fired; CMC has been added to vitreous enamel slips to prevent settling.

Carboy. A glass container for acids, etc.; it has a narrow neck and its capacity is 5 gallons or more. (See B.S. 678.)

Carburettor. The chamber of a water-gas plant, lined with refractory material and often filled with CHECKERS (q.v.) on which oil is sprayed to enrich the gas (cf. SUPERHEATER).

Carder Tunnel Kiln. A tunnel-kiln designed in about 1928 by Carder and

Sons, Brierley Hill, England, for the firing of stoneware at 1200°C.

CARES. Ceramic Analysis and Reliability Evaluation of Structures. (Formerly known as SCARE – Structural Ceramic Analysis and Reliability Examination). A Fortran 77 computer program using Weibull and Batdorf fracture statistics to predict the fast-fracture reliability of isotropic ceramics.

Carlton Shape. A tea-cup the top half of which is cylindrical, the bottom half being approximately hemispherical but terminating in a broad, shallow foot. For specification see B.S. 3542.

Carman Equation. A relationship, derived from KOZENY'S EQUATION (q.v.), permitting determination of the specific surface, *S*, of a powder from permeability measurements:

$$S = 14\sqrt{[p^3/KV(1 - p^2)]}$$

where *p* is the porosity of the bed of powder, *V* is the kinematic viscosity of the flowing fluid, and *K* is a constant. In his original application of this equation, Carman used a simple apparatus in which liquids were used as permeating fluids. (P.C. Carman, *J. Soc. Chem. Ind.*, **57**, 225, 1938.)

Carnegieite. $Na_2O.Al_2O_3.2SiO_2$, m.p. 1526°C. It is formed when NEPHELINE (same composition) is heated above 1248°C and is sometimes found in fireclay refractories that have been attacked by Na_2O vapour.

Carolina Stone. A CHINA STONE (q.v.) used to some extent in the US pottery industry.

Carrara Porcelain. Term sometimes applied to PARIAN (q.v.).

Carrousel. A four-wheeled bogie fitted with a rotating framework which carries two sets of STILLAGES (q.v.) for the handling of bricks from a dryer to a Hoffmann type of kiln. (From the French word for a merry-go-round.)

Carry. Term sometimes used in Scotland for OVERBURDEN (q.v.).

Car Top. See DECK.

Cascading. The behaviour of the charge in a ball mill when it spills from the highest point without an appreciable proportion passing into free flight.

Case Mould. See under MOULD.

Cased Glass. Glass-ware with a superimposed layer of another glass having a different composition and usually coloured. The thermal expansions of the two glasses must be carefully matched (cf. PLY GLASS).

Casella Counter. A device for the automatic counting and size-determination of particles from 1 to 200mm. (Casella Electronics Ltd., York, England.)

Casement Wall. See BREAST WALL.

Casher Box. A metal box used to catch a glass bottle after it had been severed from the blow-pipe in the old hand-blown process.

Casing Wall. See BREAST WALL.

Cassel Kiln. See KASSEL KILN.

Cassette. A design of tile CRANK based on a rectangular refractory module. The tiles are supported and separated without building up a structure of posts and pins.

Castable. BS 1902, Pt 7 defines a castable as a mixture of graded refractory aggregate and other materials, which, after mixing with water, will set owing to the formation of a hydraulic and/or a chemical bond. Castables are subdivided into two categories according to their bulk density. Insulating castables are those which when cast, cured and dried at 110 °C have a bulk density 1600 kg/m^3. All other castables are known as dense castables. BS 1902 Pt 7 specifies tests for UNSHAPED REFRACTORIES (q.v.). ASTM C401 defines REGULAR CASTABLE REFRACTORIES as those containing hydraulic setting cements and having a lime content greater than 2.5%. Normal strength castables have a modulus of rupture of at least 2.07 MN m^{-2} (300 psi); high-strength types have a modulus at least 4.14 MN m^{-2}. They are further classified on the basis of volume stability. Insulating castables are classified on the basis of bulk density. See also LOW-CEMENT CASTABLE; NO-CEMENT CASTABLE.

Casting. Shaping a fluid material (which subsequently solidifies) by pouring it into a mould. The process is used in pottery manufacture and in glass-making. To cast pottery-ware, a SLIP (q.v.) is poured into a plaster mould which absorbs a proportion of the water so that the body builds up on the walls of the mould. In the production of thin-walled ware, e.g. tableware, excess slip is poured out of the mould when the required thickness of ware has been formed. For thick ware, e.g. sanitary fireclay, the SOLID CASTING process is used, with an inner plaster core; slip is in this case poured into the space between mould and core and the body is allowed to build up without any slip being poured off. ASTM C 866, a test for the filtration rate of whiteware clays, is relevant to casting behaviour. *Glassware* can be cast by pouring molten glass into a mould. (A telescope mirror blank 6 m diameter was thus made). Television tubes and radomes are made by centrifugal casting. The glass flows upwards in a spinning mould, producing a uniform wall thickness. Complex shapes are made by *slip-casting* ground glass powder in plaster of Paris moulds, then firing.

Casting-pit Refractories. Specially shaped refractories (usually fireclay) for use in the casting of molten steel. The individual items included in the term are: LADLE BRICKS, ROD COVERS, STOPPERS, NOZZLES, MOULD BRICKS, TRUMPETS, GUIDE TUBES, CENTRE BRICKS, RUNNERS, CONES and MOULD PLUGS (q.v.). ASTM C435 classifies nozzles, sleeves and ladle bricks each into type A, B or C depending on PCE, porosity, reheat linear change and, for ladle bricks, modulus of rupture.

Casting Spot. A fault that sometimes appears on cast pottery as a vitrified and often discoloured spot on the bottom of the ware or as a semi-elliptical mark on the side. It occurs where the stream of slip first strikes the plaster mould and is attributable to local orientation of platy particles of clay and mica in the body. The fault can be largely eliminated by adjusting the degree of deflocculation of the slip so that it has a fairly low fluidity. The fault is also sometimes called FLASHING.

Castle; castling. Local term for the setting of bricks on a dryer car, two-on-two in alternate directions.

Catalyst Support. Many catalysed chemical reactions require the catalyst to present a high surface area to the reagents. One way of achieving this is for the catalyst to be supported on an inert substrate itself of high surface area – for example a ceramic honeycomb.

Catalytic Converter. A device to remove noxious components from the hot exhaust gases of diesel and petrol engines, by passing them over a suitable catalyst to promote their breakdown. Cordierite ceramic honeycombs are the usual catalyst carriers.

Cataphoresis. The movement of colloidal particles in an electric field; this forms the basis of the purification of clays by the so-called ELECTRO-OSMOSIS (q.v.) method.

Catenary Arch. A sprung arch having the shape of an inverted catenary (the shape assumed by a string suspended from two points that are at an equal height from the ground). The stress pattern in such an arch is such that there is no tendency for any bricks to slip relative to one another.

Cateracting. Term used in ball-milling for the state of the charge when a large proportion falls in free flight. This usually occurs at mill speeds higher than that needed for CASCADING (q.v.).

Cathedral Glass. Rolled flat-glass textured on one side to resemble old window glass (cf. ANTIQUE GLASS).

Cathode Pickling. See ELECTROLYTIC PICKLING.

Cationic Exchange. See IONIC EXCHANGE.

Cat Scratch. Surface flaws on glassware resembling the marks of a cat's claws.

Cat's Eye. (1) Glass tubing of the SCHELLBACH (q.v.) type for use in spirit-levels. (2) A fault, in glass-ware, in the form of a crescent-shaped blister which may contain foreign matter.

Cauchy Formula. A formula proposed by A. L. Cauchy, a 19th-century French mathematician, relating the refractive index, *n*, of a glass to the wavelength, λ, of the incident light:

$$n = A + B\lambda^{-2} + C\lambda^{-4}\ldots$$

A closer relationship is the HARTMAN FORMULA (q.v.).

Caveman. An oddjob man around (frequently under) a glass furnace.

Cavity Wall. A wall built so that there is a space between the inner and outer leaves, which are tied together at intervals by metal or other ties. Such a wall has improved thermal insulation and damp-proofness.

CBN. Cubic Boron Nitride (q.v.)

C/B Ratio. A term that has been used for the SATURATION COEFFICIENT (q.v.).

C & D Hot Top. A HOT-TOP (q.v.) designed by W. A. Charman and H. J. Darlington (hence the name 'C & D') at the time, about 1925, when they were both employed by Youngstown Sheet and Tube Co., USA. The hot top, which is fully floating, consists of a cast-iron casing lined with fireclay or insulating refractories; a refractory bottom ring is attached to the lower end of the casing to protect the latter from the hot metal.

C-D Principle. The Convergence-Divergence principle used in the FRENKEL MIXER (q.v.).

c.e.c. Cation exchange capacity. See IONIC EXCHANGE.

Celadon. An artware glaze of a characteristic green colour, which is obtained by introducing a small percentage of iron oxide into the glaze batch and firing under reducing conditions so that the iron is in the ferrous state. The name was used by the first Josiah Wedgwood for his self-coloured green earthenware.

Celeste Blue. A ceramic colour made by softening the normal cobalt blue by the addition of zinc oxide.

Celite. The name given to one of the crystalline constituents of portland cement clinker by A. E. Törnebohm (*Tonindustr.* **58** *Ztg.* **21,** 1148, 1897). This constituent has now been identified as a solid solution of $4CaO.Al_2O_3.Fe_2O_3$ and $6CaO.2Al_2O_3.Fe_2O_3$.

Cell. One of the spaces in a hollow clay building block. According to the US specification (ASTM C212) a cell must have a minimum dimension of at least 0.5 in. (12.5 mm) and a cross-sectional area at least 1.5 in^2 (950 mm^2).

Cell Furnace. A glass-tank furnace in which glass in the melting end and auxiliary chambers is heated electrically.

Cellular Brick or Block. A clay building brick or block which shall not have holes, but may have frogs or cavities (holes closed at one end) exceeding 20% of the volume of the brick or block. (B.S.

3921). Cellular bricks are usually made by pressing whereas PERFORATED BRICKS (q.v.) are made by extrusion.

Cellular Concrete. See AERATED CONCRETE.

Cellular Glass. See FOAM GLASS.

Celsian. Barium feldspar, $BaO.Al_2O_3.2SiO_2$; m.p. 1780 °C. There are two crystalline forms, resulting in a non-linear thermal expansion curve. Celsian refractory bricks have been made and have found some use in electric tunnel kilns.

Celsius. The internationally approved term for the temperature scale commonly known as Centigrade. The fixed points are 0°C, (the triple point of water) and 100°C (the steam over boiling water). See °C, °F.

Cement. See ACID-RESISTING CEMENTS; CIMENT FONDU; HIGH EARLY-STRENGTH CEMENT; HYDRAULIC CEMENT; KEENE'S CEMENT; PORTLAND CEMENT; RAPID-HARDENING PORTLAND CEMENT; REFRACTORY CEMENT, SILICATE CEMENT; SOREL CEMENT, SULPHATE-RESISTING CEMENT; SULPHO-ALUMINATE CEMENT; SUPERSULPHATED CEMENT. For methods of testing cement see BS 4550, Pts 0 to 6, which specifies chemical analysis, strength and consistency tests, setting times, soundness, heat of hydration, and standard sands and coarse aggregates for mortar and concrete. BS 6432 specifies tests for glass-fibre reinforced cement.

Cementation. A process to promote the absorption and diffusion of one substance into the surface of another. (Originally, producing a hard carbon steel coating on wrought iron by packing it in charcoal and heating strongly for a long period).

Cement Bacillus. This name has been applied to the compound $3CaO.Al_2O_3.3CaSO_4.31H_2O$, which is formed by the action of sulphate solutions on portland cement and concrete.

Cemented Carbide. Metal carbides, chiefly tungsten carbide WC and titanium carbide TiC, used for cutting tools, in which the carbide grains are bonded together by a thin layer of metal, usually cobalt, though nickel has also been used. The object is to provide a degree of ductility to an inherently brittle material.

Cemented Oxide. See TOOL (CERAMIC).

Cement-kiln Hood. The head, which may be mobile or fixed, of a rotary cement kiln; through the hood the burner passes and within the hood the clinker discharges from the kiln to the cooler.

CEN. Comité Européen de Coordination des Normes, Boite 5, B-1000, Brussels. EN. Abbreviation for European Standards (Euronorm)

Centigrade. See °C.

Centre Brick. A special, hollow, refractory shape used at the base of the GUIDE TUBES (q.v.) in the bottom-pouring of molten steel. The Centre Brick has a hole in its upper face and this is connected via the hollow centre of the brick to holes in the side faces (often six in number). The Centre Brick distributes molten steel from the trumpet assembly to the lines of RUNNER BRICKS.

Centre Pip. A fault in tableware made by the ROLLER-HEAD (q.v.) process. A small raised area forms at the centre of the ware in firing. The orientation of clay particles introduced by pugging persists across the thickness of the ware at the centre. Elsewhere the shaping process reorients the particles along a diameter. The result is a central region of minimum drying and firing shrinkage and a raised centre.

Centreless Grinding. The cylindrical workpiece to be ground is supported on a workable and rotated by a 'regulating wheel' (q.v.). It is also sometimes known as a CROWN BRICK or SPIDER.

Centrifugal Casting. Tubes and other rotationally synmetric shapes are cast on the inside of moulds spun at high speed to produce large forces which increase consolidation rates. A dense cast results. No central core is needed to cast hollow articles. The moulds may be of plastics or metal and need not be porous. Cone-shaped glassware such as television tubes is made in this way.

Centrifugal Disc Mill. A horizontal disc within a chamber is rotated at high speed by an electric motor. Clay is fed in near the axis, and comminuted by the centrifugal action of the mill, at constant moisture content.

Centrifuge. A device in which the rate of settling of particles from a liquid suspension is accelerated by subjecting them to a high centrifugal force; this is done by charging the suspension into carefully-balanced 'buckets' at opposite ends of a diameter of a rapidly spinning disk or shaft. Such equipment is used industrially in the treatment of raw materials; in the ceramic laboratory, the centrifuge is used as a method of particle-size analysis, particularly for particles below about 1mm.

Centura Ware. Glass-ceramic tableware that has been glazed, and sometimes decorated. (Trade name: Corning Glass Works, USA).

CERABULL. A computer database and electronic bulletin board operated by the American Ceramic Society.

Ceramel. Obsolete term for CERMET (q.v.).

Ceramesh. 0.1 mm pore-size metal mesh coated with zirconia.

Ceramic Coating (on Metal). Defined (ASTM – C286) as: 'An inorganic essentially non-metallic, protective coating on metal, suitable for use at or above red heat.'

Ceramic Colour Standards. This set of 12 ceramic tiles provides standard colours for the calibration of colour measuring instruments used in all industries. Specially developed colours and glazes and rigorously precise firing schedules give stable and reproducible colours. There are 3 neutral grey standards with flat spectral curves to check linearity of response; 7 strongly coloured standards with steep spectral curves to check responsivity or wavelength and waveband errors; 2 colour-difference standards which 'pair' with the mid-grey and the green standards, giving a direct check on colour difference measurements. A black and a white standard are available, plus 'specials' to individual requirements. (The development of these tiles, in collaboration with the National Physical Laboratory, earned Ceram Research, Stoke-on-Trent, its second Queen's Award for Technological Achievement).

Ceramic Filter. A ceramic characterized by an interconnected pore system, the pores being of substantially uniform size. Such ceramics are made from a batch consisting of pre-fired ceramic, quartz or alumina together wlth a bond that, during firing, will vitrify and bind the surfaces of the grains together. The pore size of different grades varies from about 10 to 500mm; tests for pore size are described in B.S. 1752. The filters are commonly available as tiles or tubes, the latter sometimes being known as CANDLES; special shapes can be made as required. Uses include filtration, aeration, electrolytic diaphrams and air-slides (cf. FILTER BLOCK; SINTERED FILTER).

Ceramic Process. The ASTM definition is: 'The production of articles or coatings from essentially inorganic, non-metallic materials, the article or coating being made permanent and suitable for utilitarian and decorative purposes by the action of heat at temperatures sufficient to cause sintering, solid-state reactions, bonding or conversion partially or wholly to the glassy state.' See also APPENDIX A.

Ceramic-to-metal Seal. The joining of a metal to a ceramic is generally accomplished by METALLIZING (q.v.) the ceramic surface and then brazing-on the metal component. Ceramic-to-metal seals are used in electrically insulated and vacuum-tight 'lead-throughs', especially for high-power h.f. devices; components so made are more rugged and resist higher temperatures than those having a glass-to-metal seal, thus permitting a higher bake-out temperature and use in a nuclear environment.

Ceramics and Ceramic Materials. Non-metallic inorganic or carbon solids, artificially produced or shaped by a high-temperature process, and inorganic composites wholly or essentially comprising such materials. Ceramics are usually taken to be shaped articles. (For a discussion of the problems in defining ceramics, and for official and quasi-official definitions in various contexts, see APPENDIX A).

Ceramic Whiteware. This term is defined in the USA (ASTM C242) as: 'A fired ware consisting of a glazed or unglazed ceramic body which is commonly white and of fine texture; the term includes china, porcelain, semivitreous ware and earthenware.'

Ceramix. Tradename for a flow control device developed by Ideal Standard Inc for hot/cold water mixer taps. Two alumina discs, one solid, the other pierced by three holes, rotate against each other to open or close the holes to provide for the passage of water. The device is wear-resistant. (UK Pat. 2136544A and 2136545A, 1984)

Ceramming. The conversion of a glass to a glass-ceramic.

Ceramography. The study of microstructures of ceramics at magnifications greater than 50 .

Ceramoplastic. An electrically insulating composite of synthetic mica bonded to glass.

Ceramtec. Tradename of the Ceramic Development Co for pyrophyllite, but often used in a context which implies that it is a synthetic ceramic.

Ceranox. A collective term for binary compounds of B, C, N and Si.

CeraPearl. A castable apatite ceramic, for dental application. Its tensile strength (150 MPa) is more than twice that of feldspathic dental porcelain.

Cercoms. Composites with at least one ceramic component.

Cercon. Tradename. Lightweight honeycomb ceramics (Corning Glassworks).

Ceria. See CERIUM OXIDE.

Cerium Borides. CeB_4: m.p.2100 °C; sp. gr. 5.7. CeB_6: thermal expansion (25–1000 °C) 6.7×10^{-6}

Cerium Carbides. CeC_2: m.p. 2500 °C; sp.gr. 5.2; decomposed by water. Ce_2C_6 has also been reported.

Cerium Nitride. CeN; produced by the action of N2 on Ce at 800C or NH_3 on Ce at 500C.

Cerium Oxide. CeO_2; m.p. 2600C; sp. gr. 7.13; thermal expansion (20–800C), 8.6×10^{-6}. A rare earth derived from the monazite of India, Brazil, Florida, Australia and Malagasy. Used in the polishing of optical glass; it is also effective both as a decoloriser and as a colouring agent for glass.

Cerium Sulphides. There are three sulphides: CeS, m.p. 2450 °C ± 100 °C;

Ce_3S_4, m.p. 2050 °C ± 75 °C; and Ce_2S_3, m.p. 1890°C ± 50 °C. Special ceramic crucibles have been made of these sulphides, but they can be used only *in vacuo* or in an inert atmosphere; such crucibles are suitable as containers for molten Na, K, Ca and other highly electropositive metals. The thermal-shock resistance of CeS is good, that of Ce_2S_3 poor, and that of Ce_3S_4 intermediate.

Cerlite. An Italian body of clay, chamotte and feldspars, which is claimed to combine the strength of vitreous china with the lower deformability of fireclay, whereby it is suitable for making large objects.

Cermet. A material containing both ceramic and metal. According to one definition the proportions by weight lie between 15% and 85%. Cermets appear first to have been made by M. Hauser in Switzerland in 1926. Experiments have been made with a wide range of ceramics (oxides and carbides) and metals (iron, chromium, molybdenum, etc.) in an attempt to combine the refractory and oxidation-resistant properties of ceramics with the thermal-shock resistance and tensile strength of metals. Success has been only partial. The most promising cermets are TiC/Alloy (the alloy containing Ni, Co, Mo, W, Cr in various proportions), Al_2O_3/Cr, Al_2O_3/Fe and ThO_2/Mo. The principal use for cermets would be for the blades of gas turbines and other high-temperature engineering units. Special cermets are also currently being used in brake and clutch linings. (Name first proposed by A. L. Berger – see *J. Am. Ceram. Soc.*, **32**, 81, 1949; for general information on the subject see J. R. Tinklepaugh and W. B. Crandall *Cermets,* Reinhold, 1960.) Note also that C.A.Calour and A.Moore claimed (*J. Mat. Sci.*, **7**, (5), 543, 1972) that above 700 °C metal matrix cermets, reinforced with ceramic fibres or whiskers, were not viable. If the fibre reacts sufficiently well with the metal matrix to form a bond, its lifetime is limited by corrosion; if it is sufficiently unreactive not to corrode, an inadequate bond is formed.

Cerulean Blue. Cobalt stannate $CoO\textit{n}SnO_2$, a light blue pigment.

Chain Hydrometer. A type of hydrometer that is operated at constant depth by a chain loading device similar to that used on some analytical balances. It has been used for the determination of the particle-size distribution of clays.

Chain Silicates. See SILICATE STRUCTURES.

Chair. (1) A wooden chair of traditional design used by a glass-blower.
(2) A team of workmen producing hand-made glass-ware.

Chairman. See GAFFER.

Chalcedony. A cryptocrystalline, fibrous form of quartz; flint, for example, is chalcedonic.

Chalcogenide Glass. A glass containing a high proportion of tellurium, selenium or sulphur. Such glasses are used in electronic current-switching devices, as their conductivity differs along different crystal panes, but their switching rate has proved too slow for modern computer applications. Current research centres on their high infra-red transmission, and applications in solar cells.

Chalk. A soft rock consisting of fine-grained calcium carbonate used as an alternative material to limestone in cement manufacture, and in the production of WHITING (q.v.).

Chalkboard Enamel. BLACKBOARD ENAMEL (q.v.)

Chalked or Chalky. The condition of vitreous enamelware that, in consequence of abrasion or chemical attack, has lost its gloss and has become powdery.

Chamber Dryer. A type of dryer in which shaped clayware is placed in chambers in which the temperature, humidity, and airflow can be controlled; the ware remains stationary during the drying process. The Keller Dryer is of this type, its distinctive feature being the system of handling the bricks to be dried by means of STILLAGES (q.v.) and FINGER-CARS (q.v.).

Chamber Kiln. An ANNULAR (q.v.) kiln of the type in which the setting space is permanently subdivided into chambers; examples include the SHAW KILN (q.v.) and the MENDHEIM KILN (q.v.).

Chamber Oven. A refractory-lined gas-making unit; the capacity of such an oven may vary from about 1 to 5 tons of coal.

Chamotte. A W. European word (originally German) which has now also been adopted in the UK, denoting refractory clay that has been specially fired and crushed for use as a non-plastic component of a refractory batch; it is generally made by calcination of lumps of the raw clay in a shaft kiln or in a rotary kiln (cf. GROG)

Champlevé Enamelware. One type of vitreous-enamel artware: a pattern is first cut into the base-metal but, where the pattern requires that enamels of different colours should meet, a vertical fin of metal is left so that the colour boundary will remain sharp, the two enamels not running into one another when they are fused. (French word meaning 'raised field'.)

Characterisation. The description of those material variables that are sufficient to permit reproduction of material performance in service, in a test or in relation to material processing. The term is used loosely for the measurement of microstructural, textural and mineralogical properties.

Charge. As applied to a ball mill this term means the material being ground and the grinding media.

Charge Pad. Localised thicker or better quality refractory in a metallurgical furnace, to withstand the impact of hot metal or scrap entering the furnace.

Charge Shaft. *See* FIRE PILLAR.

Charlton Photoceramic Process. A positive print is formed on ceramic ware by a process involving the application of emulsion to the ware, exposure in contact with the negative, and development. (A. E. Charlton, *Ceramic Industry,* **56,** No. 4, 127, 1951.)

Charp. Trade-name for Calcined High-Alumina Refractory Powder; it is made from Ayrshire bauxitic clay.

Charpy Test. An impact test for fracture resistance. A V-notched specimen, fixed at both ends, is struck behind the notch by a pendulum. The energy lost by the pendulum in fracturing the test-piece gives a measure of the impact strength of the material. See FRACTURE TOUGHNESS TESTS.

Chaser Mill. Name occasionally used for an EDGE-RUNNER MILL (q.v.) particularly of the over-driven fixed-pan type.

Chatelier. See LE CHATELIER.

Chatter Sleek. = FRICTIVE TRACK (q.v.)

Check, Crizzle, Vent. A surface crack in glassware. (See also SMEAR.)

Checkers; Checker Bricks. Refractory bricks or special shapes set in a regenerator in such a way as to leave passages for the movement of hot gases; waste gases passing from a furnace through the checkers give up heat which, on reversal of the direction of gas flow in the furnace, is subsequently transferred to the combustion air and fuel.

Checkerwork, Chequerwork. A construction made with many CHECKER

BRICKS (q.v.) to provide a large surface area for heat transfer to gases.

Checking. (1) A defect in cast-iron enamelware, raised lines appearing on the surface as a result of cracks in the ground-coat. (2) The term has been used in USA to signify surface cracking or crazing of clayware.

Cheeks. The refractory side-walls of the ports of a fuel-fired furnace.

Chemically Bonded Refractory Cement. A jointing cement for furnace brickwork that may be one of two types: *Air-setting refractory cement* – finely ground refractory material containing chemical agents, such as sodium silicate, which ensure that the cement will harden at room temperature; *Air-hardening refractory cement* – finely ground refractory material containing chemical agents that cause hardening at a temperature below that at which vitrification begins but above room temperature.

Chemical Machining. Selective etching or dissolution of material to shape a component. See LEACHABLE GLASS; LITHIUM SILICATE GLASSES.

Chemical Porcelain. US term for CHEMICAL STONEWARE (q.v.).

Chemical Stoneware. A vitreous type of ceramic product for use in the chemical industry and in other locations where resistance to chemical attack is sought. A typical body composition is: stoneware clay or plastic fireclay, 25–35%, ball clay, 30–40%, feldspar, 20–25%; grog, 20–30%. The grog may be crushed stoneware or porcelain, fused alumina, or silicon carbide, depending on the combination of physical properties sought in the finished product. The firing temperature is usually 1200–1300 °C. Chemical stoneware has a crushing strength of nearly 300 MN m^{-2}; the coefficient of linear expansion (20–100 °C) is 4×10^{-6}. In the USA, this type of ware is termed 'Chemical Porcelain'.

Chemical Surface Coating (CSC). This low-temperature chemical modification technique has three stages: activation of the substrate; formation of the new ceramic precursor by chemical action; heat treatment to convert the new precursor into a new ceramic coating. For instance, a silicon nitride coating can be produced by activating a silica gel substrate by chemisorption of trichorosilane, reacting with ammonia and heating to 500 °C.

Chemical Vapour Deposition (CVD). Chemical reactions between gaseous reagents at or near a heated surface are used to deposit a material coating on that surface. Process temperature are typically above 700 °C. The technique is widely used both to prepare ceramic materials, and to apply ceramic coatings to other material. It was first developed to fabricate microelectric devices. Compared to other coating methods it has higher deposition rates and offers better control over the microstructure of the coating by adjusting the deposition parameters.

Chemical Vapour Infiltration (CVI). Vapour phase reactants infiltrate a porous preform. Their chemical reactions deposit product within the pores to produce a 2-phase composite.

Chequers; Chequer Bricks. See CHECKERS; CHECKER BRICKS.

Chert. A general term for a cryptocrystalline siliceous rock that may occur in either nodular or tabular form. Chert stones were used as pavers and runners in the old paddle type of mill for grinding potters' materials; trimmed chert blocks are now used as a lining material for ball mills.

Chest Knife. A tool used in hand-blown glass-making for removing the MOIL (q.v.)

from the blowing-iron; the moils are allowed to crack off while the blowing-irons are in a receptacle called a 'chest'.

Chevenard Dilatometer. An apparatus for the measurement of thermal expansion; it depends on the recording, by means of an optical lever, of the differential expansion of the test-piece and that of a standard. It finds use in Western Europe for the testing of ceramic products. (P Chevenard *Compte Rend.,* **164,** 916, 1917.)

Chevron Notch Bend Test. See FRACTURE TOUGHNESS TESTS.

Chevroning. Fault-like planes in rotary kiln linings. The bricks between adjacent fault planes are tilted out of alignment with a V or chevron end to the displaced region. Ovality of the kiln structure is one cause.

Chill Mark. A surface defect on glass-ware characterized by its wrinkled appearance; also known as FLOW LINE (q.v.). It is caused by uneven contact in the mould prior to the forming.

Chimney Linings. Refractory concretes are used line steel chimneys, by gunning or casting. BS 4207 provides a code of practice for their formulation and installation. See also FLUE LINER.

Chimney Pot. The relevant British Standard is BS 1181 (1961); cf. FLUE LINER.

China. BS 5416 specifies this to be pottery with water absorption 0.2% and tranlucency 0.75% (assessed by comparison with standard test pieces) See also BONE CHINA. In the USA, however, ASTM-C242 defines the word as any glazed or unglazed vitreous ceramic whiteware used for nontechnical purposes, e.g. dinnerware, sanitary-ware, and art-ware, provided that they are vitreous. The COMBINED NOMENCLATURE (q.v.) equates this term with PORCELAIN (q.v.).

China Clay (Kaolin). A white-firing clay consisting essentially of KAOLINITE (q.v.). Major deposits occur in Cornwall (England); Georgia, S. Carolina and Florida (USA); Germany, France, Czechoslovakia, and elsewhere. Annual output in USA is 2 500 000 tons; in England, 1 500 000 tons. Of these tonnages only about 20% is used in the ceramic industry (pottery and refractories). Typical composition (per cent): SiO_2, 47; Al_2O_3, 38; Fe_2O_3, 0.8; Alkalis, 1.7; loss-on-ignition, 12.

Chinaflow. A specialist distribution service for china and glassware, operated by National Carriers.

Chinagraph. Specially formulated pencils (Royal Sovereign Pencil Co) for marking pottery, glass and similar smooth, hard surfaces.

China Process. A US term defined (ASTM – C242) as the method of producing glazed whiteware in which the body is biscuit fired and glaze is then applied followed by glost firing at a lower temperature. (In the UK this process is used in the manufacture both of earthenware and of bone china.)

China Sanitary-ware. US term for vitreous ceramic sanitaryware.

China Stone. Partly decomposed granite, consisting of feldspathic minerals and quartz, it is used as a flux in pottery bodies. Examples in the UK are Cornish Stone and Manx Stone. The former is available in various grades, e.g. Hard Purple, Mild Purple, Hard White and Soft White; the feldspars are least altered in the Hard Purple, alteration to secondary mica and kaolinite being progressively greater in the Mild Purple, Hard White, and Soft White; the purple stones are so coloured by the small amount of fluorspar present. Manx Stone (from Foxdale, Isle of Man) is virtually free from fluorine.

Chinese Blue or Mohammedan Blue. The mellow blue, ranging in tint from sky-blue to greyish-blue, obtained by the early Chinese and Persian potters by the use of impure cobalt compounds as colorants.

Chinese Red, Chrome Red. Red and orange colours made from various proportions of lead oxide PbO and lead chromate $PbCrO_4$.

Chipping. The chipping of vitreous enamelware is often attributable to the enamel coating being too thick, or to the curvature of enamelled edges being too sharp. Chipping of the edges of ceramic tableware is also accentuated by poor design.

Chittering. A fault that may appear as a series of small ruptures along the edge or rim of pottery-ware. True chittering is caused by incorrect fettling.

Choke. A glass-making fault appearing as a constriction in the neck of a bottle.

Choke Crushing. The method of running a pair of crushing rolls with the space between the rolls kept fully charged with clay; with the rolls operated in this way much of the crushing is achieved by the clay particles bearing on each other which gives a finer product than with FREE CRUSHING (q.v.).

Chrome-Alumina Pink. A ceramic colour consisting principally of Cr_2O_3, Al_2O_3 and ZnO; when used as a glaze stain, the glaze should contain ZnO and little, if any, CaO. It is recommended that, for use under-glaze, the glaze should be leadless. The colour depends on diffusion of Cr into the insoluble Al_2O_3 lattice and is normally stable up to 1300 °C.

Chrome Green. See VICTORIA GREEN.

Chromel. Tradename. Nickel-chrome alloys used in *chromel-alumel* thermocouples, and kiln accessories and furniture.

Chrome-Magnesite Refractory. A refractory material made from chrome ore and dead-burned magnesite, the chrome ore preponderating so that the fired product contains 25–55% MgO. Such refractories may be fired or they may be chemically bonded; they are frequently metal-cased and find most use in the steel industry. A typical chemical analysis is (per cent): SiO_2, 3–7; Fe_2O_3, 10–14, Al_2O_3, 12–18; Cr_2O_3, 25–30; CaO, 1–2; MgO, 35–40. See also MAGNESITE-CHROME REFRACTORY for ASTM classification.

Chrome Ore. An ore consisting primarily of chrome spinels, (Fe, Mg)O. $(Cr, Fe, Al)_2O_3$; it occurs in ultrabasic igneous rocks and in rocks derived from them by alteration, e.g. serpentines. The chief sources are in S. Africa, Rhodesia, Turkey and Russia; there are other important deposits in Yugoslavia, Greece, India, Philippines and Cuba. The composition varies widely; a typical refractory-grade ore contains (per cent) Cr_2O_3, 40–45; SiO_2, 3–6; and Al_2O_3, FeO, and MgO, 15–20 of each. The principal use in the refractories industry is in the manufacture of chrome-magnesite bricks for the steel industry

Chrome Red. See CHINESE RED.

Chrome Refractory. A refractory brick made from chrome ore without the addition of other materials; the Cr_2O_3 content is 25%. and the MgO content is 25%. Chrome refractories are neutral, chemically, and find use as a separating course of brickwork between silica refractories and basic refractories; their R.u.L. q.v.) is considerably lower than that of chrome-magnesite refractories. Chemical composition (per cent): SiO_2 4–7; Fe_2O_3, 12–15; Al_2O_3, 16–20; Cr_2O_3, 38–42; CaO, 1–2; MgO, 15–20.

Chrome Spinel. A SPINEL (q.v.) in which the trivalent metal is Cr.

Chrome-Tin Pink. A colour for ceramic glazes; it was first used in England in the 18th century and was originally called 'English Pink'. The mechanism by which the colour is formed is precipitation of fine particles of chromic oxide on the surface of tin oxide in an opaque glaze; lime must also be present.

Chrome-Zircon Pink. About 70% of the SnO_2 used in CHROME TIN PINK (q.v.) can be replaced by zircon without impairing the colour or stability.

Chromic Oxide. Cr_2O_3; m.p. approx. 2270 °C; sp. gr. 5.21; thermal expansion (0–1200 °C) 7.4×10^{-6}. A source of green colours for the ceramic industry but usually added in the form of a chromate, e.g. potassium dichromate. The oxide is highly refractory and has been used to a limited extent to improve the bond in CHROME-MAGNESITE REFRACTORIES (q.v.)

Chromite. Ferrous chromite. $FeO.Cr_2O_3$; sp. gr. 4.88; thermal expansion (0–1200 °C) 8.3×10^{-6} Chromite is a major constituent of many CHROME ORES (q.v.).

Chromium Borides. At least seven chromium borides have been reported. The compound that has received most attention is the diboride, CrB_2; m.p. approx. 1900°C; sp. gr. approx. 5.5; thermal expansion 5.0×10^{-6}. When heated in air, a coating of B_2O_3 is formed and prevents further oxidation up to the limiting temperature at which B_2O_3 itself ceases to be stable. It has been used for the flame-spraying of combustion-chamber linings.

Chromium Carbides. At least two Cr carbides exist. Cr_3C_2; m.p. 1890 °C; sp. gr. approx. 6.7; thermal expansion (20–1000 °C) 10×10^{-6}. CrC: m.p. 1550°C but oxidizes at 1100 °C. These carbides are hard, refractory, and chemically resistant. As such they have found limited use in bearings and corrosion-resistant nozzles, etc.

Chromium Dioxide. CrO_2. A ferromagnetic; CURIE POINT (q.v.) 116 °C; magnetic coercive force 20–30 oersteds; saturation magnetization 130 emu/g.

Chromium Nitrides. There are two compounds, Cr_2N and CrN; the latter decomposes at 1500°C.

Chromium Oxide. See CHROMIC OXIDE; CHROMIUM DIOXIDE.

Chromium Silicides. Several compounds exist e.g. Cr_3Si, Cr_2Si, CrSi, Cr_3Si_2 and $CrSi_2$. The m.p. varies from approx. 1550°C for CrSi and $CrSi_2$ to approx. 1700°C for Cr_3Si. Hardness varies from 76–89 Rockwell A. These silicides resist oxidation up to about 1000°C; they have good resistance to thermal shock but low resistance to impact.

Chromophore. Atoms or ions whose arrangement of electrons leads to selective absorption of particular wavelengths of light, and so imparts colour to their compounds, which may be used as PIGMENTS (q.v.). Typical chromophores are multivalent metals such as Co, V. In *lattice colours* the chromophore forms part of the crystal lattice. See also ENCAPSULATED COLOURS.

Chrysotile. The principal mineral of the ASBESTOS (q.v.) group. When pure it has the composition $3MgO.2SiO_2.2H_2O$.

Chuff Brick. A soft, underfired, salmon pink brick (USA).

Chün Glaze. A thick, high-temperature opalescent glaze, often with red or purple streaks.

Chum. A shaped block of wood or plaster on which a bat of pottery body can be roughly shaped before it is placed in the mould for jolleying in the making of large items of hollow-ware; a piece of flannel is placed over the chum before it

is used and this subsequently slips off the chum when the shaped clay is removed.
Chunk Glass. Rough pieces of optical glass obtained when a pot of glass is broken open.
CICS. Ceramic Industry Certification Scheme. The body which grants accreditation certificates for QUALITY ASSURANCE under BS 5750, for the ceramics industry.
CIE System. The name derives from the initials of the Commission Internationale de l'Eclairage. It is a trichromatic system of colour notation that is used, for example, in the glass industry.
Cinpres. Gas assisted injection moulding, a plastics manufacturing technique.
Ciment Fondu. Trade-name: aluminous hydraulic cement. (Lafarge Aluminous Cement Co. Ltd, 73 Brook St., London, W1) For general properties of this type of cement see HIGH-ALUMINA CEMENT. See also GORKAL CEMENT.
Cimita. A natural mixture of clay and feldspar occurring in parts of Chile. The composition is not uniform but a typical analysis (per cent) is: SiO_2, 58; Al_2O_3, 33; Fe_2O_3, 1; Alkalis, 4; H_2O, 4.
Cinder Notch. See SLAG NOTCH.
Circle Brick. A brick with two opposite larger faces curved to form parts of concentric cylinders (cf. RADIAL BRICK, and see Fig. 1, p39).
Circle System of Firing. See under ROTARY-HEARTH KILN.
Cire-perdue. French words for LOST WAX (q.v.).
Citadur. Trade-name: a HIGH-ALUMINA CEMENT (q.v.) made in Czechoslovakia.
Citric Acid. An organic acid sometimes used for the treatment of steel prior to enamelling, or in some metal release tests.
Cladding. An external non-load bearing facing, e.g. of bricks, faience or tiles, used in frame-type buildings (cf. VENEERED WALL). ASTM C1088 is a specification for thin veneered brick units (maximum thickness 1 in., 25.4mm) made from clay or shale.
Cladless Hot Isostatic Pressing, See SINTER-HIP.
Clamming. Local name for the brick and fireclay filling of the WICKETS (q.v.) of an old-type pottery kiln; sometimes spelled CLAMIN.
Clamp. A kiln constructed, except for the permanent foundations, of the bricks that are to be fired, together with combustible refuse and breeze. STOCK BRICKS (q.v.) in the London area were formerly fired exclusively in clamps but these produced only about 50% of first-quality bricks and were replaced by annular or tunnel kilns.
Clark Circle System. See under ROTARY-HEARTH KILN.
Clash. Local Scottish term for a thin slurry of clay and water.
Classification. The subdivision of a powder according to its particle size distribution or its separation into fractions of differing particle sizes. See also SIEVES, AIR SEPARATIONS, etc.
CLAW. Control of Lead at Work Regulations, 1980.
Clay. A natural material characterized by its plasticity, as taken from the clay-pit or after it has been ground and mixed with water. Clay consists of one or more clay minerals together with, in most cases, some free silica and other impurities. The common clay mineral is kaolinite; most clays consist of kaolinite in various degrees of atomic disorder. See also BRICK CLAYS, CHINA CLAY, FIRECLAY and the clay minerals HALLOYSITE, KAOLINITE, MONTMORILLONITE, etc.
Clay Building Brick. A brick for normal constructional purposes; such bricks can be made from a variety of BRICK CLAYS

(q.v.). Relevant British Standard is: BS 3921. The US Standards are ASTM – C62 (Building Brick); ASTM – C216 (Facing Brick); ASTM – C67 (Sampling and Testing); ASTM – C126 (Glazed Bricks).

Clay Gun. The modern preferred term (BS 3446 Pt3 1990) for TAPHOLE GUN (q.v.)

Clay Lath. BS 2705 describes this as copper-finished steel wire mesh at the intersections of which suitably shaped unglazed clay nodules have been bonded by a firing process. Clay lath provides a stable, well-keyed base to cover the whole of a surface with the minimum number of joints; it is supplied in rolls or mats.

Clay Shredder. A unit for the preliminary preparation of plastic clay. The machine consists of a hopper with a flat or conical base; adjustable knives operate from a vertical, central, rotating shaft. The clay falls from the shredder through slots in the casing.

Clayite. A term proposed by J. W. Mellor (*Trans. Brit. Ceram. Soc.*, **8**, 28, 1909) 'for that non-crystalline variety of the hydrated aluminium silicate- $Al_2O_3.2SiO_2.2H_2O$ – which occurs in china clay and in most clays yet examined. Kaolinite is crystalline clayite'. When X-ray analysis was subsequently able to demonstrate the crystallinity of all the clay minerals except ALLOPHANE (q.v.), the term was discarded.

Clayspar. Trade-name for a siliceous raw material occurring in Scandinavia and containing approx. 95% SiO_2, 2.5% Al_2O_3, 1.5% K_2O, 0.5% Na_2O.

Cleaner. A hot solution of alkalis (strength about 5%) used to remove grease and dirt from the base-metal before it is enamelled.

Clear Clay. A clay such as kaolin that is free from organic matter and so does not give rise to bubbles if used in a vitreous enamel; such clays are used in enamels when good gloss and clear colours are required.

Clear Frit. A frit that remains non-opaque when processed into a vitreous enamel.

Clear Glaze. A ceramic glaze that is transparent.

Cleavage, Cleavage Plane. Cleavage is the FRACTURE (q.v.) of a crystal along specific crystallographic planes known as *cleavage planes.* See also CRYSTAL STRUCTURE.

Cleavage Crack. Typically, a plastically deformed groove with median and lateral cracks emanating from it, caused by a hard sharp object crossing the (glass) surface. In crystalline materials, cleavage cracks are those proceeding across the grains.

Clinker. (1) Hydraulic cement in the unground state as it issues from the rotary or shaft kiln in which it was made. (2) Dead-burned basic refractory material, e.g. stabilized dolomite clinker. (3) Lumps of fused ash from coal or coke as formed, for example, in the fireboxes of kilns. (4) For the similar word as applied to continental engineering bricks see KLINKER BRICK.

Clinkering Zone. That part of a cement kiln which is in the temperature range (1350–1600 °C) in which the constituents react to form the CLINKER (q.v.).

Climbing Temperature Programme. A systematic method to determine the optimum conditions for HOT PRESSING (q.v.). Full pressure is applied at room temperature. The temperature is then increased slowly, and densification monitored (by observing the press piston movement) until it stops for a preset time interval, when full densification is assumed. Density is plotted against time, at constant pressure. An elongated S-shaped curve results, corresponding to the three stages of densification. The isothermal pressing temperature is found by using regression techniques to fit the approximately linear second stage.

Clino-enstatite. See ENSTATITE.

Clinotherm System. A prefabricated low thermal mass kiln.

Clip. A brick which has been cut to length (USA).

Clip Tile. A ceramic unit designed to fit around the base of a steel I-beam.

Clobbering. Decorating another producer's ware without permission (USA).

Closed Chip. A fractured corner of edge where the material is still in place.

Closed Fraction. A controlled distribution of particle size, between upper and lower set limits.

Cloisonné Enamelware. One type of vitreous-enamel artware: a pattern is outlined on the base-metal by attaching wire fillets to the metal; enamels of different colours are applied within the partitions thus made; the ware is then fired and the surface is polished. (From French word meaning 'partitioned'.)

Closed-circuit Grinding. A size-reduction process in which the ground material is removed either by screening or by a classifier, the oversize being returned to the grinding unit. Typical examples are a dry-pan with screens, dry-milling in an air-swept ball mill and wet-milling in a ball mill with a classifier.

Closed Porosity. See under POROSITY.

Closed Pot. See under POT.

Closer. See Pup

Closet Suite. A suite of ceramic sanitary-ware including the closet and the flushing cistern.

Closure. US equivalent of CLOSER (q.v.)

Clot. Roughly shaped clay or body ready for a final shaping process.

Clot Mould. The mould, in some types of stiff plastic brick making machines, into which a clot of clay is extruded and from which it is then ejected prior to the final re-pressing.

Clunch. A name used in some parts of England for a tough coarse, clay or for a marly chalk.

CMAA. Concrete Masonry Association of Australia. See BENDER ABRASION METHOD.

CMC. (1) Abbreviation for CARBOXYMETHYLCELLULOSE (q.v.). (2) Critical Moisture Content. (3) Ceramic Matrix Composite – see COMPOSITES.

CMZP. Calcium magnesium zirconium phosphate $(Ca_{1-x} Mg_x)Zr_4(PO_4)_6$. The material is chemically stable to 1500 °C, has thermal conductivity 1–2 W/mK from 200 to 1200°C, bulk thermal expansion of 5×10^{-7}/K from 25 to 750 °C.

CNB. Chevron Notch Bend. See FRACTURE TOUGHNESS TESTS.

CNC. Computer Numerical Control.

CO_2 Process. See CARBON DIOXIDE PROCESS.

Coade Stone. A vitreous ware, used for architectural ornament, made in London by Mrs Coade from 1771 until her death in 1796; manufacture finally discontinued in about 1840. The body consisted of a kaolinitic clay, finely ground quartz and flint, and a flux (possibly ground glass).

Coated Dolomite Grain. Also known as unfired semi-stable dolomite refractory, this comprises calcined dolomite bonded with tar or oil.

Coating. See METAL PROTECTION; REFRACTORY COATING.

Cob Mill. A disintegrator for breaking down agglomerates of the raw materials used in the compounding of vitreous-enamel batches.

Cobalt Aluminate. As used as a stain in the pottery industry, this is of indeterminate composition but it is more stable and gives a better blue than most other cobalt stains. When used to whiten a pottery body, up to about 0.1% is required.

Cobalt Carbonate. $CoCO_3$; sp. gr. 4.1. Sometimes used as a source of COBALT OXIDES (q.v.) in ceramic colours.

Cobalt Chloride, $CoCl_2.6H_2O$;

Cobalt Nitrate, $Co(NO_3)_2.6H_2O$. Soluble cobalt compounds sometimes used in the ceramic industry for the purpose of neutralizing the slightly yellow colour caused by the presence of 'iron' impurities; the principle is the same as the use of a 'blue bag' in the laundry.

Cobalt Chromate, $CoCrO_4$. Finds some use, in admixture with ZnO and Al_2O_3, as a green or blue colour for vitreous enamels.

Cobalt Oxides. Grey or 'prepared' oxide, CoO, m.p. approx 1800°C; cobaltic oxide, Co_2O_3; black oxide, Co_3O_4. Used to give a blue colour to glass and pottery ware, and added (together with nickel oxide) to ground-coat enamels for steel to impove their adherence. The tint of cobalt oxide colours can be modified by adding other oxides (see MAZARINE BLUE, WILLOW BLUE and CELESTE BLUE). With the addition of the oxides of manganese and iron a black colour can be produced.

Cobalt Sulphate, $CoSO_4.7H_2O$. A soluble cobalt compound used for the same purpose as COBALT CHLORIDE (q.v.). It is also sometimes used in vitreous enamels for mottling single-coat grey-coloured ware.

Cobble Mix. U.S. term for concrete containing aggregate up to 150mm diameter. (Cf. CYCLOPEAN).

Coble Creep. CREEP (q.v.) in which the creep rate is controlled by diffusion at the grain boundaries, and is proportional to the cube of the grain size. (R.L. Coble, *J. Appl. Phys.* **32** (5) 1961 p787).

Cobo Process. The cold bonding of other materials to bricks.

Cock-spur. See SPUR.

Coconut Piece. A special shape of ceramic wall tile (see Fig 7, p350).

COD. Crack Opening Displacement. (q.v.)

Cod Placer. See under PLACER.

Coefficient of Scatter. A term used in the testing of vitreous enamelware and defined (ASTM 286) as: 'The rate of increase of reflectance with thickness at infinitesimal thickness of vitreous enamel over an ideally black backing; for method of test see ASTM C347.'

Coesite. A form of silica produced at 500–800°C and a pressure of 35 kbar; sp.gr. 3.01; insoluble in HF. Named from L. Coes who first obtained this form of silica *(Science,* **118**, 131, 1953).

Co-generation. The combined generation of heat and electric power, using a diesel or gas-engine generator. Land-fill gas (methane) is a frequently used fuel.

Coil Building. A primitive method of shaping clay vessels by rolling clay into a 'rope', which is then coiled to form the wall of the vessel; the inner and outer surfaces of the roughly shaped ware are finally smoothed.

Coin Brick. See JAMB BRICK. (a solecism for QUOIN BRICK).

Coke Oven. A large, refractory-lined structure consisting of a series of tall, narrow chambers in which coal is heated out of contact with air to form coke. Silica refractories are used for most of the coke-oven structure. (See BS 4966) (BS 6886 specifies alumino-silicate refractories for coke-ovens).

Cokilite. A refractory castable based on silicon nitride.

Coking. See CARBONIZATION.

Colburn Process. The production of sheet glass by vertical drawing for about 4 ft and then bending over a driven roller so that the cooling sheet then travels horizontally. The process was invented in USA by I. W. Colburn in 1905 and was subsequently perfected by the Libbey-Owens Company.
Coldclay. A self-hardening clay, tradename of Fulham Pottery. Cf. DUOCLAY.
Cold Crushing Strength. See CRUSHING STRENGTH.
Cold-end Handling, Operations in the manufacture of glass when the glass is cool after processing.
Cold Joint. The surface between a run of concrete which has passed to its final set, and a second run poured alongside it but which is no longer able to blend into it.
Cold-process Cement. Older name for SLAG CEMENT (q.v.).
Colemanite. Naturally occurring calcium borate, $2CaO.3B_2O_3.5H_2O$; there are deposits in Nevada and California. It is used in some pottery glazes.
Collar. A short fireclay section used to join the main (silica) part of a horizontal gas retort to the metal mouthpiece.
Collaring. The final process in THROWING (q.v.) a vase: both bands are used to compress the clay and thus reduce the size of the neck.
Collet. A flange for the mounting of an abrasive wheel on a spindle.
Colloid. Material in a form so finely divided that, when dispersed in a liquid, it does not settle out unless flocculated by a suitable electrolyte. The finest fraction of plastic clays is in this form. The colloidal size range is approx. 0.001–0.20mm.
Colloid Mill. A high-speed dispersion unit yielding a suspension of particles of the order of 1mm size. The original mill of this type was designed by H. Plauson (Brit. Pat. 155 836 and subsequent patents).
Colloidal Colours. Colloidal-sized particles in glass scatter light, the wavelength of maximum scatter being dependent on particle size. Coloured effects are thus produced (as in sunsets). Colloidal gold, silver or copper are most used. Red filter glasses are made by precipitating coloured crystals such as CdS or CdSe. The particles are dispersed in the glass by controlled nucleation and growth. The metal to be precipitated is dissolved in the glass along with a reducing agent (tin oxide or antimony oxide). The molten glass is cooled to a temperature at which the nucleation rate is high, then heated to promote crystal growth.
Colorimeter. An instrument for COLOUR MEASUREMENT (q.v.).
Colour. See also ENCAPSULATED COLOURS; LATTICE COLOURS; COLLOIDAL COLOURS; TRISTIMULUS VALUES; TRICHROMATIC PRINTING; MUNSELL SYSTEM; METAMERISM. Colour is important in the ceramics industry because of its aesthetic value. Most ceramic ‘colours’ are solid coloured particles (PIGMENTS q.v.) suspended in a glass. Such DECORATION (q.v.) must be durable, non-toxic and stable. It is applied by a wide range of processes, as designs or coloured glazes. *Colour measurement* is important, particularly for glazed tiles, sanitaryware and vitreous enamelware. These may have to match eachother and other coloured materials in e.g. the bathroom. ASTM C609 specifies a photometric method for the measurement of small colour differences between tiles. There are two basic methods of colour measurement: spectrophotometry, and the Tristimulus Filter Method typified by the Colourmaster, Hilger and Hunterlab

instruments. See CERAMIC COLOUR STANDARDS. BS 5252, 1976 specifies a wide range of colours used in industry. The Methuen *Handbook of Colour* 1983 provides colour charts, the names of a wide range of colours, and their technical co-ordinates. See also ISCC-NBS COLOUR SYSTEM and RAL COLOUR REGISTER. See also *Colour. Its measurement, computation and application.* Chamberlin and Chamberlin, Heyden 1980. For an elementary discussion of colour measurement in the ceramic industry, see W. Ryan and C. Radford *Whitewares.: Production, Quality and Quality Control,* Institute of Ceramics, 1987.

Colour Retention. The colour retention of red, orange and yellow vitreous enamels containing cadmium sulphide, or sulphoselenide, when exposed to atmospheric corrosion is measured by treating the enamel surface with cupric sulphate solution; details of the test are given in ASTM – C538.

Colour Twist. Twisted coloured glass rods as a form of decoration within a wine-glass stem.

Columbium. Name used in USA for NIOBIUM (q.v.).

Comb Rack. A bar of acid-resisting metal, e.g. Monel, used to support and to separate metal-ware in a PICKLING (q.v.) basket, while it is being prepared for the application of vitreous-enamel slip. The term is also applied to a burning tool for vitreous enamelling.

Combed. A surface texture of narrowly spaced lines produced on clay facing bricks by fixing wires or plates above the extruding column of clay so that they 'comb' its surface.

Combined Nomenclature of the European Communities. Published by the EC Commission in Luxembourg, 1987, this document replaced the BRUSSELS NOMENCLATURE. It classifies ceramic products in Chapter 69, according to use, dealing with refractories, bricks, tiles and china and other tableware. Ceramics for electrical applications are classified in Chapter 85. The 'Explanatory Notes' to the classification give definitions of different types of ceramics, for international tariff and trade purposes. See APPENDIX A.

Comeback. In the vitreous-enamel industry this term denotes the time that elapses before a batch-type furnace regains its operating temperature after a fresh load of ware has been placed in it.

Comminution. A term covering all methods of size reduction of materials, whether by crushing, impact, or attrition.

Common Bond. Term sometimes used in USA for AMERICAN BOND (q.v.).

Common Pottery. A term used in the COMBINED NOMENCLATURE (q.v.) for 'products obtained from ferriginous and calcareous clay (brick earth) which, when fractured, present an earthy, matt and coloured (generally brown, red or yellow) appearance. Particles, inclusions and pores representative of their structure are greater than 0.15mm and visible to the naked eye. Water obsorption (measured by a specified test) is 5wt %.'

Commons. Clay building bricks that are made without attention to appearance and intended for use in the inner leaf of cavity walls or for internal walls; for the UK, dimensions are specified in BS 3921. The crushing strength of such bricks varies from about 10 to 40 MNm^{-2}, the water absorption from about 10 to 30 wt%.

Compacting Factor. A factor indicating the workability of any concrete made with aggregate not exceeding 40 mm, but of particular value for assessing concretes of low workability. The factor is the ratio of the bulk weight of the concrete when compacted by being allowed to fall through a specified height, to that when fully compacted. For details see BS 1881.

Compact Tension. See FRACTURE TOUGHNESS TESTS.

Compaction. The production of a dense mass of material, particularly of a powder, in a mould. Ceramic powders may be compacted by dry-pressing, tamping, vibration, isostatic pressing, or explosive pressing. A.R. Cooper and L.E. Eaton (*J. Am. Ceram. Soc.* **45**, 1962, p897) proposed a quantitative model for the compaction of ceramic powders. This equated the reduction in volume of the compact to the statistically distributed and pressure-activated processes of filling large and small voids. Experimental rate constants for these two processes reveal whether compaction is dominated by primary particles or by agglomerates. Such analysis neglects interaction between the powder and the die walls (see DIE WALL FRICTION). More efficient bonding results if the particles are subjected to shear forces as well as compressive forces. (D. Train, *Trans Inst Chem Engrs*, **35**, 1957, p258).

Compass and Wedge. Term sometimes used for a brick that has a taper both on the side and on the face, e.g. a 9 in. (228 mm) brick tapered 4.5/3.5in. (114/89 mm) and 2.5/2 in. (63/50 mm).

Compo. A siliceous, highly grogged fireclay composition, with crushed coke or graphite sometimes added, for use as a general ramming material in the casting-pit of a steelworks or in a foundry. A typical composition is (per cent): crushed firebrick, 65–70; chamotte, 15–20; siliceous fireclay, 14; foundry blacking, 1. The batch is made up with about 10% water. (The word is an abbreviation for 'Composition'.)

Composites. Materials comprising two or more components with different properties. The components may exist as continuous phases, or one may be continuous and another discrete. In ceramic matrix composites the continuous phase is ceramic – the discrete phase metallic or a second ceramic. In metal matrix composites, the continuous phase is metallic. See also CERMETS.

Composite Refractory. A refractory brick, different parts of which have different compositions and/or textures, e.g. an insulating refractory having a dense working face, or a metal-cased basic refractory having internal partitions filled alternately with magnesite and chrome-magnesite.

Composition Brick. Scottish term for a common building brick made by the stiff-plastic process from clay and colliery waste; characteristically, it has a black core.

Compound Rolls. Two or more pairs of CRUSHING ROLLS (q.v.) arranged above one another, the upper pair acting as a primary crusher and the lower pair as a secondary crusher. Compound rolls find use in the size-reduction of brick clays.

Compressibility. The compressibility of ceramic fibre is the percentage relaxation after 5min of the pressure which gives a 50% reduction in thickness. Cf. RESILIENCE. BS 1902 Pt section 8 specifies tests.

Compression Precracking. See FRACTURE TOUGHNESS TESTS.

Compressive Strength. The maximum load per unit area, applied at a specified rate, that a material will stand before it fails. BS 3921 specifies a compressive strength test for building bricks, specifying a strength of 5 N/mm^2 for all bricks, but 50 or 70 N/mm² for Class B and Class A ENGINEERING BRICKS. ASTM E447 gives two tests for masonry prisms, one for comparative laboratory tests with different bricks and mortars; the other for use on-site. The prisms are single-wythe, laid in stack bond, containing at least two morter joints, and cured for 28 days.

Concrete. A mixture of hydraulic cement, fine aggregate (e.g. sand) and coarse aggregate, together with water; it is placed *in situ* or cast in moulds and allowed to set. The ratio of the three solid constituents is usually expressed in the order given above, i.e. '1: 2: 5: concrete' means 1 part cement, 2 parts fine aggregate and 5 parts coarse aggregate. The properties are largely determined by the cement and 'fines', but, for normal constructional purposes, the coarse aggregate must be graded to give dense packing.

Concrete Aggregate. Normal (as opposed to lightweight) concrete aggregate includes sand and gravel, crushed rock of various types and slag. The nomenclature is given in BS 812; the mineralogical composition is dealt with in ASTM C294 and C295.

Concrete Block. The properties required of precast masonry units, both dense and lightweight, are specified in the UK in BS 6073. In USA the properties required of a solid concrete building block are specified in ASTM C.90, C936 for concrete paving blocks and C 744 for units faced with resin or cement plus filler; the properties of hollow concrete blocks are specified in ASTM C90 and C129; for methods of sampling and testing see ASTM C140, C426 and C1006. C139 specifies special units for manhole construction. There are some 15 more ASTM specifications for specialised pre-cast concrete products.

Concrete Brick. A building brick made from portland cement and a suitable aggregate. ASTM – C55 specifies two types: Type I. Moisture-Controlled Units; Type II. Non-moisture-Controlled Units. There are three grades of each type, these being classified according to crushing strength and water absorption.

Concrete Pipes. Pipes for water and drainage are made in large sizes from pre-cast concrete units. BS 5911 specifies the pipes, fittings for joints, inspection chambers etc., including special requirements for porous, reinforced or glass composite concrete pipes. The Annual Book of ASTM Standards, Vol 4.05 lists some 25 standards relating to the requirements and test methods for concrete pipes, fittings and joints, this number almost doubling if metric equivalents are counted separately.

Condenser. The condenser attached to a horizontal zinc retort for the cooling of the zinc vapour and its collection as metal is made of fireclay.

Conditioning Zone. (1) The part of a tank furnace for flat glass where the temperature of the glass is adjusted before it flows into the forehearth or drawing chamber.
(2) That part of the feeder, away from the wall of a glass-tank furnace, in which the temperature of the molten glass is adjusted to that required for working.

Conduit (for Electric Cable). See CABLE COVER.

Cone. (1) The usual term for a PYROMETRIC CONE (q.v.) or a TEST CONE (q.v.); for nominal squatting temperatures of Pyrometric Cones see Appendix 2.
(2) A special fireclay shape sometimes used in the bottom pouring of molten steel; it has a hole through it and conveys the steel from runner brick to ingot mould. The cone is frequently

made as an integral part of a runner brick, in which case the latter is known as a RISER BRICK or END RUNNER (q.v.).

Cone and Plate Viscometer. A cone with wide (c. 174°) apical angle is placed on a horizontal flat plate. The space between is filled with the liquid to be studied and the cone or the plate rotated about the axis of the cone.

Cone Crusher. A primary crusher for hard rocks, e.g. quartzite used in making silica refractories; it consists of a hard steel cone that is rotated concentrically within a similarly shaped steel casing (cf. GYRATORY CRUSHER).

Cone Hip. See HIP TILE.

Cone-screen Test. A works' test for the fineness of milled enamels. A standard volume of enamel slip is washed through a conical sieve; the amount of oversize residue is read from a graduated scale along one side of the sieve.

Cone Wheel. A small cone-shaped abrasive wheel of the type frequently used in portable tools.

Congruent Melting. The melting point of a solid is said to be congruent if the material melts completely at a fixed temperature and without change in composition. (cf. INCONGRUENT MELTING).

Coning and Quartering. A method of obtaining a random sample of a powder by pouring it out to form a cone, which is then divided into four by two diametral 'cuts'. One quarter is selected, and the process may be repeated.

Consistency. The flow or mechanical displacement properties of cements, mortars and refractory castables. See KELLY BALL TEST and VICAT NEEDLE for cements; BALL-IN-HAND TEST for castables, FLOW TABLE TEST for mortars and concretes.

Consistodyne. Trade-name; a device for attachment to the barrel of a pug for controlling the workability of the clay; *(Ceramic Age,* **77**, No. 11, 34, 1961).

Consistometer. Term used in the vitreous-enamel industry for various instruments designed for the evaluation of the flow properties of enamel slips. The earliest consistometer was that used by R. D. Cooke (*J. Amer. Ceram. Soc.,* **7**, 651, 1924); this was of the capillary type and was developed into an instrument suitable for works' control by W. N. Harrison *(ibid.,* **10,** 970, 1927). Other instruments for controlling the properties of vitreous-enamel slips include the GARDNER MOBILOMETER *(q.v.).*

Consolidation. Heat treatment of POROUS GLASS and RECONSTRUCTED GLASS (q.v.) to eliminate porosity.

Constant Moment Test. See FRACTURE TOUGHNESS TESTS.

Constringence. See ABBE NUMBER.

Contact Stress. The tensile stress component in a (glass) surface near the contact area of force due to an applied object.

Container Glass. One of the major divisions of the glass industry – jars, bottles and other containers. ASTM C147 tests their internal pressure strength; C225 their resistance to chemical attack; C148 the quality of annealing; C149 their thermal shock resistance. British standards specify requirements for different types of containers for various applications, including bottles for beer and cider (BS 6118); soft drinks (BS 6119,7367 and 7488); wine (BS 6117); medicine (BS 1679); milk (BS 6106 and 3313); food (BS 5771); honey (BS 1777); drinking glasses for wine tasting BS 5586.

Containerless Hot Isostatic Pressing. See SINTER-HIP and GAS-PRESSURE SINTERING.

Continental Porcelain. The true, feldspathic, hard-paste PORCELAIN (q.v.) made in Western Europe.

Continuous Casting. Molten metal flows continuously into a shallow, water-cooled mould and the solidified ingot is withdrawn at a constant rate between cooled moving belts or rollers. Refractories for such a process must have long lives.

Continuous Chamber Kiln. See TRANSVERSE-ARCH KILN.

Continuous Kiln. A kiln in which the full firing temperature is continuously maintained in one or other zone of the kiln. There are two types: ANNULAR KILN (q.v.) and TUNNEL KILN (q.v.).

Continuous Vertical Retort. A type of gas retort, built of silica or siliceous refractories. Coal is charged into the top of the retort, coke is extracted from the bottom, and town gas is drawn off, the whole operation being continuous (cf. HORIZONTAL RETORT). Continuous vertical retorts are also used in the zinc industry, in which case they are built of silicon carbide refractories.

Contrast Ratio (C80). A ratio related to the reflectance of a vitreous enamel coating and defined in ASTM C347 as: The ratio of the reflectance of an enamel coating over black backing to its reflectance over a backing of reflectance 0.80 (80%).

Contravec. Trade-name: a system for the blowing in of air at the exit end of a tunnel kiln to counteract the normal convection currents. (Gibbons Bros. Ltd., Dudley, England)

Conversion. A change in crystalline structure on heating that is not immediately reversible on cooling. The most important example in ceramics is the conversion of quartz at high temperature into cristobalite and tridymite (cf. INVERSION).

Converter. A refractory-lined vessel supported on trunnions and used for the production of steel by oxidation of the impurities in the molten pig-iron that forms the charge; this is done by a blast of air or oxygen. In the original BESSEMER CONVERTER the blast passed through the converter bottom, the basic refractory lining being pierced with tuyeres for this purpose. In the TROPENAS CONVERTER the blast strikes the surface of the molten metal via tuyeres passing through the refractory wall – hence the alternative name SIDE BLOWN CONVERTER. In the oxygen process of steelmaking, TOP-BLOWN CONVERTERS are used; they are lined with tarred dolomite or tarred magnesite-dolomite refractories.

Cooke Elutriator. A short-column hydraulic elutriator for sub-sieve sizes designed by S. R. B. Cooke (U.S. *Bur. Mines, Rept. Invest.,* No. 3333, 1937).

Cooler. As used in the portland cement industry, the term 'cooler' refers to the ancillary unit of a cement kiln into which hot clinker is discharged to cool before it is conveyed to the grinding plant.

Cooling Arch. A furnace for the annealing of glassware, which is placed in the furnace and remains stationary throughout the annealing (cf. LEHR).

Co-ordinating Size. The size of a co-ordinating dimension (ISO 1803). E.g. that for a BRICK (q.v.) is 225 × 112.5 × 75mm. The target size to which the product is actually manufactured is called *work size*, and is less, by the nominal thickness of the joints (10mm for bricks, 1mm for tiles). Neither of these is the *actual size* (obtained by measurement) which must be within a given range of the specified work size. The *nominal size* is the size used to describe the product, and may again differ.

Copacite. Trade-name; a Canadian SULPHITE LYE (q.v.).

Coping. The top course of masonry in a wall. BS 5642 Pt2 specifies materials, sizes and design of concrete, stone and clayware copings.

Copper Carbonate. The material is the basic carbonate, $CuCO_3.Cu(OH)_2$; it is used as a source of copper for coloured glazes, particularly for those glazes fired under reducing condions.

Copper Enamel. A vitreous enamel specifically compounded for application to copper, the composition is essentially lead silicate with small additions of alkalis, arsenic oxide and (sometimes) tin oxide.

Copperhead. A fault (reddish-brown spots) liable to appear in the ground-coat during vitreous enamelling; the spots are exposed areas of oxidized base metal. Causes include boiling from the base-metal, inadequate metal preparation, very thin application of enamel, and the presence of acid salts in the enamel slip.

Copperlight. A glass window pane, 6 mm thick and up to 100 mm^2 size, fitted in a special copper frame and used as a 'fire-stop'.

Copper Oxide. Cupric oxide (black), CuO; cuprous oxide (red) Cu_2O. Used as colouring agents in pottery and glass. Copper oxide normally gives a green colour but under reducing conditions it gives a red, due to the formation of colloidal copper, as in *rouge flambé* and *sang de boeuf* art pottery and copper ruby glass. Mixed with other metal oxides it yields black colours for glazes and for vitreous enamels.

Copper Ruby Glass. See RUBY GLASS.

Copper Titanate. $CuTiO_3$; a compound sometimes added in amounts up to 2% to $BaTiO_3$ to increase the fired density.

Co-precipitation. Precipitation of one compound is assisted by the addition of a second compound which precipitates with the same reagent. Co-precipitation can be used to prepare ceramics or their precursors with a mixed composition of carefully controlled proportions.

Coquille. Thin glass with a radius of curvature of 90 mm; used in the production of sun glasses; from French word meaning 'shell' (see also MICOQUILLE).

Coral Red. A ceramic colour. One form of coral red consists of basic lead chromate; this compound is unstable and the decorating fire must be at a low temperature.

Corbel. Brickwork in which each course projects beyond the course immediately below.

Cord. A fault in glass resulting from heterogeneity and revealed as long inclusions of glass of different refractive index from that of the remainder of the glass.

Cordierite. A magnesium alumino-silicate $2MgO.2Al_2O_3.5SiO_2$; part of the Mg can be replaced by Fe or Mn. Cordierite exists in three crystalline forms but only the α-form is commonly encountered; it melts incongruently at 1540 °C. There is a large deposit of cordierite in Wyoming, USA, but it is usually synthesized by firing a mixture of clay, steatite and alumina (or an equivalent batch) at 1250 °C or above. The mineral has a low and uniform thermal expansion (2.8×10^{-6} between 0° and 1200 °C) and cordierite bodies therefore resist thermal shock. The electrical resistance (10^8–10^{10} ohm-cm at 300 °C) is adequate for many purposes when high thermal shock resistance is also required, e.g. arc chutes, electric fire-bars, fuse cores, etc. The dielectric constant (1 MHz) is 5.0 and the power factor (1 MHz) 0.004–0.008. Cordierite honeycombs are used as supports for automobile engine exhaust catalysts.

Core. (1) The central part of a plaster mould of the type used in SOLID

CASTING (q.v.). (2) The central part of a sand-mould as used in foundries. (3) A one-piece refractory or heat-insulating shape for use at the top of an ingot mould and serving the same purpose as a HOT-TOP q.v.); this type of core is also sometimes called a DOZZLE.

Cored Brick. A brick that is a least 75% solid in any plane parallel to the load-bearing surface.

Corex Process. The direct reduction of iron ore using non-coking coal. Introduced by Iscor Ltd, S. Africa, and formerly known as the KR process.

Corhart. Trade-name for various types of electrically fused refractories: Corhart Standard is a fused mullite-corundum refractory (18–20% SiO_2, 72–74% Al_2O_3); Corhart Zac is a similar product containing zirconia (8–12% SiO_2, 65–75% Al_2O_3, 15–20% ZrO_2). These electrocast refractories are used chiefly in glass-tank furnaces. Corhart 104 is a fusion-cast chrome-magnesite refractory for use in the steel industry. (Corhart Refractories Co. Louisville 2, Kentucky, USA; English agent--Electrocast Ltd., Greenford, Middlesex.).

Cornelius Furnace. A type of glass-melting furnace in which the glass is heated by direct electrical resistance. The design was introduced in Sweden by E. Cornelius (Brit. Pat. 249 554; 23/3/25; 303 798; 8/1/29).

Corner Wear. The wear of an abrasive wheel along one or both of its circumferential edges.

Cornish Crucible. A small (e.g. 90 mm high, 75 diam.) clay crucible of a type used for the assaying of copper; these crucibles are made from a mixture of about equal parts of ball clay and silica sand.

Cornish Stone. See CHINA STONE.

Corridor Dryer. Term sometimes used for a CHAMBER DRYER (q.v.).

Corrosion. Wear caused by chemical action (cf. ABRASION and EROSION (q.v.). See also SLAG ATTACK. ASTM C 621 and C 622 describe tests for the resistance of refractories to corrosion by molten glass.

Corundum. The only form of alumina that remains stable when heated above about 1000 °C; also known as α–Al_2O_3; m.p. 2050 °C; hardness 9 Moh; sp. gr. 3.98; thermal expansion (20–1000 °C) 8.5×10^{-6}; sp. heat 0.18 (20 °C), 0.22 (100 °C). It occurs naturally, but impure, in S. Africa and elsewhere but is generally produced by extraction from bauxite followed by a firing process at high temperature. Corundum is used as an abrasive and as a refractory and electroceramic, e.g. in sparking plugs (see also ALUMINA).

COSHH. Control of Substances Hazardous to Health Regulations, 1988. See HEALTH AND SAFETY.

Co-spray Roasting. A process for making FERRITES (q.v.) in which a solution of mixed Mn and Fe chlorides is sprayed into a large roaster. A mixture of Mn and Fe oxides is formed, and HCl recovered. Zinc oxide is added, and MnZn ferrites produced.

Cotac. Cobalt-bonded Tantalum Carbide. Tradename, (ONERA, France)

Cottle. Term used in the N. Staffordshire potteries for the material, e.g. stiff canvas, used to form the sides of a plaster mould while the plaster is being poured in and until the plaster has set. (Probably from N. country. *Cuttle* a layer of folded cloth.)

Cotto. Traditional Italian tiles, extruded as split tiles from weathered shaly red or yellow clays of Tuscany or Emilia. The tiles are given a rustic finish by wirebrushing or sandpapering after drying, then fired set upright using special steel kiln furniture. Tiles up to 61 cm square are produced.

Coulter Counter. A high-speed device for particle-size analysis designed by W. H. Coulter *(Proc. Nat. Electronics Conf.,* **12**, 1034, 1956) and now made by Coulter Electronics Inc., Chicago, USA. A suspension of the particles flows through a small aperture having an immersed electrode on either side, with particle concentration such that the particles traverse the aperture substantially one at a time. Each particle, as it passes, displaces electrolyte within the aperture, momentarily changing the resistance between the electrodes and producing a voltage pulse of magnitude proportional to particle volume. The resultant series of pulses is electronically amplified, scaled and counted.
Counter Blow. In the BLOW-AND-BLOW (q.v.) process of shaping glass-ware, the operation during which the parison is blown out.
Counterflow Kiln. A tunnel kiln with two parallel tunnels, through which the ware moves in opposite directions. The resultant air-flow pattern in the linked tunnels leads to fuel economy.
Coupe. A design of plate in the form of a very shallow bowl, lacking any flat rim at the edge.
Course. By convention, a course of brickwork includes one layer of mortar as well as the bricks themselves.
Cove Skirting. A special shape of ceramic wall tile (see Fig. 7, p350).
Cover. An item of KILN FURNITURE (q.v.). The cover is the flat refractory shape forming the top of a CRANK (q.v.); it protects the top piece of ware and at the same time holds the PILLARS (q.v.) in position.
Covercoat. (1) The final coat applied to vitreous enamelware resulting in the top surface. Normally there is a ground-coat and a cover-coat, but some enamels are now sufficiently opaque for single-coat application to appropriate grades of base-metal.
(2) The clear layer covering the ink design of a transfer. Commercial Decal Inc USA, incorporated a low-melting glass into the covercoat, to protect the colour layer during firing. This also fortuitously served to reduce METAL RELEASE (q.v.) from colours containing lead or cadmium. (Br. Pats. 1120486, 1968 and 1426219, 1976).
Covered Pot. See under POT.
Covering Power. The ability of a glaze or vitreous enamel to cover, uniformly and completely, the surface of the fired ware.
Cowper Stove. See HOT-BLAST STOVE.
CPDA. Clay Pipe Development Association.
Crack Arresters. Macroscopic design features to prevent or delay CRACK GROWTH (q.v.) Examples are holes drilled near sharp corners to arrest cracks which may originate there, by reducing stress concentrations.
Crack, Cracking. See FRACTURE.
Crack Branching. When a single crack divides into two or more cracks, the rate of crack growth is slowed, because the available energy must be shared over the production of a greater number of new fracture surfaces (see FRACTURE, CRACK PROPAGATION). This effect can be induced by the pre-existence of microcracks or zones of weakness (e.g. at grain boundaries) throughout the ceramic. The TOUGHNESS (q.v.) of ZIRCONIA (q.v.) ceramics can be enhanced by this mechanism. Crack branching is also known as FORKING in FRACTOGRAPHY (q.v.).
Crack Bridging. A mechanism leading to higher TOUGHNESS (q.v.) in ceramics and composites. Intact material behind the crack tip imparts closure forces on the crack. The crack is 'bridged' behind

the tip either by reinforcing fibres or whiskers (*fibre-bridging*) or by larger grains in the matrix (*grain bridging*). As the crack opens under load (i.e. the *crack opening displacement* (COD) increases) the 'bridges' are subject to stresses which rapidly increase, until the whiskers or grains fracture or pull out from the matrix. Whiskers will support stresses up to a maximum crack opening displacement comparable to a fraction of the whisker radius, and given fibre-bridging stress which decays rapidly as the COD increases. Large grains will support only much lower bridging stresses, but the COD required to pull out a matrix grain is of the order of the grain size, much larger than whisker diameters. The grain bridging zone then extends (e.g. in aluminium with ~10μm grains) to several hundred microns behind the crack tip, and the (smaller) grain-bridging stress decays correspondingly slowly as the crack opens. This leads to greater fracture resistance with increased crack length (R-CURVE (q.v.) behaviour) in ceramics such as larger-grained alumina.

Crack Deflection. A method for increasing TOUGHNESS (q.v.) in which a dispersed second phase prevents the direct growth of a crack, diverting it and lengthening its path.

Cracking off. The severing of shaped glass-ware from the MOIL (q.v.).

Crackle. (1) A multiply crazed or cracked surface on art pottery or glass. To produce the effect on pottery the glaze is compounded so as to have a higher thermal expansion than the body; the craze pattern is sometimes emphasized by rubbing colouring matter, such as umber, into the fine cracks. With glass, the ware is cracked by quenching in water; it is then reheated and shaped. (2) A crackled vitreous enamel – the surface appearing to be wrinkled due to its mottled texture – can be produced by the wet process of application.

Crack Nucleation. New cracks form in ceramics when an external stress forces DISLOCATIONS towards barriers such as GRAIN BOUNDARIES, impurities or other POINT DEFECTS (see CRYSTAL STRUCTURE). This build-up of dislocations forms a *microcrack*, which grows initially by this PLASTIC DEFORMATION PROCESS. See CRACK PROPAGATION.

Crack Opening Displacement. The linear separation of the two surfaces formed by cracking. It significantly affects the diminution in crack propagation due to CRACK BRIDGING (q.v.)

Crack Propagation. The movement of a developing crack through a solid. See FRACTURE; CRACK NUCLEATION. A newly-formed microcrack grows initially by plastic deformation of the material just ahead of the crack tip. This absorbs much energy, so that initial crack growth is slow. It also tends to blunt the crack tip, reducing the stress concentration effect. On the other hand, small cavities form in the deformed area, which link to the tip of the crack, which extends. In ceramics, the nature of the material limits the amount of plastic deformation; crack tip blunting is limited and the local stress quickly comes to exceed the fracture stress of the material. The interatomic bonds are broken and the crack propagates rapidly and catastrophically. See FRACTURE; BRITTLE FRACTURE; GRIFFITH-OROWAN CRITERIA; TOUGHNESS; STRESS INTENSITY FACTORS.

Crank. This word is used in the pottery industry in two related senses: (1) A thin refractory bat (Fig. 4, p177) used as an item of KILN FURNITURE (q.v.) in the glost-firing of wall tiles. A number of

cranks, each supporting one or more tiles, are built up to form a stack; the cranks are kept apart by refractory distance-pieces known as DOTS.
(2) A composite refractory structure for the support of flatware during glost- and decorating-firing; the crank is designed to prevent the glazed surfaces of the ware from coming into contact with other ware or kiln furniture.

Craquelé. See CRACKLE.

Crater. A clearly-defined hole in a refractory brick or block whose diameter and depth are measurable.

Crawler. Local term for an apron-feeder to a pan mill used in brickmaking.

Crawling. (1) A defect that sometimes occurs during the glazing of pottery, irregular areas that are unglazed, or only partially glazed, appearing on the fired ware. The cause is a weak bond between glaze and body; this may result from greasy patches or dust on the surface of the biscuit ware, or from shrinkage of the applied glaze slip during drying.
(2) A similar defect liable to occur in vitreous enamelling when one coating of enamel is fired over another coating that has already been fired. Causes include a too-heavy application of enamel, poorly controlled drying, and the use of enamel that has been too finely ground.

Crazing. The formation of a network of surface cracks. A typical example is the crazing of a glaze; this is caused by tensile stresses greater than the glaze is able to withstand. Such stresses may result from MISMATCH (q.v.) between the thermal expansions of glaze and body, or from MOISTURE EXPANSION (q.v.) of the body; ASTM specifies a water-quench test from 149 °C for vitreous whitewares (C554) and an autoclave test (C424) for nonvitreous or semivitreous bodies for which moisture expansion may be a factor. In the special case of glazed tiles fixed to a wall, a third cause is movement of the wall or of the cementing material between the tile and the wall. Crazing of vitreous enamelware (particularly cast dry-process enamel) may also occur, the system of fine cracks penetrating through the enamel to the base-metal. The crazing of cement and concrete is due partly to natural shrinkage and partly to volume changes following surface reaction with CO_2 present in the atmosphere. (For CRAZING TESTS see under AUTOCLAVE; HARKORT TEST; PUNCH TEST; RING TEST; SINGER'S TEST, STEGER'S TEST; TUNING-FORK TEST).

Crazing Pot. Popular name in the pottery industry for an AUTOCLAVE (q.v.).

Creep. The slow deformation that many materials undergo when continuously subjected to a sufficiently high stress. With most ceramics, creep becomes measurable only when the stress is applied at a relatively high temperature. A 20–50 h creep test for refractories is described in ASTM C832. BS 1902 Pt 4.10 describes a compressive creep test up to 1600 °C for refractories. See also SLOW CRACK GROWTH.

Creeper. Correcting, with a mortar wipe, a tendency for brick rings to drift outward axially in the upper part of an arch.

Cremer Kiln. A German design of tunnel kiln that can be divided into compartments by a series of metal slides to permit better control of temperature and atmosphere. The fired ware is cooled by air currents through permeable refractory brickwork in the kiln roof or by water-cooling coils. (G. Cremer, Brit. Pats. 697 644, 30/9/53; 740 639, 16/1/55; 803 691, 29/10/58.)

Crescent Crack. See FRICTIVE TRACK.

Crespi Hearth. A type of open-hearth steel furnace bottom characterized by the fineness of the particles of dolomite used for ramming; after it has been burned-in, the hearth is very dense and resistant to metal penetration. (G. B. Crespi, Brit. Pat. 507 715. 10/8/38.)

Crimping. The production of a rolled or curled edge to the base metal prior to vitreous enamelling.

Cristobalite. The crystalline form of silica stable at high temperatures; its m.p. is 1723 ± 5 °C. Cristobalite is formed when quartz is heated with a mineralizer at temperatures above about 1200 °C and is itself characterized by a crystalline inversion, the temperature of which varies but is generally between 200 °C and 250 °C; this inversion is accompanied by a change in length of about 1 %. Cristobalite is a principal constituent of silica refractories, causing their sensitivity to thermal shock at low temperatures; it is also present in many pottery bodies and is synthesized for use as a refractory powder in the investment casting of metals, for which purpose its high thermal expansion is advantageous.

Cristobalite Squeeze. The process of putting ceramic glazes into compression, by taking advantage of the INVERSION (q.v.) of cristobalite at 220 °C. On cooling from the glost kiln, by 573 °C the glaze has solidified, and the body contraction, due to the β to α cristobalite phase change, compresses the glaze.

Critical Moisture Content. The moisture content of a clay at which the clay particles touch, the water occupies the voids between the particles, and no shrinkage occurs on further drying. It corresponds roughly to the LEATHER-HARD (q.v.) state. Drying can be more rapid below the critical moisture content, without risk of distortion.

Critical Speed. (1) The maximum safe speed of rotation of an abrasive wheel; at higher speed vibration becomes dangerous. (2) In ball-milling, the speed of rotation above which the balls remain against the casing, as a result of centrifugal force, throughout a revolution of the mill; this speed is given by the equation: $N\sqrt{R} = 54.18$ – where N is the rev/min and R is the radius of the interior of the mill, less the radius of the ball, in feet.

Critical Stress Intensity. See STRESS INTENSITY FACTOR.

Crizzle. See CHECK.

Crockery. A popular term for ceramic tableware, sometimes restricted to EARTHENWARE (q.v.).

Crocodile Kiln A type of TOP-HAT KILN (q.v.) developed by Keller GmbH. The hood has a swivel bearing at one end, and so opens at an inclined angle, rather than being raised vertically.

Crookes Glass. A glass, usually containing cerium, that absorbs ultra-violet light and is used for protective goggles, etc. This glass resulted from the work of Sir W. Crookes, in 1914, for the Glass Workers' Cataract Committee of the Royal Society.

Crop. To cut replacement refractory bricks roughly flush with surrounding worn lining bricks, to equalise behaviour when returned to service.

Cross-bend Test. (1) Term sometimes used for TRANSVERSE STRENGTH TEST (q.v.).
(2) A test to determine the resistance of vitreous enamelware to cracking when it is distorted.

Cross-fired Furnace. A glass-tank furnace heated by flames that cross the furnace perpendicular to the direction of flow of the glass; the furnace has several pairs of ports along its melting end (cf. END-FIRED FURNACE).

Crouch Ware. Light-coloured Staffordshire salt-glazed stoneware of the early 18th century; it was made from a clay from Crich, Derbyshire, the word 'crouch' being a corruption.
Crowding Barrow. A hand-barrow for bricks; it has a base and front, but no sides.
Crown. A furnace roof, particularly of a glass-tank furnace.
Crown Blast. The procedure of blowing air at roof level into the exit end of a tunnel kiln to counteract the natural flow of gases in this part of the kiln.
Crown Brick. See KEY BRICK and CENTRE BRICK.
Crown Glass. Glass of uneven thickness and slightly convex (thus producing some optical distortion), hand-made by blowing and spinning (cf. OPTICAL CROWN GLASS).
Crucible Test. See SLAG ATTACK TESTS.
Crush. A defect of flat glass – a lightly pitted dull-grey area.
Crush Dressing. Shaping the face of an abrasive wheel to a required contour by means of steel rolls.
Crushing Rolls. A unit frequently used for the size reduction of brick clays, etc. Two steel rolls are arranged horizontally and adjacent, with a gap between them of a uniform width equivalent to the maximum size of particle allowable in the crushed product; the rolls are rotated in opposition, and sometimes at different speeds, so that the clay is carried downwards and 'nipped' between the approaching surfaces of the rolls (cf. EXPRESSION ROLLS).
Crushing Strength. The maximum load per unit area, applied at a specified rate, that a material will withstand before it fails. Typical ranges of values for some ceramic materials are:
Fireclay and silica refractories: 14–35 MN m^{-2}
Common building bricks: 14–42 MN m^{-2}
Engineering bricks Class 'A': 70 MN m^{-2}
Sintered Alumina: 350 MN m^{-2}
Now usually referred to as COMPRESSIVE STRENGTH (q.v.). BS 1902 specifies cold crushing strength of shaped insulating refractories (Pt. 4.1) and dense refractories (Pt. 4.3) and ULTIMATE COMPRESSIVE STRENGTH (q.v.) of dense refractories (Pt. 4.2). BS 1902 Pt. 7.6 specifies cold crushing strength tests for unshaped refractories. ASTM C133 and C93 specify tests for refractory bricks and insulating firebricks respectively; C 773 is a test for whitewares; E447 for masonry prisms.
Crusilite. A silicon carbide heating element. (Trade-name: Morganite Electroheat Ltd, London.).
Cryogenic. Relating to (the production of) low temperatures. See SUPERCONDUCTIVITY.
Cryolite. Natural sodium aluminium fluoride, Na_3AlF_6; m.p. 980 °C; sp. gr. 2.95. Because of its low m.p. and its fluxing action, it is used in the manufacture of enamels and glass and in the ceramic coatings of welding rods. Opal glass is often made from batches containing about 10% cryolite; a similar preparation is sometimes used in white cover-coat enamels.
Crypto System. An impulse system of oil firing, more particularly for the top-firing of annular kilns. Trade-name: R. Aebi & Cie., Zurich.
Cryptocrystalline. Crystalline, but with sub-microscopic grain size.
Cryptofluorescence. Term proposed by W. A. McIntyre and R. J. Schaffer (*Trans. Brit Ceram. Soc.*, **28**, 363, 1929) for soluble salts that have crystallized in the interior of a clay building product

and are therefore hidden; the term has not gained general currency.

Crystal Glass. A popular, but misleading, name for a type of decorative glass that is usually deeply cut so that the brilliance resulting from its high refractive index is fully displayed. For British Standard definitions see FULL LEAD CRYSTAL GLASS and LEAD CRYSTAL GLASS.

Crystalline. (1) (of solids) having the atoms in a definite long-range order. See CRYSTAL STRUCTURE.
(2) (of particles in a powder) having the geometric shape of a freely-developed crystal. See MORPHOLOGY.

Crystalline Glaze. A glaze containing crystals of visible size to produce a decorative effect. Typical examples are glazes containing zinc silicate crystals and the AVENTURINE (q.v.) glazes.

Crystal Structure. Crystals are solids whose constituent atoms have a regular repetitive arrangement in space.

The basic unit of a crystal structure is the *unit cell*, which is the smallest group of atoms which has the characteristics and symmetry of the whole crystal. It is the volume formed by joining adjacent *lattice points*, those points at which the arrangements of atoms begins to repeat itself. The lines along the edges of the unit cell are the *lattice vectors*. (The *lattice constants* are the lengths of these lines and the angles between them). The *space lattice* is the regular, repetitive and symmetric three-dimensional array formed by placing repeating unit cells alongside one-another. The space lattice describes the geometric arrangement of the atoms in a crystal. There are only 14 such symmetrical arrangements possible, the BRAVAIS LATTICES (q.v.) (See fig. 2, p81. See also RECIPROCAL LATTICE, X-RAY CRYSTALLOGRAPHY, SPACE GROUP). While it is possible to prepare *single crystals* in which such symmetric arrangements of atoms make up the whole of the material, most crystalline solids actually consist of small crystals packed together at random, and separated by *grain boundaries*. The MICROSTRUCTURE (q.v.) of these *polycrystalline* materials has an important influence on their properties. Moreover, the arrangement of the atoms in crystals is not perfect. *Defects* occur. These may be *point defects*, affecting single atoms, or *line defects*, affecting lines and planes of atoms. The various types of *point defect* are *Substitutional Defect* – an atom in the lattice is substituted by a foreign atom of similar size. *Vacancy* – an atom is missing from a lattice site. *Interstitial* – a (small) foreign atom fitting in between the lattice atoms. *Frenkel Defect* – the displacement of an atom from its position in the lattice (leaving a *vacancy*) to an *interstitial* position. *Schottky defect* - a pair of *vacancies* in an ionic crystal, the two missing atoms being ions of opposite electric charge. All such *point defects* strain the lattice, and affect the movement of *dislocations*, the *line defects*. Solid-state DIFFUSION (q.v.) occurs by movement of atoms into neighbouring vacancies. In general, the numbers of point defects increase as the temperature is raised. There are two types of *line defect*, or *dislocation*: *Edge dislocations* occur when an extra plane of atoms is present on one side of a perpendicular plane (the *slip plane*) but not on the other. Such dislocations are relatively easy to move by low stresses, making the crystals weaker and more susceptible to plastic flow. *Screw dislocations* mark the boundary between slipped and unslipped crystal parallel to the direction of slip, as if a cut had been made into the solid crystal and the planes of atoms sheared one atom space

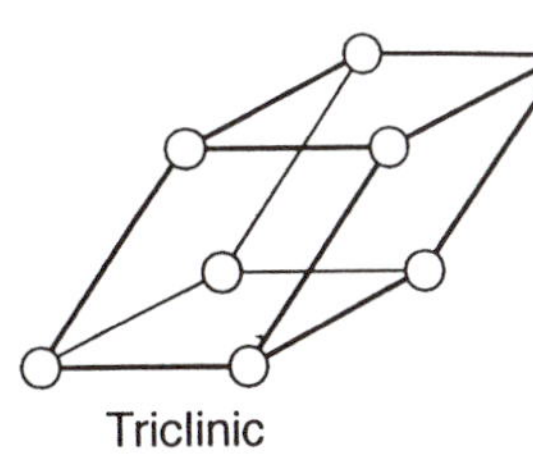

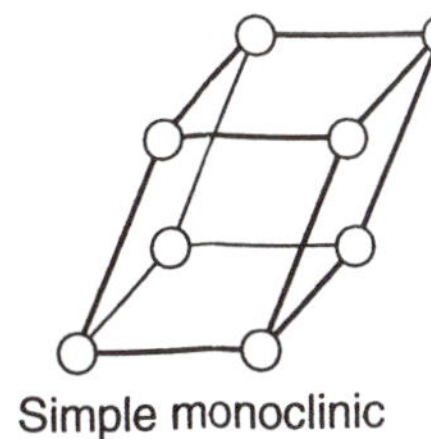

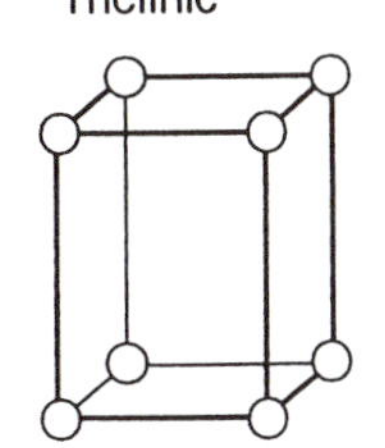

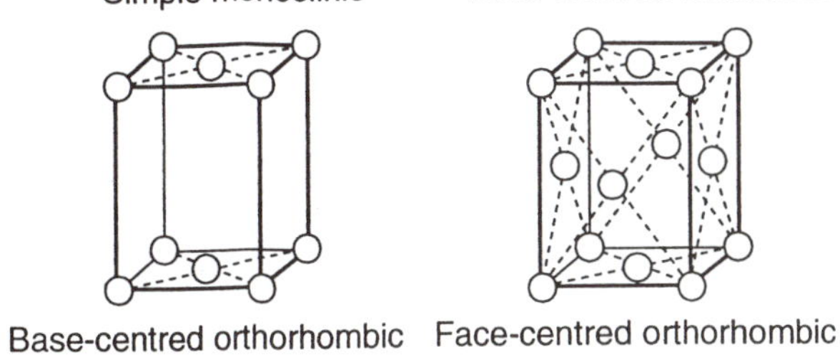

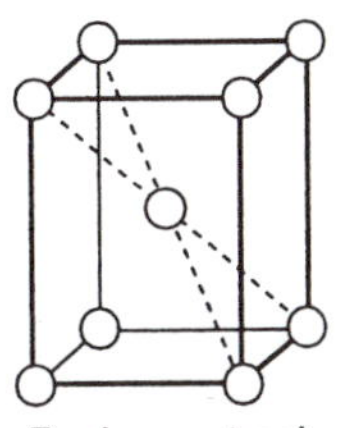

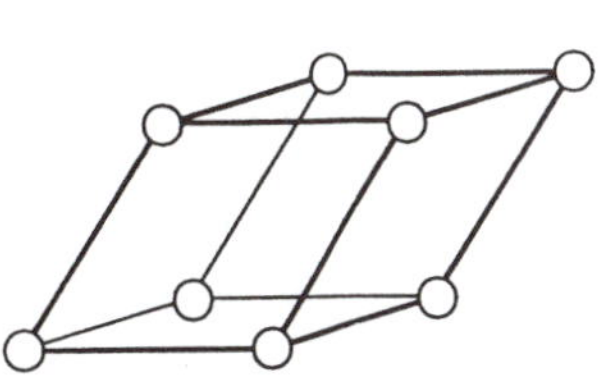

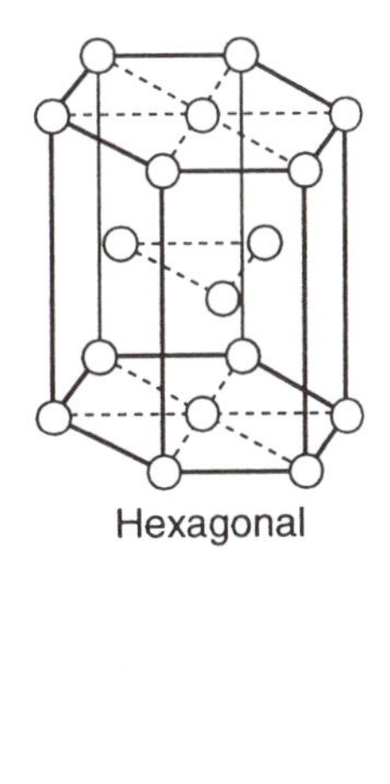

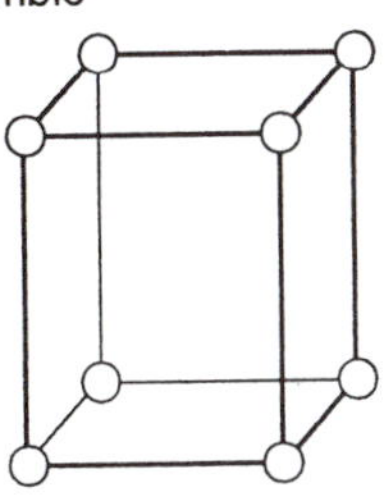

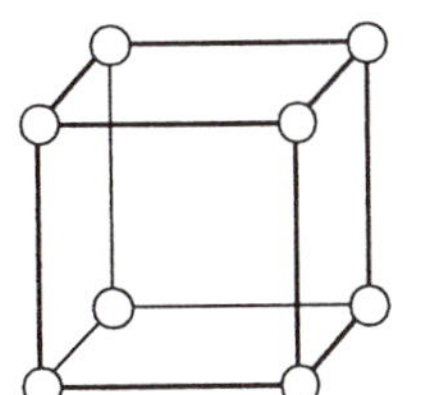

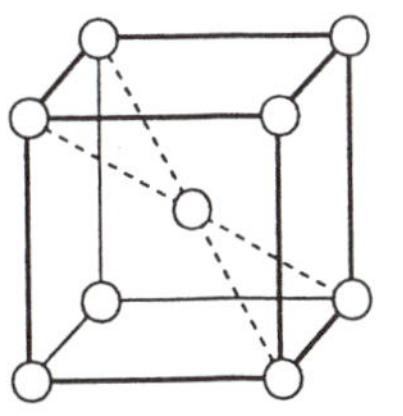

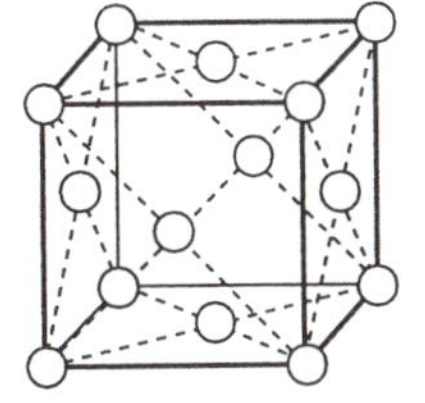

Fig. 2 Bravais Lattices

parallel to the edge of the cut. This transforms successive atom planes into the surface of a helix, hence the name. The *Burger's Vector* indicates the direction of displacement due to dislocations, which are usually combined edge and screw dislocations. This is the direction of *plastic deformation* when shear forces are applied. The Burger's vector is normal to an edge dislocation, parallel to a screw dislocation. Point defects and grain boundaries stop the movement of dislocations, and thus affect strength as well as other properties.

Crystar. Tradename. A form of RECRYSTALLISED SILICON CARBIDE (q.v.) in which SiC and electronic grade Si react at over 2300 °C, and the grains recrystallize to form a continuous network of SiC. Crystar is used to produce thermal shock resistant kiln furniture as well as tubes and other refractory shapes. (Norton Co, USA).

C-Scan Acoustic Microscopy, C-SAM. Focussed transducers from 10 to 100 MHz are coupled to the test piece in a water-immersion tank so that the test-piece is located at the desired depth below the test-piece surface. (The reflections from that surface itself are cut out electronically). Scanning the transducer over the surface produces an image of internal flaws at that depth in the specimen.

CSZ. Cubic Stabilised ZIRCONIA (q.v.)

C.T. Compact Tension – See FRACTURE TOUGHNESS TESTS.

C.T. Nozzle. Trade-name: a refractory nozzle for steel-pouring designed to give a *Constant Teeming* rate (hence the name). The nozzle consists of an outer fireclay shell and a refractory insert of different composition. Strictly speaking, the term refers to a particular type of insert developed for the teeming of free-cutting steels. (Thos. Marshall & Co. (Loxley) Ltd., Brit. Pats. 832 280, 6/4/60; 904 526, 29/8/62.)

Cubing Rolls. CRUSHING ROLLS (q.v.) having projections and used for breaking down hard 'slabby' clays into a cube-like product that is more suitable for feeding to a secondary grinding unit.

Cuckhold. (1) An iron tool for cutting off lumps of prepared clay, from a pug, ready for the hand-moulding of building bricks. (2) A two-pronged fork with a taut wire joining the extremities of the prongs; it is used for cutting clay from a WASHBACK (q.v.) for STOCK BRICK (q.v.) manufacture

Cuckle. See CUCKHOLD.

Cull. US equivalent of the English WASTER (q.v.).

Cullet. Broken glass that can be recharged to the glass furnace. The word is derived from the French *collet* the little neck left on the blowing iron when bottles were hand blown; these 'collets' were returned to the glass-pot and remelted. FACTORY CULLET or DOMESTIC CULLET is from the same glassworks at which it is to be used; FOREIGN CULLET is from a different glassworks.

Cullet Cut. Scratches on glassware due to particles of cullet in the polishing felt.

Culm. A Carboniferous shale used for brickmaking in the Exeter area. (Dialect word for coal dust or soot.)

Cummings' Sedimentation Method. An approximate method of particle-size analysis having the merit of giving a weight/size distribution directly. (D. E. Cummings, *J. Industr. Hyg. Toxicol.,* **11**, 245, 1929.)

Cup Gun. A spray gun, particularly as used for touching-up vitreous enamelware, with a container for the enamel slip forming an integral part of the gun.

Cup Wheel. An abrasive wheel shaped like a cup; such wheels are used, for example, in the grinding of flat surfaces. Diamond cup-wheels are employed in the grinding of tungsten carbide tools.
Cupel. A small, refractory, tapered cylinder (broad end up) with a shallow depression in the top; cupels are made from bone ash or calcined magnesia and are used for the assay of non-ferrous ores.
Cupola. A shaft furnace used in a foundry for the melting of iron. Cupolas are generally lined with fireclay refractories covered with a ganister-clay mixture. For the production of cast iron with a low sulphur content, a basic lining is sometimes used; the lining is in this case built of chrome-magnesite or dolomite refractories, or it may be rammed with a monolithic basic refractory composition.
Cupola Brick. See KEY BRICK.
Cupping. A process in which vitreous enamel slip is poured over selected areas of a piece of ware while it is being drained, to ensure that the overall thickness of application shall be uniform.
Curb Bend. A special shape of wall tile (see Fig. 7, p350).
Curie Law and Curie-Weiss Law. Magnetic susceptibility $\chi = C/T$ where C is the Curie constant. and T is the absolute temperature. A FERROMAGNETIC material may be regarded as a paramagnetic material in which the atomic magnetic moments are aligned by an internal interaction the 'Weiss field' (P.Weiss, 1907) proportional to the magnetization. Above the CURIE POINT, T_c, the (then) paramagnetic behaviour of ferromagnetic materials is described by the CURIE–WEISS LAW, $\chi = C/(T - T_c)$.
Curie Temperature. Originally discovered by Pierre Curie in 1895 as the tcmperature at which a material changed from ferromagnetic to paramagnetic behaviour. Now generally applied to the temperature at which ordered arrangements of atoms, which give rise to a variety of electric and magnetic properties, break down.
Curing. The process of keeping freshly placed concrete moist to ensure complete hydration so that maximum strength is developed. Compounds are available for spraying on concrete to retard loss of moisture during the curing period (see ASTM C156 and C309). Pre-cast concrete units are often STEAM CURED (q.v.).
Curling. A defect in vitreous enamelling that is similar to CRAWLING (q.v.).
Curtain Arch. An arch of refractory brickwork that supports the wall between the upper part of a gas-producer and the gas uptake.
Curtain Wall. See SHADOW WALL.
Curtains. An enamelling defect, in the form of dark areas having the appearance of drapery, liable to occur in sheet-steel ground-coats. The probable cause is boiling or blistering whcn the ground-coat is being fired.
Cushion Edge. Defined in BS 6431 as: 'A slight convex radiusing of the periphery of the glazed surface of a tile and/or tile fitting where the glazed surface meets the edges'.
Cut-back. The careful removal of damaged or worn refractory surrounding a lining failure site, until sound refractory capable of being patched is reached.
Cut Glass. Glass-ware into which a pattern has been ground by means of an abrasive wheel; the grinding is followed by polishing (cf. BRILLIANT CUT).
Cut Glaze. A faulty glaze, spots or patches being bare or only very thinly covered. The common cause is

contaminated areas on the biscuit-ware, i.e. patches of oil, grease, dust, or soluble salts. A fault resulting in a similar appearance is KNOCKING (q.v.)

Cut-off Scar. Marks on the base of a glass bottle made by the Owen's suction machine; the 'scar' is largely caused during the final blowing operation, however.

Cutlery-marking. See SILVER-MARKING.

Cutting-off Table or Cutter. A frame carrying a tightly stretched wire, or a system of such frames and wires, that operates automatically at a short distance from the mouthpiece of a pug or auger to cut off clots or finished bricks or pipes from the extruded column.

Cutting-off Wheel or Parting Wheel. A thin abrasive wheel of the type used for cutting-off or for making slots. Such wheels generally have an organic bond.

Cutting Tools. See TOOL TIPS.

Cutty Clay. A variety of English ball clay that was formerly used for making tobacco pipes.

CVR. Abbreviation for CONTINUOUS VERTICAL RETORT (q.v.).

Cyanide Neutraliser. See under NEUTRALISER.

Cyanite. Obsolete spelling of KYANITE (q.v.).

Cyclic Fatigue. The breakdown of a material by repeated application of a regularly increasing and decreasing stress.

Cyclone Classifier. A dust-laden gas enters a cylindrical or conical chamber tangentially and leaves through a central opening. The dust particles, moving in a helical path, experience a force towards the centre of M*v*/r, where M is the particle mass, *v* the particle velocity and r the radius of the chamber. Thus particles are separated according to size, by a force which can be 5 to 2500 times that due to gravity in a settling tank. The detailed mechanics and flow patterns in practical cyclone classifiers are complex.

Cyclopean. Concrete containing aggregate large than 150mm, used in dams and other large structures.

Cylinder Process. An old method of making flat glass by blowing molten glass to form a cylinder, which is then cracked open and flattened in a special furnace known as a flattening kiln. In Belgium , France and Germany, the cylinder was made of a length to correspond to that of the glass sheet and the circumference corresponded to the width. Bohemian practice was the reverse of this and only small sheets (about 3 ft square) could be made.

Cylinder Test. See THERMAL SHOCK TESTS.

Cylindrical Screen Feeder. One type of feeder for plastic clay. It consists of a vertical cylindrical screen though which clay is forced by blades fixed to a vertical shaft that rotates within the cylinder. This machine not only feeds, but also mixes and shreds the clay.

Czochralski Technique. A method of growing single crystals of the refractory oxides, and of other compounds, by pulling from the pure melt; the compound must melt congruently. (J Czochralski *Z. Phys. Chem.,* **92**, 219, 1917.)

Dam Ring Significant local thickening of the lining of a rotary kiln in an axial direction, to expand the upstream feed surface area to increase residence time, or for other control purpose.

Damp Proof Course. An impervious layer below the surface of a floor, or through the width of a wall, designed to prevent moisture rising from the ground below. The usual materials are bituminous, but a course of ENGINEERING BRICKS (q.v.) is common in older buildings. B.S.3921 specifies

WATER ABSORPTION limits for bricks for damp-proof courses of 4.5 (class 1) or 7.0 (class 2) the latter being recommended for external works.

Danielson–Lindemann Deflection Test. A procedure for assessing the ability of vitreous enamelware to suffer a small degree of bending without the enamel cracking (R. R. Danielson and W. C. Lindemann, *J. Amer. Ceram Soc.*, **8**, 795, 1925). The procedure has been standardized by the American Ceramic Society (*Bull. Amer. Ceram. Soc.*, **7**, 360, 1928; **9**, 269, 1930).

Danneberg Kiln. An ANNULAR KILN (q.v.) in which the products of combustion leave each firing chamber through holes in the hearth and through numerous small interconnecting flues.

Danner Process. A method for the continuous production of glass tubing invented by Edward Danner in the USA in 1917. Glass flows from a tank furnace on to a mandrel, which is inclined and tapered and slowly rotates. The mandrel is hollow and air is blown through it to maintain a hole through the glass, which is continuously drawn from the lower end of the mandrel as tubing, 1.5 to 60mm dia., at a rate of up to 250 kg/h. (c.f. VELLO PROCESS.)

Danny. An open crack at the base of the neck of a bottle.

Dapple. External or internal surface irregularity in a glass container.

Darcy's Law; Darcy. Darcy's Law states that the rate of flow of a fluid, subjected to a low pressure difference, through a packing of particles is very nearly proportional to the pressure drop per unit length of the packing. This Law forms the basis of methods for the determination of the permeability of ceramics. The DARCY is the c.g.s. unit of permeability: a material has a permeability of 1 darcy if in a section 1 cm^2 in area perpendicular to the flow, 1 ml of fluid of unit viscosity flows at a rate of 1 cm/s under a pressure differential of 1 atm.; 1 Darcy = 1.013×10^{-9} PERM (q.v.) (The Law was propounded by H. P. G. Darcy, when designing the fountains at Dijon in 1856).

Datolite. A boron mineral approximating in composition to $CaO.B_2O_3.2SiO_2.H_2O$; it occurs in Russia and elsewhere. Trials have shown that it is a suitable flux for use in glazes for structural clay products.

Datapaq Trade-name. An automatic recording system of travelling thermocouples, which can pass through e.g. a tunnel kiln and provide a record of the firing schedule.

Davis Revergen Kiln. The word 'Revergen' is a trade-mark. A gas-fired tunnel kiln of the open-flame type; the flame does not come in actual contact with the ware. The combustion air is preheated by regenerators (hence the name) below the kiln. The design was introduced by Davis Gas Stove Co. Ltd, Luton, England; this firm was absorbed by Gibbons Bros Ltd, Dudley, England.

Day Tank. A periodic glass-tank furnace, usually consisting of a single chamber, from which glass is worked out by hand; the furnace is operated (charging, melting, and working) on a 24-h cycle.

DCB Double Cantilever Beam. See FRACTURE TOUGHNESS TESTS.

DCCA Drying Control Chemical Additives. Chemicals such as oxalic acid, tetramethylorthosilicate (TMOS) and dimethyl formamide, which improve drying and produce larger pores in gels. See SOL-GEL PROCESS.

DCDC Double Cleavage Drilled Compression. See FRACTURE TOUGHNESS TESTS.

DCL Fusion-cast Refractory. A US fusion-cast refractory, e.g. glass-tank block, made by a process that largely eliminates the cavities liable to occur as a result of shrinkage during cooling; the mould is L-shaped and is tilted while it is being filled so that the shrinkage cavities concentrate in the smaller leg of the L (the 'lug'), which is then sawn off and discarded. (DCL = Diamond Cut Lug; c.f. RO, RT and SR).
DCMA Number. See Dry Colour Manufacturers' Association
Dead-burned. Term applied to a refractory raw material, and especially to magnesite, after it has been heated at a sufficiently high temperature for the crystal size to increase so that the oxide becomes relatively unreactive with water. Magnesite is deadburned in shaft kilns or rotary kilns at a temperature of 1600–1700°C.
De-airing. The removal of air from plastic clay or body, from the moist powder in dry-pressing, from casting slip, or from plaster during blending. There are various devices íor submitting these materials to a partial vacuum during their processing. De-airing is most commonly practised in extrusion, shredded plastic clay being fed to the pug, or auger, via a de-airing chamber. The original patent was that of R. H. Staley (US Pat., 701, 957, 1902.)
Debinding The removal of a binder from a green ceramic compact. c.f. BURNOUT.
Debiteuse. A refractory block having a vertical slot; it is used in the FOURCAULT PROCESS (q.v.) of sheet-glass manufacture, being depressed below the molten glass which is drawn upward through the slot. (French word for a feeding device.)
Debonding The breakdown, under thermal, mechanical or chemical stress, of the bonds between the components of a composite.
Debye-Scherrer Technique. See X-RAY CRYSTALLOGRAPHY.
Decal. The American term for a ceramic *transfer* or *litho* which is now also current in the UK. Decals are used to apply designs to ceramic tableware, ornamental ware and tiles, and to glass containers. The decal comprises three layers: the colour, or image, layer which comprises the decorative design; the covercoat, a clear protective layer, which may incorporate a low-melting glass; the backing paper on which the design is printed (by SCREEN PRINTING OR LITHOGRAPHY q.v.). There are various methods of transferring the design while removing the backing-paper, some of which are suited to machine application. (See HEAT-RELEASE DECAL, HEAT-ACTIVATED DECAL, SLIDE-OFF DECAL).
Decalcomania (USA). A particular type of transfer printing, now known in England as LITHOGRAPHY (q.v). The term is derived, via the French, from two Greek words: *decal* (off the paper), and *mania,* this form of printing having had a short, but extreme, popularity with young ladies in 1860/65.
Deck. The refractory top of a car used in a tunnel kiln or bogie kiln.
Decking. The stacking of vitreous enamelware in several layers ready for firing.
Decolorizer. A material added to glass to counteract the colour imparted by impurities such as iron; the decolorizer may be an oxidizing agent, removing the colour by chemical action, or it may counteract the colour already present by introducing the complementary colour. The materials used include the oxides of As, Ce, Co, Mn, Nd and Se.
Decorating Firing or Enamel Firing. The process of firing pottery-ware after the

application of coloured or metallic deeoration; the temperature is usually 700–800°C and this fixes the decoration and makes it durable.

Decoration. The application of colours and designs to ceramic ware. Bodies and glazes may be coloured or enhanced by crystalline effects, and coloured decorations applied on or under-glaze by printing or by applying transfers. Gold and other precious metal is applied to more expensive ware, and may require additional firing. Glassware may also be decorated by cutting or engraving. See also COLOUR; DECAL; MURRAY-CURVEX MACHINE; SCREEN PRINTING; BANDING; GLAZE; GOLD DECORATION; LUSTRE; ENGOBE; ENGRAVING.

De-enamelling. The removal of vitreous enamel from the base metal; this can be done by sand-blasting or by solution in alkali.

Deep Cut. Alternative name for CUT GLASS (q.v.).

Deflecting Block or Spreader Block. A block of refractory material, triangular in cross-section, that is built into a coke-oven below a charging hole; the sharp edge of the block is uppermost and this deflects or spreads the stream of descending coal so that it comes to rest more uniformly in the oven.

Deflocculation. The dispersion of a clay slip by the addition of a small amount of suitable electrolyte, e.g. sodium silicate and/or sodium carbonate.

Defluorinated Stone. CHINA STONE (q.v.) from Cornwall, England, from which the small amount of fluoride present has been removed by flotation.

Deformation Eutectic. The composition within a ceramic system (e.g. china clay, flint and feldspar) which, when heated under specified conditions, deforms at a temperature lower than that required to produce deformation in any other composition within the system. This term is used more particularly in USA.

Deformation Temperature. The temperature at which, when a ceramic material is heated under specified conditions, the rate of subsidence becomes equal to the rate of thermal expansion. With glass, this temperature corresponds to a viscosity of 10^{11}–10^{12} poises.

Dégourdi. The preliminary low-temperature (800–900°C) firing of feldspathic porcelain, as practised in Europe; the second (glost) firing is at approx. 1400°C. (French word meaning 'warming' as distinct from the high-temperature-*grand feu*-glost firing.)

Delft Ware. An early type of porous earthenware covered with a tin-opacified glaze and decorated in blue before the glost fire; named from Delft, Holland, but the process was already in use in England in the 16th century. In USA the term is defined (ASTM C242) as a calcareous earthenware having an opaque white glaze and monochrome on-glaze decoration.

De-hacking. The removal of ware from a kiln setting. The term is usually used when this is done mechanically, particularly in brick kilns.

Deko-Press. A process intended to reduce the production costs of tableware by combining the dry-pressing of spray-dried powder with decoration and single-firing in a roller-hearth kiln. Developed by Hutschenreuther AG. (Keram Z **36** (1984) 551.)

Dekram. Trade-name. A multi-colour printing machine for ceramic tableware (Brit.Pat. 1247001, 1971, Service Engineers Ltd).

Dela-Drum. Powder and spray are fed into the lower end of an inclined rotating drum. As the moist powder rises along the axis of the drum it agglomerates, and particles of uniform size are discharged. (Dela Gesellschaft für Granuliertechnik m.b.H)

Della Robbia Ware. Terra-cotta artware with white or brightly-coloured glaze.

Deltek An advanced ceramic developed by FOSECO (FS) Ltd and Birmingham University. It has good thermal shock resistance, high refractoriness and strength, and can be fabricated to complex shapes with good surface finish and dimensional tolerances. Its applications are to investment casting, aluminium die casting and other metal casting processes.

Demijohn. A glass container for wine or spirits; it has a narrow neck and a capacity of over 10 litres. The name is derived from the French *Dame Jeanne*, a popular 17th-century name for this type of large bottle.

Dendritic see MORPHOLOGY

Dense. When applied to structural clay or refractory products the term generally signifies 'of low porosity'; when applied to a glass it means 'of high refractive index' (in this context the term is sometimes expanded to OPTICALLY DENSE).

Densification The processes during shaping and firing which lead to closer packing of particles, the removal of pores, and a denser ceramic compact or article.

Density. The determination of the various densities ascribed to refractory materials is specified in B.S.1902 Pts. 3.4 to 3.8. See APPARENT SOLID DENSITY; BULK DENSITY; DENSITY FACTORS; PACKING DENSITY; TRUE DENSITY.

Density Factors for Glass. Factors for calculating the density of a glass from its composition; the original set of factors was that of A. Winkelmann and O. Schott *(Ann. Physik. Chem.*, **51**, 730, 1894). Numerous amendments to these have since been put forward; probably the most reliable are those of M. L. Huggins and K. H. Sun, *J. Amer. Ceram. Soc.*, **26**, 4, 1943.

Dental Ceramics. DENTAL PORCELAIN (q.v.) has been supplemented by other BIOCERAMICS, in particular APATITES and ALUMINA, for general maxillofacial restorative work, using crowns and implants. Metallising, metal-ceramic bonding and colour matching are important in dental ceramic work. BS 3365 specifies dental silicate and silico-phosphate cements; BS 6039 glass-ionomer cements.

Dental Porcelain. Feldspathic porcelain, shaped, tinted and fired for use as false teeth; the firing is sometimes carried out in a partial vacuum to remove small air bubbles and thus ensure maximum density and strength. X-ray diffraction studies have shown fired dental porcelain to be some 80% glassy phase. Dental porcelain jacket crowns are specified by BS 5612, denture teeth by BS 6817.

Deox. A masonry cleaning solution (National Chemsearch (UK) Ltd).

Derby Press. Trade-name; a machine for the re-pressing of wire-cut building bricks. (Bennett & Sayer Ltd., Derby, England.)

Design. Traditionally, ceramic design was governed by aesthetic considerations. The design of modern ceramic engineering components is governed by the need to avoid sharp corners and re-entrant shapes (both for ease of shaping and to avoid stress concentration, leading to crack propagation), and to take advantage of the high compressive strength of ceramics.

Detergent. For resistance of the decoration of tableware see DURABILITY OF ON-GLAZE DECORATION.

Detonation Gun (D-gun). A D-gun comprises a water-cooled barrel about 1m long and 25mm diameter. An oxygen/acetylene mixture ignited in the barrel drives a charge of powder (particle diameter 45 μm) at a velocity of c. 800 m/s, temperatures c. 4000°C being reached. After ignition, the barrel is purged with nitrogen and the cycle repeated 4 to 8 times per second, each detonation depositing overlapping 25mm circles of powder some 20μm thick, on the target as the gun is carefully traversed.

Detonation Forming. See EXPLOSIVE PRESSING.

Detonation Plating. Term suggested (*Iron & Steel, London*, **39**, (2), 63, 1966) to replace FLAME PLATING (q.v.).

Devil's Tongue A starch plant from whose tuberous root Konjaku Meal is prepared. This has been used to improve the filterability of slips.

Devitrification. The change from the glassy to the crystalline state; it may occur either as a fault or by controlled processing to produce a devitrified ceramic; see GLASS-CERAMIC.

Devitrified Glass. The original term for GLASS-CERAMIC (q.v.).

Devitrite. A crystalline product of the devitrification of many commercial glasses; the composition is $Na_2O.3CaO.6SiO_2$; its field of stability in the ternary system is small and far removed from its own composition When heated to 1045°C devitrite decomposes into wollastonite and a liquid. First named by G. W. Morey and N. L. Bowen (*Glass Industry*, **12**, 133, 1931).

De-watering. The removal of water from a ceramic slip, e.g. by filter-pressing.

DF Stone. Abbreviation for DEFLUORINATED STONE (q.v.).

D-gun. See DETONATION GUN.

Diagonal Bond. Brickwork in which headers are laid to form a pattern of diagonals.

Diamantini. See GLASS FROST.

Diamond Pyramid Hardness (DPH). A hardness test based on the measurement of the depth of indentation made by a loaded diamond; for details see B.S.427. As applied to the testing of glass, this procedure affords a measure of the yield point of the glass structure.

Diamond Wheel. An abrasive wheel consisting of graded industrial diamonds set in a ceramic, metal, or resinoid bond.

Diaphragm Wall. A wide-cavity brick wall with the two leaves bonded together by cross-ribs of brickwork replacing the normal wall ties. The resulting connected box structure gives a wall of greater resistance to lateral and vertical loading, suitable for tall single-storey structures.

Diaspore. One of the monohydrates of alumina; its equilibrium temperature with corundum and water is 360°C. It occurs, mixed with a certain amount of clay (and thus more properly termed DIASPORE CLAY), in Missouri (USA) and in Swaziland (S. Africa). Diaspore, after strong calcination to eliminate firing shrinkage, is used as a raw material for the manufacture of high-alumina refractories.

Diatomite; Diatomaceous Earth. A sedimentary material formed from the siliceous skeletons of diatoms, which are minute vegetable organisms living in water (both fresh and marine). Large deposits occur in USA, Denmark, and France; there are smaller deposits in Ireland and elsewhere. Because of its cellular nature (porosity about 80%) and the fact that it can be used up to a temperature of about 800°C, diatomite is used as a heat- and sound-absorbing material; it is employed either as a powder for loose-fill, as shaped and fired bricks, or as an aggregate for lightweight concrete.

Dice. The small, roughly cubical, fragments produced when toughened glass is shattered.

Dice Blocks. See THROAT.

Dichroic Glass. A glass which transmits some colours, and reflects others.

Dickite. $Al_2O_3.2SiO_2.H_2O$. This is the best crystallized of the kaolin minerals, the crystals consisting of regular sequences of two basic kaolin layers. Dickite is comparatively rare; it is occasionally found in sandstones.

Dicor. Tradename. A castable, machineable glass-ceramic based on mica (tetra-silicic fluormica) with up to 7% added ZrO_2 to improve chemical durability and translucency for dental applications. (P.J. Adair, Dentsply International and Corning Glass. U.S. Pat, 4431420 1984)

Didier-March Kiln. A coal-fired tunnel kiln; typically, there are four fireboxes, two on each side.

Didymium. A mixture of rare earth oxides, chiefly the oxides of LANTHANUM, NEODYMIUM and PRASEODYMIUM (q.v.).

Die. (1) An attachment at the exit of an extruder designed to give the final shape to an extruded clay column. A brick die usually has an internal set of steel platès arranged to permit lubrication of the internal surface presented to the clay column.

(2) In dry-pressing and plastic shaping, this term is often applied to the metal mould into which the moist powder or plastic clay is charged prior to pressing.

Die Cast Brick Machine. A machine developed by Machinery Inc. USA to make special shapes from the extruded slug, so not disrupting the production process.

Die Pressing. Term used in some sections of the industry for DRY PRESSING (q.v.).

Dielectric. A material that is capable of sustaining an electrical stress, i.e. an electrical insulator. Electroceramics of high dielectric constant include the titanates, stannates, and zirconates; with suitable compositions the dielectric constant can attain 20 000, the power factor varying from 1×10^{-4} to 500×10^{-4}. They are used in high-capacity condensers at radio-frequencies.

Dielectric Heating. The heating of a dielectric by energy loss within the material when it is exposed to a high-frequency electric field. Trials have been made with this form of heating, at radio-frequencies, as a method of drying clayware but it is generally uneconomic.

Die-Wall Friction. The interaction between the wall of the die and the powder compact in uniaxial dry-pressing, which leads to a non-uniform pressure distribution across the compact. Special lubricants may be used to reduce this effect.

Differential Thermal Analysis (DTA). A method for the identification and approximate quantitative determination of minerals. In the ceramic industry, DTA is particularly applied to the study of clays. The basis of this technique is the observation, by means of a thermocouple, of the temperatures of endothermic and/or exothermic reactions that take place when a test sample is heated at a specified rate; in the differential method, one junction of the thermocouple is buried in the test sample and the other junction is buried in an inert material (calcined Al_2O_3) that is heated at the same rate as the test sample. In the DTA of a clay, the major effect is the endotherm resulting from the evolution of the water of constitution. The temperature of the peak of this endotherm varies according to the particular clay mineral present; the area of the endotherm (as measured on DTA curve) affords a means of assessing the quantity of thc mineral present.

Differential Scanning Calorimetry (DSC). The difference in energy inputs to a substance and a reference material is measured as a function of temperature while both are subjected to a controlled temperature programme.

Differential Sintering. If two separate powder compacts of different green densities are sintered to be fully densified, then the lower density compact must shrink more. However, if the compacts are joined as one piece, the region of lower green density is mechanically constrained from shrinking more than the adjacent higher density region, and so does not sinter to full density. This phenomenon is known as *differential sintering*.

Differential Thermogravimetry (DTG). A technique for the study of the changes in weight of a material when heated; it has been applied, for example, in following the dehydration process of clay minerals.

Diffusion. The transport of matter by the movement of atoms, ions or molecules. The process is driven by differences in concentration of chemical species in different parts of the substance, with thermally activated movement. In solids, atoms migrate through the lattice either by the diffusion of INTERSTITIALS, or of VACANCIES moving in the opposite direction (see CRYSTAL STRUCTURE). Diffusion is important in SOLID-STATE SINTERING. (q.v). It may occur within the grains of a polycrystal (volume diffusion) or along the GRAIN BOUNDARIES (q.v.). (Surface or grain boundary diffusion.)

Diffusion Bonding. A group of joining techniques in which strong bonds are achieved through solid state diffusion without melting of the base materials. The techniques are usually used to join dissimilar materials, either directly, or using thin interlayers. In diffusion brazing a melting filler metal is interposed between the components to be joined. In diffusion welding the interlayers do not usually melt. (For a review, see O.M. Akselsen, *J. Mater. Sci.* **27**, 569, 1992).

Diffusion Sintering. See SOLID-STATE SINTERING.

Diffusivity. See THERMAL DIFFUSIVITY.

Digs. Deep, short scratches on glass.

Dilatancy. The behaviour exhibited by some materials of becoming more fluid when allowed to stand and less fluid when stirred; dilatancy is shown by some ceramic bodies that are deficient in fine (<2μm) particles. (cf RHEOPEXY and THIXOTROPY).

Dilatometry. The measurement of thermal expansion, strictly, as a function of temperature. In particular, a continuous record can reveal phase changes in materials.

D-line Cracks. Fine, closely spaced weathering cracks near the edges of concrete.

Dimming Test. To determine the durability of optical glass the surface is subjected to the action of air saturated with water vapour at a definite temperature (usually 80°C) for a specified period. Any dimming of the surface is then observed.

DIMOX. The Directed Metal Oxidation process (q.v).

Dimple. (1) A fault in vitreous enamelware appearing as a small shallow depression. Causes include: (i) Oil or water spots from spray line. (ii) Contamination of slip by other enamel

frit. (iii) Contamination of dry biscuit by dust of another enamel. (iv) Contamination of dry biscuit by iron or other dust from dryer or racks. (v) Incomplete healing of a burst blister. (vi) Incorrect spraying due to suspension of slip being too highly set, or to the presence of large frit particles.
(2) A fault in a pottery glaze resulting from the failure of the glaze to heal completely over the depression formed during the bursting of a large bubble during the glost firing.

DIN. Abbreviation for Deutsche Industrie-Norm (German Industry Standard). These standards are published by the Deutscher Normenausschuss (abbreviation DNA) which has its offices in Berlin, W.15.

Dinas Brick. The original name for a silica refractory, so called from Dinas, in S. Wales, where the silica rock was quarried. The name is preserved in the German *dinas-stein* and the Russian equivalent.

Dinnerware. An American term (ASTM C242) for ceramic whiteware made in a consistent pattern and a full range of articles to make up a dinner service.

Dinosaur. A precision setting machine which automatically cuts, faces and sets green bricks. (Pearne & Lacy, USA; Craven Fawcett, Leeds).

Diopside. $MgO.CaO.2SiO_2$; m.p. 1392°C; sp. gr. 3.3; thermal expansion (0–1200°C) 8.8×10^{-6}. There is a deposit in New York State. Trials have been made with synthetic diopside as a high-frequency electro-ceramic. It is formed as a devitrification product of sodalime glass if the CaO is partially replaced by MgO; it is also formed when siliceous slags attack dolomite refractories.

Dip Mould. A one-piece glass forming mould. The molten glass enters at the top, and the finished ware is removed through the same hole.

Dipping. (1) BISCUIT-FIRED POTTERY (q.v.) is dipped into a suspension of the glaze ingredients in water; the dipped ware is then dried and GLOST-FIRED (q.v.)
(2) In vitreous enamelling, the base-metal can be dipped in slip and drained (wet process) or it can be first heated and then dipped in powdered frit (dry process).

Dipping Weight. See PICK-UP.

Direct-arc Furnace. See under ELECTRIC FURNACES FOR MELTING and REFINING METALS.

Direct-bonded ceramics. Ceramics in which there is no change in chemical composition at the grain boundaries, but only of particle orientation, so that the bonds across the grain-boundries are of the same nature as those in the bulk of the ceramic. Such ceramics have high mechanical strength.

Direct Coagulation Casting. In this NEAR NET SHAPE technique for forming complex shapes, an aqueous electrostatically stabilized suspension of low viscosity is cast into a non-porous mould and then coagulated by changing the pH or by a delayed reaction catalyzed by enzymes, to form a stiff, wet green body.

Directed Metal Oxidation Process. Ceramic reinforcement is incorporated in a ceramic matrix formed by the reaction of molten metal with an oxidant.

Direct Firing. The firing of pottery or vitreous enamelware in a fuel-fired kiln or furnace without protection of the ware from the products of combustion.

Directional Solidification. This is a technique for preparing lamellar or fibrous composites of ceramics and/or metals by growth from a molten eutectic composition, using techniques normally used for growing single crystals. The solidification front is made to advance slowly in one direction, by minimizing convection in the melt, as in some zone-melting or crucible techniques. Directionally solidified eutectics have low porosity, stable microstructures up to the eutectic temperature, and good bonding between the uniformly distributed phases. (R. L. Ashbrook, *J. Amer. Ceram. Soc.* **60** (9/10), 428, 1977).

Direct Teeming or Top Pouring. The transfer of molten steel from a ladle, through one or more refractory nozzles, directly into the ingot moulds.

Dirty Finish. See FINISH.

Dirty Ware. Foreign matter that occasionally disfigures potteryware as taken from the kiln; potential sources of the 'dirt' include the atmosphere, both in the factory and in the kiln, the placers' hands, the kiln lining, the kiln furniture. See also WHITE DIRT.

Disappearing-filament Pyrometer. An optical pyrometer consisting of a small telescope with an electrically heated filament placed in its focal plane. A hot surface within a kiln or furnace is focused through the telescope and the current supplied to the filament is adjusted until the apparent temperature of the filament and furnace coincide, the filament then disappearing in the general colour of its background. The corresponding temperature is read from a scale on the instrument.

Disappearing-highlight Test. A test to determine the degree of attack of a vitreous-enamelled surface after an acid-resistance test; (see ASTM – C282).

Disc. A thin refractory support used to protect other KILN FURNITURE from glaze attack during glost firing.

Discharge-end Block. See NOSE-RING BLOCK.

Disintegration Index. A measure of the durability of a hydraulic cement proposed by T. Merriman (*Engng. News Record*, **104**, 62, 1930). The test involves shaking with a lime-sugar solution followed by titration of one aliquot against HCl with phenolphthalein as indicator and another with methyl orange as indicator. The Disintegration Index is the difference between the two titrations. The test was superseded by the test now known as the MERRIMAN TEST (q.v.).

Disintegrator. A machine used for the size reduction of some ceramic materials. A rotor is rapidly revolved within a casing, both rotor and casing having fixed hammers which impact on the material being ground (cf. HAMMER MILL).

Disk Feeder. A type of clay feeder for attachment to the base of a storage bin. There are various types. In one of these there is a short fixed cylinder with a side outlet; below the cylinder is a revolving horizontal disk. In another design the disk is stationary, the clay being discharged by moving arms inside the cylinder.

Disk Wheel. An abrasive wheel of a type that is usually mounted on a plate so that grinding can be done on the side of the wheel.

Dispex. Trade-name; ammonium polyacrylate, sometimes used as a deflocculant in clay slips. (Allied Colloids Ltd., Bradford, England.)

Dislocations. See CRYSTAL STRUCTURE.

Disordered clays. Clays in which there is substantial ISOMORPHOUS SUBSTITUTION (q.v.). Disordered clays have fine particle size, high plasticity, high unfired strength and high drying shrinkage. In general, ball clays are disordered, china clays not.

Disorp. A process developed by Didier-Werke, Germany, to absorb HF from the hot flue gases from the firing of bauxite or fireclay, by filtration through a bed of limestone granules.

Dispersion Strengthening. Very fine particles of a second phase, widely dispersed throughout the crystal lattice of the first phase, inhibit the movement of dislocations, increasing strength and fracture toughness. The technique is used in cements and to strengthen metals by introducing e.g. thoria. If the dispersion is produced by solid state reactions which precipitate the second phase chemically, the technique is known as *precipitation hardening*. See TOUGHNESS.

Dissector. A person employed to classify defective potteryware according to the nature of the fault.

Disthene. Obsolete name for KYANITE (q.v.).

Diver Method. A technique for the determination of particle size by sedimentation. The specific gravity at a given depth in a sedimenting suspension is determined by means of small loaded glass 'divers' of known specific gravities in a range between the specific gravity of the dispersion medium and that of the homogeneous suspension. If a 'diver' is placed under the surface of a sedimenting suspension it will descend to a level where its weight is equal to the weight of suspension displaced; it will then continue to descend at the same rate as the largest particles at the level of its geometrical centre of gravity and at a greater rate than all the particles in the suspension located above that level. (S. Berg. *Ingeniorvidenskabelige Skrifter*, No. 2, 1940.)

Division Wall. A wall of refractory bricks between two adjacent settings in a bench of gas retorts.

D-line Cracks. Fine, closely spaced weathering cracks near the edges of concrete.

D-load. The strength of a concrete pipe in a 3-edge bearing test, expressed as the load per unit length as a fraction of internal diameter or horizontal span. The load may be the maximum the pipe will support, or may be that load which will produce a 0.001 in (0.025mm) crack throughout a length of 1 ft (305mm) of the pipe.

DNA. See under DIN.

Dobbin. A type of dryer used in the tableware section of the pottery industry; the ware, while still in the plaster mould, is placed on horizontal turntables within the drying cabinet; the turntables can be rotated about a vertical axis so that the ware moves from the working opening into the interior of the dryer where moisture is removed from the mould and the ware by means of hot air.

Dobie. A hand-shaped clay building brick before it has been fired; from ADOBE (q.v.). In USA a 'dobie' is a lump of raw fired clay.

Docking. The immersion of building bricks in water as soon as they are taken from the kiln; this is done only when the bricks are known to contain lime nodules and is a method for the prevention of LIME BLOWING (q.v.).

Doctor Blade. (1) A thin, flexible, piece of steel used for smoothing a surface, e.g. for cleaning excess colour from the engraved copper plate used in printing on pottery.

(2) A blade used for parting thin ceramic sheets or wafers of the type used in miniature condensers.

Document Glass. A glass that absorbs ultraviolet rays and thus protects documents from deterioration.

Dod Box. An old device for extruding rods or strips of a pottery body for use in the making of cup handles or of basket-ware. The term may be a corruption of WAD BOX (q.v.) or it may be from DOD the old name for the Reed Mace or Bulrush.

Dod Handle. A cup- or jug-handle made by the old DOD BOX (q.v.) method.

Dodecacalcium Heptaluminate. $12CaO.7Al_2O_3$; cubic; sp. gr. 2.68 (anhydrous), 2.73 (mono-hydrate); m.p. 1392°C; thermal expansion (0–1200°C) 8.0×10^{-6}. A constituent of high-alumina hydraulic cement. This compound has not been prepared in a strictly anhydrous form; when sorption of water has reached a max. the composition is $12CaO.7Al_2O_3.H_2O$.

Dodge-type Jaw Crusher. A jaw-crusher with one jaw fixed, the other jaw being pivoted at the bottom and oscillating at the top; the output is low but of uniform size (cf. BLAKE-TYPE JAW CRUSHER).

DOFP. Direct-On Finish Process of vitreous enamelling (US abbreviation).

Dog-house. In an open-hearth steel furnace, the arched refractory area through which a metallic burner (for oil-firing) is inserted; in a glass-tank furnace, the refractory-lined extension into which batch is fed.

Dog's Teeth or Dragon's Teeth. A fault sometimes found on the edges of a rectangular extruded column of clay, the greater friction at the corners of the die holding the clay back relative to the centre of the extruding column; if this corner friction is too great it results in a regular series of tears along the edges of the column. Methods for curing the fault are increasing the moisture content of the clay, improving the lubrication of the die, or enlarging the corners of the die at the back of the mouthpiece.

Dolly. A gathering iron with a refractory tip used in the making of glass-ware in semi-automatic machines.

Dolly Dimples. A slight defect in cast-iron vitreous enamelware, blisters in a leadless enamel having almost completely healed.

Doloma. Calcined dolomite, i.e. a mixture of the oxides CaO and MgO; the term was introduced by the Basic Furnace Linings Committee of the Iron and Steel Institute (I.S.I. Spec. Rept. 35, 1946).

Dolomite. The double carbonate of calcium and magnesium, $(Ca,Mg)CO_3$. Dolomite occurs abundantly in many countries: in England it extends as a belt of rock from Durham to Nottinghamshire; in Wales the Carboniferous Limestone has been dolomitized locally. Typical analysis *(%)*: MgO, 20; CaO, 30; CO_2, 46; impurities, 4. It is used as a source of magnesia in glass production but the principal use is as a refractory material, for which purpose it is calcined. Because of the free lime present calcined dolomite rapidly 'perishes' in contact with the air; it may be tar-bonded to give it partial protection, or it can be stabilized by firing it, mixed with steatite or other siliceous material, so that the lime becomes combined as one or more of the calcium silicates. Stabilized dolomite refractory bricks find some use in the lining of electric steel furnaces and rotary cement kilns, tarred dolomite bricks or blocks (sometimes with dead-burned magnesite added) are much used in the newer oxygen steelmaking processes. See also TEMPERED DOLOMITE BLOCK.

Domain. In a ferroelectric or ferromagnetic crystal, e.g. barium titanate, a 'domain' is a small area within which the polarization is uniform. If the crystal is exposed to a high electric or magnetic field, those domains in which the polarization is in a favourable direction will grow at the expense of other domains. A domain structure gives rise to HYSTERESIS (q.v.).

Dome Brick. A brick in which both the large and the side faces are inclined towards each other in such a way that, with a number of these bricks, a dome can be built. (See Fig. 1, p39.)

Dome Plug. A refractory shape, usually made of aluminous fireclay or of a refractory material of still higher alumina content, used in the top of the dome of a HOT-BLAST STOVE (q.v.).

Doping. The addition of small quantities of other materials to markedly affect the properties of the main component. Electrical properties of ceramics are particularly susceptible to modification in this way.

Dorfner Test. A test for stress in glazed ware proposed by J. Dorfner (*Sprechsaal*, **47**, 523, 1914): a cylinder of the ware is partly glazed and the shrinkage of the glazed portion is noted.

Dorr Mill. A TUBE MILL (q.v.) designed for operation as a closed-circuit wet-grinding unit.

Dorry Machine. Apparatus for testing the abrasion resistance of a ceramic; the flat ends of cylindrical test-pieces are abraded under standardized conditions by movement in contact with a specially graded sand.

Dot. A small refractory distance-piece for separating CRANKS (q.v.) and SETTERS (q.v.).

Dottling. The setting of pottery flatware horizontally on THIMBLES (q.v.).

Double Cantilever Beam. See FRACTURE TOUGHNESS TESTS.

Double Cleavage Drilled Compression. See FRACTURE TOUGHNESS TESTS.

Double Draining. A further period of flow of slip from dipped vitreous enamelware after the initial draining has finished and the enamel appears to have set. A possible cause of this trouble is excessive alkalinity of the slip caused by the solution of alkalis from the frit; alternatively, the amount of electrolyte added to the slip may have been incorrect.

Double Embossing. Producing three shades of decoration on glassware, by producing a design with acid, followed by two additional acid treatments.

Double-face Ware. Vitreous enamelware that has a finish coat on both sides.

Double-frit Glaze. A glaze containing two frits of different compositions. As an example, a glaze may contain a lead frit and a leadless frit; the glaze is thus rendered highly insoluble by the inclusion in the second frit of those constituents liable to increase lead solubility.

Double-layer Extrusion. Making bricks by extruding two layers of clay simultaneously, the outer layer usually forming a thin surface layer of enhanced appearance.

Double Layer Theory. Ceramic particles generally contain atoms that are easily polarized, with resulting strong interparticle attraction. AGGLOMERATION (q.v.) can be prevented by mixing the powders in a liquid with dissolved ions, whereby a "*double-layer*" of dissolved ions forms round each particle. The double-layer theory describes the electrostatic repulsion between two particles whose electric double-layers overlap. This overlap causes a local increase in the electrolyte concentration. The resulting osmotic pressure counterbalances the attractive electrostatic van der Waals forces, and prevents the particles coming together in agglomerates.

Double-roll Verge Tile. A single-lap roofing tile having a roll on both edges so that verges on the two sides are similar.

Double Roman Tile. See under ROMAN TILE.

Double-screened Ground Refractory Material. A US term defined as: A refractory material that contains its original gradation of particle sizes resulting from crushing, grinding, or both, and from which particles coarser and finer than two specified sizes have been removed by screening (cf. SINGLE-SCREENED).

Double Standard. A brick (particularly a refractory brick) that is twice as wide as a standard square, e.g. 9 × 9 × 3 in (229 × 229 × 76 mm).

Double-thread Method. A procedure for determining the coefficient of thermal expansion of a glass by forming a thread by fusing a fibre of the glass under test to a fibre of a glass of known expansion; from the curvature of the double-thread, when cold, the coefficient of expansion of the glass under test can be calculated. (M. Huebscher, *Glashütte*, **76**, 57, 1949).

Double Torsion. See FRACTURE TOUGHNESS TESTS.

Double-wing Auger. An AUGER (q.v.) with two discharge screws.

Down-cast. Local term for negative pressure in the atmosphere of some zones of an annular kiln.

Down-draught Kiln. A kiln in which the hot gases from the fireboxes first rise to the roof, then descend through the setting and are finally withdrawn through flues in the kiln-floor.

Down-draw Process. The production of glass tubing by continuously drawing molten glass downward from an orifice. See also overflow process.

Downtake or Uptake. One of the two vertical passages, built of refractory bricks, leading from the ports to the slag-pockets of an open hearth furnace. As such a furnace operated on the regenerative principle, the direction of gas-flow being periodically reversed, the identical passages at the two ends of the furnace alternately served as Downtake for the waste gases leaving the furnace and Uptake for the hot air for combustion and (in gas-fired furnaces) the fuel gas.

Downward Drilling. Wear on glass-tank furnace bottoms.

Dozzle. See under CORE.

DPH. Abbreviation for DIAMOND PYRAMID HARDNESS (q.v.).

DPT – Diffuse Phase Transition - see RELAXORS.

Drag-ladle or Dragade. To make quenched CULLET (q.v.) by ladling molten glass into water.

Drag-line. A type of mechanical excavator often used in the winning of brick-clays; a ‘bucket’, suspended from a boom, is lowered on to the clay and is then dragged towards the excavator by a wire rope, thus filling the ‘bucket’ with clay. An advantageous feature of a drag-line is its ability to work clay below the level of the excavator itself.

Dragged. A surface texture on clay facing bricks produced by a tightly stretched wire contacting thc column of clay as it is extruded from the pug in the wire-cut process: this texture is also known as rippled.

Dragon’s Teeth. See DOG’S TEETH. Also large, tooth-like formations of clinker near the outlet of the sintering zone of a rotary kiln.

Drain Casting or Hollow Casting. Terms used (more particularly in USA) for the slip-casting process for making hollow-ware, the excess slip being drained by inversion of the mould.

Drain Lines. Lines or streaks liable to appear in badly drained wet-process vitreous enamelware after it has been fired.

Drain Tile. US term for an unglazed field-drain pipe. The properties of clay drain tiles are specified in ASTM – C4 and C498 (for perforated drain tile); of concrete drain tiles in ASTM – C412.
Draw. See LOAD.
Draw Bar. In the PITTSBURGH PROCESS (q.v.) of drawing sheet glass, the position of the sheet is defined by a refractory block (the draw-bar) submerged in the molten glass.
Drawing Chamber. The part of a tank furnace for flat glass from which the sheet of glass is drawn.
Drawn Stem. See STEMWARE.
Dredging. A dry process of vitreous enamelling in which powdered frit is sifted on to the surface of the hot base-metal.
Dresden Green. A ceramic colour for use up to about 1100°C. It consists of (%): CoO, 24; Cr_2O_3, 52; ZnO, 24.
Dressing. The process of removing, from the face of an abrasive wheel, those grains that have become dulled during use.
Dressler Kiln. The first successful muffle-type tunnel kiln was that built by Conrad Dressler in 1912. The name is now applied to a variety of kilns designed and built by the Swindell-Dressler Corp., Pittsburgh, USA.
Dri-lok. Tradename. A method for the quick and precise construction of kiln cars using interlocking insulating bricks without mortar. Developed by MPK Insulation, Colchester, the Dri-lok bricks are 230 × 114mm or 230 × 230mm, up to 114mm thick.
Drip Course. A protruding or recessed course inside the crown of a glass tank, to prevent molten material running down and corroding the breast walls.
Drip Test. See SLAG ATTACK TESTS.
Drop Arch. An auxiliary brick arch projecting below thc general inner surface of the arched roof of a furnace, brick conduit or like structure.
Drop Forward. The projection of isolated bricks from the general lining surface of a newly commissioned rotary kiln.
Drop-machine Brick. US term for a refractory brick made by dropping, from a height of about 5 m, a clot of the prepared batch into a mould. The process aims to imitate hand-moulding.
Drop Monkey. A system of ropes for moving kiln cars.
Dropper. A drop of glaze that has formed, by condensation of glaze vapour, on the inside of the roof of a glost kiln and has subsequently fallen on the ware that is being fired.
Dropping. See SAGGING.
Dropping Ball Test. See KELLY BALL TEST. BS 4551 includes a dropping ball test for the consistence of mortars.
Drop Throat. A glass-tank THROAT (q.v.) below the level of the floor of the melting tank.
Drum Dryer. A rotating drum, heated to dry tumbling raw materials.
Dry-bag Process. See ISOSTATIC PRESSING.
Dry Colour Manufacturers' Association. This body has devised a scheme for the classification of pigments, based on the chemical and structural identity of the colouring substance. It excludes single metal oxides and non-oxide PIGMENTS (q.v.) but includes all mixed metal oxides in one of 14 crystal classes and over 50 pigment categories, each having a specific *DCMA Number* (E.g. Crystal class XIII is *spinel* and DCMA No 13–26–2 refers to cobalt aluminate blue spinel.
Drum. (1) Term sometimes applied to the mouth of a port in a glass-tank furnace.

(2) A wooden former of the type that was used in making the side of a saggar by hand.

Drum-head Process. A process used in Europe for the shaping of flatware; it was developed on account of the 'shortness' of the feldspathic porcelain body. A slice of the pugged body is placed on a detachable 'drum-head' which fits on the BATTING-OUT (q.v.) machine. The'drum-head', with the shaped disk lying on it, is then removed and inverted over the jigger-head, the bat then being allowed to fall on the mould for its final jiggering.

Dry Body. An unglazed stoneware type of body. The term has been applied, for example, to CANE WARE, JASPER WARE and BASALT WARE (q.v.).

Dry Edging. A fault sometimes occurring in pottery manufacture as a result of insufficient glaze application: it is shown by rough edges and corners.

Dry Gauge. See DRAG-LADLE.

Dry Laying. A rapid form of refractory construction, chiefly for brick rings in rotary kilns. Mating faces of bricks are placed in direct contact, and mortar used sparingly, if at all, for local adjustment.

Dry Lining. See DRY LAYING.

Dry Mix. See DRY PROCESS.

Dry Pan. An EDGE-RUNNER MILL (q.v.) used for grinding relatively dry material in the refractories and structural clayware industries. The bottom has a solid inner track on which the mullers rotate and outer perforated grids through which the ground material is screened, oversize being ploughed back to the grinding track.

Dry Pressing. The shaping of ceramic ware under high pressure (up to 100 $MN.m^{-2}$), the moisture addition being kept to a minimum (5–6%) or, with some materials, eliminated by the use of a plasticizer, e.g. a stearate. Dry pressing is used in the shaping of wall and floor tiles (when it is often referred to as DUST PRESSING), most high-grade refractories, abrasive wheels, the Fletton type of building brick (the moisture content for pressing is in this case 19–20%), and many articles in the electroceramic industry. The process is also sometimes referred to as SEMI-DRY PRESSING.

Dry Process or Dry Mix. (1) Term used in the US whiteware industry and defined (ASTM – C242) as the method of preparation of a ceramic body by which the constituents are blended dry; liquid may then be added as required for subsequent processing.

(2) The process of cement manufacture in which the batch is fed to the kiln dry.

Dry Process Enamelling. In this method of vitreous enamelling, the base metal is preheated to a temperature above that at which the enamel to be used will mature (usually 850–950°C); the finely powdered enamel is then applied to the hot ware which is then fired to complete the maturing process.

Dry-rubbing Test. A test to determine the degree of attack of a vitreous-enamelled surface after an acid-resistance test; (see ASTM – C282).

Dry Shake. A dry mixture of cement and fine aggregate, spread over a concrete floor before final finishing, to provide a wear-resistant surface.

Dry Spray. Faulty spraying of vitreous enamel, leading to a rough, sandy surface.

Dry-stacked walls. See MASONRY.

Dry Strength. The mechanical strength of a ceramic material that has been shaped and dried but not fired; it is commonly measured by a transverse strength test.

Dryer. See CHAMBER DRYER; DOBBIN; HOT-FLOOR; MANGLE; TUNNEL DRYER.

Dryer Scum. See SCUM.

Drying Shrinkage. Ceramic ware (and particularly clayware that is shaped from a moist batch shrinks during drying; the drying shrinkage is usually expressed as a linear percentage. e.g. the drying shrinkage of china clay is usually 6 to 10%; that of a plastic ball clay is 9 to 12%. To produce ware (e.g. electroceramics or refractory bricks) of high dimensional accuracy, the drying and firing shrinkages must be low; this is achieved by reducing the proportion of raw clay and increasing the proportion of nonplastic material in the batch, which is then shaped by dry-pressing, for example.

DT. Double Torsion. See FRACTURE TOUGHNESS TESTS.

DTA. DIFFERENTIAL THERMAL ANALYSIS (q.v).

DTG. DIFFERENTIAL THERMOGRAVIMETRY (q.v)

Duck-nesting. Localised pockets of lining scour sometimes affecting high-alumina brickwork.

Ductile Reinforcement. See CRACK BRIDGING; STRAIN HARDENING; FIBRE REINFORCEMENT.

Dulling. A glaze fault characterized by the ware having poor gloss when drawn from the kiln; the cause is surface devitrification, which may result from factors such as SULPHURING (q.v.) or too-slow cooling.

Dummy. A foot-operated device for manipulating paste moulds when blowing glass by mouth.

Dumont's Blue. Alternative name for SMALT (q.v.).

Dumortierite. A high-alumina mineral, $8Al_2O_3.6SiO_2.B_2O_3$; it occurs sufficiently abundantly in Nevada, USA, for use in the ceramic cores of sparking plugs although bodies of still higher alumina content are now more generally used for this purpose.

Dump. An item of KILN FURNITURE (q.v.) designed for use in a RING – a bottomless saggar – for the support of large hollow-ware, e.g. basins. Dumps may also be used as spacers in a CRANK (q.v.).

Dunite. A rock consisting essentially of olivine but sometimes also containing chromite; it occurs in many parts of the world and is used in the manufacture of forsterite refractories.

Dunnachie Kiln. A gas-fired chamber kiln designed by J. Dunnachie (Brit. Pat. 3862; 1881). The first such kiln was built at Glenboig, Scotland, in 1881 for the firing of firebricks. Important features are the solid floor and the space between the two lines of chambers.

Dunt, Dunting. A crack, or the formation of cracks (which may be invisible), in ware cooled too rapidly after it has been fired.

Duo-clay. Trade name for a highly plastic modelling clay, which can be fired and glazed normally, or may be air dried at room temperature to a hard state in which it may be decorated with special cold glazes. Developed by Podmore & Sons, Stoke-on-Trent.

Duplex Ceramics. Spherical pressure zones 10–50μm diameter are homogenously dispersed in a ceramic matrix, to toughen it and to enhance its thermal shock resistance. The zones contain a high proportion of unstabilized zirconia. Spontaneous or stress-induced tetragonal/monoclinic transformation of the zirconia produces compressive stresses in the zones, and radial compressive and tensile hoop stresses in the nearby matrix. (H.E. Lutz and N. Claussen, *J. Eur. Ceram. Soc.* **7** (1991) 209)

Dupré's Equation. A relationship between surface energies in ceramic-metal bonding.

$W = \Delta G = \gamma_c + \gamma_m - \gamma_{cm}$

where W is the work of adhesion (the energy required to separate unit area of the bonded surface); ΔG is the change in surface energy; γ_c, γ_m and γ_{cm} are respectively the surface energies of the ceramic, the metal, and of the interface formed by joining, assuming pristine, clean surfaces.

Durability of On-glaze Decoration. A test for the resistance of decorated tableware to attack by detergents is provided in ASTM – C556; the reagent used is a solution of 5 g. Na_2CO_3 per litre of water; the test is for 2, 4 and 6 h at 99 ± 1.1°C. ASTM – C676 is a related test for glass tableware, C675 for alkali attack on ceramic decorations in bottles; C1203 a boiling test for the resistance to alkalis of glassware decorated with fired enamels.

Durapatite. A highly translucent hydroxyapatite of near theoretical density, made by sintering a wet compact of recently precipitated hydroxyapatite.

Duravit. Tradename. A mechanized (battery) slip casting system for sanitaryware, developed by Netzsch (Germany).

Dust. A fault, in electric lamp bulbs or valves, resulting from local concentrations of seed or finely-divided foreign matter; also known as SPEW. See also DUSTING.

Dust. Dust is defined by B.S.2955, 1958 as particles which are or have been AIRBORNE, with particle size (200 mesh BS sieve). RESPIRABLE DUST (q.v.) is a hazard in the ceramics industry, causing SILICOSIS (q.v.) and other diseases. Workers wear protective clothing; machines are fitted with dust extractors, and dust concentrations in the working environment routinely measured. See DUST SAMPLING.

Dust Coat. A coating of vitreous enamel that has been sprayed thinly and relatively dry.

Dust Pressing. See DRY PRESSING.

Dusting. (See also DUST) (1) Spontaneous falling to a powder, particularly of material containing a large amount of CALCIUM ORTHOSILICATE (q.v.), which suddenly expands when it is cooled from red heat.
(2) In dry-process vitreous enamelling, a synonym of DREDGING (q.v.)
(3) In wet-process vitreous enamelling, a defect during spraying resulting in localized concentrations of almost dry slip.
(4) The cleaning of an applied coating of vitreous enamel slip after it has dried, preparatory to firing.

Dust Sampling. RESPIRABLE DUST (q.v.) concentrations are routinely measured in the workplace. Measurements may be taken over a long period in the general environment (see GRAVIMPINGER) or by personal dust samples such as the Casella, carried by the operative. Dust samples collect specimens in filters, and the collected samples are weighed and analysed. (see THRESHOLD LIMIT VALUE). The methods for sampling are specified in the UK by MDHS 14, 1993 (q.v.) and in the USA by NIOSH (q.v.) *Manual of Analytical Methods*, DHEW Pub. No. 84–100, revised 1987. The dust is collected on special filters (see below), and weighed and analysed.

Dust Sampling, Filters. Glass fibre filters are adequate for measurements of dust concentration, but not if the dust is to be analysed. Also fibre loss in handling can be significant if dust is collected. Silver or membrane filters are then used. The Gelman DM800 is a proprietary membrane filter exemplified in MDHS 14. (Gelman Sciences Ltd). Millipore Ltd filters are of a range of types. The Gelman DM800 is made from a copolymer of acrylonitrile and PVC. Such membrane filters and those made from mixed esters of cellulose do not show excessive weight charge due to moisture absorption (as do cellulose nitrate filters), nor (as do pure PVC) do they show excessive static build-up.

Dutch Bond. Brickwork with alternate header and stretcher courses and alternate stretchers in vertical alignment.

Dutch Kiln. An early type of up-draught intermittent kiln for the firing of bricks; it had a number of small chimneys in the roof.

Dutch Oven. A simple furnace of small size and usually fired with solid fuel; it can be constructed outside a newly-built furnace, for example, and used as an air-heater for drying-out and warming up (cf. DUTCH KILN).

Dwell Mark. A fracture surface marking resembling a pronounced RIPPLE MARK OR WALLNER LINE, indicating a pause in crack propagation at that point.

Dye Absorption or Dye Penetration. A test for porosity in ceramic products that are nominally non-porous. It is applied, for example, to porcelain insulators for which B.S.137 stipulates that there shall be no sign of dye penetration after a fractured specimen has been immersed for 24 h in a 0.5% solution of fuchsine in alcohol under a pressure of 14 MNm^{-2}.

Dyer Method. A procedure for shaping the socket of a clay sewer-pipe proposed by J. J. Dyer (*Brick Clay Record* **105** (3), 27, 1944).

E.B. Gun. Trade name; a cement gun of a type designed for use in repairing the refractory lining of gas retorts. (E. B. Refractory Cement Co. Ltd., Stourbridge, England.)

'E' Glass. (Electrical.) A glass of low alkali content (<1%Na_2O), used to make electrically insulating glass fibre.

Earthenware. Non-vitreous, opaque, ceramic whiteware. The COMBINED NOMENCLATURE (q.v.) defines earthenware products as those made from selected clays, sometimes mixed with feldspars and with varying amounts of chalk, characterized by a white or light-coloured fragment (slightly greyish, cream or ivory). The fragment, which has a fine grain, is homogeneous; the diameter of the non-homogeneous elements (particles, inclusions, pores) representative of the structure of the general mass should be less than 0.15 mm: these elements are therefore not visible to the naked eye. In addition their porosity as measured by the method specified in the Nomenclature, (6912 00 10) (coefficient of water absorption) is 5% or more by weight. (See also FINE POTTERY, which is coloured.) The general body composition is (per cent): china clay, 25; ball clay, 25; calcined flint, 35; china stone, 15. The biscuit firing temperature is 1100 to 1150°C.

Easing Air. The air that is admitted through the feed-holes of an annular kiln at one stage in the firing of FLETTON (q.v.) bricks; the purpose is to check the rapid rise of temperature consequent on the ignition of the organic matter present in such bricks.

Easy Fired. Clayware, particularly earthenware, is said to be easy fired if it has been fired at too low a temperature and/or for too short a time.

Eaves Course; Eaves Tile. A course of special-size roofing tiles (*eaves-tiles*) for use at the eaves of a roof to obtain the correct lap.

Ebonex. Titanium suboxide material for electrodes for electrochemistry, and for use as a high temperature lubricant. Specific properties depend on the proportions of suboxides present. See MAGNELI PHASE.

Eccospheres. Tradename for MICROBALLONS (q.v.) made by Emerson-Cuming Corp, USA.
Ecosmalt. A process to apply vitreous enamel by a two-coat, one-fire process, in which the enamel is applied by electrostatic powder coating in two successive stages.
Edge Bowl. A hollow bowl about 7 in. deep and containing the slot through which glass is drawn in the PITTSBURGH PROCESS (q.v.).
Edge-defined film growth. A modification of the CZOCHRALSKI TECHNIQUE (q.v.) for growing single crystals. In the EFG technique a new die with a capillary hole is immersed in the melt, which is wicked to the top of the die by surface tension. A seed crystal is touched to its surface there, and slowly pulled away. A film of the melt spreads over the top of the die, and the growing crystal assumes the cross-sectional shape defined by the sharp edges of the die. Shapes similar to extruded cross-sections can be produced.
Edge Lining. The painting, by hand or machine, of a coloured line round the edge of pottery.
Edge-runner Mill. A crushing and grinding unit depending for its action on heavy mullers, usually two in number, that rotate relative to a shallow pan which forms the base; the pan bottom may be solid or perforated (cf. END-RUNNER MILL).
Edging. (1) The removal of dried vitreous enamel cover-coat from the edge of ware, to reveal the underlying coating of enamel.
(2) The application of differently coloured slip around the rim of enamel-ware.
Edinburgh Sink. A domestic or industrial sink having an overflow and with its top edge projecting outwards; cf. BELFAST SINK and LONDON SINK.
EDS Energy Dispersive Spectroscopy.
Efflorescence. A deposit of soluble salts that sometimes appears on the surface of building bricks after they have been built into a wall. If the salts are derived from the bricks themselves they consist chiefly of $CaSO_4$, $MgSO_4$, K_2SO_4 and Na_2SO_4; soluble sulphates present in the raw clay can be rendered insoluble by the addition of $BaCO_3$ to the clay while it is being mixed; this precipitates the sulphate as insoluble $BaSO_4$. Efflorescence may arise, however, from soluble salts in the mortar or, if a wall has no damp course or is backed by soil, from the soil itself. A test for the likelihood of bricks to effloresce is given in B.S.3921, which categorizes efflorescence 'nil', 'slight' (area marked), 'moderate' or 'heavy' (50% covered).
Efflorwick Test. A test for the likelihood of the formation of EFFLORESCENCE (q.v.) on a clay building brick. A 'wick', made by shaping and firing a red clay known to be free from soluble salts, is allowed to absorb any soluble salts dissolved by distilled water from the crushed sample to be tested; the 'wick' is then dried and examined for efflorescence. The conditions of the test have been standardized by the New York State College of Ceramics (*Brick & Clay Record*, **101**, No. 5, 25, 1944).
EFG. Edge-defined Film Growth (q.v.)
Eggshell Porcelain. A very thin, and hence highly translucent, porcelain originally made by the Chinese and Japanese for the European market. Bone China has also for long been available of 'eggshell' thinness.
Eggshelling. (1) A glaze fault resulting in potteryware coming from the glost-kiln with an egg-shell appearance. The fault is caused by gas bubbles that have burst

on the surface of the glaze, which has subsequently failed to heal; the glaze is too viscous at the firing temperature used.
(2) A similar surface fault in vitreous enamelware (unless this surface appears on the ground-coat, when it can be an advantage as providing a good base for the cover-coat).
Egyptian Blue. A colour resulting from the formation of a complex copper silicate in an alkaline glaze or glass; the colour was much used by the ancient Egyptians and subsequently by the Persians.
Ehecoat. Tradename. A high-emissivity refractory coating of silicon carbide, made by MPK Insulation, Colchester.
Einecs. European Inventory of Existing Commercial Chemical Substances. As well as listing thousands of individual substances. the Inventory includes general definitions. Those relevant to the ceramics industry include:
2660434 Portland Cement
2660455 High Alumina Cement
2660460 Glass, oxide
2660476 Frits
2663409 Ceramic materials and wares (for the text of this definition, see Appendix A).
From 2690474 to 2691059, and 2701852, 2702086, 2702107, 2704654 various colours and pigments, mainly spinels.
2704235 Bone Ash
2706599 Portland cement flue dust
2706667 Feldspar group minerals
2957317 Nonoxide glasses.
The purpose of the Inventory is to establish whether a substance is 'existing', or 'new' and so subject to wide ranging health and safety testing. The Inventory is published in several volumes. The general definitions are in Advance Edition Vol VIII, Substance Definitions Index, EINECS, Commission of the European Communities, Luxembourg 1987.
Eirich Mixer. An under-driven wet-pan mixer. The original design was that of two Germans L. Eirich and J. Eirich (Brit. Pat. 379265; 25/8/32).
Elastic After-effect. When glass and certain ceramic materials are subjected to stress for a long period they remain partly deformed when the stress is removed: the elastic after-effect is the ratio of the deformation remaining after a given time to the deformation immediately after removal of the stress.
Elastic Fractionation. Soft aggregate particles are separated from hard by projecting them against a steel plate, the harder rebounding further.
Electret. An electrical analogue of a permanent magnet: a material that is 'permanently' electrified and exhibits electrical charges of opposite sign at its extremities. In order to retain their charge for a long period (days or weeks) ceramic electrets must be polarized at high temperature; materials that have been treated in this way include the titanate dielectrics.
Electrical Conductivity. A measure of the ease with which an electric current can be made to flow in a material by an applied voltage.
Electric current is the flow of charge, which in ceramics can be carried by positively charged, mobile ions (*ionic conductivity*), or by electrons or holes. Electrons in a solid can have one of a range of energies. These available energies are in bands, the bands being separated from each other by quite wide *energy gaps*. If all the available energy states in a *conduction band* (so called) are full, electrons must acquire sufficient energy to cross the gap to the next higher conduction band, before they can

move freely and carry charge. A material exhibiting this behaviour is called an *electrical insulator*. If about half the available energy states in a conduction band are occupied, most of the electrons in that band can easily reach a more energetic state, and are thus able to move freely. A material of this type is an *electrical conductor* – typically a metal. If the conduction band is almost empty, only a few electrons can easily acquire more energy (the rest must cross the energy gap to the nearly empty band above them). Conduction is thus limited, and the material is an *intrinsic semiconductor*. If the conduction band is nearly full, a few electrons can reach the vacant higher energy states, leaving 'holes' behind them. These behave as positive charge carriers otherwise like electrons. (*Hole conduction*). Again the material is an *intrinsic semiconductor*. The number of such mobile electrons and holes increases as the temperature rises, so the electrical conductivity of semiconductors increases with temperature. The number of mobile electrons (or holes) can be increased by DOPING (q.v.) the material with another of higher or lower valency respectively, producing n-type (negative) or p-type (positive) *doped* or *extrinsic semiconductors*.

ASTM D1829 is a test for the electrical conductivity of ceramics at elevated temperatures.

Electrical Discharge Machining. EDM. The cutter (a profile tool or cutting wire, forms the anode and the workpiece the cathode, about 40mm apart, separated by a fluid dielectric such as paraffins in the EDM machine. Sub-micron solid impurities in the fluid migrate to form a bridge; a 'spark' passes through the fluid, and material is removed from the cathode by melting, evaporation and by thermal spalling. The fluid concentrates the spark plasma into a very small volume, with energy densities up to 3 J/mm^2 and plasma temperatures of 40000°C. The rate of removal of material depends on its electrical and thermal properties and not on its hardness or brittleness. Mirror finishes can be achieved.

Electric Furnaces for Melting and Refining Metals. Several types of electric furnace are used in the metallurgical industries, both ferrous and non-ferrous: all these furnaces are lined with refractory materials, the larger furnaces generally being bricked, the smaller furnaces usually having a monolithic refractory lining which is rammed into place. The chief types of such furnaces are DIRECT ARC, in which the electric current passes through the charge: INDIRECT ARC, in which the arc is struck between the electrodes only; INDUCTION FURNACE, in which the metal charge is heated by eddy-currents induced in it. Induction furnaces may be operated at high frequency (h.f. induction furnaces) or at low frequency (l.f. induction furnaces).

Electrical Porcelain. Porcelain made for use as an electrical insulating material. A typical batch composition is 18% ball clay, 22% china clay, 30% quartz, 15% china stone, 15% feldspar. Low-tension porcelain for the insulators used for normal supply lines and high-tension porcelain for the high voltage grid are of essentially the same composition, but the latter is generally made to a lower porosity. Large insulators may be jolleyed or, where necessary, thrown and turned; some types of insulator for suspension lines are warm-pressed. Relevant British Standards include: B.S.16, 137, 223, 3288, 3297.

Electrocast Refractory. A refractory material that has been made by FUSION-CASTING (q.v.).

Electroceramics. A general term for ceramics specially formulated for electrical and electronic applications. B.S.6045 gives eight classes:
Group C.100 PORCELAINS (alkaline alumino-silicates)
C.200 STEATITE, FORSTERITE (magnesium silicates)
C.300 high permittivity ceramics: TITANIA, TITANATES, STANNATES and NIOBATES.
C.400 alkaline earth alumino-silicates.
C.500 porous aluminium and magnesium silicate.
C.600 MULLITE
C.700 HIGH ALUMINA ceramics
C.800 OXIDE CERAMICS
The standard also defines six classes of glass insulating materials, based on composition: G.100 alkali-lime-silica; G.200 and G.300, borosilicates (chemically and electrically resistant respectively) G.400 alumina-lime-silicates; G.500 lead oxide alkali silica; G.600 barium oxide alkali silica.
B.S.4789 specifies ceramic envelopes for electron tubes.

Electrode Ring or Bull's Eye. Special refractory shapes, in the roof of an electric arc steel furnace, forming an opening through which an electrode is inserted.

Electrofusion. The process of fusion in an electric furnace. See FUSION-CASTING.

Electroless Plating. The formation of metal coatings by autocatalytic chemical reduction of metal ions from solution. The metal being deposited itself catalyses the continous chemical reduction of the metal ions, the electrons are supplied by the reducing agent, and not by an external source of electric current, and uniform coatings of theoretically unlimited thickness can be deposited at a constant rate solely on the article to be plated. The process is used to apply Ni, Co, Cu or noble metal coatings to ceramics, as well as to plastics and metals.

Electrolyte. A compound which, when dissolved in water, partially dissociates into ions, i.e. into electrically charged atoms, molecules, or radicals. Electrolytes are added to clay slips and to vitreous-enamel slips to control their flow properties.

Electron Beam Perforation. Holes are melted in ceramics by precise localised heating with an electron beam focussed with a magnetic lens.

Electronic Ceramics. Ceramics whose applications in a wide range of electronic devices, use such properties as semiconduction, piezoelectricity, ionic conduction, as well as the more widely known electrical insulating properties of traditional electroceramics.

Electron Microscopy. Electrons from a cathode can be focused by electric and magnetic fields, to form an enlarged image of the cathode on a fluorescent screen. In TRANSMISSION ELECTRON MICROSCOPY, the electrons pass through the specimen to be studied; its magnified image can then be examined. The resolving power is some 1000 times that obtainable with optical microscopes. In SCANNING ELECTRON MICROSCOPY, the electron beam is focussed on the specimen, and made to perform a raster scan (as in a TV tube) across its surface point by point. This causes secondary electrons to be emitted point by point. These are processed electronically to produce an image of the surface features on a cathode ray tube screen.

Electrolytic Pickling. A method (not much used) for the preparation of the base-metal for vitreous enamelling; chemical PICKLING (q.v.) is assisted (ANODE PICKLING) or replaced (CATHODE PICKLING) by electrolysis.

Electro-osmosis. The de-watering and partial purification of clay by a process of ELECTROPHORESIS (q.v.) first proposed by Elektro-Osmose A. G. (Brit. Pats., 135815–20 25/6/18). The process has had only limited application because of its high cost; it has been used at Karlovy Vary (formerly Karlsbad) in Czechoslovakia, and at Grossalmerode and Westerwald (Germany).

Electrophoresis. The movement of fine particles in a suspension as a result of the application of an electric field. Use is made of this effect in the electrical lubrication of the dies in some clayworking machinery, the migration of the clay particles leaving a concentration of water between the clay and the metal die. Clay particles become negatively charged and migrate towards the anode. Casting is thus possible. See ELEPHANT and VOLUPHANT.

Electro-optic Materials. See OPTOELECTRONICS, FERPICS.

Electrostatic Spraying. A process in which particles that are to be sprayed are given an electrostatic charge opposite to that on the ware to be sprayed; this attracts the sprayed particles to the ware. Although technically applicable to vitreous enamelling, this method of spraying has so far been little used in the ceramic industry.

Electrostriction. A second-order electromechanical coupling effect, usually a much smaller effect than PIEZOELECTRICITY (q.v.) Some lead magnesium niobates have significant electrostrictive strains (0.1%) with near absence of hysteresis. (L.E. Cross *et al*, *Ferroelectrics* **23**, 1980, p187).

Elephant. An electrophoretic casting machine which produces a single continuous band of clay. Two cylindrical anodes rotate in opposite directions. The cathode is placed in the nip between them, and slip fed in from above. A clay deposit is built up on the cylinders. When the deposits are thick enough, they unite to form a continuous band, whose width depends on the length of the cylinders and whose thickness depends on their speed of rotation. (Karl Händle, Germany. *Interceram* **27**, 1978, p 33)

Elevator Kiln. A kiln into which a setting of ware is raised from below; the ware is set (outside the kiln) on a refractory base which is subsequently elevated by jacks into the firing position. Kilns of this type have been used, for example. in the firing of abrasive wheels.

Elutriation. The process of separation of particles, according to their size and/or density, by submitting them to an upward current of water, air or other fluid (Cf. SEDIMENTATION).

Embossing. The decoration of pottery by means of a raised pattern (flowers, figures, etc.); the effect is usually obtained by depressions in the plaster mould in which such ware is made (Cf. SPRIGGING).

Emery. A naturally occurring, impure, CORUNDUM (q.v.); used as an abrasive.

Emissivity. A surface property, being the ratio of its emissive power for heat to that of a BLACK-BODY (q.v.) for a given wavelength and at the same temperature. Some reported values for refractory materials are:

Type of Refractory	1000°C	1500°C
Fireclay	0.8	0.7
Silica	0.85	0.7
Sillimanite	0.55	0.6
Sintered alumina	0.4	0.4
Chrome-magnesite	0.85	0.7

Emley Plastometer. An instrument designed primarily for assessment of the plasticity of building plaster; it has also been used for the testing of clay. The material to be tested is placed on a porous disk which is mounted on a vertical shaft; as the shaft revolves it rises, pressing the sample against a conical metal disk, the motion of which is resisted by a lever. Equilibrium is reached when the force of the sample under test against the metal disk is equal to the stress acting through the lever; the average relative tangential force for the first 5-min. period is taken as an index of plasticity. (W. E. Emley, *Trans. Amer. Ceram. Soc.*, **19**, 523, 1917.)

Emperor Press. Trade-name. A dry-press brickmaking machine of the rotary-table type. (Sutcliffe Speakman Ltd., Leigh, Lancs., England.)

Enamel. See ENAMEL COLOUR; GLASS ENAMEL; VITREOUS ENAMEL.

Enamel-back Tubing. Glass tubing, the back half of which (the tube being held vertically) is seen to consist of white or coloured PLY GLASS (q.v.)

Enamel Colour. A ceramic colour for the on-glaze decoration of pottery.

Enamel Firing. In the British pottery industry this term is synonymous with DECORATING FIRING (q.v.)

Enamelled Brick. A hard, smooth brick with a fired (usually coloured) wash coating.

Enamelling Iron. Cold-rolled sheet specially made from steel of very low carbon content for the vitreous-enamel industry

Encapsulated Colours. Ceramic colours in which the CHROMOPHORE (q.v.) is physically trapped or *encapsulated* in the crystal matrix, not forming a part of the crystal lattice itself (cf. LATTICE COLOURS). The pink cadmium sulphoselenide *inclusion pigments* (Degussa, Br. Pat. 1403470, 1975) were of this type, with the pigment trapped in the ZIRCON (q.v.) lattice. Later developments have led to means to entrap much greater concentrations of chromophore to enhance colour saturation, producing bright yellow, orange and red cross colours. (British Ceramic Research Ltd UK Pat. 2106530A, 1983). The technique can also be used to impart stability to pigments which would be difficult or impossible to use, either by reducing flow problems caused by reaction with glazes (cobalt blue colours) or increasing thermal stability of the pigment - grey colours based on carbon black pigment encapsulated in zircon.

Encapsulated HIPing. The ceramic powder is contained in a thin glass envelope during HOT ISOSTATIC PRESSING, excluding the furnace atmosphere from the compact. The technique has proven particularly valuable in HIPing high-T_c ceramic superconductors. It allows the oxygen chemistry to be controlled and avoids an additional treatment stage in an oxidizing atmosphere.

Encapsulation. The sealing of an electronic component, particularly of a semi-conductor, generally with a ceramic sealing compound (cf. POTTING MATERIAL).

Encaustic Tiles. Ceramic tiles in which a pattern is inlaid with coloured clays, the whole tile then being fired.

End. A brick shape used for the construction of arches and sprung roofs; the large faces are inclined towards each other in such a way that one of the end faces is smaller than the other. (See Fig 1, p39.)

End Feather. See FEATHER BRICK.

End-fired Furnace. A type of glass-tank furnace in which the ports are in the back wall (cf. CROSS FIRED FURNACE).

End Runner. See RISER BRICK.

Endoprep. A process for electrostatic powder enamelling which requires no metal preparation.

End-runner Mill. A small grinding unit, primarily for laboratory use, operating on the principle of the pestle and mortar; the runner is set eccentrically in the mortar, which is mechanically driven (cf EDGE-RUNNER MILL).

End Skew. A brick (particularly a refractory brick) with one end completely bevelled at an angle of 60°. This bevel can be towards a large face (*end skew on flat*) or towards a side face (*end skew on edge*). Both types of brick are used in the springing of an arch (See Fig 1, p39)

End Wall. (1) The vertical refractory wall, furthest from the furnace chamber, of the downtake of an open-hearth steel furnace.
(2) One of the two vertical walls terminating a battery of coke ovens or a bench of gas retorts; it is generally constructed of refractory bricks and heat-insulating bricks with an exterior facing of building bricks.
(3) cf. GABLE WALL.

Endell Plastometer. See GAREIS-ENDELL PLASTOMETER.

Endellite. Obsolete name for the clay mineral HALLOYSITE (q.v.); some authorities, however, would preserve the name for those halloysites containing an excess of water.

Endothermic Reaction. A chemical reaction that takes place with absorption of heat. The dehydration of kaolinite is a reaction of this type.

Enduro. Tiles made by the FIRESTREAM process (q.v.) *Fliesen u. Platten* **36** (4) 14 1986.

Energy Dispersive Spectroscopy. X-rays emitted by an element struck by a beam of electrons are detected by a Li-doped silicon crystal, in which electron-hole pairs are generated, the number of pairs being proportioned to the energy of the X-rays. The resulting electrical pulses are displayed as a spectrum of X-ray intensity vs. energy.

Enforced-order Mixer. See FRENKEL MIXER.

Engine-turning Lathe. A lathe having an eccentric motion and used to incise decorations on pottery-ware before it is fired.

Engineering Bricks. Building bricks that have been shaped from a clay such as an Etruria Marl which, when fired at a high temperature, will vitrify to produce a brick of great strength and low water absorption. If firing is under reducing conditions, blue bricks are produced; with oxidizing firing the product is a red engineering brick. In either case the brick must conform, in the UK, to B.S.3921 which sub-divides them into two classes: Class A – min. crushing strength 70 MNm^{-2}, max. water absorption 4.5 wt%. Class B – min. crushing strength 50 MNm^{-2}, max. water absorption 7 wt%.

Engineering Ceramics. Simply, ceramics used for applications in engineering. More precisely, the term usually excludes ceramics based in clay, and is particularly applied to mechanical engineering applications in which the thermo-mechanical properties of the ceramics are important. Typical engineering ceramics now widely used are alumina, zirconia, silicon carbide, and silicon nitride.

English Bond. A brick wall built with alternate header and stretcher courses.

English Crystal Glass or English Full Crystal. Older name for FULL LEAD CRYSTAL GLASS (q.v.).

English Garden Wall. A brick wall in which three stretcher courses are followed by one header course, the bond pattern then being repeated.

English Kiln. A transverse-arch chamber kiln with a system of flues and dampers above the chambers permitting any two chambers to be connected. It was designed by A. Adams in 1899 for the firing of building bricks made from highly bituminous clays.
English Pink. See CHROME TIN PINK.
English Translucent China. Ceramic tableware, etc., introduced in 1959 by Doulton Fine China Ltd, Burslem, England. In contrast to bone china it is feldspathic, but differs from Continental porcelain in that it is biscuit fired at a higher temperature than the glost fire.
English and Turner Factors. See THERMAL EXPANSION FACTORS FOR GLASS.
Engobe. A coating of slip, white or coloured, applied to a porous ceramic body to improve its appearance; a glaze is sometimes applied over the engobe, as in sanitary fireclay. A typical engobe for sanitary fireclay consists of 10% ball clay, 40% china clay, 20% flint and 30% china stone; some of the china stone may be replaced by feldspar and considerable variation is possible in the proportions of the other constituents. (French word.)
Enloc. A system for keeping plates warm. A special conductive decoration on the underside of the plate is heated inductively by a generator beneath the table.
Epitaxy. Growth of one crystal layer onto another with the same crystal orientation.
Engraving. A method of decoration. For application to pottery, the pattern is engraved on a copper plate or roller; the incised pattern is then filled with specially prepared colour which is transferred to the ware by transfer paper. As a method of decorating glass-ware, the pattern is cut directly into the glass surface by means of copper wheels; the depth of cut is shallower than in intaglio work.
Enslin Apparatus. This apparatus for the determination of the water-absorption capacity of clays was originally designed by O. Enslin (*Chem. Fabrik.*, **6**, 147, 1933) for testing soils. It consists of a U-tube, one arm of which is connected via a 3-way tap to a calibrated horizontal capillary tube; the other arm ends in a funnel with a sintered-glass base on which is placed a weighed sample of clay. Water is allowed to contact the sample and the amount absorbed is read from the capillary tube. The result is expressed as a percentage of the weight of the dry clay.
Enstatite. Magnesium metasilicate $MgSiO_3$: m.p. 1550°C. Enstatite (orthorhombic) is stable up to about 1050°C when it inverts to proto-enstatite (orthorhombic) which is the stable high temperature form. Clino-enstatite (monoclinic) is formed as a metastable phase when proto-enstatite is cooled rapidly; the transition temperature is 865°C. Thermal expansion (300–700°C): enstatite 12×10^{-6}; proto-enstatite 10×10^{-6} clino-enstatite 13.5×10^{-6}, Proto- and/or clino-enstatite are major constituents of steatite ceramics: clino-enstatite is found in magnesite re-fractories that have been attacked by siliceous slags.
Envelope Kiln. Alternative name, particularly in USA, for TOP-HAT KILN (q.v.) or SHUTTLE KILN (q.v.).
Equilibrium Diagram. See PHASE DIAGRAM.
Equivalent Particle Diameter or Equivalent Free-falling Diameter. A concept used in evaluating the size of fine particles by a sedimentation process; it is defined as the diameter of a sphere that has the same density and the same free-falling velocity in any given fluid as

the particle in question (cf. PARTICLE SIZE).

Eriometry. The measurement of particle size by using laser-generated diffraction patterns.

Erosion. Wear caused by the mechanical action of a fluid, e.g of molten steel flowing through the refractory nozzle in a ladle or of waste gases flowing through the downtake of an open-hearth furnace (cf. ABRASION and CORROSION).

Erosion-resistant Castable. A term used in the petrochemical industry to describe high-strength dense castables used to make erosion-resistant linings.

Erz Cement. A ferruginous hydraulic cement formerly made in Germany; it has now given place to FERRARI CEMENT (q.v.).

ESSE A. An Italian clay building block, 19 × 40 × 29cm, made with pore-forming additives. The arrangement of perforations is varied to emphasis loadbearing, acoustic or thermohygrometric properties.

Estuarine Clays. Clays that were deposited, during the course of geological time, in estuaries and deltas. Estuarine clays of the Middle Jurassic occur in Yorkshire and Lincolnshire and are used as raw material for making building bricks.

E.T.C. ENGLISH TRANSLUCENT CHINA (q.v.).

Ethyl Silicate. See SILICON ESTER.

Etruria Marl. A brick-clay occurring in the Carboniferous System and used for the manufacture of bricks and roofing tiles, particularly in the English Midlands and North Wales. These clays have a high iron content; they fire to a red colour under oxidizing conditions but under reducing conditions they fire to the blue colour of the well-known Staffordshire engineering brick.

Etruscan Ware. BASALT WARE (q.v.) having an encaustic decoration, mainly in red or white in imitation of early Italian Etruscan pottery.

Ettringite. $Ca_6Al_2(SO_4)_3(OH)_{12}.26H_2O$. This mineral is present during the early stages of the hardening of Portland cement; it is an essential constituent of set supersulphated cement, contributing to its high flexural strength.

Eucryptite. A lithium mineral, $Li_2O.\ Al_2O_3.\ 2SiO_2$. When heated, the α-form changes to the β-form at about 970°C; the β form expands in one direction and contracts in another direction. Eucryptite is a constituent of special ceramic bodies having zero (or even negative) thermal expansion.

Eurite. A feldspathic mineral occurring on the island of Elba and used locally as a ceramic raw material, especially as a flux.

Eurolit. A dry screed of tongue-and-grooved clay tiles, on top of which floor tiles are fixed.

Euronorms. EUROPEAN STANDARDS (q.v.).

Europresse. A continuously operating extruder developed by Rieterwerke, Konstanz, Germany, in which the conventional auger is replaced by a cylindrical body with annular grooves, with only a small clearance between the cylinder and the outer casing. The grooves are filled uniformly by a feed rotor on an axis parallel to the cylinder, and emptied by a comb-like roller with teeth arranged to fit the grooves.

Eutectic. An invariant point on an equilibrium diagram. In the Al_2O_3-SiO_2 system (see Fig. 5) there is a eutectic between silica and mullite and a eutectic between alumina and mullite. The eutectic temperature is that at which

a eutectic composition solidifies when cooled from the liquid state. The eutectic composition has been defined as 'that combination of components of a minimum system having the lowest melting point of any ratio of the components; in a binary system it is the intersection of the two solubility curves.'
Eutectic Joining. The components of a eutectic mixture are present in such proportions that the mixture melts at a lower temperature than mixtures of slightly different proportions. Eutectic joining uses such mixtures to form the bond between ceramics, or between ceramics and metals.
Eutit. Cast basalt made by Czechoslovak Ceramics.
Exadur. An extra hard, smooth plaster made by H & E Börgardts KG, Germany.
Excavator. See DRAG-LINE; MECHANICAL SHOVEL; MULTIBUCKET EXCAVATOR; SHALE CUTTER or SHALE PLANER; SINGLE BUCKET EXCAVATOR.
Excess Air. The amount of air in a combustion process that is greater than that theoretically required for complete conbustion.
Exchangeable Bases. See IONIC EXCHANGE.
Exfoliation. The property of somc hydrous silicates, notably VERMICULITE (q.v.). of permanently expanding concertina-wise when rapidly heated to a temperature above that at which heat is evolved; cf. BLOATING and INTUMESCENCE.
Exhaust catalyst. See CATALYTIC CONVERTER.
Exothermic Reaction. A reaction that takes place with evolution of heat. Such a reaction occurs at 900 to 1000°C when most clays are fired.
Expanded Clay. See LIGHTWEIGHT EXPANDED CLAY AGGREGATE.
Expansion Cards. Cardboard inserts, or cards pre-attached to refractory bricks to provide allowance for thermal expansion.
Expansive Cement. A high sulphate and alumina-containing cement, which shrinks during drying, but then expands on hardening to compensate.
Explosive Forming: A process for compaction by the blast of an explosion within the mould containing the powder to be compacted; the rate of energy release has been stated to be approx. 33 kJs^{-1} and densities as high as 97% of the theoretical have been attained.
Exposed Finish Tile. US term (ASTM – C43) for a hollow clay building block the surfaces of which are intended to be left exposed or painted: the surface may be smooth, combed or roughened.
Expression Rolls. A pair of steel rolls which when rotated will force a clay column through a die or along a cutting-off table (as in the shaping of bats for roofing-tile manufacture); cf. CRUSHING ROLLS.
Extra-duty Glazed Tile. US term for a ceramic tile with a glaze that is sufficiently durable for light-duty floors and all other surfaces inside buildings provided that there is no serious abrasion or impact.
Extrusion. The forcing of clay or other material through a die; this is done by a continuous screw on a shaft rotating centrally in a steel cylinder (an 'auger'), or by a series of knives obliquely mounted on such a shaft (a 'pug'), or, occasionally, by means of a piston (a 'stupid'). Ceramic material may be extruded either to mix and consolidate it (as in the extrusion of pottery body prior to jiggering) or to give the body its final shape (as in the extrusion of bricks and pipes).
Eye. An opening in the SIEGE (q.v.) of a pot furnace for glass melting; gas and air

for combustion enter the furnace through the eye.

Eykometer. An instrument for the determination of the yield point of clay suspensions; details have been given by A. G. Stern *(U.S. Bur. Mines, Rept. Invest.* 3495, 1940).

°F. Degrees Fahrenheit. T°F = 5/9 (T – 32)°C. See °C.

Face. The part of the abrasive wheel against which is applied the work to be ground.

Faced Wall. A wall in which the facing and backing are bonded in such a way that the wall acts as a unit when loaded (cf. VENEERED WALL).

Fabrication Traces. Anomalous markings on crack surfaces due to weaknesses originating during processing.

Facing Brick. A clay building brick having an appearance, and weather-resistance, suitable for use in the outside leaf of a wall or in an external brick panel. Such bricks are made in a variety of textures and colours; physical properties range from a crushing strength of about 7 to 100 MNm^{-2} and a water absorption of about 5 to 30 wt.% Properties are specified in B.S. 3921, and in ASTM C216.

Faggot. Defined in the S. African specification for building bricks as 'A facing brick of the normal face dimensions but with the smallest bed dimension less than the height'.

Fahrenheit. See °F.

Faience. Originally the French name for the earthenware made at Faenza, Italy, in the 16th century; the ware had a tin-opacified glaze and in this resembled Maiolica and Delft ware. The meaning has now changed. In France, *faience* is any glazed porous ceramic ware; *faience fine* is equivalent to English EARTHENWARE (q.v.). In England, the term 'faience' now refers to glazed architectural ware, e.g. large glazed blocks and slabs *(not* dust-pressed glazed tiles). In USA, *faience ware* signifies a decorated earthenware having a transparent glaze.

Faience Tiles and Mosaic. US terms defined (ASTM – C242) as follows: FAIENCE TILE – glazed or unglazed tiles, generally made by the plastic process, showing characteristic variations in the face, edges, and glaze that give an 'art' effect; FAIENCE MOSAIC – Faience tiles that are less than 6 in^2 (3871 mm^2 in facial area and usually 8 to 9mm thick.

Fairlight Clay. A Cretaceous clay of the Hastings (England) area that finds use in making building bricks.

Falling Slag. Blast-furnace slag that contains sufficient CALCIUM ORTHOSILICATE (q.v.) to render it liable to fall to a powder when cold. Such a slag is precluded from use as a concrete aggregate by the limits of composition specified in B.S. 1047.

False Set. Premature, but temporary, stiffening of portland cement resulting from overheating of the added gypsum during grinding of the clinker (cf. FLASH SET).

False Header. A half-brick used to complete a course of brickwork.

Falter Apparatus. Apparatus designed by A. H. Falter *(J.Amer. Ceram. Soc.,* **28**, 5, 1945) for the determination of the SOFTENING POINT (q.v.) of a glass by the fibre-elongation method as defined by J. T. Littleton *(J.Soc. Glass Technol.* **24**, 176, 1940).

Famille Rose. Red colours based on gold and tin, obtained from PURPLE OF CASSIUS (q.v.)

Famille Verte. Green colours based on chromic oxide, developed in the 18th century.

Fantail. The opening in the refractory brickwork that separated the slag pocket from the regenerator of an open-hearth steel furnace.

Faraday Effect. The rotation of the plane of polarization of a beam of polarized light as it passes through glass that is located in a magnetic field; first observed by Michael Faraday in 1830.

FARE-gun. Fuel-Air Repetitive Explosion coating deposition process.

Farren Wall. A cavity wall (4-in. cavity) for house construction introduced in USA; *(Brick & Clay Record,* **95**, No. 4, 38, 1939).

Fat Clay. Term sometimes used for a clay that is highly plastic.

Fastfire Kiln. A kiln for rapid firing. The preferred term according to BS 3446 Pt2.

Fast Firing. see RAPID FIRING

Fast Fracture. See FRACTURE

Fast ion conductors. Solid electrolytes with high electrical conductivites due to charge transport by mobile ions. See IONIC CONDUCTIVITY.

Fatigue. The failure of a component after prolonged exposure to a stress significantly less than the stress which would produce immediate failure (*static fatigue*); or to repeated cyclic stresses at level below the immediate fracture stress (*dynamic fatigue*). A silicon oxynitride surface film greatly increases the fatigue resistance of silica fibres. See also STRESS CORROSION; THERMAL CYCLING TEST.

Fat Oil. A thick oil of turpentine made by heating turpentine in air at 145°C and then subjecting it to steam distillation. It is used as a vehicle, particularly for gold, in the decoration of pottery.

Faugeron Kiln. A coal-fired tunnel kiln of a design proposed in 1910 by E. G. Faugeron (French Pat. 421 765; 24/10/1910); the distinctive feature is the division of the tunnel into a series of chambers by division-walls on the cars and drop-arches in the roof. Such kilns have been used for the firing of feldspathic porcelain.

Fayalite. Ferrous orthosilicate, $2FeO.SiO_2$; m.p. 1205°C. This low-melting mineral is formed when ferruginous slags attack, under reducing conditions, aluminosilicate refractories.

Feather. (1) A fault, in glass, of feather-like appearance and caused by SEED (q.v.) produced by foreign matter picked up by the glass during its shaping.
(2) A fault in wired glass resulting from bending of the transverse wires.
See also FEATHERING.

Feather. A STRIATION looking like a feather.

Feather Brick. A specially moulded brick of the shape that would be produced by cutting a standard square diagonally from one edge. This could be done in three ways to produce a FEATHER END (or END FEATHER) a FEATHER-END-ON-EDGE, or a FEATHER SIDE (or SIDE FEATHER) depending on whether the diagonal terminates, for a 9 × 4½ × 3 in. brick, at a 4½-in. edge, a 3-in. edge or a 9-in. edge. (See Fig. 1, p39).

Feather Combing. A method of decoration sometimes used by the studio potter: the ware is first covered with layers of variously coloured slips and a sharp tool is then drawn across the surface while it is still moist.

Feather End. See FEATHER BRICK.

Feather Side. See FEATHER BRICK.

Feathering. A glaze fault caused by devitrification. It is particularly liable to occur in glazes rich in lime. To prevent the fault the initial rate of cooling in the kiln, after the glaze has matured, should be rapid. (See also FEATHER.)

Feed Shaft. See FIRE PILLAR.

Feeder. (1) A device for supplying raw material, e.g. clay, to a preparation machine.

(2) A mechanical system for the production of gobs of glass for a forming machine.
Feeder Channel. The part of the forehearth of a tank furnace, producing container-glass or pressed glass-ware, that carries the molten glass from the working end to the feeder mechanism.
Feeder Gate or **Feeder Plug**. A shaped refractory used to adjust the rate of flow of molten glass in the feeder channel.
Feeder Sleeve or **Feeder Tube.** A cylindrical tube that surrounds the feeder plunger in a glass-forming machine.
Feeder Spout or **Feeder Nose.** The part of the feeder in a glass tank furnace containing an opening in which the orifice ring is inserted; it forms the end of the forehearth.
Feeing. See FEYING.
Feed Rate. The rate of advance of a cutting tool into the work (in distance per revolution).
Feine Filter. A FILTER-PRESS (q.v.) filter, made from parallel strings instead of cloth.
Feldspar. The internationally agreed spelling is feldspar; previously the usual spelling in Britain was *felspar.* The feldspars are a group of minerals consisting of aluminosilicates of potassium, sodium, calcium, and, less frequently, barium. The most common feldspar minerals are:
Orthoclase and Microcline: $K_2O.Al_2O_3.6SiO2$
Albite:$Na_2O.\ Al_2O_3.\ 6SiO_2$
Anorthite:$CaO.\ Al_2O_3.\ 2SiO_2$
These minerals rarely occur pure, isomorphous substitution being common; most potash feldspars contain sodium, and soda feldspars usually contain both potash and calcium. There is a continuous series of solid solutions, known as plagioclase feldspars, between albite and anorthite. Feldspars are the chief components of igneous rocks and occur in economic quantity, and in adequate purity for ceramic use, in USA (N. Carolina, S. Dakota, New England), Canada (Ontario, Quebec, Manitoba), Scandinavia, France, and many other countries. Over half the world's output is consumed by the glass industry: in USA and Europe it is a component (20–30%) of practically all pottery bodies, Feldspar constitutes 25–50% of most vitreous enamel batches.
Feldspar Convention. See under RATIONAL ANALYSIS.
Felite. The name given to one of the crystalline constituents of portland cement clinker by A. E. Törnebohm (*Tonindust. Ztg.* **21** 1148, 1897). Felite is now known to be one form of $2CaO.SiO_2$.
Felspar. See FELDSPAR.
Fender. A grate in the floor of a chamber kiln immediately in front of the inter-chamber wall: it may consist of fireclay blocks or firebars.
Felvation. The classification of powders by fluidisation, elutriation and seiving.
Feolite. Tradename. A refractory for storage heaters, based on ferric oxide, developed by the Electricity Council, UK.
Feret's Law. States that the strength (S) of cement or concrete is related to its mixing ratio by the equation $S = K\,[c/(c+w+a)]^2$ where *c, w,* and *a* are the absolute volumes of the cement, water and air in the mix. This relationship was proposed by R. Feret at the beginning of the century.
Ferrari Cement. A sulphate-resistant cement consisting principally of $3CaO.SiO_2$, $2CaO.SiO_2$ and $4CaO.Al_2O_3.Fe_2O_3$. The sulphate resistance results from the formation of a protective film of calcium ferrite around the calcium aluminate crystals formed by

hydrolysis of the brownmillerite. (F. Ferrari, *Cemento Armato,* **38**, 147, 1941.)

Ferpic. **Fer**roelectric **pic**ture devices. These OPTO-ELECTRONIC CERAMICS (q.v.) were developed by Bell Telephone Laboratories, USA, to provide low cost, solid state displays. The ceramic rotates the plane of polarisation of light shone through it, the degree of rotation depending on the wavelength of the light and the electric field applied to the ceramic. It is possible to transmit light of a single colour, to vary this colour by varying the applied field, and to produce various colours at different points in the ceramic surface.

Ferric Oxide. Fe_2O_3, the mineral hematite; m.p. 1565°C, but when heated in air it loses oxygen below this temperature to form magnetite, Fe_3O_4; sp. gr. 5.24; thermal expansion (0–1000°C)12×10^{-6}. Ferric oxide is present in the surface layers of blue engineering bricks: it gives the characteristic sparkle to aventurine glazes.

Ferrielectric. A mixed phase material, often a solid solution, which can be regarded as an incompletely compensated antiferroelectric, the atomic dipoles of the two phases being unequal. A permanently polarized state results, in which antiferroelectric and ferroelectric phases coexist, each with its own CURIE POINT (q.v.).

Ferrimagnetic. A material in which the atomic magnetic moments are aligned antiparallel in an unbalanced fashion so that a spontaneous magnetization occurs in the absence of an applied field. They have many properties in common with ferromagnetics, but their saturation magnetization is much less that the sum of the atomic moments.

Ferrites. Specifically, this term refers to a group of electroceramics having FERRO-MAGNETIC (q.v.) properties, combining high magnetic permeabiiity (up to 4300) and low coercive force (0.1–4 oersteds). Attention was first drawn to these materials by S. Hilpert *(Ber Deuts. Chem. Ges.* **42**, 2248, 1909). The ferrites are synthesized from Fe_2O_3 and the oxide, hydroxide, or carbonate of one or more divalent metals, e.g. Ba, Zn, Mn, Ni, Co, Mg or Cu. The firing process must be carefully controlled to maintain the required state of oxidation. Mg-Mn ferrites have a square hysteresis loop; $BaO.6Fe_2O_3$ is a 'hard', or permanent, magnetic material.

Ferroconcrete. Concrete reinforced with iron or steel, usually as rods or mesh.

Ferroelectric. A dielectric material that shows a net spontaneous polarization below the electric CURIE POLNT (q.v.) The dielectric constant reaches a maximum value at this temperature. An applied electric field governs the direction of the polarization, and there is hysteresis and a non-linear relation between the polarization and the applied field. An important property of ferroelectric ceramics is their PIEZOELECTRIC (q.v.) nature.

Ferromagnetic. A material in which the atomic moments are aligned parallel, giving rise to spontaneous magnetization in the absence of an applied field. The magnetization increases to a saturation value equal to the sum of the atomic moments if an increasing external field is applied. The materials show magnetic hysteresis, and have high susceptibilities and permeabilities. Their magnetization decreases with increasing temperature until at the CURIE POINT (q.v.) the ferromagnetic behaviour disappears and the materials become paramagnetic. For a discussion of ceramic ferromagnetics, see FERRITES.

Ferrous Oxide. FeO; m.p. 1370°C; sp. gr. 5.7. This lower oxide tends to be formed under reducing conditions; it will react with SiO_2 to produce a material melting at about 1200°C, hence the fluxing action of ferruginous impurities present in some clays if the latter are fired under reducing conditions.

Ferroxdure. Tradename. Ceramic permanent magnets made by N.V. Philips Holland, of $BaFe_{12}O_{19}$. They have high coercivity, low remanance and low saturation magnetisation.

Ferroxplana. Ceramics with the general chemical formula $MO.6Fe_2O_3$ where M is Ba, Sr or Pb have the *magnetoplumbite* structure – hexagonal with a unique '*c*' axis. In the basic magnetoplumbite structure, the '*c*' axis is the easy direction of magnetization. In some of the more complex substituted structures, with interleaved magnetoplumbite and spinal layers in various proportions, the '*c*' axis becomes the difficult direction of magnetization. The easy direction is one in the plane perpendicular to the '*c*' axis. Such materials are called *ferroxplana*.

Fettling. The removal, in the 'clay state', and usually by hand, of excess body left in the shaping of pottery-ware at such places as seams and edges (cf. SCRAPPING). For the purposes of COSHH (q.v.) this potentially dusty process includes towing, scalloping, sand-papering and sand-sticking. *Damp fettling* is done either with a wet sponge, or whilst the ware is still damp and no dust is given off.
(2) The process of repairing a steel-furnace hearth, with deadburned magnesite or burned dolomite, between tapping and re-charging the furnace.

Fettling Refractory. Refractory grain, relatively rich in fluxes, for maintaining furnace hearths.

Feying. A local term in the English brick industry for the removal of overburden or for the cleaning up of a clay-pit floor after an excavator has been at work.

FGM. Functionally Gradient Materials.

Fiberfrax. Tradename. Aluminium silicate fibres, and a wide range of textile products made therefrom. (Carborundum Co., USA)

Fiberglas. Trade name: Owens-Corning Fiberglas Corp., USA.

FFF. Finnish Flotation Feldspar, a flux for pottery bodies, beneficiated by a flotation process.

Fianites. Yttria-stabilized zirconia single crystals.

Fibre Bridging. See CRACK BRIDGING.

Fibre (Ceramic). A mass of thread-like ceramic material, cf. FILAMENT and WHISKER. Typically, these fibres are made from a batch consisting of alumina and silica (separate or already combined as kaolin or kyanite) together with a borosilicate flux; zirconia may also be present. Other types of ceramic fibre are made from fused silica and from potassium titanate. These fibres are used in the production of lightweight units for thermal, electrical, and sound insulation; they have also been used for high-temperature filtration, for packing, and for the reinforcement of other ceramic materials. BS 1902 Pt6 specifies requirements and tests. BS 6466 is a code of practice for the design and installation of ceramic fibre furnace linings. See also GLASS FIBRE.

Fibreglass. Though originally a trade name (of Fibreglass Ltd., St. Helens, England – cf FIBERGLASS, Owens Corning Fiberglass, USA) these words are now generally used for GLASS FIBRE (q.v.) products.

Fibrekote. A sprayed refractory lining, comprising a mixture of alumina-silica fibres with an inorganic binder, stable to 1600°C.

Fibre Pullout. The DEBONDING mechanism in fibre-reinforced composites. If a small crack develops in the matrix, the load is transferred by shear forces to the still intact fibres, becoming greater than the bond strength. The bond fails and a cylindrical crack propagates down the fibre. Test methods aim at measuring the forces required to pull out fibres. As this may be practically difficult to achieve, *fibre pushout* tests, using indenters aligned with the fibres, are sometimes substituted. See also CRACK BRIDGING.

Fibre Reinforcement. Incorporating *continuous fibre* reinforcement into a ceramic matrix leads to composites with very high TOUGHNESS (q.v.), and essentially ductile behaviour. The fibres and the matrix may separately be brittle (fracture toughness of a few MPa√m) but in combinations with appropriately mismatched properties and careful microstructural control, CRACK PROPAGATION (q.v.) may be virtually completely checked. The fibre elastic modulus is chosen to be higher than that of the matrix, to promote load transfer to the fibre. The fibre should also have higher thermal expansion, so that on cooling from processing temperatures, the matrix is in compression. *Medium-strength* fibre-matrix bonding allows for crack redirection, crack branching and crack energy dissipation by propagation along the fibre-matrix interface. The fibres then remain intact and bear the loads until they fracture separately. Extra energy is then absorbed in fracturing the fibres and in FIBRE PULLOUT (q.v.) from the matrix. A fibrous FRACTURE SURFACE (q.v.) results, rather than the smooth continuous surface which would result from BRITTLE FRACTURE (q.v.) of both components were the fibre-matrix bond strength high. Most continuous fibre reinforcement is unidirectional, and strengths anisotropic, being low (~20MPa) normal to the fibre direction. Two-dimensional composites have fewer fibres and lower strengths than unidirectional composites tested in their strongest direction (parallel to the fibres), but are strong in two directions.

Fibretech. Refractory castables reinforced with steel fibres.

Fibrous. See MORPHOLOGY.

Fiburloc. Trade name of U.S. Concrete Pipe Co., Cleveland, Ohio, for vitrified clay pipes with joints comprising fibreglass/polyester sockets and urethane spigots.

Fick's Law. The rate of diffusion is proportional to the negative of the concentration gradient of the diffusing species (Adolf Fick, 1855).

Fidler-Maxwell Kiln. A straight tunnel kiln designed to be fired with gas, coal, or oil; a distinctive feature was the use of cast-iron recuperators in the cooling zone. (F. Fidler and J. G. Maxwell, Brit. Pat. 141 124 and 141 125, 6/1/19.)

Field-drain Pipe. An unglazed, fired clay pipe, generally 75 or 100 mm dia. and about 305 mm long, for the drainage of fields; occasionally these pipes have a flattened base, or longitudinal ribs, to facilitate alignment during laying. For specification (which includes pipes from 63.5 to 305 mm dia) and joints see B.S. 1196.

Figuier's Gold Purple. A tin-gold colour, produced by a dry method; it has been used for porcelain decoration.

Figured Glass. Flat glass with a pattern on one or both surfaces.

Filament. A single long thread of ceramic material which may, or may not, be a single crystal. cf. FIBRE (CERAMIC) and WHISKER.

Filler. A general term for a solid material that is chemically inert under the conditions of use and serves to occupy space and improve physical properties; an example is the use of china clay in some paints and plastics. In vitreous enamelling a filler (or plugging compound) consisting of a mixture of inorganic compounds is used to fill any holes in iron castings, thus providing a uniform surface on which to apply the enamel.

Film Strength. The mechanical strength, determining its resistance to damage by impact or abrasion, of the dried coating on vitreous enamel slip prior to firing.

Filling Packet. In a glass-tank, the DOG-HOUSE (q.v.).

Filter Block. The properties required of a clay filter-block for use in trickling filters in the chemical industry are specified in ASTM–C159 (cf. CERAMIC FILTER).

Filter Candle. A porous ceramic tube, which may be rounded and closed at one end, made with a high porosity and of substantially uniform pore size.

Filter Cloth. Cloths form the actual filtering medium in the FILTER PRESSES (q.v.) used to de-water clay slips. Originally, these cloths were made of cotton, subsequently of twill with a backing cloth of jute; cotton and jute cloths need to be proofed against mildew, usually by cuprammonium treatment. Nylon or terylene filter cloths have the advantage of being rot-proof and are stronger than cotton; they are now much used in the ceramic industry.

Filter Press. Equipment for the dewatering of a slurry by compression between two filter cloths, which are supported by metal frames; a number of such filtering compartments are set side by side, the compression being applied through the whole series of units. Filter presses are much used in the pottery industry, in which it is customary to blend the constituents of the body in slip form. ASTM C866 is a test for the filtration of whiteware clays.

Fin. A fault, sometimes occurring in pressed or blown glass-ware, in the form of a thin projection following the line between the parts of the mould. (Also called FLASH).

Final Set. The time required for a hydraulic cement to develop sufficient strength to resist a prescribed pressure. In the usual VICAT'S NEEDLE (q.v.) test this stage of the setting process is defined as that at which the needle point will, but its circular attachment will not, make a depression on the surface of the cement. In B.S. 12, the conditions of the test are 14.4 to 17.8°C (58–64°F) and $\nless$90% relative humidity; for normal cements, the Final Set must not exceed 10 h (the usual time is 3–4 h).

Findlings Quartzite. A compact, cemented quartzite of a type occurring in Germany as erratic blocks – hence the name, which is the German word for 'foundling'. This type of quartzite, which is used as a raw material for silica-brick manufacture, is composed of about 60% of quartz grains set in 40% of a chalcedonic matrix. In Germany the term is being displaced by the more informative term 'cemented quartzite'.

Fine Ceramics. The term is used for high-quality ceramics, particularly whitewares, but latterly for engineering ceramics, for which it is the usual term in Japanese literature.

Fine China. A term used, without being precisely defined, to describe ware as high-quality CHINA (q.v.). Fine china is not necessarily BONE CHINA, and the term may aptly be applied to ware (e.g. E.T.C.) which does not contain bone ash.

Fine Grinder or Pulverizer. A machine for the final stage of size reduction, i.e. to –100 mesh. Such machines include ball mills, tube mills, and ring-roll mills.

Fine Pottery. Products similar to EARTHENWARE (q.v.) defined similarly by the COMBINED NOMENCLATURE (q.v.) except that the body is made from a mixture of selected clays, to produce fine-grain, homogeneous porous body of water absorption > 5 wt %, coloured from yellow to brown or reddish brown.
Fineness Modulus. Defined in USA as one-hundredth of the sum of the cumulative values for the amount of material retained on the series of Tyler or US sieves (excluding half sizes) up to 100 mesh. Example:

On 4 mesh 2%
On 8 mesh 13
On 16 mesh 35
On 30 mesh 57
On 50 mesh 76
On 100 mesh 93
Sum = 276 ÷ 100 = 2.76 Fineness Modulus

Finger-car. A small four-wheeled bogie having two uprights from which project pairs (usually 10 in number) of 'fingers'; these can be raised or lowered by a lever and cam. Finger-cars are used in the KELLER SYSTEM (q.v.) of handling bricks.
Finger Dip Test. See SLAG ATTACK TESTS.
Finial. An ornamental piece of fired clayware formerly much used to finish the ends of roof ridges.
Fining. See REFINING.
Finish. The top section of a glass bottle. The finish may be DIRTY (rough or spotted) or SHARP; other faults include: BULGED FINISH – distended so that the bung will not fit; FLANGED FINISH – having a protruding fin of glass; OFFSET FINISH – unsymmetrical relative to the axis of the bottle.
Finish Mould. See NECK MOULD.
Finite Element Analysis. A method for the numerical approximation of a continuous variable. The system is considered as an array of discrete small elements, interconnected at separate node points. The technique is used for numerical calculations of heat flow, stresses and electrical behaviour.
Fin Wall. A piered cavity wall in which the piers are narrow and deep 'fins', instead of being wide shallow structures which merely increase the wall thickness locally. The fins increase the structural strength of the wall by converting it into a series of 'T' sections. (cf. DIAPHRAGM WALL).
Fireback. The refractory shape, sometimes made as two or more sections to improve thermal-shock resistance, that forms the back of an open domestic fire. A typical batch composition is 60–70% fireclay and 30–40% grog. B.S. 1251 specifies four sizes 350, 400, 450 and 500 mm; and a minimum P.C.E. of Cone 26.
Firebox. One of the small refractory-lined chambers, built wholly or partly in the wall of a kiln, for combustion of the fuel.
Firebrick. Popular term for a FIRECLAY REFRACTORY (q.v.).
Firebridge. See BRIDGE.
Fire-Chek Keys. Trade-name: pyrometric cones made by Bell Research Inc., E. Liverpool, Ohio, USA.
Fireclay; Fireclay Refractory. A clay, commonly associated with the Coal Measures, that is resistant to high temperatures; it normally consists of kaolinite together with some free silica, other impurities rarely exceeding a total of 5%. The most important deposits in the UK are in the Central Valley of Scotland and Ayrshire, Northumberland and Durham, in the Stourbridge district, S. W. Yorkshire, and the Swadlincote area. In these districts fireclays of various grades are worked for the manufacture of

refractories; in Scotland, S.W. Yorkshire, and Swadlincote the fireclays are also used for making sanitary-ware and glazed sewer-pipes; BS 1206 specifies fireclay sinks. In a few other areas fireclays are also worked for less exacting purposes. The P.C.E. of fireclay, and of refractories made from it, generally increases with the alumina content from about 1 550°C for the poorer qualities to 1750°C for grades containing 42% Al_2O_3 when fired; B.S. 3446 defines a fireclay refractory as containing less than 78% SiO_2 and less than 38% Al_2O_3, those grades containing 38–45% Al_2O_3 being termed Aluminous Fireclay Refractories. The principal uses of fireclay refractories are in blast-furnace linings, casting-pit refractories, boiler-furnace linings, and many general purposes where the conditions are not too severe.

Fireclay Plastic Refractory. A US term defined as: A fireclay material tempered with water and suitable for ramming into place to form a monolithic furnace lining that will attain satisfactory physical properties when subjected to the heat of furnace operation.

Fire Crack; Firing Crack. A fire crack is a crack in glass-ware caused by local thermal shock: a firing crack is a crack caused in clay-ware, or in a non-clay refractory, by too-rapid firing.

Fire Finished. Glass-ware that has received its final surface gloss by heating the ware, usually in a flame. See FIRE POLISH.

Fire Mark. A defect, in the form of a minute indentation, that may occur in vitreous enamelware if it is over-fired.

Fire Pillar. One of the vertical shafts, beneath each firehole, left in a setting of bricks in a top-fired kiln. (Also known as a CHARGE SHAFT, FEED SHAFT or FIRE SHAFT).

Fire Polished. The final surface gloss applied to glass by heating, in a flame or as the last stage in high temperature production. See FIRE FINISHED.

Firepower. Tradename. A small cross-section LTM tunnel kiln for tableware. (N. Barsby, Br. Ceram. Rev. (90), 1979, p14. Cf. QUICKFIRE KILN).

Fireproofing Tile. US term for a hollow fired-clay building block for use as a protection for structural members against fire.

Fire Resistance. This term has at times been used indiscriminately to denote the resistance of a material to ignition or to the spread of flame. In the relevant British Standard (B.S. 476) the meaning is restricted to the performance of complete elements of a building structure without regard to the performance of the materials of which they are composed. In USA, fire tests for building construction and materials are the subject of ASTM – El19.

Firesand. An intermediate product in the manufacture of silicon carbide.

Fire Shaft. See FIRE PILLAR.

Firestone. A siliceous rock (80–90% SiO_2) suitable for use, when trimmed, as a furnace lining material in its natural state. Firestone has been worked in England at Mexborough and Wickersley, and in Co. Durham, but is now little used.

Firestream Process. Unglazed tiles are fired in a roller hearth kiln, and glaze powder is sprayed on to the glowing hot tiles. The process was developed by Marazzi, Sassuolo, Italy. (UK Pat Appl. GB 2193205A, 1988).

Fire Travel. The movement of the zone of highest temperature around the gallery of an annular kiln. A typical rate of fire travel is one chamber per day, often a little faster.

Fire Trough. A long trough, about 20 in. broad, on which, instead of on a grate, solid fuel is burned in some annular kilns.

Fired-on. Decoration fused into the surface of glazed pottery or glass-ware.

Firing. The process of heat treatment of ceramic ware in a kiln to develop a vitreous or crystalline bond, thus giving the ware properties associated with a ceramic material.

Firing Crack. See FIRE CRACK.

Firing Expansion. The increase in size that sometimes occurs when a refractory raw material or product is fired; it is usually expressed as a linear percentage expansion from the dry to the fired state. Firing expansion can be caused by a crystalline conversion (e.g. of quartz into cristobalite, or of kyanite into mullite plus cristobalite), or by BLOATING (q.v.). Cf. AFTER EXPANSION.

Firing Range. See FIRING TEMPERATURE.

Firing Schedule. The intended relationship between temperature and time during the firing of a ceramic product.

Firing Shrinkage. The decrease in size that usually occurs when ceramic ware is fired; it is usually expressed as a linear percentage contraction from the dry to the fired state. Firing shrinkage always occurs with shaped products containing plastic clay and often amounts to 5–6% (cf. AFTER CONTRACTION).

Firing Temperature. Typical ranges of firing temperature of traditional ceramic wares are as follows:

	°C
Aluminium enamels	430–540
Wet-process cast-iron enamels	620–760
Decorating firing of pottery	700–800
Sheet-iron cover-coat enamels	730–840
Sheet-iron ground-coat enamels	760–870
Dry-process cast-iron enamels	820–920
Building bricks	900–1100
Glost firing of pottery	1050-1100
Engineering bricks	1100–1150
Earthenware biscuit	1100–1150
Salt-glazed pipes	1100–1700
China biscuit	1200–1250
Fireclay refractories	1200–1400
Hard porcelain	1300–1400
Silica refractories	1400–1450
Basic refractories	1550–1750

Firsts. Pottery Ware that has been selected as virtually free from blemishes (cf. SECONDS, LUMP).

Fisher Sub-Sieve Sizer. An air-permeability apparatus for the measurement of particle sizes between 0.2 and 50 μm. (Fisher Scientific Co., USA).

Fisheye. A bubble on the outer surface of a glaze or vitreous enamel.

Fish-scaling. A fault sometimes occurring in vitreous enamelware, small semicircles ('fish scales') of enamel becoming detached – often some days after the ware has been fired. The source of this defect is always the ground-coat and the ultimate cause is release of hydrogen followed by a build-up of pressure; the hydrogen may originate in the base metal, in the water used in the preparation and application of the enamel, or in the furnace atmosphere during the process of firing.

Fittings (UK); **Trimmers** (USA). Special sizes and shapes of wall and floor tiles (see Fig. 7, p350). For another meaning of Fittings see ODDMENTS.

Fitz Mill. A type of fine-grinding unit used, for example, in the preparation of the body for sparking plugs; trade name – W. J. Fitzpatrick Co., Chicago 7, USA.

Fixing Block; Fixing Brick. A building unit that may be made of clay, light-weight concrete or breeze, and that is sufficiently soft to permit nails to be driven in but sufficiently non-friable to hold the nails firmly in position. Clay fixing blocks are made to have a high porosity, the pores being of controlled size.

Flabby Cast. A fault sometimes encountered in the casting of pottery-ware. The article appears satisfactory as cast but subsequently deforms, either as a result of a thixotropic effect or because the interior of the cast is still fluid; if the cause is thixotropy, the amount of Na silicate and Na_2CO_3 used as deflocculant should be increased; the second cause is most common in the casting of thick ware and is prevented by increasing the casting rate.

Flake enamel. Vitreous enamel frit in the form of thin fragments.

Flaking. The detachment of thin patches of refractory from the silica lining of a gas retort. It is probable that flaking is initiated during the working period of the retort and that the removal of SCURF (q.v.) may be a contributory, but not the principal, factor.

Flambé. See ROUGE FLAMBÉ.

Flaky. See MORPHOLOGY.

Flame Gunning. The application of refractory powder through a high temperature burner to deposit a fused mass onto a worn area of a refractory lining.

Flame Plating. This term has been proposed for the application of a thin coating of refractory material to a surface by the introduction of the plating powder, oxygen and acetylene into a chamber where the explosive gas mixture is detonated, the plating powder thus being melted and projected on the inner surface of the chamber and on any object within it. The process was developed in USA by Linde Air Products Co., the original patent being in the name of R. M. Poorman (US Pat. 2 714 563. 2/8/55).

Flame-spraying. The process of coating a surface (of metal or of a refractory) by spraying it with particles of oxides, carbides, silicides or nitrides that have been made molten by passage through an oxy-acetylene or oxy-hydrogen flame; the coating material can be fed into the flame either as a powder or as a continuous rod. The object is to provide a thin protective coating usually to prevent oxidation – as in the flame-spraying of alumina on to steel. ASTM C63 specifies a test for the 'Adhesion and Cohesive Strength of Flame-Sprayed Coatings'.

Flameless Combustion. Term sometimes used for SURFACE COMBUSTION (q.v.).

Flanged Finish. See FINISH.

Flare Bed. The refractory-lined duct that conveys gas from the producers to the combustion chambers in a setting of horizontal gas retorts.

Flaring Cup. An abrasive wheel shaped like a cup but with a larger diameter at the outer (open) edge than at the bottom of the cup, the bottom being the disk that fits on the spindle.

Flash Calcination. Cold powdered clay is passed through a gas or oil flame, then quenched with cold air. Steam is generated within the clay particles faster than it can escape by diffusion, and the particle structure is disrupted. Flash calcined kaolin is of lower density than normally calcined kaolin, and does not contain mullite or cristobalite.

Flash, Flashing. (1) The formation, by surface fusion or vitrification, of a film of different texture and/or colour on clay products or on glass-ware. In the firing of clay products flashing may occur unintentionally; it is then a defect because of its uncontrolled nature. Bricks that are intentionally flashed make possible pleasing architectural effects. Flashed glass-ware is made by

fusing a thin film of a different glass (usually opaque or coloured) on the surface of the ware.
(2) In structural brickwork, a sheet of impervious material secured over a joint through which water might otherwise penetrate.
(3) A fault in glass-ware (see FIN).
(4) The fin of excess body formed during the plastic-pressing of ceramic ware, e.g. electrical porcelain; it is removed by an auxiliary process.
(5) Alternative name for CASTING SPOT (q.v.).
Flash Method. See THERMAL DIFFUSIVITY.
Flashing. See FLASH.
Flashover. A discharge of electricity across the surface of an insulator. Damage may result.
Flash Reaction Hot Pressing. Material is pressed in a mould which has been preheated to a temperature above that at which the material decomposes. The more rapid temperature rise (compared to reaction hot pressing) produces almost simultaneous decompositions in mixtures of materials with differing decomposition temperatures. The process has been used to produce rare earth silicate glasses, ferrites and other mixed ceramics. (D.B. Black *J. Canad. Ceram. Soc.* **45,** 47, 1976).
Flash Set. Premature and permanent stiffening of portland cement that has not been adequately retarded; cf. FALSE SET.
Flash Wall. A continuous wall of refractory brickwork built inside a downdraught kiln in front of the fireboxes; its purpose is to direct the hot gases towards the roof of the kiln and to prevent the flames from impinging directly on the setting.
Flask. Local term for the mould, made of wood or cast iron, used in the shaping of crucibles for steel melting.
Flat Arch. See JACK ARCH.
Flat Drawn. Sheet glass made by the vertical drawing process.
Flat glass. One of the major divisions of the glass industry – windows, automobile glass, and other forms of sheet and plate glass. The structural performance of glass sheets can be assessed destructively (ASTM E997, an air pressure test) or non-destructively (ASTM E998, observation of deflections and strains under a standard pressure).
Flatter. See STONER.
Flatting. A process for trueing-up hand-made fireclay refractories while they are still only partially dried. Handmaking is now little used except for some special shapes.
Flat-ware. Plates, saucers, dishes, etc. (cf. HOLLOW-WARE).
Flaunching. The weathering, formed in mortar, at the top of a chimney or at the base of a chimney pot.
Flemish Bond. A brick wall built with alternate headers and stretchers in each course. Single Flemish Bond is so bonded on one face of the wall only; Double Flemish Bond is bonded in this way on both faces of the wall.
Fletton. An English building brick originally made at Fletton, now a suburb of Peterborough; centres of Fletton brickmaking also include Bedford and Bletchley. The bricks are made by the semi-dry process from Oxford Clay; this clay is shaly and contains much organic matter, which assists in the firing process. The crushing strength varies from about 14 to 35 MNm^{-2} and the water absorption from about 17 to 25 wt%.
Flexmesh. A type of HEXMESH, constructed to be flexible on one axis, rigid on the other.
Flexmould. Tradename. Flexible polyvinyl chloride or room temperature vulcanizing silicone rubber is used for

the block mould. The flexibility of the material allows sharp undercuts to be designed into the mould. (Flexmould Inc. USA).

Flexural Strength. See MODULUS OF RUPTURE.

Flint. Nodular chalcedonic silica from the chalk deposits of W Europe and elsewhere. These flint pebbles are calcined and ground for use in earthenware and tile bodies. In USA the term 'flint' is often applied to other finely ground high-silica rocks used in whiteware manufacture. See also CALCINED FLINT.

Flint Clay. A non-plastic refractory clay of a type found in Missouri (USA), Southern France, and a few other localities. Flint clays are kaolinitic, but laterization (See LATERITE) has occurred so that the Missouri deposit, for example, may also contain diaspore whereas the French may contain boehmite. Flint clays contain (fired) 40–45% Al_2O_3; P.C.E. 34–35.

Flint Glass. 'White Flint' is a term applied to many colourless glasses; 'Optical Flint' is any glass of high dispersion for optical equipment (cf. OPTICAL CROWN). The introduction of lead oxide into the batch to produce the original flint glass, in the 17th century was an English contribution to glass-making; silica was introduced into the batch in the form of crushed flints.

Flintlime Brick. CALCIUM SILICATE BRICKS (q.v.) in which part or all of the aggregate is crushed flint.

Flint Shot. US term for the sharp sand sometimes used in sandblasting the base metal as part of the preparation process for vitreous enamelling.

Flintless Stoneware. Defined in the UK Pottery (Health and Welfare) Special Regulations of 1950 as: Stoneware the body of which consists of natural clay to which no flint or quartz or other form of free silica has been added.

Flip-flop. A glass decanter with a thin base.

Float Glass Process. A process for making sheet glass introduced in 1959 by Pilkington Bros. Ltd, at St. Helen's, England. A ribbon of glass is floated on molten tin, the product being sheet glass with truly parallel surfaces, both fire-polished. The process is limited to soda-lime-silica glasses but can produce 50000m^2 (500–800 ton) of glass per day. It was improved in 1975 by the PPG delivery system. The stream of molten glass is of the required production width (c. 4m) from the melting furnace. This reduces optical distortion in the finished glass arising from flow problems when the glass stream passed from melting furnace to the molten tin through a restricted opening.

Floater. A refractory shape that is allowed to float on the surface of molten glass in a tank furnace in order to hold back any scum that may be present (cf. RING).

Floating. (1) Slight displacement of a pattern applied to potteryware; it is caused by the presence of grease or moisture on the glaze at the time when the decoration was applied.
(2) The smoothing of freshly-placed concrete to give a good surface finish.

Floating Agent. In the vitreous-enamel industry, the US equivalent of the English term SUSPENDING AGENT (q.v.).

Floating-die Press. The punches, die set and core rods can be controlled to move during the press cycle. Such multiple synchronised movements can produce complex shapes, and minimize density gradients in the pressed parts. Both are of value in producing technical ceramics.

Flo-con. Tradename. A SLIDING-GATE VALVE originally developed by United

States Steel Corp. (U.S. Pat 3685707, 1972) (Now Flo-Con Corp. USA; Flogates, Sheffield).

Floc Test or Water Test. A test for the durability of hydraulic cement: 1g of the cement is shaken with 100 ml of water in a test-tube which is then placed on its side and allowed to stand for 7 days; if the amount of floc formed is very small the cement is considered to be durable. The test was proposed by I. Paul *(Proc. Assoc. Highway Officials, N. Atlantic States,* **12**, 140, 1936).

Flocculation. The addition of a suitable electrolyte to a colloidal suspension to agglomerate the particles and so hasten their settling; the sediment thus formed is usually more readily removed from the containing vessel than is the sediment from a deflocculated suspension. An example is the addition of up to 0.05% $CaCl_2$ to flocculate slop glaze.

Floe Rock. US term for quartzite that occurs as loose boulders, e.g. TUSCARORA QUARTZITE (q.v.).

Flood Casting. Term used in the British sanitary-ware industry for the process of slip casting in which excess slip is removed from the mould by draining. In other sections of the pottery industry the process is referred to merely as 'casting'; in USA the process is known as 'Drain Casting'.

Floor Brick. A brick having mechanical, thermal, and chemical resistance to the conditions to which it is likely to be exposed when used in an industrial floor. Properties are specified in ASTM-C410: 'Industrial Floor Brick'.

Floor Quarry. (UK) **or Quarry Tile** (USA). A heavy ceramic flooring material. Floor quarries (as distinct from FLOOR TILES q.v.) are usually made by a plastic process. They are relatively thick and are hard fired to produce a body resistant to heavy abrasion and to attack by most industrial liquids – hence their wide use for factory floors. For sizes, tolerances and trueness of shape, see B.S. 1286.

Floor Tile. (1) Ceramic tiles, normally unglazed, for flooring. It is difficult to draw a sharp distinction between floor tiles and FLOOR QUARRIES (q.v.) but the former are always dry-pressed, and they are relatively thin and do not normally exceed 152×152 mm in size. For standard sizes see B.S. 6431. Concrete floor tiles and fittings are specified in BS 1197.
(2) One of the refractory shapes used in the construction of a gas retort; a group of these tiles is laid horizontally to brace the retorts of a vertical setting and to limit the combustion flues.
(3) In the USA the term is used for a hollow fired-clay block for use in the construction of floors and roofs; two grades, FT1 and FT2, are specified in ASTM – C57.

Floridin. An older name for ATTAPULGITE (q.v.); now a tradename of the Floridin Co., New York, USA.

Flotation. A process for the concentration of a mineral by crushing the rock in which it is dispersed, suspending the crushed material in water and causing a froth to form one of the constituents of the mineral adhering to the air bubbles so that it can be floated off, leaving the remainder to settle out. Three types of reagent are required: frothers, collectors, and controllers. Flotation has been applied industrially to the preparation of the following ceramic raw materials; feldspar, talc, magnesite, kyanite, glass-sands, lithium ores.

Flow Blue. A deep cobalt blue which was used for under-glaze printing on pottery. As the name indicates, the colour tended to flow into the glaze, giving a blurred effect; this result was obtained by placing FLOW POWDER (q.v.) in the saggar containing the ware. Chlorine evolved from the powder and combined with some of the cobalt, thus rendering it slightly soluble in the glaze.

Flow-button Test. See FUSION-FLOW TEST.

Flow Coating. The application of vitreous enamel slip by allowing it to flow over the metal and then to drain.

Flow Cup. An orifice-type VISCOMETER, of which there are several patterns: FORD, REDWOOD, SAYBOLT, ZAHN. Their method of use is specified in BS EN 535, which superseded BS 3900 and the earlier BS 1753. Not all the early types are now specified.

Flow Line. Streaks of different reflectance from the surrounding area, arising in glass and in injection moulded parts, from the flow pattern of material during forming the part; see CHILL MARK; WELD MARK.

Flow Machine. A machine used in glass-making; molten glass flows into it from a feeder under the action of gravity.

Flow Powder. A mixture formulated to evolve chlorine at the temperature of the glost firing of pottery and used in the production of FLOW BLUE (q.v.). For ware covered with a lead glaze, a suitable composition is (per cent): NaCl, 22, white lead, 40; $CaCO_3$, 30; borax, 8. For use with a leadless glaze a suitable mixture is (per cent): NaCl, 15; $MgCl_2$, 55; KNO_3, 15; $CaCO_3$, 15.

Flow Process. A method of producing flat glass by allowing molten glass to flow continuously from a tank furnace between rollers. The term is sometimes also applied to a process for making hollow-ware, but this is more correctly referred to as the GOB PROCESS; (q.v.).

Flow Table Test. A test for the consistency of concrete in terms of its tendency to spread when placed on a metal table and jolted under specified conditions. (ASTM C124). The test is also applicable to refractory concretes (ASTM C860) and to tests on hydraulic cement for mortars (ASTM C230). BS 1902 Pt 11 describes its use for refractory mortars, BS 4551 its use for testing bricklaying mortars, wherein the flow table is described.

Flowers. See MOTTLING (of silica refractories).

Flue Liner. A fireclay shape for use in the flues and chimneys of domestic heating appliances; BS 6461 is a code of practice for the design and installation of domestic masonry chimneys and flues. See also CHIMNEY LININGS.
ASTM C315 is the US standard, and ASTM E835 gives metric sizes related to a 100-mm module.

Fluid-energy Mill. A size-reduction unit depending for its action on collisions between the particles being ground, the energy being supplied by a compressed fluid (e.g. air or steam) that enters the grinding chamber at high speed. Such mills will give a product of 5μm or less; they have been used for the fine grinding of frits, kaolin, zircon, titania, and calcined alumina, but the energy consumed per ton of milled product is high.

Fluidity of liquids, see VISCOSITY; of powders, see MOHR-COULOMB FAILURE LAW.

Fluidized Bed. A bed of solid particles maintained in suspension in a gas, and thus behaving rather like a fluid; the powder to be fluidized is supported on a porous base, e.g. a trough of special porous ceramic material, through which the gas (usually air) is fed from below and under pressure. The principle is used as a method of conveying powders along a slightly inclined porous ceramic trough; the powder can be simultaneously dried and/or calcined.

Flume. Local term for the alkali vapour volatilized from the glass in a glass-tank furnace; it causes corrosion of the furnace roof, downtakes, and regenerator refractories. (A corruption of 'Fume'.)
Fluorapatite. The fluorine-containing modification of APATITE (q.v.)
Fluorspar. CaF_2; m.p. 1420°C; sp. gr. 3.1. The largest deposits of this mineral are in Mexico, China, USA and USSR, but it is also worked in England and most other European countries. It is used as an opacifier in glass and vitreous enamel. Fluorspar crucibles have been used in the melting of uranium for nuclear engineering.
Flux. A substance that, even in small quantities, lowers the fusion point of material in which it is naturally present (e.g. alkalis in clays) or of material to which it has been added (e.g. borax added to glazes). The term is also used for the prepared low-melting glasses that are added to ceramic colours to fuse them to the ware on which they are being used as decoration.
Flux Factor. A factor for assessing the quality of steelworks grade silica refractories. It is defined in ASTM-C416 as the percentage of Al_2O_3 in the brick plus twice the total percentage of alkalis; for first-quality (Type A) bricks the flux factor must not exceed 0.50.
Flux Line. The level of the molten glass in a tank furnace; this level is generally marked by a horizontal line of maximum attack on the refractories. The term is also sometimes applied to the boundary between unmelted batch and molten glass in a tank furnace.
Flux-line Block. A refractory block for use in the upper course of the walls of a glass-tank furnace. The flux line is the surface level of the molten glass and attack on the refractories is more severe at this level than beneath the molten glass.
Fly-ash. The ash carried by the waste gases in a furnace or kiln (cf. PULVERIZED FUEL ASH).
Flying Arch. In a modern glass-tank furnace the double-walled bridge built across the furnace to separate the working end from the melting and refining end; the flying arch is independent of the general furnace structure.
Flying Bond. See MONK BOND.
Foam Glass. Cellular glass, in the form of blocks, usually made by mixing powdered glass with a gasifying agent (e.g. carbon or a carbon compound) the mixture then being heated for a short time to fuse the glass and trap the evolving gas bubbles. Foam glass is used as a structural heat-insulating material (cf. BUBBLE GLASS).
Foamed Ceramic. Highly porous ceramics made by generating a foam of bubbles in a ceramic slip, using a chemical agent, then drying and firing. Foams differ from honeycomb in having a random varied pore-size distribution, rather than regular uniform holes.
Foamed Clay. Lightweight cellular clayware for heat and sound insulation. Foam is generated in a clay slip, either mechanically or by a chemical reaction that evolves gas bubbles, and the slip is then caused to set. Some insulating refractories are made by this process.
Foamed Concrete. See AERATED CONCRETE.
Fold. See LAP.
Folded Foot. The foot of a wine glass is said to be folded if the outer edge has been raised by folding it back on itself.
Font. (1) An alternative term for a VOID. (2). A reservoir above the mould for fusion-casting refractories; molten material from the font helps to fill the PIPE (q.v.).

Foot-boards. Wooden boards, hinged together, for hand-shaping the foot of glass stem-ware.
Ford Cup. An orifice-type viscometer. It has been used to a limited extent in the testing of the flow properties of ceramic suspensions. The aperture sizes, length of orifice and method of use were given in B.S. 1733 and ASTM-D333. This viscometer was introduced by the Ford Motor Co., England, in 1925.
Forehearth. (1) An extension to the bottom of a CUPOLA (q.v.) serving as a reservoir for molten iron.
(2) An extension of a glass-tank furnace from which glass is taken for forming.
Fork. A metal device for placing vitreous enamelware in, and subsequently removing it from, a box furnace.
Forking. The separation of a propagating fracture into two or more new fractures at acute angles. See CRACK BRANCHING.
Forshammer Feldspar. A Swedish feldspar used as a flux in tableware bodies, to replace DF Stone, compared to which it was higher in alkalis and lower in silica.
Forsterite. Magnesium orthosilicate, $2MgO. SiO_2$: sp. gr. 3.2; m.p. 1890°C; thermal expansion (0–1200°C) 12.2×10^{-6} In economic amounts forsterite occurs naturally only in association with FAYALITE (q.v.) in the mineral OLIVINE (q.v.). Forsterite refractories are made from olivine or SERPENTINE (q.v), dead-burned magnesite being added to the batch to combine with the fayalite and produce the more refractory mineral MAGNESIOFERRITE (q.v.). Pure synthetic forsterite is used as a low-loss dielectric ceramic; it is particularly suitable (on account of its relatively high thermal expansion) for making vacuum-tight seals with metallic Ti.
Fotalite. Tradename of Corning Glass for a PHOTOSENSITIVE GLASS (q.v.)
Fotoform Process. Tradename. An image is nucleated and crystallized within a lithium silicate glass, doped with an appropriate sensitizer. The pattern can then be preferentially etched away to produce holes, patterns, and internal channels, with an accuracy of a few microns.
Founder. See TEASER.
Founding (of glass). See REFINING.
Fourcault Process. A method of making sheet glass invented in 1902 by E. Fourcault of Lodelinsard (Belgium). The glass is drawn vertically and continuously from the glass-tank through a DEBITEUSE (q.v.).
Four-colour Printing. See TRICHROMATIC PRINTING.
Fourier Transform Specroscopy. Two beams of radiation are used to produce an interference pattern, and the spectrum is produced by taking its Fourier transform – a mathematical technique for building up a complex pattern as the sum of a set of waves of different harmonic wavelengths.
Fractal. 'A shape made of parts similar to the whole in some way' (B.B. Mandelbrot '*The Fractal Geometry of Nature*' W.H. Freeman, San Francisco, 1982).
More mathematically strictly, a fractal is a set such that two standard definitions of a dimension give different answers. For example, a 1-dimensional curve such as a coastline may have one length according to one definition, but may have arbitrarily small irregularities giving rise to a different, larger length by another definition. This is a *fractal curve.* The *fractal dimension* D_F is defined by $D_F = [(\log L_0 - \log L)/\log S] + D$ where L_0 is a constant, L is a measured length, S is the scale of measurement and D is the number of dimensions (1 for a line, 2 for a plane). Fractal curves of surfaces

are self-similar, in the sense that a magnification of their small-scale features looks like their larger scale features. Fractal geometry can be used to study particle morphology, or aid in fracture mechanics through the study of fracture surfaces, which have been shown experimentally to be fractal surfaces. Their fractal dimensions are related to the FRACTURE TOUGHNESS (q.v.) of the materials. For example, the area of the FRACTURE SURFACE (q.v.), which is one determinant of the energy required to produce FRACTURE (q.v.) is scale-dependent. If L and L_0 are the irregular and straight path lengths of the crack respectively, the real area of the fracture surface is L/L_0 times the area used if fractal effects are omitted. By the equation defining fractal dimension D_F, $L/L_0 = S^{1-D_F}$ and values for (e.g.) critical crack extension forces and energies required to produce fracture should be correspondingly increased by amounts deducible from the fractal geometry.

Fractional Porosity. See POROSITY.

Fractography. The study of the features of the fracture surfaces of brittle materials, to elucidate the causes of the cracks. ASTM C 162–92 and ASTM STP 827 define many terms.

Fracture. Cracking. The physical separation of parts of a solid subjected to mechanical stress. When a crack develops (see CRACK NUCLEATION) two new surfaces are formed. The energy to break the interatomic bonds to do this is supplied by the external stress. It is equal to twice the product of the surface energy of the solid and the area of the new fracture surface. Mechanisms which absorb this energy restrain cracking (see TOUGHNESS). CRACK PROPAGATION (q.v.) is controlled by conditions at the crack tip: macroscopically, by the stress distribution which is concentrated at the crack tip; also by the microstructure and, on the atomic scale, by the CRYSTAL STRUCTURE (q.v.) See also FRACTOGRAPHY; SLOW CRACK GROWTH; FRACTALS

Fracture Mechanics. The study of the ways in which crack geometry, material strength and TOUGHNESS (q.v.) and stress patterns affect the FRACTURE (q.v.) of materials. The objective is to determine the critical size of crack or defect which will cause the crack to propagate rapidly and cause catastrophic failure. See also FRACTALS; WEIBULL DISTRIBUTION; BATDORF THEORY; GRIFFITH OROWAN CRITERIA; SHETTY'S CRITERION; R-CURVE.

Fracture Mirror. A very smooth, reflective fracture surface indicating that the fracture was moving relatively slowly along a single spreading front.

Fracture Modes. Mode I – the *opening mode* – the stress is perpendicular to the crack plane.

Mode II – the *sliding mode* – shear stress between the crack faces in the direction of crack propagation.

Mode III – the *tearing mode* – shear stress between the crack faces acting in a direction parallel to the crack front.

See also SHETTY'S MIXED MODE EQUATION.

Fracture Surface. The newly-formed surface within a solid formed by cracking. See also FRACTOGRAPHY.

Fracture Toughness. A measure of the ease of crack propagation through a material. The units are $MPa.m^{0.5}$. Values for ceramics range from below 1 for glasses and most single crystals ; 1 to 3 for glass-ceramics, clay-based ceramics and MgO; 2.5. to 4 for early engineering ceramics (alumina, boron carbide, silicon carbides, RBSN); 4 to 6 for hot-pressed silicon nitride and SRBSN and for transformation-toughened alumina composites; 6 to 12 for sialons, PSZ, TZP and low binder content hard metals; over 15 for Ceria-stabilized TZP, fibre-reinforced engineering ceramics and high binder content hardmetals. See FRACTURE TOUGHNESS.

Fracture Toughness Tests. There are numerous techniques but no standard specimen. The choice is set by the type of information required; the nature of the material; the test geometry compared to the component geometry; the fracture energy around large or small flaws; the test temperature; service environment. See FRACTURE; STRESS CORROSION; J-BASED FRACTURE TESTING.

Most tests involve the preparation of test-pieces of more or less complex geometry, containing cuts or notches of prescribed size and shape. Some require the careful introduction of a sharp starter crack tip in the appropriate location – a matter of some delicacy. Cyclic compression can produce a controlled fatigue flow through the specimen thickness. The specimen can then be tested in flexure by the single-edge notched beam technique. Direct measurements on the FRACTURE SURFACE (q.v.) and other techniques of FRACTOGRAPHY (q.v.) can yield information on fracture toughness. Pertinent features of specifically designed tests are as follows.

The ***Compact Tension*** (CT) test applies a separating force to the blind end of a sawn notch by tensile forces applied to bolts in holes either side of the notch. A short, stable crack is first established with minimum possible force, then the increased crack lengths due to repeated applications of the force are measured. The test is simple, and applicable to non-porous materials at ambient temperature for slow rates of loading.

The ***Constant Moment*** (CM) test open up a V-shaped notch in the specimen by forces parallel to the long axis of the V on moment arms to either side. The technique is applicable to porous materials at ambient temperature with slow loading rates, and can measure critical fracture toughness as a function of crack growth rate.

In the ***Double Torsion*** (DT) test upward forces are applied on either side of a prepared notch, with a pair of downward forces at a greater distance on either side. The test is a common one, with K_I independent of the crack length. The loading arrangements are simple but a large volume of material is required, and the crack front is curved, with non-uniform stress intensity. The test is useful for slow crack growth studies on both porous and non-porous materials. It can be used at high temperature in severe environments and will give K_{IC} as a function of crack growth rate.

The ***indentation fracture*** (IF) test is simple and requires only small volumes of material. Cracks develop around a Vickers indentation in a flat ceramic surface in proportion to its fracture toughness. K_{IC} can be estimated from crack lengths. The ratio of force to crack length is independent of load and material hardness, and so is a useful parameter. The test is suitable for dense materials (97% theoretical) at ambient temperature.

The ***indentation strength*** (IS) method induces Vickers indentations into flexure strength bars with smooth surfaces. The critical stress intensity at fracture is measured, and K_{IC} calculated. It is not necessary to measure crack lengths, but results are dependent on the indenter load.

In the ***Chevron Notched Bar*** (CNB) test, a chevron or M-shaped notch is sawn in

a short bar or rod and forces applied either side as in the CM test. The test is applicable at high temperature, requires no knowledge of material properties, but calculates K_{IC} from the load and specimen geometry. Specimen preparation is complex, requireing careful machining of the sawn 'M' notch.
In the ***Double Cantilever Beam*** (DCB) test, tensile forces are applied at right angles to a sawn notch cut parallel to the long axis of a bar. K_{IC} is derived from the load and the geometry. K_I is independent of crack length for constant moment loading. The test procedure and the sample preparation are simple, but the specimen must be very carefully pre-cracked to produce a sharp crack tip from the blunt sawn notch. The test is applicable to non-porous materials at ambient temperature, and can give K_{IC} as a function of crack growth rate. Slow loading rates are used.
The ***Tapered Cantilever Beam*** (TCB) test is a variant of the DCB requiring a more complex, wedge-shaped specimen with notches running the length of both tapered faces of the wedge. The test is applicable to porous materials at slow loading rates at ambient temperature, and will give K_{IC} as a function of crack growth rate.
The ***Single Edge Notched Beam*** (SENB) test can be used for rapid rates of loading. It is simple. A narrow notch cut across the width of a long specimen is easier to make and measure than a sharp crack. The test is performed in 4-point bend, with the inner pair of forces applied to the opposite face to the notched one. It is applicable to non-porous materials at both ambient or high temperature. The results are sensitive to notch size, and they are improved by precracking the specimen at the end of the notch.
The ***Single Edge Precracked Beam*** (SEPB) test is thus a refinement of the SENB, to introduce a precrack. The specimen is held in a loading fixture which generates a Vickers indentation or a single notch at the centre of the tensile surface of the specimen, which is then loaded in 4-point bend to failure, as in the SENB. The specimen must be made with accurately parallel faces and squared corners.
ASTM C1212 describes how to make reference specimens from SiC and Si_3N_4, containing known discontinuities or seeded voids, by making surface depressions with small spheres in pre-pressed compacts and then isostaticaly pressing them.

Framework Silicates. See SILICATE STRUCTURES.

Franchet Lustre. A LUSTRE (q.v.) for the decoration of pottery. Metal salts are incorporated with the glaze and the ware is then fired reducing at 550–600°C. (L. Franchet, La Fabrication Industrielle des Emaux et Couleurs Ceramiques, Paris, 1911.)

FRCI. Fibrous Refractory Concrete Insulator. A lightweight refractory concrete incorporating ceramic fibres to act as a thermal insulator.

Free-blown. Alternative term for OFF-HAND (q.v)

Free Silica. The amount of silica present in a ceramic body or raw materials mix, as SiO_2 uncombined with any other element. Free silica may present a respirable dust hazard, and may give rise to thermal expansion changes in a body at the temperatures at which silica is subject to phase changes.

Free Crushing. The method of running a pair of crushing rolls so that virtually all the crushing occurs as the particles pass between the rolls, only a small proportion of the particles being crushed by bearing on one another. When crushing rolls are operated in this way, the product contains less fines than with CHOKE CRUSHING (q.v.).

Freehand Grinding or Off-hand Grinding. The method of grinding in which the object to be ground is held by hand against an abrasive wheel.

Freeze-casting. A process for making intricate shapes of special ceramic material, e.g. turbo-supercharger blades. Refractory powder, with a small proportion of binder, is made into a thick slip, which is cast into a mould and then frozen; the cast is then dried and sintered.

Freeze-Drying. A method for preparing fine powders. A liquid solution of a soluble precursor salt is sprayed into a cryogenic medium. The resulting fine dispersion can be calcined or reacted to produce the required product, after heating to drive off the frozen solvent.

French Chalk. See STEATITE.

Frenkel Defect. See CRYSTAL STRUCTURE

Frenkel Mixer. A screw-type, 'enforced order', mixer depending on interlayer displacement of the material between machined screw surfaces; it operates on the convergence divergence (C-D) principle. (M. S. Frenkel, Brit. Pat. 888 864, 7/2/62).

Fresnel Formulae. A beam of light striking a glass surface is partly reflected. If *i* and *r* are the angles of incidence and refraction respectively, for an incident beam of unit amplitude polarized in the place of incidence, the reflected wave amplitude B is given by $-B = \tan(i-r)/\sin(i+r)$ for an incident beam polarized at right angles to the place of incidence, the reflected wave amplitude b is given by $-b = \sin(i-r)/\sin(i+r)$. (and if A and a are the respective amplitudes of the refracted waves, $A^2 + B^2 = 1$ and $a^2 + b^2 = 1$ by the Law of Conservation of Energy).

Freudenbergite. $Na_2Al_2Ti_6O_{16}$. A SYNROC (q.v.) phase, for immobilizing sodium.

Frey Automatic Cutter. Trade name: a machine for cutting an extruding column of clay into bricks by one or more horizontal wires that cut downwards while the clay is moving forward. (G. Willy A.G., Chur, Switzerland.)

Friability. The tendancy for particles of a powder to break down in size under light forces during handling and storage.

Friction Element. Some clutches and brakes for use in severe conditions are now lined with CERMETS (q.v.) Amongst the materials used are corundum and sillimanite as the ceramic component, and Mo, Cr, Fe, and Cu alloys as the metallic bond.

Friction Press or Friction-screw Press. A machine for dry-pressing: a plunger is forced into the mould by a vertical screw, the screw shaft being driven by friction disks or rollers – downwards for pressing by one disk and upwards for release of pressure by a second disk on the opposite side of the driving wheel. This type of press is used for special shapes of tiles and sometimes for making silica refractories, etc.

Friction Welding. A method of joining ceramics in which the components are heated by rubbing together at high speed.

Frictive Track. A series of crescent-shaped cracks, lying along a common axis parallel to the direction of rubbing contact.

Frigger. A small whimsically shaped piece of hand-blown glass ware – a glass-blower's bagatelle.

Fringe. A thin layer of material on one face of a brick, projecting beyond the edge.

Frischer Ring. A type of TOWER PACKING.

Frit. A ceramic composition that has been fused, quenched to form a glass, and granulated. Frits form an important part of the batches used in compounding enamels and glazes; the purpose of this pre-fusion is to render any soluble and/or toxic components insoluble by causing them to combine with silica and other added oxides.

Frisket. A stencil (usually paper) which defines the area of pottery to be glazed, or decorated. (USA).

Fritted Glass. See SINTERED GLASS.

Fritted Glaze. A glaze in which some of the constituents have been previously fused together to form a frit, the constituents so pretreated include lead oxide, which fritting converts into an insoluble silicate, and other constituents that would otherwise be soluble.

Fritted Porcelain. Alternative name for SOFT-PASTE (q.v.).

Fritting Zone. See SOAKING AREA.

Frizzling. A fault liable to develop during the firing of potteryware that has been decorated with lithographic transfers – if the varnishes are burned away too rapidly in the early stages of the enamel fire, the colour is liable to crack and curl up. To prevent this fault, the layer of size should be thin and the rate of firing between 200° and 400°C should not greatly exceed 1°C/min.

Frog. A depression in the bedding face of some pressed building bricks to decrease the weight and improve the keying-in of the mortar. The term is probably derived from the same word as applied to the similar depression in the centre of a horse's hoof. BS 3921 specifies that the volume of the frog(s) shall not exceed 20% of the volume of the brick.

Frost Glass. Glass whose surface has been treated chemically, sandblasted lightly, or decorated with very thin crushed glass like tinsel, (*frost*) to scatter light and produce decorative and obscurative effects.

Frost's Cement. An early form of hydraulic cement patented in England in 1811 by James Frost; it was made from two parts chalk to one part clay.

Frosting. See ACID FROSTING.

Frost Resistance. Porous building materials may be damaged by the pressures generated by the freezing of moisture which has penetrated the pores. Frequent freeze-thaw cycles are more likely to cause damage than prolonged low temperatures. Waterproofing treatments which seal most of the pores but leave some route for the ingress of moisture can worsen the damage. Frost resistance tests are based on (i) indirect appraisal from properties such as strength and/or porosity; (ii) long-term exposure tests or experience; (iii) accelerated simulation. BS 3921, 1974 was based on (i) and (ii); ASTM C62–75 on (i) and (iii). Accelerated tests may be omni-directional, in which all faces of the bricks or tiles are exposed to the freeze-thaw conditions. Only a simple freezing cabinet is then required. In uni-direction tests, only one face is exposed to the low temperature, a more realistic simulation of service conditions. The BCRA Panel Freezing Test subjects a panel of ten courses of three bricks soaked for 7 days, to a 100-cycle freeze-thaw regime, each cycle comprising 132 min freezing at –15°C, followed by 20 mins thawing by radiant heaters at 25°C, with 5 min water-spray and drain treatment to replace evaporated water. BS 3921 classifies bricks into 3 categories of frost resistance. BS 6431 specifies a cyclic freeze-thaw test for tiles.

Froth Flotation. See FLOTATION.

FSCS. **F**ibre **S**andwich **C**onstruction **S**ystems. In the construction of a LOW THERMAL MASS KILN (q.v.) ceramic fibre is sandwiched between a thin refractory hot-face tile and the kiln casing. The tiles are held in place by a pin which rotates 90° to lock into a stainless steel socket welded to the casing. (Shelley Furnaces, UK)

FSZ. Fully stabilized ZIRCONIA.

FTIR. Fourier Transform Infra-Red (spectroscopy)

Fuch's Gold Purple. A tin-gold colour, produced by a wet method; it has been used in the decoration of porcelain.

Fugitive Wax Slip-casting. A process for shaping complex ceramic components. A positive model of the component is machined from a water insoluble/organic soluble wax. The positive model is repeatedly dipped into liquid wax, and then dissolved out. The wax mould is placed on a plaster mould which draws the water from the slip, forming a unidirectional casting. The wax mould is dissolved in an appropriate organic solvent when the ceramic cast has solidified. (Ford Motor Co, U.S. Pat 4067943 1978).

Full Lead Crystal Glass. A CRYSTAL GLASS (q.v.) specified in B.S. 3828 as containing >30% PbO and >12% alkali; cf. LEAD CRYSTAL GLASS.

Fullerenes, Fullerides. Molecular structures of atoms arranged on the surface of a sphere or a cylinder, in hexagonal and pentagonal arrays, in configurations similar to R. Buckminster Fuller's geodesic domes. Nanometer-sized tubes are being developed based on cylindrical Fullerenes. 28 to 540 carbon atoms in such a spherical arrangement form the most symmetric possible molecules – *Buckminster Fullerene* or *'Buckyballs'*.

Fuller's Earth. A type of clay formerly used in the cleansing of cloth. The clay minerals present are of the montmorillonite and/or beidellite types. It is used for decolorizing oils and has also been used as a bond for foundry sands.

Fuller's Grading Curve. A method of graphical representation of particle-size analysis: the grain size (in fractions of an inch) is shown on the abscissa and the cumulative percentage on the ordinate. In the original paper (W. B. Fuller and S. E. Thompson, *Trans. Amer. Soc. Civil Engrs,* **59**, 67, 1907) the concept of 'ideal' grading curves was introduced, these being selected to be ellipses with straight lines tangential to them; more strictly, the 'ideal' curves are parabolas having the form $d = P^2D/10\,000$, where d is any selected particle diameter, D is the diameter of the largest particles and P is the percentage finer than d.

Fumed Silica. A pure colloidal form of silica produced by reacting silicon tetrachloride in a flame. Particle size is some 12nm, and BET surface area 230 m^2/g.(Tradenames include Cab-O-Sil M5 and Degussa OX-50).

Functionally Gradient Materials. Composites with graded properties. See GRADED CERMETS.

Furring Brick. An American term for a hollow brick with surface grooves to retain plaster.

Furring Tile. US term for a hollow fired-clay building block for lining the inside of walls and carrying no superimposed load.

Fused Grain Refractory. Refractories made from raw material which has been melted, destroying its natural crystal structure; solidified; crushed to an appropriate particle size and sintered.

Fused Quartz. This term is often applied to *Transparent Vitreous Silica* (See under VITREOUS SILICA) that has been obtained by the fusion of quartz crystal electrically or in a flame.

Fused Silica. In the UK this term has, in the past, often been applied to *Translucent Vitreous Silica;* this has caused confusion because the same term has frequently been used in the literature (especially in USA) when referring to the transparent variety. For preferred terms see VITREOUS SILICA.

Fusion-cast Basalt. An abrasion-resistant material made by fusing natural basalt and casting the molten material into moulds to form blocks. The hardness is 8–9 Mohs; crushing strength, 500 MNm^{-2}. These blocks can be used for industrial flooring and for the lining of bunkers, chutes, and other equipment where abrasion is severe.

Fusion Casting. A process for the manufacture of refractory blocks and shapes of low porosity and a high degree of crystallinity; the refractory batch is electrically fused and, while molten is cast into a mould and carefully cooled. The usual types of fusion-cast refractory are those consisting of mullite, corundum, and zirconia in various proportions; such refractories find considerable use as tank blocks for glass-melting furnaces. Trials have been made with fusion-cast refractories in steel furnaces.

Fusion-flow Test. A method for the evaluation of the fusion flow properties of a vitreous enamel or of a glaze. As used for testing enamel frits, the method is standardized in ASTM – C374.

Fuzzy Texture. A fault sometimes occurring in vitreous enamelware, the 'fuzziness' resulting from minute craters and bubbles in and near the enamel surface. Also known as GASSY SURFACE.

G Stone. A name that has been used for PYROPHYLLITE (q.v.).

G-Value. The basis of a method of calculation for compounding slips and glazes, the 'G-value' being the grams of suspended solids per cm^3 of suspension: G = SP/100, where S is the sp. gr. of the suspension and P is the percentage of solids in the suspension. (K. M. Kautz, *J. Amer. Ceram. Soc.*, **15**, 644, 1932).

Gabbrielli Penetrometer. The pressure required to penetrate an unfired body with a flat-headed needle parallel to its surface is measured. This provides a rapid estimate of porosity as a control test on the effectiveness and uniformity of pressing dust-pressed tiles. (Gabbrielli s.r.l., Italy).

Gable Tile. A roofing tile that is half as wide again as the standard tile. Gable tiles are used to complete alternate courses at the VERGE (q.v.) of a tiled roof.

Gable Wall. The refractory wall above the tank-blocks at the end of a glass-making furnace; it is also sometimes referred to as the END WALL or BACK WALL.

Gadget. A tool, used in the hand-made glass industry, for holding the foot of a wine glass while the bowl is being finished.

Gadolinium Oxide. Gd_2O_3; m.p. approx. 2350°C; sp. gr. 7.41; thermal expansion (20–1000°C), 10.5×10^{-6}. A white crystalline solid, with properties similar to MgO and CaO. It is a neutron absorber, and is used in nuclear applications 'diluted' with alumina or magnesia, as a 'burnable poison'. Such materials control the flow of neutrons in a nuclear reactor by absorbing them, being converted to a higher isotope of the same material. Gadolinium has a high neutron capture cross-section but is used as a minor constituent of the ceramic control material to avoid excessive degradation of mechanical properties as the neutrons are absorbed and the $^{157}_{54}Gd$ transformed to $^{158}_{54}Gd$.

Gaffer. The foreman of a SHOP (q.v.) making glass-ware by hand; also called a CHAIRMAN.

Gahnite. See ZINC ALUMINATE.

Gaize. A siliceous rock, containing some clay, found in the Ardennes and Meuse Valley (France). It has been used as a POZZOLANA (q.v.).

Galena. Natural lead sulphide, PbS; formerly used in the glazing of pottery.
Gall. Molten sulphates sometimes formed on the surface of glass in a furnace; the cause is inadequate reduction of the salt cake to the more reactive sodium sulphite. Other aggregations of unfused material floating on molten glass are also sometimes referred to as gall, e.g. batch gall.
Galleting. Small pieces of roofing tile bedded in the top course of single-lap tiles to give a level bedding for the ridge tiles.
Galleyware or Gallyware. Term for the early (16th century) tin-glazed earthenware; the name derives either from the importation of the ware in Mediterranean galleys or from the use of the tin-glazed tiles in ships' galleys.
Gallium Arsenide. GaAs. A semiconductor used in electro-luminescent devices, photocells.
Gallium Nitride. GaN; m.p. 800°C. A special ceramic of high electrical resistivity 4×10^8 ohm.cm at 20°C)
Gallium Oxide. Ga_2O_3; m.p. 1800°C: sp. gr. 6.4.
Gamut. See TRI-STIMULUS VALUES for colours.
Gangue. The impurities surrounding the important constituent(s) in an ore, e.g. the silicate matrix in chrome ore.
Ganister. Strictly, this term should be reserved for a silica rock formed by deposition of dissolved silica in the siliceous seat-earth of a coal seam; the true ganister was typically found as the seat-earth of some of the coal seams of the Lower Coal Measures in the Sheffield district of England and was the raw material on which the manufacture of silica refractories in that area was based. The term 'ground ganister' is used to denote the highly siliceous patching and ramming material used in cupolas and foundry ladles. In some parts of the USA, 'ganister' alone is used for such materials, including crushed firebrick, possibly mixed with clay and/or silica rock.
Gap Grading. See GRADING.
Gardner-Coleman Method. A method for determining the absorption of oil by powders. (ASTM D1483/60 Pt 20).
Gardner Mobilometer. An instrument for the evaluation of the flow properties of vitreous-enamel slips. It consists of a plunger ending in a disk, which may be solid or may have a standardized system of perforations; the plunger is inserted in a tall cylinder containing slip and is loaded so that it descends through the slip; the time taken to fall through a specified distance is a measure of the mobility of the slip. (H. A. Gardner Laboratory, Bethesda, Maryland, USA.)
Gareis-Endell Plastometer. Consists of two disks between which a cylinder of clay is squeezed; the upper disk is rotated while the lower disk is slowly raised by a revolving drum; a stress deformation curve is recorded. (F. Gareis and K. Endell, *Ber. Deuts. Keram. Ges.*, **15**, 613, 1934.)
Garnet. (1) A group of minerals crystallizing in the isometric system and of general composition $3RO.R_2O_3.3SiO_2$, for example: $3CaO.Al_2O_3.3SiO_2$ (grossularite), $3MgO.Al_2O_3.3SiO_2$ (pyrope) $3CaO.Fe_2O_3.3SiO_2$ (andradite). Natural garnet occurs in economic quantity in USA, Canada, India, S. Africa, and elsewhere: it is used as an abrasive, particularly in the woodworking industry and for the lapping of bronze worm wheels.
(2) Mixed oxides with the general formula $A_3B_2X_3O_{12}$, with body-centred cubic structure. Yttrium and rare earth garnets are FERRIMAGNETIC MATERIALS, (q.v.) of which YTTRIUM IRON GARNET (q.v.) is the most important.

Garnex. A lining system of preformed boards, for continuous casting tundishes. (*Steel Times* Feb 1977, p165).
Garspar. Finely-ground quartz and glass, a by-product of polishing plate glass, used as a substitute for feldspar.
Gas Adsorption Method. A technique for the determination of SPECIFIC SURFACE (q.v.); variants of the method include the BRUNAUER, EMMETT AND TELLER METHOD (q.v.) and the HARKINS AND JURA METHOD (q.v.).
Gas Bubbling Brick. A special porous brick shape, set into the base lining of a LADLE (q.v.) or GLASS TANK for the injection of gases for *bubbling* or *stirring* to improve the homogeneity of the steel or control teeming temperature. Adjoining bubbler blocks form a *bubbler strip.*
Gas Concrete. See AERATED CONCRETE.
Gas-fire Radiant. The radiants for gas fires are made of refractory material having good resistance to thermal shock. Thc usual composition is a mixture of clay and crushed fused silica; this is shaped and then fired at about 1000°C.
Gas Pickling. A method of preparing sheet steel for vitreous enamelling by treatment, while hot, with gaseous HCl.
Gas-pressure Bonding. ISOSTATIC PRESSING (q.v.) at high temperature with gas (e.g. argon) as the fluid for transmission of the pressure (cf. EXPLOSIVE PRESSING).
Gas Pressure Sintering. The specimen is first sintered until all pores are closed pores, under vacuum or at low pressure using a gas which readily diffuses through the material (e.g. oxygen for oxides). Then the external gas pressure is raised (to c. 10MPa) using a gas which does not easily diffuse through the material. The technique does not require cladding and decladding of the products, and combines sintering with low-pressure containerless HIPing in one continuous operation. cf. SINTER-HIP, (q.v.) which uses pressures of 70 to 200 MPa.
Gas Retort. A refractory structure used for the conversion of coal into coke with the simultaneous distillation of town gas. There are two types: CONTINUOUS VERTICAL RETORT (q.v.) and HORIZONTAL RETORT (q.v.). In the UK the refractories used must meet the specifications issued by the Gas Council (1 Grosvenor Place, London, S.W.1) in collaboration with the Society of British Gas Industries and the British Coking Industry Association.
Gas Turbine. A device for the conversion of the energy of hot gases, derived from internal combustion, into rotary motion of a machine element. The efficiency increases with operating temperature and is at present limited by the safe temperature at which heat-resisting alloys can be used. There has been much research on the possible use of cermets and other special ceramics in these turbines, particularly in the blades.
Gassy Surface. See FUZZY TEXTURE.
Gate. See STOPPER.
Gate Scar. The U.S. term for a VOID (2).
Gater Hall Device. See BARRATT-HALSALL FIREMOUTH.
Gather. To take molten glass from a furnace for shaping; the amount of glass so taken (gathered) is called a 'gather' also.
Gathering Hole. An opening in the working-end of a glasstank furnace, or in the wall of a pot furnace, to permit the gathering of molten glass.
Gator-Gard Process. A development of PLASMA SPRAYING (q.v.) in which a constricted throat in the gun accelerates the powder particles before they join the plasma. Tradename.
Gatorizing. A superplastic forging technique used to attach ceramic blades to a metal turbine disc.
Gaudin's Equation. An equation for the particle-size distribution that can be

expected when a material is crushed in a ball mill or rod mill; it is of the form $P = 100(x/D)^m$, where P is the percentage passing a sieve of aperture x, D is the maximum size of particle, and m is a constant which is a measure of dispersion. The equation holds good only if the ratio of size of feed to size of balls is below a critical value which, for quartz, is 1:12. (A. M. Gaudin, *Trans. Amer. Inst. Min. Met. Engrs.*, **123**, 253, 1926.) cf. SCHUHMANN EQUATION.

Gauge. The exposed length of a roofing tile as laid on a roof; also sometimes known as a MARGIN.

Gauged Brickwork. Brickwork built to fine tolerances, sometimes using bricks specially made or ground to accurate sizes. (see RUBBER).

Gauging of Cement. The process of mixing cement with water. For the preparation of a cement paste of standard consistency prior to testing, B.S. 12 stipulates that the time of gauging shall be 3–5 min.

G-brick. A large French hollow block with good thermal insulating properties, due to its many perforations and thin cell walls.

Gault Clay. A calcareous clay with a short vitrification range used for making building bricks in S. E. England. Bricks made from this clay are generally porous and cream coloured, but in a few localities red bricks are made from it.

Gee Casting-Rate Meter. A metal ring is placed on top of a clean, dry standard plaster base, which has been blended to a fixed consistency. The ring is filled with slip of known density and a close-fitting cover put in place. The space within the ring is connected to a manometer via a valve in the cover. The pressure drop in that space gives a measure of the casting rate.

Gehlenite. 2CaO. Al_2O_3. SiO_2; m.p. 1593°C; thermal expansion (0–1200°C) 8.7×10^{-6} This mineral is sometimes formed by the action of lime on firebricks; it may occur in the slagged parts of blast-furnace linings.

Gelatine-pad Printing. See MURRAY-CURVEX MACHINE.

Gelman Membrane. See DUST SAMPLING, FILTERS.

Generator. In a water-gas plant, the refractory-lined chamber in which solid fuel is gasified by blowing in steam and air alternately.

Geomod. Tradename. CAD software for tableware (SDRC CAE International, Stevenage).

Geopolymer. Geopolymers are semiamorphous three-dimensional networks of polymeric sodium, potassium, lithium and magnesium silicoaluminates of the poly (sialate) type (–Si–O–Al–O–) or of the poly sialate-siloxo) type (–Si–O–Al–O–Si–). Geopolymers harden between 20 and 120°C and are similar to thermosetting organic resins, but stable up to 1200/1400°C without shrinkage. Hardening involves the chemical reaction of alumino-silicate oxides (Al^{3+} in IV-fold coordination) with alkali polysilicates yeilding polymeric Si–O–Al bonds. A wide range of alkaline resistant inorganic reinforcements were combined with geopolymer matrices (in particular SiC fibre) without high temperature, high energy processing. SiC fibre/K-poly (sialate-siloxo) matrix composites shaped and hardened at 70°C for 1 h 30 min develop flexural mean strengths of at least 380 MPa, unchanged after firing at 450°C, 700°C and 900C. (*Ceram. Eng. Sci. Proc.* **9** (7/8), 835, 1988).

Georgian Glass. WIRED GLASS (q.v.) in which the wire mesh is square.

Germanium Nitride. Ge_3N_4; decomposes at 800°C. A special electro-ceramic of high resistivity.

Germanium Oxide. GeO_2; m.p. 1115°C; sp. gr. 43. This oxide has been used to make glasses of high refractive index.

Gerstley Borate. A sodium-calcium borate produced at the Gerstley Mine, Death Valley, California. (US Borax and Chemical Corp., Los Angeles).

Getting. The actual process of digging clay, by hand or bv excavator; getting and transporting form the successive stages of winning.

Gibbsite. Aluminium trihydrate, $Al_2O_3.3H_2O$. Occurs in Dutch Guiana, the Congo, and some other areas where laterization has occurred. A typical analysis is 65% Al_2O_3, 33% H_2O and 2% impurities, but many samples contain clay. Gibbsite requires calcination at a very high temperature to eliminate all the shrinkage that results from the loss of water.

Gilard and Dubrul Factors. See THERMAL EXPANSION FACTORS FOR GLASS.

Gild. The painting of pottery with liquid gold, this is subsequently fired on at about 700°C.

Gillmore Needle. Apparatus for the determination of the initial and final set of portland cement (Q. A. Gillmore, *Practical Treatise on Limes, Hydraulic Cements and Mortars,* New York, 1864). The present form of apparatus consists of two loaded rods which slide vertically in a frame: the rod ('needle') for thc determination of initial set is 1⁄12 in. (2 mm) dia. and weighs ¼ lb (110g), the needle for the final set is 1⁄24 in. (1 mm) dia. and weighs 1lb (446g). Details are given in ASTM–C266 (cf. VICAT NEEDLE).

Ginneter. Term in the N. Staffordshire potteries for a person who grinds from china-ware, after it has been taken from the glost kiln, any adhering particles of refractory material from the kiln furniture; cf. SORTING. (From *Ginnet,* an old term for a tool used by carpenters to remove excrescences from wood.)

Glaceramic. A term that has been used in USA for devitrified glass products of the type exemplified by PYROCERAM (q.v.)

Glarimeter. See INGERSOLL GLARIMETER.

Glascol. Tradename. Acrylic copolymer used as sealers for porous glazes and as media for colours. (Allied Colloids, UK).

Glas-lok. Tradename. A glass-fibre joint for clay sewer pipes. (Pomona Pipe, USA)

Glass. A solid with no long-range order in the arrangement of its atoms. ASTM C162–92 defines a glass as an inorganic product of fusion that has cooled to a rigid condition without crystallizing, and notes that glasses are typically hard and brittle, with a conchoidal fracture. The word is used loosely as a synonym for GLASSWARE (q.v.) The Annual Book of ASTM standards Vol 15.02 lists over fifty test methods and property requirements for glass and glass products. The names of many types of glass indicate their main constituents, e.g. soda-lime glass (the presence of silica being understood); for the system of designating types of optical glass see OPTICAL GLASS CLASSIFICATION. A major advance in the understanding of the fundamental nature of a glass was made when W. H. Zachariasen (*J. Amer. Chem. Soc.,* **54**, 3841, 1932) deduced that the characteristic properties of glasses are explicable if the interatomic forces are essentially the same as in a crystal, but if the three-dimensional atomic network in a glass lacks the symmetry and periodicity of the crystalline network. Zachariasen's rules for the formation of an oxide glass are (1) the sample must contain a high proportion of cations that are surrounded by oxygen tetrahedra or by oxygen triangles; (2) these tetrahedra or triangles must share only corners with each other; (3) some oxygen atoms must be linked to only two such cations and must not form further bonds with any other cations. The ions in a glass are thus divided into NETWORK-FORMING (q.v.) and NETWORK-MODIFYING (q.v.). See also STRUCTON; VITRON. See also HALIDE GLASSES, CHALCOGENIDE GLASSES.

Glassblowing. Shaping hot glass by air pressure. This may be done by hand and mouth by a craftsman *glassblower*, or may be an element of various forming processes: see BLOW-AND-BLOW; PRESS-AND-BLOW; PUFF-AND-BLOW.

Glass-bonded Mica. See MICA (GLASS-BONDED).

Glass-Ceramic. A type of ceramic material that, while in the form of a molten glass, is shaped by one of the conventional glassmaking processes, and is subsequently devitrified in a controlled manner so that the finished product is crystalline. The crux of the process is the precipitation, during cooling of the shaped ware, of nucleating agents previously added in small amounts to the glass batch; the nucleated article is then heated to a temperature at which the nucleated crystals can grow. Devitrified-glass products can be made in a wide range of compositions; the properties can thus be varied, but typically the ware is impermeable and has high strength and good thermal-shock resistance. Uses include RADOMES (q.v.), high-temperature bearings, thread-guides and domestic ovenware. (See also NEO-CERAMIC GLASS).

Glass-coated Steel or Glass-lined Steel. A grade of vitreous enamelled ware that has good resistance to chemical attack at high temperatures and pressures; also known as Glassed Steel.

Glass Enamel. A mixture of coloured metallic oxides, or salts, and low-melting flux, e.g. lead borosilicate, used for the decoration or labelling of glass-ware; the enamel is fired-on at a red heat.

Glass-encapsulated HIPing. See ENCAPSULATED HIPING.

Glass Eye. A large unbroken blister on vitreous enamelware.

Glass Fibre. Filamentous glass made by mechanical drawing or centrifugal spinning; or by the action of a blast of air or steam to produce *Staple Fibre*. The unprocessed filaments are known as *Basic Fibre*; a number of filaments bonded together form a *Strand*. Long glass fibres are known as *'Silk'*, a fleece-like mass of fibres is *'Wool'*; felty material is *'Mat'*. Formerly glass for making fibres was produced as marbles which were remelted in an electric furnace to form filaments. Nowadays the *direct-melt method* is used. Fibres are drawn directly from holes in an alloy bush, at winding speeds of 50 m/s. The Owens STEAM-BLOWING PROCESS (q.v.) is used to make glass wool. Glass filaments are a few μm diameter, and much stronger than ordinary glass because free of surface flaws. There are over 20 British Standards for glass fibre products and plastics and cement products with glass fibre reinforcement.

Glass Frost. Very thin glass that has been crushed for use as a decorative material (cf. TINSEL).

Glass-ionomer Cements. Translucent dental cements, made by the reaction between aluminosilicate glass powders, and polymers and copolymers of acrylic acid. Developed by A.D. Wilson and B.E. Kent at the Laboratory of the Government Chemist. (*J. Appl. Chem. Biotechnol* **21**, (11), 313, 1971). For specification, see B.S. 6039.

Glassivation. Passivation of semiconductor devices by complete encapsulation in glass. An Americanism.

Glass-like Carbon, Vitreous Carbon. A monolithic, non-graphitic form of carbon, made by shaping a thermosetting organic resin and pyrolysing at 2000°C and above. The material is hard, impervious, has high compressive (700–1400 $MPam^{-2}$ and tensile (1–200 $MPam^{-2}$ strengths, retained at high temperatures, but low impact strength. Higher temperature fabrication gives higher thermal conductivity (0.012 cal/cm°C.s) at the expense of some strength. The material was developed by Lockhead Missile and Space Co. Palo Alto. AD 708314, 1969).

Glassmakers Soap. A selenium compound, or manganese dioxide, used to remove green coloration due to iron compounds.

Glass-making Sands. High purity silica sand, with low iron content especially, suitable for making clear glass. BS 2975 specifies sampling and analysis.

Glass-to-Metal Seal. Metal components varying in size from fine wires to heavy flanges are sealed to glass for many purposes, e.g. electric lamp bulbs and radio valves. Metals that have been used for this purpose include Pt, Cu, W, Mo and alloys such as Fe-Cr, Ni-Fe and Ni-Fe-Co; the SOLDER GLASS (q.v.), which is used as a powder suspended in a volatile liquid, is essentially lead borate. See Kohl. *'Handbook of Materials and Techniques for vacuum devices'* Reinhold, New York & London 1967; or the article on Glass Sealing in *'Encyclopaedia of Materials Science and Engineering'* Vol 3, M.B. Bever (ed) Pergamon 1986.

Glassware. The products of the glass industry, usually CONTAINER GLASS; FLAT GLASS. The other major branches of the glass industry produce FIBREGLASS and special OPTICAL and ELECTRICAL GLASS.

Glaze. A thin glassy layer formed on the surface of a ceramic product by firing-on an applied coating, by firing in the presence of alkali vapour (as in SALT GLAZING (q.v.), or as a result of slag attack on a refractory material. A glaze may, however, be partially crystalline (see CRYSTALLINE GLAZE and MATT GLAZE). The term 'glaze' is also applied to a prepared mixture, which may be either a powder or a suspension in water (SLOP GLAZE q.v.) ready for application to ceramic ware by dipping or spraying.

Glaze Fit. The matching of the thermal expansion of a glaze to that of the body on which it is held. To prevent CRAZING (q.v.) the glaze must be in compression when the ware has been cooled from the kiln to room temperature; to achieve this, the thermal expansion of the glaze must be less than that of the body.

Glazing. (1) The process of applying a glaze to ceramic ware; the latter may be unfired or in the biscuit state.
(2) The formation of a glazed surface on a refractory material as a result of exposure to high temperature and/or slagging agents. Occasionally, a glaze is deliberately applied to a refractory to seal the surface pores.
(3) The dulling of the cutting grains in the face of an abrasive wheel, generally as a result of operating at an incorrect speed.
(4) Inserting glass sheets into windows, doors etc.

Glenboig Fireclay. A fireclay occurring in the Millstone Grit in the region of Glenboig, Lanarkshire, Scotland. A typical per cent analysis (raw) is: SiO_2, 50–51; Al_2O_3, 33; Fe_2O_3, 2.5; alkalis 0.5. The P.C.E. is 32–33.

Glide Planes. SLIP PLANES (q.v.)

Globar. A silicon carbide heating element. (Trade-name: Carborundum Co., USA and England.)

Glory Hole. An opening in a furnace used in the glass industry for reheating glass-ware preparatory to shaping.
Gloss. A surface property, for example of a glazed ceramic or of vitreous enamelware, related to the reflectivity. For a quantitative definition see SPECULAR GLOSS.
Gloss Point. When a layer of glaze powder is heated, a temperature is reached at which the surface changes its appearance from dull to bright; this temperature has been termed the 'gloss point'. (R. W. P. de Vries, *Philips Tech. Rev.*, **17**, 153, 1955.)
Glost. This word, meaning 'glazed', is used in compound terms such as 'Glost ware', 'Glost firing' and 'Glost placer'.
Glost Box. A form of CRANK for totally enclosing flatware during glost firing.
Gluing. The construction of a rotary kiln lining in axial strips, alternatively glued and seated normally on the kiln shell. Individual brick rings are keyed on the side, between glued strips.
Glut Arch. A brick arch below the firemouth of a pottery BOTTLE OVEN (q.v.) for the admission of primary air and the removal of clinker.
Gob. A lump of hot glass delivered, or gathered, for shaping.
Gob Process. A method of making hollow glass-ware, gobs of glass being delivered automatically to a forming machine.
Gold Decoration. See BURNISH GOLD and BRIGHT GOLD.
Gold Ruby Glass. See RUBY GLASS.
Gold Scouring. Alternative term for burnishing; see under BURNISH GOLD.
Goldstone Glaze. An AVENTURINE (q.v.) glaze; a quoted composition is (parts): white lead, 198.7; feldspar, 83.4; whiting, 8; ferric oxide, 11.2; flint, 41.4. This glaze matures at cone 04.
Gold-tin Purple. A mixture of gold chloride and brown tin oxide, used as a pigment.
Gonell Air Elutriator. A down-blast type of ELUTRIATOR (q.v.) designed by H. W. Gonell *(Zeit. Ver. Deutsch. Ing.,* **72**, *945, 1928);* it has found considerable use in Europe for assessing the fineness of portland cement.
Gorkal Cement. The Polish equivalent of Ciment Fondu (q.v.)
Gooseneck. See under BUSTLE PIPE.
Goskar Dryer. A chamber dryer for bricks and tiles designed by T. A. Goskar *(Trans. Brit. Ceram. Soc.,* **37**, 62, 1938). Each chamber has a false floor and a false roof; air enters the chamber via the space above the false roof and is withdrawn via the corresponding space below the false floor.
Gossler Process. The drawing of glass fibres from streams of molten glass issuing from holes at the bottom of a melter. (O. Gossler, Br. Pat. 354 763, 1930).
Gothic Pitch. See under PITCH.
Gottignies Kiln. The original electric MULTI-PASSAGE KILN (q.v.); it was introduced in 1938 by two Belgians, R. Gottignies and L. Gottignies (Belgian Pat. 430 018).
Gouging Test. A procedure for the evaluation of the resistance of a vitreous-enamelled surface to mechanical wear. In the procedure laid down in a Special Bulletin issued by the Porcelain Enamel Institute (Washington, D.C., USA), a small steel ball is rolled on the enamel surface under various loads.
Gouy-Freundlich Double Layer. See DOUBLE-LAYER THEORY.
Grade. As applied to an abrasive wheel this term refers to the tenacity of the bond for the abrasive grains, i.e. the resistance to the tearing-out of the abrasive grains during use.

Grading. (1) The relative proportions of the variously sized particles in a batch, or the process of screening and mixing to produce a batch with particle sizes correctly proportioned. A batch with a grading for low porosity will contain high proportions of coarse and fine particles and a low proportion of intermediate size; if a particular particle size, e.g. the medium size, is excluded from the batch this is said to be a GAP GRADING.
(2) In the abrasives industry, the process of testing to determine the GRADE (q.v.) of a wheel; testing machines are available for this purpose.
Graded Cermet. A CERMET (q.v.) in which the proportion of metal to ceramic in continuously, gradually and uniformly varied along the length of the article, so that one end is pure metal, the other the pure ceramic, or any section from the length of such a cermet.
Gradient Kiln. A kiln in which the temperature changes uniformly along its length.
Grain. (1) Abrasive material, dead-burned magnesite, or similar materials that have been size graded. (2) A crystal or microcrystal in a polycrystalline material.
Grain Boundary. The narrow, disordered regions separating individual grains in a polycrystalline material. They may comprise glassy material of this is formed during SINTERING (q.v.) Grain boundaries provide a second route (as well as through the grains) for DIFFUSION processes, and are important to the related processes taking place in SINTERING and in CREEP. Grain boundaries interface with the movement of DISLOCATIONS and CHARGE CARRIERS, and are this important in determining how the MICROSTRUCTURE (q.v.) affects mechanical and electrical properties. See also CRYSTAL STRUCTURE.
Grain Bridging. See CRACK BRIDGING.
Grain Density. B.S. 1902 Pt 3.6 defines *grain density* as the BULK DENSITY (q.v.) of granular materials with grain size larger than 2mm. It specifies four techniques for its determination (mercury with vacuum; arrested water absorption; mercury in burette; mercury in Amsler volumeter) of which the first is the referee method, but the others are preferred on health grounds for routine testing, as they do not involve large quantities of mercury in glass containers.
Grain Growth. In general, in polycrystalline materials, the larger grains tend to grow at the expense of the smaller. Grain growth is more marked at higher temperatures, which facilitate atomic mobility. Grain growth may be by DIFFUSION processes and the movement of GRAIN BOUNDARIES, or by selective dissolution processes such as OSTWALD RIPENING (q.v.).
Graining; Graining Oxide; Graining Paste. The production on vitreous enamelware of a patterned surface resembling grained wood; this is done by transferring, by means of a metal roller, GRAINING PASTE (ALSO KNOWN AS GRAINING OXIDE) from an etched plate to the enamel surface. This paste consists of very finely ground colouring oxides and a flux; it is made to the required consistency with an oil, e.g. clove oil.
Grain-size Analysis. See PARTICLE-SIZE ANALYSIS.
Grain Spacing. A term in the US abrasives industry equivalent to STRUCTURE (q.v.).
Granite Ware. This term has had various meanings wilhin the vitreous-enamel industry but the definition given in ASTM – C286 would restrict the meaning to: a one-coat vitreous-enamelled article with a mottled pattern

produced by controlled corrosion of the metal base prior to firing.

Granito. The Italian word meaning 'granite' applied to a vitrified 'porcelain-stoneware' tile body.

Granny Bonnet. Term sometimes used for a Bonnet Hip Tile – see under HIP TILE.

Granular. See MORPHOLOGY.

Granulate. Components of a powder which behave as units of a given size distribution, although each may be made up of many smaller particles adhering together. For example, spray-dried granulate may be in the form of roughly spherical or toroidal grains, of uniform size, each made up of large numbers of much smaller platelike particles.

Granulometry. The measurement of particle size distribution, or of grain size.

Graphite. A crystalline form of carbon occurring naturally in Korea, Austria, Mexico, Ceylon, Madagascar, and elsewhere; it is produced when amorphous carbon is heated, out of contact with air, at a high temperature. The m.p. exceeds 3500°C, the electrical and thermal conductivities are high. Graphite is used (under reducing conditions) as a special refractory for electrodes and electrical heating elements; mixed with fireclay it is used in the manufacture of plumbago crucibles and stoppers for use in the metal industry. See also CARBON REFRACTORIES. ASTM Annual Book of Standards Vol 15.01 lists some twenty-five standards and tests relating specifically to graphite and manufactured carbon products.

Grappier Cement. A cement made from fine-ground underburned or overburned slaked lime.

Gravé. Decoration on glass, by sandblasting to various depths.

Gravimpinger. A simple gravimetric dust sampler developed at British Ceramic RA. The dust is collected by high-velocity impingement in a fluid which wets clay particles well and has a low vapour pressure minimizing evaporation over the sampling period of a week. (*Annals Occup. Hyg.* **9** (1) 29 1966).

Gravity Process. See GOB PROCESS.

Green. Of ceramics, shaped but not dried or fired.

Green House. A heated room in which pottery-ware is inspected and stored between the shaping process and the firing process.

Green Spot. A fault that occasionally becomes serious in the manufacture of sanitary fireclay and glazed bricks. The green spots are comparatively large and frequently of an intense colour. The usual causes are the presence of chalcopyrite ($CuFeS_2$) in the raw clay or accidental contamination by a particle of copper or copper alloy, e.g. a chip off a bronze bearing.

Green Strength. The mechanical strength (usually measured by a transverse test) of ceramic ware in the GREEN (q.v.) state.

Green Vitriol. Ferrous sulphate $FeSO_4.7H_2O$, produces a red colour when used as a source of iron in pigments.

Grey Stock. A clamp-fired STOCK BRICK (q.v.) that is off-colour.

Griffith Cracks. Minute cracks assumed to be present in the surface of a piece of glass; their existence was suggested by A. A. Griffith (*Phil. Trans., A,* **221**, 163, 1920) to explain the low strength of glass compared with that calculated from the forces of molecular cohesion.

Griffith-Orowan Criteria. When crack tip stresses are accommodated only by elastic deformation, the fracture stress is inversely proportional to the square root of the flaw size, which is often the size of

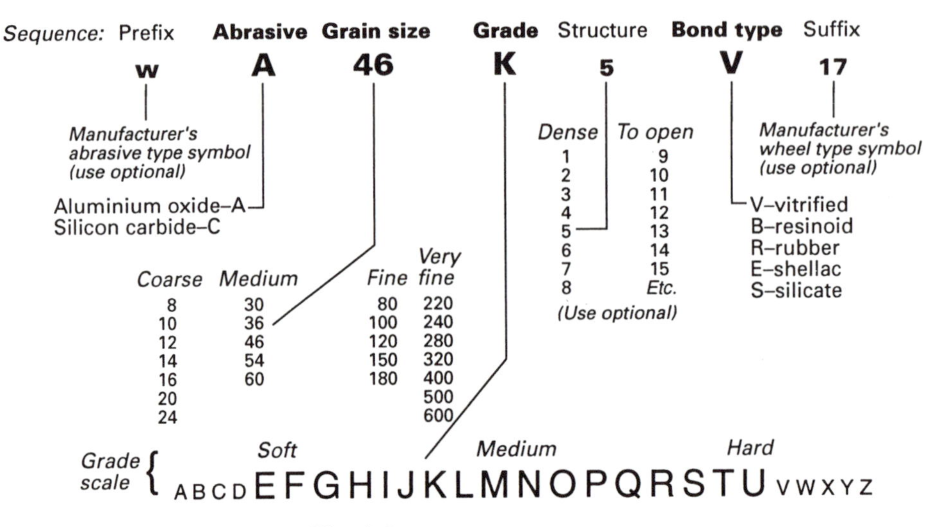

Fig. 3 See GRINDING WHEEL

the largest most highly stressed defect. See FRACTURE, CRACK PROPAGATION.

Grin. **G**radient **R**efractive **In**dex lens. Instead of by variation in thickness, the light path through the lens is determined by controlled variations in refractive index obtained by ion exchange. The usual form is a cylinder or rod with the refractive index higher on the outside.

Grinding. (1) Particle-size reduction by attrition and/or high-speed impact; in the size reduction of ceramic raw materials coarse crushing usually precedes fine grinding.
(2) Final shaping to close tolerance, e.g. of an electroceramic component, by means of abrasive wheels.

Grinding Aid. An additive to the charge in a ball mill or rod mill to accelerate the grinding process; the additive has surface-active or lubricating properties. Grinding aids find particular use in the grinding of portland cement clinker, but in the UK their use is precluded by the conditions laid down in B.S. 12.

Grinding Wheel. A disk, or comparable symmetrical shape, of bonded abrasive material. The abrasive is either alumina or silicon carbide; the bond may be of the vitrified ceramic type, or it may consist of sodium silicate (here called a SILICATE BOND), resin, rubber, or shellac. A standard marking system for grinding wheels was adopted many years ago by the Grinding Wheel Manufacturers' Association of America; in 1952 this system was also adopted in the UK as British Standard 1814. This marking system is reproduced in Fig. 3. BS 1814 is now withdrawn and replaced by BS4481 Pt1.

Grinstead Clay. A Cretaceous clay for brickmaking in parts of Sussex, England.

Grip-coat. See GROUND-COAT.

Grisaille. A method of decorating, at one time used on pottery vases, etc., in which different shades of grey were used to produce the effect of low relief; from French word meaning grey shading.

Gritting. The process of forming a smooth surface on blocks of marble, or other natural stones, by means of abrasive blocks known as 'rubbing blocks .

Grizzle. A clamp-fired STOCK BRICK (q.v.) that is soft and suitable only for internal walls. The term went out of use before 1940.

Grizzly. The name used in the English structural-clay products industry for a simple, stationary, screen consisting merely of a frame and perforated metal plate, or wire mesh; the screen is set at an angle, the inclination depending on the angle of repose of the material being screened.

Grizzly Crusher. A machine with moving bars which crush material and sparate it by size.

Grog. Firebrick, which may already have been used in a furnace lining or elsewhere, crushed to a size suitable for incorporation in a batch, either for remaking into bricks or for use in a refractory cement, ramming, or patching material. The word is probably derived from the French *gros grain* (coarse grain) via the English 'grogram' (coarse cloth) (cf. CHAMOTTE).

Grog Fireclay Mortar. A US term defined as: raw fireclay mixed with calcined fireclay, or with broken fireclay brick, or both, all ground to suitable fineness.

Grossalmerode Clay. A refractory clay from Grossalmerode, about 10 miles (16 km) E.S.E. of Kassel, West Germany. These clays are of Tertiary origin. They have been used for making glass-pots since A.D. 1503. The typical Grossalmerode clay is worked at the Faulbach and at the Lengemann deposits and contains (unfired) 70% SiO_2 and 18% Al_2O_3; the P.C.E. is 28–29. An aluminous clay (P.C.E. 33–34) is worked at the Salzmann claypit.

Grotthuss Chain Reaction. In absorbed water layers, $H_3O^+$and OH^- ions are present, with the fraction dissociated being as much as 1%. (10^6 times that in liquid water). The H_3O^+ ion releases a proton to a neighbouring water molecule, which then releases a proton to another. The charge transport which results provides a conduction mechanism in the surface layers of humidity sensitive oxides.

Ground-coat. The coating on vitreous enamelware that is applied to the metal. The principal constituents of a sheet-iron ground-coat are feldspar, borax, and silica which, together, constitute 75–90% of the batch; in addition, there is about 10% Na_2CO_3+$NaNO_3$, 5% fluorspar, 0.5% cobalt oxide and 1.5% MnO_2. The chief constituents of ground-coats for cast iron are also feldspar, borax. and silica, but the batches are more variable than for ground-coats on sheet iron.

Ground-hog Kiln. US term for an art-potters kiln (usually fired with solid fuel) partly buried in a convenient hillside to support the roof and conserve heat.

Ground Laying. A process for the application of a uniform coating of colour to pottery-ware by painting with oil the area to be coloured and then dusting powdered colour over the ware; the colour sticks only where oil has been first applied. The process is now used only for the decoration of some expensive types of china and porcelain.

Ground Mass. See MATRIX.

Grout. Mortar or cement mixed to a fluid consistency for use in filling the joints between brickwork or tiles.

Grouting. (1) Filling tile joints with GROUT (q.v.) (2) Pumping refractory material from outside the furnace to seal localised voids formed in the lining in service, or to make a new lining. (3) Filling the voids in soils to make firm, massive foundations, by injecting a hardenable compound or concrete.

Growan. A term in the Cornish china clay industry for incompletely and unevenly decomposed granite.

Gruneisen's Law. relates thermal conductivity γ, linear thermal expansion coefficient α, compressibility K and specific heat C_v, for a solid of volume V, by $3\alpha V = \gamma K C_v$.
Guard Ring. An arrangement in thermal-conductivity apparatus designed to ensure that heat shall flow through the sample actually under test, in a direction perpendicular to the hot and cold faces, i.e. no heat flows through the sides of the test-piece.
Guide Tube. A fireclay tube having a spigot and socket, for use in the TRUMPET (q.v.) assembly in the bottom-pouring of molten steel. Two sizes are specified in B.S. 2496: 4⅝ in. (117 mm) o.d. and 2⅞ in. (73 mm) i.d. for tubes up to 12 in. long (305 mm); 6 in. (152 mm) o.d. and 3¹⁵⁄₁₆ in. (100 mm) i.d. for tubes 12–15½ in. (305-393 mm) long.
Gull Wings. Gull wings and WAKE HACKLE form when the fracture front encounters an inclusion, which overlap slightly on rejoining. The mark then appears as a TWIST HACKLE (called a WAKE HACKLE if so formed) with two outspreading strong WALLNER LINES, the GULL WINGS.
Gum. Various natural gums are used as BINDERS (q.v.) in the ceramics industry. e.g. *Gum Arabic* (or *Gum Acacia, Kordofan gum*); *Gum Tragacanth.*
Gum Set. Erratic quick setting of cement in concrete.
Guncrete. Refractory concrete applied by GUNNING. (q.v.)
Guniting. Hot spraying of refractory concrete.
Gunning. The placing of a refractory powder or slurry by means of a cement gun; most commonly, gunning is practised as a method of repairing furnace linings without the need to cool them to room temperature.
Gunning Mix. A CASTABLE, MOULDABLE OR RAMMING MIX that has been designed to be installed by pneumatic projection. (B.S. 1902 Pt7).
Guthrie Kiln. A variant of the BELGIAN KILN (q.v.), a trough replacing the transverse grate; the design was patented by H. Guthrie in 1877.
GVC. Abbreviation for GLAZED VITRIFIED CLAY: term applied to glazed clay pipes.
Gypsum. Natural hydrated calcium sulphate, $CaSO_4 . 2H_2O$, from which PLASTER of PARIS (q.v.) is produced. In England gypsum occurs in the Newark and Tutbury zones of the Keuper Marl, and in the Purbeck Beds of Sussex. It occurs abundantly in USA, Canada, France, USSR and elsewhere.
Gyratory Crusher. A primary crusher for hard rocks such as the quartzite used in silica brick manufacture. The material to be crushed is fed into the space between a vertical steel cone and a similarly shaped steel casing; the cone rotates eccentrically to the casing (cf. CONE CRUSHER).
H Cassette. A support for handling and firing tiles, which is shaped so that each tile is individually supported, with the tile camber moulded into the cassette shape, to minimise twist during firing. (CERIC, France)
Habit. The form and structure characteristic of a particular crystalline compound.
Habla Kiln. A ZIG-ZAG KILN (q.v.) that may be archless or with a permanent flat roof; top-fired with fine slack; output 25 000–50 000 bricks/week. It was designed by a Czech, A. Habla (Brit. Pats. 242 051 and 311 884).
HAC. High Alumina Cement (q.v.).
Hack. A double row of building bricks, with a 1-ft (305 mm)space between them and 8–9 bricks high, set on boards in the open to dry; the top of a completed hack is protected from rain by a 'cap'. Side

protection, when necessary, is provided by LEE BOARDS usually pronounced LOO or LEW BOARDS. Hack drying has now largely been replaced by artificial drying under more controlled conditions.

Hackle. A finely structured FRACTURE SURFACE of rough or matt appearance. Also termed frosted, gray, matte, mist or stippled areas.

Hackle Marks. Fine ridges on the FRACTURE SURFACE of glass, parallel to the direction of crack propagation. See also RIB MARKS.

Hafnium Borides. HfB: m.p. 3060°C: sp. gr. 12.8. HfB_2 m.p., 3150°C; sp. gr. (theoretical), 11.2; thermal expansion (25–1000°C), 5.5×10^{-6}; resistivity (20°C), 10–12 µohm.cm.

Hafnium Carbide. HfC; m.p. 3890°C; theoretical density, 12.6 g/ml.; hardness, 2533 (K100); modulus of rupture, 200–350 MNm^{-2} (20°C.); 170 MNm^{-2} (1000°C.); 105 MNm^{-2} (1300°C.). Modulus of elasticity 315 GNm^{-2} (20°C); thermal expansion, 8.8×10^{-6} (25–2500°C).

Hafnium Nitride. HfN; m.p. 3310°C; theoretical density, 14.0 g/ml. This special ceramice can be prepared by sintering Hf in dry N_2 at 1000°C.

Hafnium Oxide. HfO_2; m.p. approx. 2800°C; sp. gr. 9.68; thermal expansion (20–1000°C) 5.8×10^{-6}; (20–1500°C) 6.5×10^{-6}. The low-temperature (monoclinic) form changes to the high-temperature (tetragonal) form at 1700–1900°C.

Hafnium Silicate. A compound analogous to zircon, hence the suggested name HAFNON. It can be synthesized from the oxides at 1550°C. Thermal expansion (150–1300°C), 3.6×10^{-6}.

Hafnium Titanate. Special refractory compositions have been made by sintering mixtures of HfO_2 and TiO_2 in various proportions. The m.p. of these sintered bodies is approx. 2200°C; there appears to be a phase change at about 1850°C. Some of the compositions have negative thermal expansions.

Hafnon. See HAFNIUM SILICATE.

Hager Process. Glass fibre is made from molten glass dropped on to a horizontal revolving disc, from which fibres are thrown off radially. (US Pat. 2 234 087, 1941).

HAGS. Hot Anode Gas Shrouded thermal spray gun.

Haigh Kiln. A US type of CHAMBER KILN (q.v.) designed by H. Haigh; it is U-shaped with an open space between the two lines of chambers, permitting the chambers to be fired from both sides. In one such kiln, as used at a US brickworks, the chambers were 12–15 ft wide, 16 ft long and 10 ft high; (4–4.5×5×3m); this particular kiln was gas-fired at 1150°C.

Hair Cracks appear in concrete whose surface has dried more rapidly than the interior.

Hair Line. (1) Fine CORD (q.v.) on the surface of glass-ware; see also AIR LINE.
(2) A fault sometimes seen in vitreous enamelware. The hair lines are a series of fine parallel, or nearly parallel, cracks that have healed in the later firings (thus differing from crazing). The cause is excessive stress in the biscuit enamel resulting from slight movement in the metal arising from non-uniform heating or poor design.
(3) A line of separation sometimes found near the centre of thick ceramic ware that has been shaped by SOLID CASTING (q.v.).

Hair-pin Furnace. A type of furnace used in vitreous enamelling, see U-TYPE FURNACE.

Half-bat. A building brick of half the normal length; also called a SNAP-HEADER.

Half Crystal. This term has been applied to CRYSTAL GLASS (q.v.) containing only about 15% PbO.

Half-penny Crack. A circular or semi-circular flaw developed from an indentation crack when sub-surface lateral cracks nearly parallel to the surface reach the surface and the material within the ring is removed. See FRACTURE TOUGHNESS TESTS.

Halide Glasses. Glasses containing heavy metal fluorides have good infra-red transmission properties, and high thermal expansion coefficients (up to 20×10^{-6}/K).

Halifax Hard Bed. A siliceous fireclay of the Lower Coal Measures extensively worked in Yorkshire, England. It contains (fired) 68–74% SiO_2, 20–25% Al_2O_3, 2.5–3.0% Fe_2O_3 and 1.0–1.5% alkalis.

Hall Effect. A magnetic field applied at right angles to an electric current flowing in a conductor generates a force in the third mutually perpendicular direction. This deflects the charge carriers. A net surface charge density builds up until its electrostatic field counterbalances the force due to the magnetic field. The sign of the surface charge density gives the sign of the charge carriers.

Hall's Factors. For calculating the thermal expansion of a glass; see THERMAL EXPANSION FACTORS FOR GLASS.

Halls Kiln. An annular kiln with permanent walls dividing the chambers; there are openings at the ends of these partition walls to give a zig-zag fire travel, but there are also large trace holes through these walls. The kiln was designed by G. Zehner of Wiesbaden, Germany. The name 'Halls Kiln' derives from the hall-like appearance of the long narrow chambers.

Halloysite. A clay mineral of the kaolinite group approximating in composition to $Al_2O_3.2SiO_2.4H_2O$. Halloysite is a constituent of the clays of New Zealand and Brazil; it also occurs in Missouri and Utah (USA), in N. Africa, Japan, and elsewhere. The special feature of halloysite is the tubular shape of the crystals resulting from the rolling up of the fundamental sheet structure; this causes difficulty when halloysitic clays are used in ceramic manufacture because of the consequent high and non-uniform shrinkage. (Named after the Belgian O. d'Halloy, 1783–1875.)

Hamaker Constant. A measure of the electrostatic dipole attraction of two particles in a given solid liquid system (See DOUBLE LAYER THEORY).

Hamburg Blue. One of a range of iron-bearing pigments.

Hammer Mill. An impact mill consisting of a rotor, fitted with movable hammers, that is revolved rapidly in a vertical plane within a closely fitting steel casing. Hammer mills are sometimes used for the size reduction of clay shales, glass cullet, and some of the minerals used in the ceramic industry (cf. DISINTEGRATOR; IMPACT MILL).

Hammer Test. Tapping anchors or cured castable with an engineer's hammer, to detect flaws or voids by the change of sound.

Hanging Rack. A heat-resistant metal frame for suspending porcelain enamel-ware during manufacturing processes.

Hard. As applied to a glass, glaze, or enamel, this word means that the softening temperature is high; such a product, when cold, is also likely to be hard in the normal sense. See also HARDNESS.

Hard Ferrite. A FERRITE (q.v.) permanent magnet e.g. $BaO.6Fe_2O_3$. See also SOFT FERRITE.

Hard Metals. Intermetallic carbides, particularly cemented carbides (q.v.) used as cutting tools and abrasives.

Hard-paste. See HARD PORCELAIN.

Hard Porcelain. True feldspathic porcelain (CONTINENTAL PORCELAIN), *Hard-paste Porcelain* being the older term still used by collectors. The COMBINED NOMENCLATURE (q.v.) defines *hard porcelain* as made from kaolin or kaolinitic clays, quartz, feldspar and sometimes calcium carbonate. It is covered with a colourless transparent glaze fired with the body. See PORCELAIN.

Hardening-on. Biscuit ware that has been decorated prior to glazing is heated at a comparatively low temperature (c. 700°C) to volatilize the oils used in decorating and to fuse the flux in the applied colour so that the decoration remains fixed during the subsequent dipping in the liquid suspension.

Hardening-on Fire. A low-temperature firing of underglaze decoration, to burn off the organic media used in applying the colours, before glazing.

Hardinge Mill. A ball mill for continuous operation in which the casing is in the form of two conical sections joined by a short, central, cylindrical section; the material to be ground is fed into the mill through one of the hollow trunnions, being discharged through the opposite trunnion. Named after H. W. Hardinge, USA, who invented this design of ball mill in 1906.

Hardness. See KNOOP HARDNESS; MICROCHARACTER HARDNESS; MOHS' HARDNESS; ROCKWELL HARDNESS; VICKERS HARDNESS.

Harkins and Jura Method. A gas-adsorption method for the determination of the specific surface of a powder. The sample is first evacuated and then exposed to a vapour near to its saturation pressure; the wetted powder is then immersed in the liquid itself and the rise in temperature is measured. From this, the surface energy change, and hence the surface area, is calculated. (W. D. Harkins and G. Jura, *J. Chem. Phys.*, **11**, 430, 1943.)

Harkort Test. Although H. Harkort was a German, details of his so-called crazing test for pottery-ware were first published in USA *(Trans. Amer. Ceram. Soc.,* **15**, 368, 1913). The test-piece is heated to 120°C and then plunged into cold water; the cycle is repeated, with successive increases of 10°C in temperature, until crazing can be detected after the quenching. The criteria are: no cracks after quenching from 150°C – crazing at room temperature likely after 3–4 months; 160°C – 15 months; 170–180°C – no crazing after 2½ years (with a few exceptions); 190°C – no crazing after 2½ years without exception. It is now known that crazing may result from moisture expansion as well as from thermal shock. (A modification of this test forms the basis of ASTM-C554.)

Harrop Kiln. One of the variety of tunnel kilns built to the designs of Harrop Ceramic Service Co., Columbus, Ohio. The early examples of Harrop Kilns in the N. Staffordshire potteries were of the large open-flame type.

Harrow. See under WASH-MILL.

Hartford-Empire Individual Section Machine (IS Machine). This in-line machine is the industry standard for making glass containers, with up to 12 mechanised glass and mould manipulating sections (e.g. H.E. IS-12). Each section functions independently, and can handle up to three GOBS (q.v.) of glass at a time, depending on the size of the ware. Malfunction in one section does not stop the whole machine, which produces about 10 pieces/min. Narrow-necked containers are made by a *blow-and-blow* process (the PARISON (q.v.) is blown, and the second blow into the mould imparts the final shape to the ware. The only movement is opening and closing the moulds. The machine can be easily converted to make wide-mouthed containers with screw threads, using a PRESS-AND-BLOW (q.v.) process in iron moulds. A modification (HE-28) uses the press-and-blow process to make tumblers in paste moulds, the ware being rotated to avoid seams. The glasses are formed as almost closed hollow objects, cut off and fire-polished with a burn-off machine.

Hartman Formula. Relates the refractive index (n) of a glass to the wavelength (λ) of the incident light: $n = n_o + c(\lambda - \lambda_0)^{-\alpha}$ where λ_0, c and α are constant and α may have a value from 1 to 2. This is an improvement on the CAUCHY FORMULA (q.v.).

HASAWA. Health and Safety at Work, etc Act, 1975. See HEALTH AND SAFETY.

Hashin's Equations. Expressions for the bulk and shear moduli, and hence Young's modulus, of composite materials, including porous materials. (A.G.Hashin and S. Shtrikman, *J. Mech. Phy. Solids* **11,** 1963, p127–140)

Hassall Joint. A type of joint for glazed pipes designed by Wm. Hassall (of John Knowles & Co., Woodville, Burton-on-Trent) in the late 19th century. Bitumen rings are attached to the outside of the spigot and the inside of the socket of the fired pipe. A thin smear of cement is rubbed around these rings just before the pipes are laid, the spigot of one pipe then being pushed into the socket of the next. Liquid cement is finally poured through holes in the socket to complete the joint.

Hastings Beds. A series of clay and sand deposits in the Lower Cretaceous of S.E. England; the Fairlight Clays at the base of these deposits have been used for brickmaking near Hastings and Bexhill.

Haunch. The part of the furnace roof between CROWN and SKEWBACK (q.v.).

Hawk Pug. Trade name: a de-airing pug having two barrels in tandem, the first rotating around a fixed bladed shaft, whereas the second is fixed while the bladed internal shaft rotates. Each shaft is set slightly off-centre. (Service Engineers, Stoke-on-Trent, England.)

Hawse Clay. A local English term for a clay that is crumbly but becomes plastic when worked up with water.

Haydite. Trade name: a lightweight expanded clay aggregate, made in USA, named after its inventor S. J. Hayde (US Pat. 1 255 878 and 1707 395).

HB. Symbol for BRINELL HARDNESS (q.v.).

HCFS. Hypersonic Flame Spraying (q.v.)

Header. A brick laid with its length across the width of a wall, thus leaving its smallest face exposed. A course of such bricks is known as a Header Course (cf. STRETCHER). A header in a coke-oven is a refractory brick set perpendicular to the wall, separating two heating flues.

Healing. The ability of a glaze to cover any areas of the ware that may have been damaged before the glaze has been fired on; this depends on a correct combination of surface tension and viscosity at the glost-firing temperature.

Healy-Sullivan Process. See HYDROGEN-TREATING PROCESS.

Health and Safety and environmental control have increased in importance. Some British regulations affecting the ceramic industry's responsibilities to its workers, its customers and the general public include:

**Health and Safety at Work etc Act*, 1974; London HMSO, 1975 (HASAWA)

(which is explained by *Essentials of Health and Safety at Work,* London, HMSO, revised 1990).
**Control of Lead at Work Regulations*, 1980 (CLAW). London, HMSO.
**Control of Substances Hazardous to Health Regulations*, S.I. 165, 1988. London, HMSO, 1988 (COSHH).
**Control of Substances Hazardous to Health in the Production of Pottery.* Approved Code of Practice, Health and Safety Commission, (HSC) London, HMSO, 1990. ISBN 011 8855301 (ACOP).
**Potteries etc (Modifications) Regulations* 1990. (These supersede the Pottery (Health and Welfare) Special Regulations 1947 and 1950).
Some of the HSE Methods for the Determination of Hazardous Substances are relevant to the ceramics industries. See MDHS.
The Health and Safety Information Bulletin* **107 (6) 2 1984 Classification, Packing and Labelling Regulations 1984, Pt.1 gives requirements for labelling dangerous substances.
Other more specific regulations deal with noise, smoke and fume emission, guards on power presses, using abrasive wheels, lighting gas furnaces, storing gaseous and liquid fuels and a whole range of similar and related topics.
*B.S. 7750 is a specification for environmental management systems aimed at ensuring compliance with stated objectives.
In the USA, the relevant bodies are OSHA (Occupational Safety and Health Administration) and NIOSH (National Institute for Occupational Safety and Health) which publishes a *Manual of Analytical Methods* P13 85–179018, 1984.
The FDA (Food and Drug Administration) is responsible for METAL RELEASE TESTS (q.v.) in the USA. ASTM C1191 describes the safe use and handling of ceramic slips; C1192 safe spraying of ceramic glazes; C1023 the labelling of ceramic art materials for chronic adverse health hazards.
See also TA-LUFT, METAL RELEASE, DUST, RSI.

Hearth. The lower part of a furnace, particularly the part of a metallurgical furnace on which the metal rests during its extraction or refining. In a blast furnace the hearth is constructed of aluminous fireclay or carbon blocks; in a basic open-hearth furnace or basic electric-arc steel furnace, the hearth is constructed of basic refractory bricks which are covered with dead-burned magnesite or burned dolomite rammed in place.

Heat Activated Decal. The colour layer medium contains an activable resin. The transfer is applied to cold ware, the resin activated by a heat source, and the backing paper blown off.

Heat Craze. The CRAZING (q.v.) of vitreous enamelware as a result of overheating in service; thick enamel and poor design are contributory causes.

Heat Release Decal. This type of transfer is applied to hot ware. The heat activates an adhesive layer, to stick the transfer to the ware, and simultaneously melts a wax layer which held the transfer to the backing paper.

Heat-Setting Mortar. See REFRACTORY MORTAR.

Heat-work. An imprecise term denoting the combined effect of time and temperature on a ceramic process. For example, prolonged heating at a lower temperature may result in the same degree of vitrification in a ceramic body as a shorter period at a higher temperature; a similar amount of heat-work is said to have been expended in the two cases.

Heavy Magnesia. A form of magnesium oxide made (unless intended for pharmaceutical use) by extraction from sea-water by the dolomite process (see SEA-WATER MAGNESIA). The extracted hydrate is calcined at 800–1000°C and has a bulk density of 0.5–0.7. On exposure to the air, heavy magnesia takes up water and CO_2 to form a basic carbonate. It is used in making SOREL CEMENT (q.v.), as a constituent of glass batches, and in various chemical processes.

Heavy Spar. See BARYTES.

Hecht's Porcelain. A German refractory porcelain similar to MARQUARDT PORCELAIN (q.v.).

Hectorite. Originally known as HECTOR CLAY from its source near Hector, California, USA. It is a hydrous magnesium silicate related to montmorillonite and forming an end-member of the saponite series.

Hedvall Effect. A brief period of enhanced reactivity associated with a solid-solid phase transition. (J.A. Hedvall, *Chem. Rev.* **15,** 1934, p139).

Heel Tap. See SLUGGED BOTTOM.

Heliglobe. A spherical mixer, disperser and comminutor which can be operated under presure or under vacuum. (*Industrie Ceram.* (635) 921 1970).

Helton System. An Austrian system for dry (mortarless) brick masonry (see also NOVAMUR).

Hepsleve. Plastics joint rings for plain-ended fully-vitrified sewer pipes. (Tradename. Hepworth Iron Co)

Heramicor. Gold decoration on tableware, subdivided into dots to prevent the build up of electrical potential leading to destructive electrical discharges (Heraeus GmbH, Germany).

Hercules Press. Trade name: a semi-dry brickmaking machine of the rotary-table type. (H. Alexander & Co. Ltd., Leeds, England.)

Hercynite. Ferrous aluminium spinel, $FeO.Al_2O_3$; m.p. 1780°C; thermal expansion (0–1000°C) 10.3×10^{-6}. Hercynite is sometimes present in solid solution with other spinels in aluminous chrome ores; it can also be formed when ferruginous slag attacks an aluminous refractory under reducing conditions. When heated under oxidizing conditions, hercynite changes into a solid solution of Fe_2O_3 and Al_2O_3.

HERF. High Energy – Rate Forming (q.v.)

Hermansen Furnace. A recuperative pot-furnace for melting glass; the first furnace of this type was built in Sweden in 1907 and was soon afterwards introduced into England and elsewhere.

Herreshoff Furnace. A multiple-hearth cylindrical muffle roasting furnace; such a furnace has been used to calcine of fireclay.

Hessian Crucible. A type of refractory crucible for use in the small-scale melting of non-ferrous metals. These crucibles are made from the GROSSALMERODE CLAYS (q.v.).

Hertzian Crack. = PERCUSSION CONE (q.v.).

Hexite-Pentite Theory. A theory (long since discarded) for the atomic structure of clays proposed by W. Asch and D. Asch (see *Trans. Brit. Ceram. Soc.,* **13,** 90, 1914). The theory was based on the hypothesis that silicates are built up of hexagonal and pentagonal rings of hydrated silica and hydrated alumina.

Hexmesh. Steel strips set vertically, to form a grating with hexagonal apertures to anchor CASTABLE REFRACTORY linings. Hexmesh is anchored to 50mm square plates (*plate heads*) drilled and fixed to round studs.

HGMS. High-Gradient Magnetic Separation (q.v.).

Higgins Furnace. An electric arc furnace for the fusion of refractory materials, e.g. alumina. The shell is a truncated cone (small-end up) which fits over a carbon hearth; when the charge has been fused, the shell is raised so that the ingot of fused material can be removed from the hearth (A. C. Higgins, US Pat., 775,654, 22/11/04).

High-alumina Cement. (1) A special hydraulic cement made by fusing (or occasionally by sintering) a mixture of limestone and bauxite. Although the setting time is similar to that of portland cement, the final strength is virtually attained in 24 h. The composition generally lies within the following limits (per cent): CaO, 36–42; Al_2O_3, 36–42; Fe oxides, 10–18; SiO_2, 4–7.
B.S. 915 stipulates 32% ≥ Al_2O_3 and Al_2O_3.CaO (by wt) equal to 0.85–1.3. High-alumina hydraulic cement, if mixed with crushed refractory material, can be used as a refractory concrete.
High-alumina cement (though previously so used) was banned for structural work in buildings in 1974. Strength is lost if HAC is exposed to high temperature and humidity during setting or in service, or when the water/cement ratio is high. The reasons for loss of strength and durability are complex, related to crystalline state changes from a metastable to a more stable state, accompanied by increased porosity and the production of water. The strength of *in-situ* HAC cannot be measured non-destructively, but DTA can give information on the crystal changes causing loss of strength.
(2) A refractory cement (non-hydraulic) of high alumina content.

High-alumina Refractory. A refractory of the alumino-silicate type. In the UK it is defined in B.S. 3446 as containing over 45% Al_2O_3. In the USA seven classes of high-alumina refractory are defined (ASTM – C27): 50%–Al_2O_3 (P.C.E. ≥ 34; Al_2O_3 content 50 ± 2.5%); 60%-Al_2O_3 (P.C.E. ≥ 35; Al_2O_3 content 60 ± 2.5%); 70%–Al_2O_3 (P.C.E. ± 36; Al_2O_3 content 70 ± 2.5%); 80%–Al_2O_3 (P.C.E. 2 37; Al_2O_3 content 80 ± 2.5%), plus three more classes, namely 85%, 90% and 99% – Al_2O_3. ASTM C673 classifies High Alumina Plastic Refractories and Ramming Mixes into 7 grades from 60% Al_2O_3 on a similar basis, except that the last two grades are 95 (±2.5%) and 100% (97.5%) Al_2O_3. The ISO classification is: High-alumina Group I >56% Al_2O_3; High-alumina Group II, 45–56%, Al_2O_3.

High-duty Fireclay Refractories. Defined in ASTM – C27 as a fireclay brick with a P.C.E. ≥31½. There are three types: Regular, Spall Resistant, Slag Resistant. Further properties are specified for the last two types. ASTM C673 specifies a high duty fireclay plastic refractory with P.C.E. ≥31.

High Early-strength Cement. US term equivalent to the UK RAPID-HARDENING PORTLAND CEMENT (q.v.). It is Type III Cement of ASTM Specification C150, which stipulates a crushing strength 1700 p.s.i.(12 MNm^{-2}) after 1 day in moist air and ≥ 3000 p.s.i. (21 MNm^{-2}) after 1 day in moist air followed by 2 days in water.

High Energy-rate Forming (HERF). A punch is carried by a ram, and fired into a closed die by gas pressure which gives an impact velocity of about 50mph, and forming pressures of over 2 GPa.

High-frequency Heating. See INDUCTION HEATING.

High-frequency Induction Furnace. See under ELECTRIC FURNACES FOR MELTING AND REFINING METALS.

High Gradient Magnetic Separation (HGMS). Also known as *high intensity magnetic separation*. Magnetic fields of sufficient intensity (up to 15 000 gauss) are used to separate paramagnetic minerals (as well as ferromagnetic materials) from china clay. (English Clays Lovering Pochin & Co. Ltd, Br. Pat 1077242, 1967 and subsequent patents).

Highlight Test. A simple, empirical estimate of the chemical resistance of a surface (of glass, glazes etc) by noting the decrease in sharpness of reflected images in it.

High Pressure Self-Combustion Sintering. (HPCS). A variation of glass encapsulated hot isostatic pressing, in which a compact made by cold isostatic pressing is sealed under vacuum in a glass capsule heated to the glass softening point. High (100MPa) pressure inert gas is applied to the capsule, and its contents ignited electrically, to sinter and densify. This specialised method has been used to process TiB_2.

High-velocity Oxyfuel Spraying. A coating process. Fuel gas and oxygen are fed into the combustion chamber of a spray gun, and there mixed with powder carried by an inert gas. The powder is heated to 2800°C and accelerated to 800 m/s in the gun, to produce low porosity coatings of good surface finish up to 1.5mm thick.

High-velocity Thermocouple. See SUCTION PYROMETER.

Highway Bridge Loading. British Standard B.S. 5400 Pt. 2 1978 specifies loads on buried rigid pipes due to vehicles, on the basis of their wheel spacing, axle spacing and tyre contact areas.

Hillebrandite. A hydrated calcuim silicate $Ca_2SiO_3(OH)_2$ which decomposes at about 500°C to form low crystalline β-Ca_2SiO_4.

Hinckley Index. A measure of the degree of crystallinity of a material, based on X-ray diffraction peaks. The sum of the heights of the $1\bar{1}0$ and $11\bar{1}$ peaks is divided by the height of the $1\bar{1}0$ peak above the background. The first two peak heights are based on a line drawn from the trough between the 020–$1\bar{1}0$ peaks, to the background just beyond the $11\bar{1}$ peak. (D.N. Hinkley, *Clays & Clay Minerals* Proc.11th Natl. Conf. p229, 1963).

Hind Effect. The porosity of a dried cast (deflocculated) pottery body is less than that of the same body, dried, if made by the plastic (flocculated) process; the observation was first made by S. R. Hind (*J. Inst. Fuel.* **24**, 116, 1951).

HIPing. Hot isostatic pressing(q.v.).

Hip Tile. A specially shaped roofing tile for use on a descending ridge that forms the junction of two faces of a roof. Hip tiles are available in a number of patterns varying from the close-fitting hip of angular section to the broadly curved types known as BONNET HIPS and CONE HIPS.

Hispano-moreseque Ware. A type of lustre or tin-glazed pottery.

HIVF. The Hypersonic Velocity Impact Fusion thermal spray coating process.

Hob. The so-called ceramic hobs (flat surfaces for supporting cooking ware on cookers) are usually made from glass ceramic sheets.

Hob-mouthed Oven. A pottery BOTTLE OVEN (q.v.) in which the firemouths projected 0.45–0.6m outwards, coal being fed to the fire from the top.

Hoffmann Kiln. A top-fired LONGITUDINAL-ARCH KILN (q.v.) of a type introduced by F. Hoffmann in 1858; variants of this type of kiln have been widely used in the firing of clay building bricks.

Hofmeister Series. Cations arranged in order of decreasing flocculating power for clay slips: H^+, Al^{3+}, Ba^{2+}, Sr^{2+}, Ca^{2+},

Mg^{2+}, NH^{4+}, K^+, Na^+, Li^+. (F. Hofmeister, *Arch Exptl. Path. Pharmakol.*, 24, 247, 1888.)

Hog's Back. A type of RIDGE TILE (q.v.).

Hoimeter. Trade name: a device for controlling the softness of an extruded column of clay; two small sensing rollers run on the clay column, their axial positions changing with any change in softness of the clay; this change can be made to adjust the rate of water addition to the mixer and so bring the consistency of the clay back to normal. (Buhler Brothers, Uzwil, Switzerland.)

Hoi-Tronic. A plasticity and moisture control system for ceramics industry materials preparation. (Manfred Leisenberg KG).

Holdcroft Bars. Pressed and prefired ceramic bars (57 × 8 × 6.5mm) made of blended materials so proportioned that, when placed horizontally with only their ends supported, and when heated under suitable conditions, they will sag at a stated temperature. The bars are numbered from 1 (bending at 600°C) to 40 (1550°C). These bars were introduced in 1898 by Holdcroft & Co., and are now available from Harrison-Mayer Ltd., Stoke-on-Trent, England.

Holding Ring. A BEAD (q.v.) in the finish of a glass bottle for use when the parison is transferred to the blow mould: sometimes known as the BACK RING or TRANSFER RING.

Holes (in electrical conductivity). The absence of an electron from a compound or crystal structure leaves a *hole*. Holes move through the crystal or compound under the action of an electric field, in the opposite direction to that taken by electrons. They thus behave much like electrons, but carrying positive charge.

Holey Boy. A type of perforated floor in intermittent downdraught kilns firing salt-glazed pipes.

Hollow Blocks. See HOLLOW CLAY BLOCKS.

Hollow Casting. See DRAIN CASTING.

Hollow Clay Blocks. Fired clay building blocks, usually of comparatively large size, with cells (air spaces) within the block; they are used for wall construction and, with metal reinforcement, for floors and prefabricated panels. For specifications see B.S. 3921; ASTM – C34, C56, C112 and C212 (cf. HOLLOW MASONRY UNIT; PERFORATED BRICK).

Hollow Masonry Unit. Defined in ASTM – C43 as: A unit whose net cross-sectional area in any plane parallel to the bearing surface is less than 75% of its gross cross-sectional area measured in the same plane.

Hollow Neck. A fault in the neck of a bottle resulting from an insufficiency of glass in the gob from which the bottle was made; if the bottle is of the screw-necked type, the faulty neck is said to be BLOWN AWAY.

Hollow-ware. Cups, basins etc. (cf. FLAT-WARE). Recent American usage spells this HOLLOWARE, particularly when applied to glassware.

Holyboy. A BATT (q.v.) with holes for combustion gases to pass through.

Hommelaya Process. Trade name: a procedure in which cobalt is deposited on sheet steel before it is enamelled, the purpose being to eliminate the need for a ground coat; (Hommel Co., Pittsburgh, USA).

Hone. A block of fine abrasive, generally SiC; the block is appreciably longer than it is broad or wide. Hones are used for fine grinding, particularly of internal bores.

Honeycomb. Ceramic honeycombs were first developed to provide a light sheathing material for the protection of the outer surface of supersonic aircraft, missiles, and space-craft from the heat generated by air friction. A high-alumina ceramic is generally used; experiments have also been made with zirconia, thoria, and various carbides. The refractory powder is mixed with a binder and rolled into a thin flexible sheet; the core is shaped by corrugated rolls. The sheet and core are bonded with the same mixture and fired together to produce the honeycomb. Thin sheets of glass on a paper substrate can be corrugated, stacked and heated to produce glass honeycombs. Honeycombs are also used as catalyst carriers.

Hood. (1) A protective casing, with an exhaust system, for use in carrying out dusty operations (e.g. TOWING (q.v.)) in pottery manufacture.
(2) Alternative name for POTETTE (q.v.).

Hooded Pot. See under POT.

Horizontal Retort. (1) An intermittent unit for the production of town gas from coal; it is constructed of segments of silica or siliceous refractory material (cf. CONTINUOUS VERTICAL RETORT).
(2) An intermittent unit formerly used for the production of zinc. Horizontal zinc retorts were generally made from a siliceous fireclay, suitably grogged.

Horse; Horsing. (1) A 'horse' is a heavy stool with a convex rectangular top on which roofing tiles, after they have been partially dried, can be given the slight curvature (about 3m radius) necessary to ensure ventilation in a tiled roof; the process itself is termed 'horsing'.
(2) In the old bottle-ovens for firing pottery, a 'horse' was used by the placers to enable them to reach the top parts of the bungs of saggars.

Hospital. In a vitreous-enamel factory, the department in which ware is repaired that has faults, not too serious for remedy.

Hot-blast Circulating Duct. See BUSTLE PIPE.

Hot-blast Main. A duct, lined with refractory material, through which hot air passes from a HOT-BLAST STOVE (q.v.) to the BUSTLE PIPE (q.v.) of a blast furnace.

Hot-blast Stove or Cowper Stove. A unit for heating the air delivered to the tuyeres of a blast furnace. It is a cylindrical furnace, about 25m high and 6m dia., lined with fireclay refractories. There is a combustion chamber up one side; refractory checker bricks (usually special shapes known as STOVE FILLINGS (q.v.) fill the remainder of the space. The checker bricks are heated by the combustion of blast-furnace gas in the combustion chamber and then, on reversal of the direction of gas flow they deliver heat to incoming air which then passes to the blastfurnace tuyeres. These stoves were first proposed by E. A. Cowper (hence their alternative name) in 1857 and were first used at Middlesbrough, England, in 1860. The idea of heating the blast, however, had been put forward by J. B. Neilson, a Scotsman, in 1824.

Hot Extrusion. This term is best reserved for the process of extrusion in which clay that has already been prepared and has the correct water content is heated while passing through the extruder; these features distinguish the process from STEAM TEMPERING (q.v.). The Hot Extrusion process is typified in the extruders designed by the Bedeschi Company, Padova, Italy.

Hot Floor. A drying floor, heated from below by hot gases or by steam pipes, for special shapes of heavy clayware, refractories, etc.

Hot Forging. Pressure is applied to a partially-densified compact, with the material free to flow without lateral constraint, at a rate selected to avoid cracking or pore formation, by controlling the strain rate. Sintered materials are thus densified with preferred crystallographic orientation.

Hot Isostatic Pressing. Shaping and sintering by the simultaneous application of heat and isostatic pressure.

Hot-metal Ladle. A ladle for the transfer of molten iron from a blast furnace to a mixer furnace and from there to a steel furnace; alternatively, the ladle may transfer molten pig-iron direct from blast furnace to steel furnace. Such ladles are generally lined with fireclay refractories but for severe conditions high-alumina and basic refractories have been tried with some success.

Hot-metal Mixer. A large holding furnace for molten pig-iron. The capacity of these furnaces, which are of the tilting type, is up to 1400 tons. Hot-metal mixers may be ACTIVE (i.e. the pig-iron is partially refined while in the furnace) or INACTIVE (i.e. the pig-iron is merely kept molten until it is required for transfer to a steelmaking furnace). In either case the bottom and walls of the furnace are made of magnesite refractories and the roof of silica refractories.

Hot Patching. The repair of the refractory lining of a furnace while it is still hot; this is most commonly done by spraying a refractory slurry through a cement gun. See also AIR-BORNE SEALING and SPRAY WELDING.

Hot-plate Spalling Test. A spalling test designed specifically for the testing of silica refractories; it was standardized in ASTM – C439, now withdrawn, though a description still appears for information in the ASTM Annual Book of Standards.

Hot Preparation. See STEAM TEMPERING.

Hot Pressing. The simultaneous application of high temperature and high pressure in a mould, to shape and sinter a powder compact at below the temperature of normal sintering, without sintering aids. The process is applied to ENGINEERING CERAMICS which are often difficult to shape and fire conventionally to zero porosity. Furthermore, fine-grained microstructures may be developed without grain growth inhibitors. Densification takes place in three stages: particle rearrangement; viscous and plastic flow; material transport by diffusion. The optimum temperature and pressure combination can be found using the CLIMBING TEMPERATURE PROGRAMME or the ISOTHERMAL TECHNIQUE (q.v.), though the characteristics of the actual powder used, or property limitations of the die materials, usually prevent the optimum conditions being achieved for the whole of the pressing operation.

(**Hot pressing** has also an obsolete usage: JIGGERING (q.v.) with a heated profile tool. Some electrical insulators were shaped in that way).

Hot-top. A container lined with refractory or heat-insulating tiles and used at the top of an ingot mould in the casting of steel: the molten steel is allowed to fill the mould and the hot-top, the extra metal remaining molten so that it sinks into the ingot as the latter cools and shrinks, thus preventing the formation of a 'pipe'.

Hot-topping Compound. A granular, exothermic insulating refractory added to the top of an ingot after teeming. Often used in conjunction with HOT-TOPPING TILES, to prevent 'piping' – the formation of cavities at the top of castings due to contraction of the melt as it cools.

Hot-topping Tile. Refractory insulating tiles performing the function of a MOULD BRICK (q.v.), i.e. delaying solidification of the melt in a HOT-TOP.

Hot-wire method. A dynamic method for the measurement of THERMAL CONDUCTIVITY (q.v.) applied to refractories up to 1500°C (PRE/R32 1978). A linear heat source embedded in the test piece gives a power output constant along its length and in time. The temperature rise in this hot wire is a measure of the thermal conductivity of the material. The method was developed for solids and powders with thermal conductivities below 1.5 $Wm^{-1}K^{-1}$. B.S. 1902 Pts. 5.6 and 5.7 and ASTM C1113–90 now specify the hot-wire method for refractories.

Hotel China. See AMERICAN HOTEL CHINA.

Hotel-ware. Tableware of extra strength (usually achieved by extra thickness) for use in hotels and restaurants; for specification see B.S. 4034, which demands no crazing after 16 hours in an autoclave at 340 10 kPa, and a water absorption ≤0.2%.

Hough Kiln. A barrel arch kiln with internal cross walls.

Hounsfield Tensometer. A compact device, manual or motor driven, which will perform tensile, shear, compression and bend strength tests on ceramics and other materials at forces up to 20kN. (Monsanto plc, Swindon).

Hourdis. French term for large hollow blocks for ceiling or roof construction, first produced in Switzerland, and named after their inventor. Dimensions are 50 to 100 × 20 × 7cm.

Housekeeper Seals. Ductile metal seals to glass. For example, a glass bead can be applied to the thinned end of a copper tube, and then sealed to a glass tube, the ductile thinned copper yielding to stresses from thermal expansion differences, and preventing cracking. Good chemical bonds between glass and copper are required. (See W.G. Housekeeper, *J.Am.Inst.Elec. Eng.* **42,** (1923) p840).

Hovel. The large conical brick structure that enclosed the old type of pottery oven or glass-pot furnace.

Hover Kiln. A tunnel kiln through which the ware is carried on a cushion of hot air; the design was developed by Shelley Electric Furnaces Ltd. Stoke-on-Trent, England (Brit. Pat. 990,589, 28/4/65; US Pat. 3,184,224, 18/5/65).

HPCS. High Pressure Self-Combustion Sintering (q.v.).

HPSN. Hot pressed silicon nitride. See SILICON NITRIDE.

HPZ. Hydropolysilazane, a precursor for Si_3N_4 films.

HR. Symbol for ROCKWELL HARDNESS (q.v.).

HREX. Symbol for a special shape of wall tile – Round Edge External Corner, Left Hand (see Fig. 7, p350).

HTI. Abbreviation for High-Temperature Insulating refractory; see INSULATING REFRACTORY.

Hulo System. A system for the handling of building bricks from setting in the kiln to delivery at the building site. (Trade name: Van-Huet, Pannerden, Holland.)

Humboldt Rotary Kiln. A kiln designed for burning cement; the batch is fed to the kiln as a suspension in hot gases with consequent economy in fuel consumption.

Humidity Dryer. A dryer for clay-ware (particularly bricks) in which the relative humidity (r.h.) of the atmosphere in the dryer is controlled so that the initial stages of drying take place at high r.h., this being progressively reduced as drying of the clay-ware proceeds.

Humper. A warped item of tableware which rocks on its foot on a flat surface. (Cf. WHIRLER).
Huntington Dresser. See STAR DRESSER.
Hutch. Scottish term for a waggon of about 1m^3 capacity used for hauling clay.
Hutchins Furnace. An electric arc furnace for the fusion of refractory materials, e.g. alumina. The shell is a truncated cone (big-end-up) and the water-cooled hearth is made from a mixture of pitch and coke (O. Hutchins, US Pat., 1 310 341, 15/7/19)
HV. Symbol for VICKERS HARDNESS (q.v.).
h,x diagram. A chart of enthalpy v. humidity. Points on this diagram correspond clearly to the continuously changing condition of the air in a dryer, and provide a clear way of displaying the changes.
Hydrargillite. An obsolescent name for GIBBSITE (q.v.).
Hydration. Dolomite and other lime-containing refractories are attacked by atmospheric moisture. To protect them before use, they may be metal-cased; tar-impregnated, wrapped in polythene sheeting etc.
Hydration Tendency. The degree of HYDRATION in a test sample of a basic refractory exposed to moisture under specified conditions of humidity, temperature and pressure. It is measured by firing the hydrated sample to 1000°C and determining the percentage weight loss. B.S. 1902 Pt 3.14 specifies the test. ASTM specifies tests for the hydration of granular dead-burned dolomite (C492); of magnesite or periclase grain (C544); for the hydration resistance of basic brick (C456) and for the hydration resistance of pitch bearing basic refractory brick (C620).
Hydraulic Cement. A cement which sets and gains strength through chemical action with water; such a cement will set under water. PORTLAND CEMENT (q.v.) is a typical example.
Hydraulic Mining. The method used to win china clay; a high-pressure (2 MPa) jet of water is aimed at the exposed vertical clay face and the clay is washed down. The suspension is then pumped to settling tanks where the impurities (chiefly mica) settle out. The purified suspension of china clay is then concentrated, filtered, and dried.
Hydraulic Press. A machine used for dry pressing, or semi-dry pressing; the pressure (up to 100 MPa) is applied hydraulically to the top, and sometimes also to the bottom, mould-plate. Such presses are used, for example, in dry pressing basic refractories.
Hydraulic Refractory Cement. A REFRACTORY CEMENT (q.v.) containing aluminous hydraulic cement, e.g. Ciment Fondu, so that it sets at room temperature.
Hydraulicking. Old term for HYDRAULIC MINING (q.v.).
Hydrite Process. In this process bricks are fired at 775°C in the presence of water. Gas fuel is burnt in steel tubes on the sides of periodic kilns, in which low quality clays may be fired more rapidly without fluorine emission.
Hydrocasing. A form of kiln/dryer construction. The dryer and kiln are in the same tunnel, made in steel prefabricated sections. A pool of water runs the full length (c.100m) of the kiln/dryer. The kiln cars enter this water at the entrance of the dryer, and a perfect seal against gas leakage is formed, enabling the kiln to operate at higher gas pressures and provide an even flow of air around the setting (CERIC, Paris).
Hydrogen-treating Process. A method for the preparation of sheet steel for vitreous enamelling by first driving

hydrogen into the surface of the steel (this is effected electrolytically) and then removing the hydrogen by immersion of the steel in boiling water. The process was introduced by J. H. Healy and J. D. Sullivan (US Pat. 2 754 222, 10/17/56).

Hydrogenic Furnace. A hydrogen atmosphere pressure furnace.

Hydrometer. A glass instrument, shaped rather like a fisherman's float, for the determination of the specific gravity of a liquid, e.g. of clay slip or of vitreous-enamel slip.

Hydrostatic Pressing. See ISOSTATIC PRESSING.

Hydrothermal Preparation. Hydrothermal crystral growth takes place when an aqueous solution is held at high pressure and high temperature, with a temperature gradient along the container. A solute may be transported by convection from the hotter region to the cooler, where the solution becomes supersaturated and the solute is deposited The materials and solution conditions can be chosen to promote chemical reactions. There are many types of such hydrothermal processes, which may be combined with other ceramic processes such as sintering, reaction sintering, hot-pressing, isostatic pressing etc.

Hydroxyapatite. That variant of APATITE (q.v.) which contains the OH^- hydroxyl ion.

Hygrometer. An instrument for measuring the relative humidity of air, e.g. in a dryer for clayware.

Hypersonic Flame Spraying (HCFS). A high velocity chemical combustion process accelerates and melts a powder stream to 500 to 1000m/s for spraying composite powders.

Hysil. Trade name: A borosilicate glass of high thermal endurance and chemical resistance used for chemical ware. (Chance Bros., England.)

Hyslop Plasticity Diagram. A diagram relating the extensibility (E) of a clay, as determined by a penetration method, to its softness (S); the relationship is of the form $E = KS^n$, where K and n are constants. (J. F. Hyslop, *Trans. Brit. Ceram. Soc.,* **35**, 247 1936.)

Hysteresis. The degree of lag in the reaction of a material to a change in the conditions (of mechanical, electrical, or magnetic stress) to which it is exposed. Ceramic ferro-electrics, for example, exhibit hysteresis when subjected to a changing external electric field; hysteresis is also shown by plastic clay when it is stressed cyclically.

IC Silicon Carbide. Abbreviation for Impregnated-Carbon Silicon Carbide; it contains free carbon and silicon and the bulk density is comparatively low (2.60).

ICP. Inductively Coupled Plasma. An emission spectrometry technique for the chemical analysis of ceramics.

ICTA. International Committee on Thermal Analysis.

Ice Colour. A relatively coarse (100 mesh) flux used to give a decorative effect on glass-ware. The surface to be decorated is coated with a tacky oil. The flux is then applied and the ware is fired at a low temperature so that the grains of flux remain discrete.

IF. Indentation Fracture. See FRACTURE TOUGHNESS TESTS.

Ignition Arch. A refractory arch in a boiler furnace fitted with a mechanical grate; its purpose is to assist the ignition of the fuel as the latter moves under the hot brickwork.

Illite. A group name for micaceous clay minerals of variable composition. The name derives from the fact that it was from Illinois (USA) clays and shales that samples were isolated by R. E. Grim, R. H. Bray, and W. F. Bradley (*Amer. Mineral.,* **22,** (7), 813, 1937) who first proposed this name. As there has been some confusion as to the exact meaning of the term, the following statement in the original paper should be noted: '. . . the term *illite* . . . is not proposed as a specific mineral name, but as a general term for the clay mineral constituent of argillaceous sediments belonging to the mica group.'

IL/MA. Ignition-Loss/Moisture-Adsorption. The relationship between these properties was proposed as a simple method for assessing the nature and proportion of the clay mineral present in clays by P. S. Keeling *(Trans. 6th. Int. Ceram. Congr.,* 195, 1958). For the purpose of this test, IL is taken as the difference between the ignition loss under oxidizing conditions at 1000°C and at 375°C; MA is determined by subjecting samples of dried clay to 75% R.H. at 25°C for 24h); this humidity is provided by the atmosphere above a saturated solution of sodium chloride).

Ilmavit. Trade-name. (VEB Technischer Glas, Ilmenau). A machineable glass ceramic. The material is 35% sodium fluorphlogopite, and is crystallized directly from the melt without further heat treatment. Its thermal expansion is similar to that of iron alloys.

Ilmenite. Ferrous titanate, $FeO.TiO_2$; m.p. 1365°C. This mineral is the principal constituent of the heavy minerals in the beach sands of Australia and elsewhere; it is thus the chief source of titania, which is used in the ceramic industry as an opacifier and as a constituent of some ceramic dielectrics.

Image Furnace. Apparatus for the production of a very high temperature in a small space by focusing the radiation from the sun (SOLAR FURNACE) or from an electric arc (ARC-IMAGE FURNACE) by means of mirrors and/or lenses. Such furnaces have been used for the preparation and study of some special ceramics.

Image Gloss. The distinctness with which a sharply outlined character is reflected from the surface in question. This property is important in the selection of vitreous-enamelled architectural panels, which should not reflect recognisable images of neighbouring objects. A test is described in ASTM – C540.

Imbibition. The particular case of absorption or adsorption of a liquid by a solid in which the solid increases in volume. A typical example is the swelling of Na-bentonite when it takes up water.

Imbrex. See ITALIAN TILE.

Imitation Porcelain. See SEMI-PORCELAIN.

IMM Sieve. See under SIEVE; for mesh sizes see Appendix E.

Impact Box. A tile, clamped to an inner face of a 30mm cubic rotating steel box, is subjected to random impacts of 5mm and 8mm diameter steel balls contained in the box. Wear is assessed visually. The test, applied to unglazed floor tiles, quarries and brick pavers, is claimed to simulate pitting attributed to steel-tipped stiletto heels.

Impact Bruise. = PERCUSSION CONE (q.v.).

Impact Mill. A crushing unit in which a rapidly moving rotor projects the charged material against steel plates; impact mills find use in the size-reduction of such materials as feldspar, perlite, etc. (cf. DISINTEGRATOR; HAMMER MILL).

Impactopus. A pulverised coal firing system (see OCTOPUS).

Impact Resistance. The resistance of a material to impact. A test for this property (which is clearly of importance in the assessment of tableware, for example) must proceed by a series of blows – usually delivered by a swinging pendulum – of increasing severity; the major difficulties in the impact testing of ceramics are to produce homogeneous test-pieces and to determine the proportion of the impact force causing fracture that is actually used in fracturing the test-piece. In USA a standard impact test for ceramic tableware is described in ASTM – C368. B.S. 6206 is an impact test for flat safety glass.

Imperiale. A large wine bottle, 6l for table wines, 4.5l for sparkling wines.

Imperial Red. Red pigments based on ferric oxide.

Impervious. In the USA this word has a defined meaning (ASTM – C242) as applied to ceramic ware, namely, that degree of vitrification shown by complete resistance to dye penetration; the term generally implies that the water absorption is zero, except for floor tiles and wall tiles which are considered to be impervious provided that the water absorption does not exceed 0.5%.

Impulse Burners. Burners for the feeding of fuel oil to a kiln. The oil, supplied at high pressure in a ring-main, is injected into the kiln in metered quantities.

Inamori Stones. Synthetic gemstones manufactured by Kyocera Corp, Japan.

In-and-out Bond. Brickwork consisting of alternate courses of headers and stretchers.

Inclination Flowmeter. Powder flows from a funnel to form a cone on an inclined glass plate. The angle of inclination is increased mechanically to that value which causes gliding of the cone tip. The cotangent of that angle is an index of the flowability of the powder (*Powder Metallurgy Int.* **19,** (1), 40, 1987).

Inclined Flow Test. A pellet of glaze is fired on a specially made plane of biscuit body, supported at a fixed angle in the kiln. The distance the glaze flows down the plane is a measure of its fluidity. (Cf. FUSION FLOW TEST FOR VITREOUS ENAMELS.).

Inclusion. (1) A foreign atom or ion in the crystal lattice.
(2) A larger foreign particle in the matrix of the material. See also NON-METALLIC INCLUSION; STONE.

Inclusion Pigment. See ENCAPSULATED COLOURS.

Incongruent Melting. A solid melts incongruently if, at its melting point, it dissociates into a liquid and a solid of different composition; for example ORTHOCLASE melts incongruently at about 1170°C to form LEUCITE and a liquid that is richer in silica.

Increment. See PREFERRED INCREMENT.

Indenting. In structural brickwork, the omission of a suitable series of bricks so that recesses are left into which any future work can be bonded.

Indentation Fracture. See FRACTURE TOUGHNESS TESTS.

Indentation Hardness. The values of hardness obtained by using one of the variety of test methods in which a hard probe or indenter is pressed into the surface to be tested. See VICKERS, KNOOP, BRINELL hardness.

Indentation Strength. See FRACTURE TOUGHNESS TESTS.

Indialite. A modified form of CORDIERITE with hexagonal structure.

Indian Red. A red ferric oxide pigment made by calcining ferrous sulphate.

Indiana Method. A technique for the determination of the quantity of entrained air in concrete on the basis of the difference in the unit weights of a concrete sample with and without air. The name derives from the fact that air-entrained cement was first used to any extent in Indiana, USA. (*Proc. A.S.T.M.,* **47**, 865, 1947.)

Indirect-arc Furnace. See under ELECTRIC FURNACES FOR MELTING AND REFINING METALS.

Induction Furnace. See under ELECTRIC FURNACES FOR MELTING AND REFINING METALS.

Induction Heating. The heating of an electrically conducting material by the effect of induced electric currents, which may be set up by a high-frequency field in a small object or a low-frequency field in a large object (as in a large induction furnace for steel melting). High-frequency heating has been proposed (Brit. Pat. 898 647, 14/6/62) as a method for the firing of glaze on ceramic tiles.

Inductively Coupled Plasma. Stable, hot (10 000°C) highly ionized gas 'flames' are produced by induction heating with an RF generator. Materials of high melting points can be evaporated. The technique can be used as the source for a type of sophisticated emission spectrometer for chemical analysis of the vapourised material. It can also be used as a synthesis technique for high purity ultrafine particles, free of contamination by any electrode materials. When liquid reactants are used, e.g. to produce oxides from solutions of metal salts, the technique is known as *spray ICP.*

Indumix. Trade-name. A kiln atmosphere control system developed by Manfred Leisenberg KG, Laubach. A pressurised gas stream 6–10mm diameter is injected into a chamber with side openings through which the kiln atmosphere is sucked in on the injector principle. In a variant, the VARIOMIX system, external air is also introduced, in amounts adequate for burning the gas.

Industrial Floor Brick. See FLOOR BRICK.

Inertisation. Biscuit firing which drives gases from the body, without closing the pores.

Inert Plasma Spraying (IPS). See PLASMA SPRAYING.

Infra-red Drying. Drying by exposure to infra-red radiation from specially designed electric lamps or gas burners. The process has found some application in the vitreous-enamel industry and in the pottery industry.

Infrasizer. An air elutriator for the fractionation of powders into seven grades within the size range 100 mesh to 7μm. It was designed by H. E. T. Haultain *(Trans. Canadian Inst. Min. Met.,* **40**, 229, 1937; *Mine Quarry Engng.,* **13**, 316, 1947).

Infusorial Earth. An obsolete name for DIATOMITE (q.v.).

Ingersoll Glarimeter. An instrument designed primarily to measure the gloss of paper; it has been used to evaluate the abrasion resistance of a glaze in terms of loss of gloss after a specified degree of abrasion. The apparatus was designed by L. R. Ingersoll (*J. Opt. Soc. Amer.,* **5**, 213, 1921).

In-glaze Decoration. A method of decorating pottery; the decoration is applied on the surface of the glaze, before the glost fire, so that it matures simultaneously with the glaze. Such decoration can be applied by hand-painting, spraying, or silk screen.

Inglis Equation. Relates the stress at the tip of an elliptical flaw in glass to the applied stress, the flaw length and the radius of curvature of the crack tip. When the stress at the crack tip is sufficient to overcome the strength of the chemical bonds there, fracture occurs. Long, sharp cracks produce higher stresses. See FRACTURE MECHANICS.

Ingot Head Tile. See HOT TOPPING TILE.

Ingot Mould. See MOULD BRICK; MOULD PLUG.

Initial Set. The time interval between the GAUGING (q.v.) of a hydraulic cement and its partial loss of plasticity. In the VICAT'S NEEDLE (q.v.) test, the end of this period is defined as the time when the needle will no longer descend through the cement test block to within 5 mm of the bottom. In B.S. 12, the test conditions are 14.4–17.8°C (58–64°F) and ≥90% r.h.; for normal cements the Initial Set must not be less than 30 min (the usual period is 1–2 h).

Injection Moulding. A process sometimes adopted for the shaping of non-plastic ceramics, e.g. alumina. A plasticizer such as polystyrene or phenol formaldehyde composition is mixed with the ceramic powder and the batch is then warmed and injected into the die.

Ink-Jet Printing. A printing process in which the ink is applied as a succession of sprayed drops. The drop sizes and spray velocity are determined electrostatically.

Inside Colour. The reverse side or inside surface of fired gold decoration on glass. Its appearance is a check on adequate ventilation and proper firing temperature.

Instron. Trade-name. A range of machines for measuring mechanical properties at constant rates of strain. The loads and strain rates are controllable; cyclic fatigue testing is possible; accessories are available to allow tension, compressive and torsional testing of green and fired ceramics at ambient and elevated temperatures.

Insulating Refractory. A refractory material having a low thermal conductivity and used for hot-face insulation in a furnace. Such products have a porosity of 60–75%. If they are made from refractory clay this high porosity is produced by the incorporation of a combustible in the batch or, less commonly, by foaming or by chemical means. In the USA, ASTM – Cl55 classifies insulating refractories as show in the table below.

Insuloc. A system for protecting water-cooled pipes in steel reheating furnaces. An inner layer of refractory fibre is covered by a tough outer skin of interlocking refractory tiles.

Insweep. Term applied to the lower part of a glass container if the sides curve inward or taper towards the base.

Intaglio. The decoration of glass-ware by cutting a pattern to a depth intermediate between that of deep cutting and engraving (The term Intaglio Printing has been applied to the decoration of pottery-ware by transfer-printing from a copper plate.)

Integrated Thermal Process. A system developed in Italy for the firing of tiles by the kilnmaker Poppi and the tile

Insulating Refractories – Classification. ASTM C155

Group Identi-fication	Reheat Change not more than	Bulk Density not greater than
Group 16	1550°F (845°C)	34 lb/ft^3 (0.54 g/cm^3)
Group 20	1950°F (1065°C)	40 lb/ft^3 (0.64 g/cm^3)
Group 23	2250°F (1230°C)	48 Ib/ft^3 (0.77 g/cm^3)
Group 26	2550°F (1400°C)	54 Ib/ft^3 (0.86 g/cm^3)
Group 28	2750°F (1510°C)	60 lb/ft^3 (0.96 g/cm^3)
Group 30	2950°F (1620°C)	68 lb/ft^3 (1.09 g/cm^3)
Group 32	3150°F (1730°C)	95 lb/ft^3 (1.52 g/cm^3)
Group 33	3250°F (1790°C)	95 lb/ft^3 (1.52 g/cm^3)

manufacturer CISA. The tiles are heated rapidly and uniformly by radiation from plates below and above the tiles, which are moved to and fro on rollers in the firing zone. Single firing in 20 minutes is possible, and it is easy to change to double firing. The flexibility of the system allows a greater variety of more artistic tiles to be fired. (*Keram Z.* **38,** (10), 626, 1986).

Intercalary Decoration. A design trapped between two layers of glass. Small pieces of gold foil, coloured glass etc. are placed on one surface, which is then covered by the second layer of hot glass.

Interlocking Tile. A large type of roofing tile having one or more longitudinal ridges and depressions that interlock with the complementary contour of neighbouring tiles on the roof. Such tiles are commonly made in a REVOLVER PRESS (q.v.).

Intermediate Crusher. A machine of a type suitable for size reduction from about 8 to 20 mesh, e.g. a pan mill or ball mill (note, however, that a ball mill can more properly be used as a FINE GRINDER (q.v.)).

Intermediate Piece. See MATCHING PIECE.

Intermittent Kiln or Periodic Kiln. A batch-type kiln in which goods are set, fired, cooled, and then drawn. The principal types in the ceramic industry are ROUND KILNS (or BEEHIVE KILNS), RECTANGULAR KILNS, BOTTLE OVENS, BOGIE KILNS and TOP-HAT KILNS. (See separate entries under these headings).

Intersection Scarp. The locus of intersection of two portions of a crack with each other.

Interstitial. See CRYSTAL STRUCTURE.

Interstop. Trade-name for a two-plate revolving SLIDING GATE VALVE (q.v.) developed by Stopinc AG.

Intumescence. The property of some silicates, notably of PERLITE (q.v.), of expanding permanently, when heated, to form a completely vesicular structure; cf. BLOATING and EXFOLIATION.

Inverse Spinel. See SPINEL.

Inversion; Inversion Point. An instantaneous change in the crystalline form of a material when it is heated to a temperature above the INVERSION POINT; the change is reversed when the material is cooled below this temperature. An example of importance in ceramics is the α ↔ ß quartz inversion at 573°C.

Investment Casting. A process for the casting of small metal components to a close tolerance. In the usual process a wax replica of the part to be cast is coated ('invested') with refractory powder, suitably bonded, and the whole is then warmed (150°C) to melt out the wax – hence the alternative name *Lost-wax Process;* the refractory mould is then fired at 1000–1100°C. The refractory used may be powdered sillimanite or alumina, or specially prepared cristobalite; the latter is particularly used in the application of the process in dentistry. B.S 1902 Pt 10 specifies methods for the determination of resistance to deformation at elevated temperatures; permeability and standard air flow capacity at elevated temperatures; thermal profile determination, for refractories for investment casting shell mould systems prefabricated brickwork, in which reinforcement is provided by steel rods running horizontally across perforations in the blocks.

Inwall. US term for the refractory lining of the STACK (q.v.) of a blast furnace.

Iodoeosin Test or Mylius Test. For determining the durability of optical glass. The amount of free alkali in a freshly fractured surface is determined by means of iodoeosin; the surface is then exposed to moist air at 18°C for 7 days and the free alkali is again determined. Any increase in free alkali is taken as a measure of lack of durability; a decrease indicates stability. The test is not valid for many modern optical glasses. (F. Mylius, *Z. Anorg. Chem.,* **67,** 200, 1910; *Silikat Z.,* **1**, 2, 25, 45, 1913.)

Ion Beam Machining. See ELECTRON BEAM PERFORATION.

Ion Implantation. The introduction of foreign ions into solids, using accelerating energies in the range 20–500 keV. It is a non-equilibrium process, useful for preparing metastable alloys or for modifying suface properties such as friction, wear and corrosion. It can be used to modify superconductive behaviour.

Ion Plating. A vapour phase is created by evaporation, and transported through a gaseous glow discharge to the substrate. In reactive ion plating, metallic species interact with reactive gases to form oxides, nitrides or carbides. Ion plated coatings tend to have greater adherence than evaporated or sputtered coatings.

Ionic Conductor. A material in which the charge carriers are ions rather than electrons or holes. Ionic conductors usually have crystal structures containing channels, or planes through which cations can move. ß-Al_2O_3 ($NaAl_{11}O_{17}$) is an ionic conductor, used as the electrolyte in sodium-sulphur storage batteries. See ELECTRICAL CONDUCTIVITY.

Ionic Exchange. The replacement of ions on the surface, or sometimes within the lattice, of materials such as clay. The ions become adsorbed to balance a deficiency of charge in the clay structure, e.g. in a montmorillonite in which some Mg^{2+} has been replaced by Al^{3+}, or to satisfy broken bonds at the edges of the clay crystals. Ionic exchange capacity is generally expressed in milli-equivalents per 100 g: typical values are: kaolinite, 1–3; ball clay, 10–20; bentonite, 80–100. ASTM C837 tests the absorption of methylene blue dye by clays, a measure of the ion exchange capacity.

Ipro Brick. An I-shaped clay paving brick designed for use in roadmaking, particularly on soils of poor bearing capacity – as in Holland.

IPS. Inert PLASMA SPRAYING (q.v.).

Iridizing. The formation of a film of metal oxide on the surface of a ceramic, particularly of glass. The surface to be treated is heated and then exposed to the vapour, or to atomized spray, of a metal salt. Such surface films may be applied as decoration or to provide a surface that is electrically conducting, e.g. for anti-frosting windscreens.

Irising. A surface fault, in the form of stained patches, sometimes found on flat glass that has been stacked with surfaces in contact. The term originates from the interference colours that often accompany the fault. It is caused by moisture. If the glass is annealed in an acid atmosphere and adequately washed irising is unlikely to occur; separation of the stacked sheets by paper also prevents this trouble.

Iron Modulus. The Al_2O_3:Fe_2O_3 ratio in a hydraulic cement. In portland cement this modulus usually lies between 2 and 3.

Iron Notch. Alternative name for the TAPHOLE (q.v.) of a blast furnace.

Iron Ore Cement. See ERZ CEMENT.

Iron Oxide. See FERROUS OXIDE; FERRIC OXIDE; MAGNETITE.

Iron Saffron. See Indian Red.

Iron Spot. A dark, sometimes slaggy, spot on or in a refractory brick, resulting from a localized concentration of ferruginous impurities; such spots can cause carbon deposition, or even disintegration of the brick, if the latter is exposed to CO attack at 400–500°C or to hydrocarbons at 800–900°C.

Iron-Zirconium Pink. See ZIRCONIUM-IRON PINK.

Ironing. A fault that may arise during the firing of decorated ware having cobalt blue bands, etc.; the band appears dull and may have a reddish scum caused by the crystallization of cobalt silicate.

Ironstone China. See MASON'S IRONSTONE CHINA.

Irregular. See MORPHOLOGY.

Irridizing. US spelling of IRIDIZING (q.v.).

Irwin Consistometer. A simple capillary-flow viscometer designed by J. T. Irwin (*J. Amer. Ceram. Soc.,* **21**, 66, 1938) for testing vitreous enamel slips. Two capillaries are placed in the lower end of a long wide-bore tube; both capillaries have the same bore but one is twice as long as the other. The wide outer tube is filled with the slip to be tested and is then allowed to discharge through the two capillaries into separate measuring cylinders. The rate of flow through each capillary can thus be calculated and a curve can be drawn from which both the yield value and mobility can be read.

Irwin Slump Test. A works test for assessing the setting-up of vitreous enamel slips for spraying. It was first described by J. T. Irwin *(Finish, p.* 28, Sept. 1946) in the following words: 'A ground-coated plate is placed on a table and a steel cylinder, 1 13⁄16ths in. internal diameter and 2½ in. high, (46mm i.d. × 63.5mm high) is placed on the plate. The cylinder is filled with enamel to be tested. The cylinder is then lifted vertically, with a rapid motion, by means of a hook and cord attached to the top of the cylinder and passing over a pulley to a weight. When the weight is released, the cylinder is raised vertically. This action results in a 'pancake' of enamel on the test plate. The diameter of this 'pancake' is a function of the set or stiffness of the enamel slip.'

IS. Indentation Strength. See FRACTURE TOUGHNESS TESTS.

IS Machine. The HARTFORD-EMPIRE INDIVIDUAL SECTION MACHINE (q.v.).

ISCC-NBS Colour System. A system for the designation of a range of identifiable colours used industrially, colours drawn up in USA by the Inter-Society Color Council and the National Bureau of Standards. For full details see NBS Circular 553, 1955, published jointly by the ISCC and the NBS (Now the National Institute of Standards and Technology).

ISO. International Standards Organisation, 1 Rue de Varembé, Geneva.

Isoelectric Point. The pH at which particles in suspension carry zero net electric charge.

ISO-JET Burner. The burner has a combustion chamber with an aligned fuel nozzle at one end, and is arranged to direct a high velocity stream of gaseous combustion products from its other end, producing suction in the firing chamber to improve kiln atmosphere circulation. Dilution air is introduced into this gas stream at an acute angle to the flow, to reduce the gas temperature and enhance the suction effect. (Bickley furnaces Inc. Br.Pat 1461309).

Isomorphous. Literally, having the same shape. Isomorphs are different compounds with the same crystal structure. In *isomorphous substitution*, one ion in the crystal structure is replaced by a foreign ion, facilitated by the crystal structures of the two compounds being isomorphous. See also CRYSTAL STRUCTURE, POINT DEFECTS.

Isostatic Pressing. The dry powder mix or preform to be pressed is put in a flexible rubber or elastomer container, (often called the TOOLING) which is then placed in a container with hydraulic fluid and subjected to hydrostatic pressure, generally 70–140 MPa but for special purposes up to 6 GPa. With metal tooling the process can be used at high temperatures. The merit of this process is the uniform manner in which pressure is applied over the whole surface, resulting in uniform density in the shaped component. Uniform non-directional objects or tubes can be made by incorporating a central MANDRELL. In the WET BAG process the hydraulic fluid is water containing soluble oil. Small batches and large articles are made. The automatic DRY BAG process is used to mass-produce small articles such as spark plug insulators. The pressure medium is a gas. Isostatic pressing avoids problems of die-wall friction but does not provide as accurate tolerances as die-pressing. The process was invented by H. D. Madden (US Pat. 1081 618, (1913).

Isothermal Technique. A systematic method to determine optimum conditions for HOT PRESSING (q.v.). A series of hot pressings are conducted under a set of different isothermal conditions, by heating the material slowly to the required temperature (the isotherm), then applying pressure which is maintained until densification is complete. The time to achieve a given density is plotted against the temperature. Two straight lines appear, whose point of intersection gives the optimum hot pressing conditions.

Istra. Trade-name: a HIGH-ALUMINA CEMENT (q.v.) made in Yugoslavia.

Italian Red. One of the red ferric oxide pigments.

Italian Tile. An alternative name for OLD ROMAN TILE (q.v.) The term is also sometimes used erroneously, as a synonym of SPANISH TILE (q.v.).

ITO Process. Soft-mud extruded large blocks are set directly on to kiln cars. (Fuchs Ziegeleimaschinen GmbH, Austria).

ITP Process. See INTEGRATED THERMAL PROCESS.

IUPAC Rules. The International Union of Pure and applied Chemistry, *'How to name an Inorganic Substance' A guide to the use of nomenclature of inorganic chemistry*. Definitive Rules 1970, Pergamon Press 1977.

Izod Test. A method for measuring impact strength. The energy required to fracture a (usually notched) specimen is calculated from the initial height of a pendulum, and the height of the swing after striking the specimen. (Cf. CHARPY TEST, FRACTURE TOUGHNESS TESTS; EDGE CHIPPING TEST.).

J_C. Critical Current Density, above which a material ceases to be a SUPERCONDUCTOR (q.v.).

J-based Fracture Testing. A method for applying FRACTURE MECHANICS to composites. See also CRACK BRIDGING, and cf. R-CURVE. The crack opening displacement is related to the applied stress by measuring the crack tip stress intensity and the *J-integral.* This is the sum of the energies needed to form a new crack tip and to develop the fracture process zone. It depends on the specimen geometry and is found by measurements on two specimens with different crack lengths. (V.C. Li in *'Applications of Fracture Mechanics to Cementitious Composites'* and (with others) in *J. Amer. Ceram. Soc.* **77**, 1994, p1553)

Jack Arch. A sprung arch of bricks specially shaped so that the outer surface of the arch is horizontal, the inner surface being either horizontal or arched with a large radius. Also known as a FLAT ARCH.

Jack Bricks. Refractory bricks for the glass industry; they are perforated to accommodate the prongs of a fork truck and used beneath a newly-set pot.
Jacking. Using hydraulic or screw jacks to secure the brickwork when constructing refractory rings. Jacks may be used radially, to hold the partly-completed ring against the casing. *Spreader jacks* are used circumferentially in the KEY ZONE to tighten the ring before inserting closure bricks and key plates.
Jam-socket Machine. A machine for shaping the sockets in clay sewer-pipes. The pipes are extruded plain, cut to length and fed to the jam-socket machine in which a ram, having the internal profile of the socket, is forced into the end of the pipe. The machine was introduced by Pacific Clay Products Co., Los Angeles, USA (*Brick Clay Record,* **124,** No. 4, 55, 1954).
Jamb. The brickwork, or other material, forming one of the vertical sides of an opening in a wall of a building or furnace, e.g. a doorjamb.
Jamb Brick. See BULLNOSE, but in particular, specially-shaped refractory bricks to form the portal around the external door of a coke-oven.
Jamb Wall. See BREAST WALL.
Japanese Red. Alternative name for THIVIER EARTH (q.v.).
Jar Mill. A small BALL MILL (q.v.), the revolving cylinder being a vitreous ceramic jar; such mills are used in the grinding of small batches of ceramic colours and vitreous enamels.
Jardiniere Glaze. A former type of unfritted lead glaze containing PbO, K_2O, CaO, ZnO, Al_2O_3 and SiO_2. There were soft (cone 02) and hard (cone 4) types.
Jasper Ware. A fine coloured stoneware first made in 1774 by Josiah Wedgwood who referred to it as being 'peculiarly fit for cameos, portraits and all subjects in bas relief; as the ground may be of any colour throughout, and the raised figures of a pure white'. The body contains barytes and the colours include blue lavender, and sage green (typically blue with white bas relief). A quoted body composition is: 26% ball clay, 18% china clay, 11% flint, 45% barytes.
Jaw Crusher. A primary crusher for hard rocks, e.g. the quartzite used in making silica refractories. The crusher has two inclined jaws, one or both being actuated by a reciprocating motion so that the charge is repeatedly 'nipped' between the jaws. For different types see under BLAKE, DODGE and SINGLE-TOGGLE.
Jena Glass. Various types of glass (the Jena chemically resistant and optical glasses are particularly well known) made by Jenaer Glaswerk Schott and Gen., Jena, Thuringia, Germany.
Jeroboam. A 4-quart wine bottle.
Jersey Fireclay Brick. A US siliceous refractory.
Jersey Stone. A CHINA STONE (q.v.) occurring in Jersey, C.I. A stated composition is (%): SiO_2, 77; Al_2O_3, 12; CaO, 0.5; Fe_2O_3, 0.5; K_2O, 4.5 ; Na_2O, 3.
Jet Dryer. A type of dryer in which the ware is dried chiefly by the action of jets of warm air directed over the surfaces of the ware; the principle has been successfully applied in the drying of pottery.
Jet-enamelled Ware. A type of 18th century porcelain decorated with black on-glaze transfers; (cf. JET WARE).
Jet Impact Mill. See FLUID-ENERGY MILL.
Jet-Kote. A HYPERSONIC FLAME SPRAYING technique. (q.v.).
Jet Ware. Pottery-ware, chiefly tea-pots, having a red clay body and a black, manganese-type, glaze; (cf. JET-ENAMELLED WARE).
Jeweller's Enamel. A vitreous enamel fusing at a low temperature and suitable for application to copper, silver, and

other nonferrous metals in the production of jewelry or badges. Because the principal European centre for enamelled jewelry was Limoges, France, the chief types have French names: BASSE-TAILLE, CHAMPLEVE, CLOISONNE, PLIQUE-A-JOUR (for details see under each name).

Jigger; Jiggering. A jigger is a machine for the shaping (jiggering) of pottery-ware by means of a tool fixed at a short distance from the surface of a plaster mould that is mechanically rotated on the head of a vertical spindle. In the making of flatware a suitable quantity of prepared body from the pug is first shaped into a disk by a batting-out machine placed alongside the jigger; in semi-automatic jiggers the movements of the two machines are co-ordinated. The disk of body is placed by hand on a plaster mould which has been fixed on the head of the vertical spindle of the jigger; the upper surface of the mould has the contour required in the upper surface of the finished ware. The mould is then set in motion by a clutch mechanism and a tool having the profile of the bottom of the ware is brought down on the plastic body which is then forced to take the required shape between mould and tool. The essentially similar process applied to the shaping of hollow-ware, e.g. cups, is known as JOLLEYING. In both processes water is necessary as a lubricant.

Jiggerman. A workman in a glassworks whose job is to return SCULLS (q.v.) to the charging end of a glass-tank furnace. (cf. JIGGERER, the man who operates a JIGGER (q.v.) in a pottery.)

Jiki. Japanese name for porcelain.

Jockey Pot. A POT (q.v.) for glassmaking of such size and shape that it can be supported in the furnace by two other pots.

Joggle. See NATCH.

Joining. The term is particularly applied to the technology required to bond similar or dissimilar materials to form components for mechanical or electrical engineering applications. It is especially used of joining ceramics to metals in machine parts.

Joint Line. A visible line on imperfect glass-ware reproducing the line between separate parts of the mould in which the glass was made. Also known as a PARTING LINE, MATCH MARK, MITRE SEAM MOULD MARK or SEAM.

Jolley; Jolleying. Terms applied to the shaping of hollow-ware in the same senses as JIGGER (q.v.) and JIGGERING are applied to the shaping of flat-ware.

Jolt Moulding. A process sometimes used for the shaping of refractory blocks. A mould is charged with prepared batch which is then consolidated by jolting the mould mechanically; top pressure may simultaneously be applied via a mould plate. cf. TAMPING.

Journey. The period of emptying of a glass-pot before it is again filled with batch; the term is also applied to a shift in which an agreed number of pieces of glass-ware are made.

Jumbo.

(1) A type of TRANSFER LADLE (q.v.) for the conveyance of molten iron.

(2) A hollow clay building block of a type made in USA, its size is 11 ½ × 7½ × 3½ in. (292 × 191 × 89mm) and it has two large CELLS (q.v.) and a 1½-in. (38mm) SHELL (q.v.); the weight is 15–16 lb. (6.7–7.1kg).

(3) Roofing tiles (claimed to be the world's largest) made by Tuiles de Limoux are 53cm × 36cm.

Jump. The slight rise in temperature when a kiln is closed.

Jumpers. See POPPERS.

K_{IC} – See STRESS INTENSITY FACTOR.

Kady Mill. Trade-name: a high-speed dispersion unit consisting principally of a bottom-feeding propeller, a main dispersion head containing a rotor and a stator, and an upper shroud and propeller. (Steele & Cowlishaw Ltd., Stoke-on-Trent, England.)
Kaki. An opaque stoneware glaze, in which a surface layer of iron oxide crystal produces a reddish brown colour.
Kaldo Process. A process for the production of steel by the oxygen-blowing of molten iron in a rotating, slightly inclined vessel; the latter is usually lined with tarred-dolomite refractories. The name is derived from the first letters of the name of the inventor, Professor Kalling, and of the Domnarvet Steelworks, Sweden, where the process was first used.
Kaliophilite. $K_2O.Al_2O_3.2SiO_2$. Transforms at about 1400°C into an orthorhombic phase that melts at approx. 1800°C. Crystals of this mineral have been found in fireclay refractories that have been attacked by potash vapour.
Kalopor. A hollow clay block, with the voids filled with polystyrene. The geometry of the leaves is designed to maximize thermal insulation, and to provide tongue and groove connections for accurate bricklaying. (KLB Vertriebs GmbH, Germany).
Kalsilite. $K_2O.Al_2O_3.2SiO_2$ together with a small amount of Na_2O. This mineral is sometimes formed when alkali vapour attacks fireclay refractories.
Kandite. Group name for the kaolinite minerals, i.e. kaolinite, nacrite, dickite and halloysite. (*Clay Minerals Bull.*, **2**, 294, 1955.)
Kankar. An Indian word for lime nodules in clay soils.
Kanthal. High temperature resistance wire and tape for heaters (A.B. Kanthal, Sweden).
Kaolin. Name derived from Chinese *Kao-Lin,* a high ridge where this white-firing clay was first discovered. See CHINA CLAY.
Kaoline. Trade-name. A fired china clay insulator (Richard, Thomas & Baldwins, Staffs.).
Kaolinite. The typical clay mineral of china clay and most fireclays; its composition is $Al_2O_3.2SiO_2.2H_2O$. The unit lattice consists of one layer of tetrahedral SiO_4 groups and one layer of octahedral $Al(OH)_6$ groups; a kaolinite crystal (which is hexagonal) consists of a number of these alternate layers. Disorder can, and normally does, occur in the stacking of these layers so that the ceramic properties of a kaolinitic clay may vary. When heated, kaolinite loses water over the range 450–600°C; the D.T.A. endothermic peak is at 580°C. When fired at temperatures above 1100°C mullite is formed.
Kaolinization. The geological breakdown of alumino-silicate rocks, usually but not necessarily of the feldspathic type, to produce kaolinite.
Kaowool. Trade-name. Aluminosilicate ceramic fibres. (Morganite Ceramic Fibres, Neston; originally Babcock & Wilcox, USA).
Kassel Brown. A substance containing about 90% humus compounds and only 1% ash from the region of Kassel, Germany; the German name is *Kasselerbraun.* Its use as a protective colloid in the deflocculation of slip was the subject of German Patent 201,987 4/8/06.
Kassel Kiln. An old type of intermittent, rectangular, fuel-fired kiln which diminished in cross-section towards the end leading to the chimney; it originated in the Kassel district of Germany.
Kavalier Glass. An early type of chemically resistant glass characterized by its high potash content; it was first made

by F. Kavalier at Sazava, Czechoslovakia, in 1837, and the Kavalier Glassworks still operates on the same site.

Kaye Disk Centrifuge. A device for particle-size analysis (1–50 μm) by means of a beam of light passed through a liquid suspension of the particles while the suspension is rotated in a transparent disk-shaped centrifuge. The prototype was designed by B. H. Kaye *(Brit. Pat.,* 895,222, 2/5/62); now made by Coulter Electronics Ltd., St. Albans, England.

Keatite. A form of silica resulting from the crystallization of amorphous precipitated silica at 380–585°C and water pressures of 350–1250 bars; sp. gr. 2.50. Named from its discoverer, P. P. Keat *(Science,* **120**, 328, 1954).

Keene's Cement or Keene's Plaster. A mixture of gypsum, that has been calcined at a dull red heat, and 0.5–1% of an accelerator usually potash alum or K_2SO_4. Also known as PARIAN CEMENT or PARIAN PLASTER.

Keel. The refractory-lined lower section of the barrel of a HOT METAL MIXER which is always in contact with the hot metal.

Kek Mill. Trade-name: (1) A pin-disk mill depending for its action on high-speed centrifugal force; (2) a beater mill in which a four-armed 'beater' revolves horizontally at high speed between upper and lower serrated disks. (Kek Ltd., Manchester, England.)

Keller System. A method of handling bricks to and from a chamber dryer; the bricks are placed on STILLAGES (q.v.) which are lifted by a FINGER-CAR (q.v.), carried into the dryer and set down on ledges projecting from the walls of the drying chamber; when dry the bricks are carried in the same manner to the kiln. The system was patented by Carl Keller of Laggenbeck, Westphalia, Germany, in 1894. Since that date the firm has introduced much additional equipment to advance the degree of mechanization in the heavy-clay industry.

Kelly Ball Test. An on-site method for assessing the consistency of freshly mixed concrete in terms of the depth of penetration, under its own weight of 30 lb (13.5kg), of a metal hemisphere, 152 mm dia. (J. W. Kelly and M. Polivka, *Proc. Amer. Concrete Inst.,* **51**, 881, 1955.)

Kelly Sedimentation Tube. A device for measuring the rate of settling of particles from a suspension, and hence for particle-size analysis. To the lower part of a sedimentation vessel, a capillary tube is joined and is bent through 90° so that it is vertical to a level a little above that of the suspension in the sedimentation vessel; above this level the capillary tube is inclined at a small angle to the horizontal. As particles settle in the main vessel the position of the meniscus in the capillary tube moves downward, affording a means of assessing the rate of settling of the particles. This apparatus, designed by W. J. Kelly *(Industr. Engng. Chem.,* **16**, 928, 1924), has been used for the particle-size determination of clays.

Kelvin Equation. An equation useful in sorption studies for the calculation of pore size and pore size distribution. It is:

$$rRT\ln(p/p_o) = -2\gamma V \cos\theta$$

where p is the equilibrium vapour pressure of a curved surface (as in a capillary or pore) of radius r; p_o is the equilibrium pressure of the same liquid on a plane surface; R is the gas constant; T is the absolute temperature; γ is the surface tension; V the molar volume; θ the contact angle of the adsorbate. When the Kelvin equation is satisfied, vapour will condense into pores of radius r. (W.T. Thompson, *Phil. Mag.* **42**, 1871, p.448).

Kelvin Temperature Scale (K). The Centigrade scale displaced downwards so that Absolute Zero (-273.16) is 0 K; 0°C = 273.16 K; etc. Once known as the ABSOLUTE TEMPERATURE,°Abs.

KER System. A coding system for technical ceramics proposed by H.W. Hennicke, *Keram Z.* **22,** (9), 544, 1970. Oxides are coded KER OX followed by a number; silicon carbides are KER SIC, etc.

Keradur. Tradename. A resin composite for moulds. (H & E Börgardts, Germany).

Keramzit. Trade-name: a Russian expanded-clay aggregate

Keraflamm. A process for flame glazing the surface of lightweight concrete. (K. Bergmann, *Ziegelindustrie* (**9**), 330, 1975).

Keralloy. A glass-ceramic coated metal alloy, for use as a substrate for thick-film circuitry. (Ceramic Developments Ltd., Nuneaton.).

KeraLum. Trade-name. A range of glasses, some of which are luminescent, others phosphorescent. (Degussa).

Keram. Trade-name. A plaster made by H & E Borgardts, Germany.

Keram'ik. Trade-name. Oxide based tool-tips for metal cutting. (Raybestos-Manhatten Inc, USA).

Keramiton. Trade-name. A ready-to-use ceramic modelling material, which air dries, or may be fired. (Faber-Castell GmbH).

Keratin. The protein in hair. An extract obtained by the treatment of hair with caustic soda is used, under the name keratin, as a retarder to control the rate of setting of plaster – in making pottery moulds for example.

Kerb. See also curb bend. In the UK wall-tile industry the accepted spelling is 'curb'. A kerb is a refractory course laid to protect a right-angled joint between two materials.

Kerlane. Trade-name. (French) Aluminosilicate ceramic fibres.

Kerner Equation. See THERMAL EXPANSION OF COMPOSITES.

Kerr Effect and Kerr Constant. The birefringence produced in glass, or other isotropic material, by an electric field; the effect was discovered in 1875 by J. Kerr, a British physicist. The Absolute Kerr Constant has been defined as the birefringence produced by unit potential difference.

Kervit Tiles. Trade-name derived from the words 'Keramik ' and 'Vitrum', denoting the mixed nature of the material which is made by casting a ceramic slip containing about 30% of ground glass. The slip is poured on to a refractory former that is coated with a separating material, e.g. a mixture of bentonite and limestone; the tiles, while on the formers, are fired at 950–1000°C then trimmed. The process was first used in Italy; it was introduced into England in 1960. (M. Korach and Dal Borgo, Brit. Pat., 468,010; 10/6/36.)

Kessler Abrasion Tester. Apparatus designed by the National Bureau of Standards, USA, for the determination of the abrasion resistance of floor tiles and quarries. A notched steel wheel is mounted on an overhanging frame so that a definite and constant weight bears on the test-piece as the wheel revolves; No. 60 artificial corundum is fed at a specified rate between the wheel and the test-piece, which is mounted in an inclined position. (D. W. Kessler, *Nat. Bur. Stand. Tech. News Bull.*, **34,** 159, 1950.)

Ketteler-Helmholz Formula. A formula for the optical dispersion of a glass:

$n^2 = n_\infty{}^2 + \Sigma_m M_m(\lambda^2 - \lambda^2{}_m)^{-1}$

where n is the refractive index of the glass for a wavelength λ, n_∞ is the index for an infinitely long wavelength, and λ_m are the wavelengths of the absorption bands for each of which there is an empirical constant M_m.

Kettle. (1) A heated iron or steel vessel in which gypsum is partially dehydrated to form plaster-of-paris.
(2) A vessel for containing molten glass.
Keuper Marl. A Triassic clay much used for brickmaking in the E. and W. Midlands, the W. of England, and the Cardiff area. This type of brick-clay usually contains a considerable amount of lime and iron oxide; magnesium carbonate and gypsum may also be present in significant quantities. Keuper Marl is of variegated colour, hence the name, from German *Köper* (a spotted type of fabric).
Key Brick. A brick with opposite side faces inclined towards each other so that it fits the apex of an arch, see Fig. 1, p39. In furnace construction such bricks are also sometimes referred to as BULLHEADS, CUPOLA BRICKS or CROWN BRICKS.
Keybrick System. A system of bricklaying in which plastic pins hold the bricks in place, while mortar is poured around them. The bricks are made to close dimensional tolerances by repressing extruded blanks.
Key-in. To undercut all or part of the surrounding sound refractory to provide a physical key for the repair to a monolithic lining.
Keying. Ensuring a tight fit in arch or ring construction.
Key-plates. Steel plates, 3mm thick in the UK, driven into one or more radial joints in a ring of bricks to tighten and secure it.
Key Zone. The final sector installed in a ring of bricks, where fitting adjustments are made, using closure bricks and key plates.
KHN. KNOOP HARDNESS NUMBER (q.v.).
Kibbler Rolls. Toothed steel rolls of a type frequently used in the crushing and grinding of brick clays; from an old word *Kibble – to grind.*
Kick's Law. A law relating to the energy consumed in the crushing and grinding of materials. The energy required to produce identical changes of configuration in geometrically similar particles of the same composition varies as the reduction in the volume or weight of the particles. (F. Kick, *Das Gesetz der Proportionalen Widerstand und Seine Anwendung,* Leipzig, 1885.)
Kidney. See POTTER'S HORN.
Kieselguhr. German name, formerly used also in Britain but now obsolete, for DIATOMITE (q.v.).
Killas. Term used in the Cornish china-clay mines for the altered schistose or hornfelsic rocks in contact with the granite and often considerably modified by emanations from the latter.
Kilmo. Computer software for calculating heat flow through kiln walls, and heat flow and air flow in the kiln. (Donald Shelley, UK).
Kiln. A high-temperature installation used for firing ceramic ware or for calcining or sintering. Kilns for firing ceramic ware are of three main types: INTERMITTENT, ANNULAR and TUNNEL; for calcining and sintering: SHAFT KILNS, ROTARY KILNS and MULTIPLE-HEARTH FURNACES are used (see entries under each of these headings).
Kiln Car. A car for the support of ware in a tunnel kiln or bogie kiln; the car has four wheels and runs on rails. In a tunnel kiln, through which a number of these cars are moved end-to-end, the metal undercarriages are protected from the hot kiln gases by continuous sand-seals. The car deck is constructed of refractory and heat-insulating material.
Kiln Furniture. General term for the pieces of refractory material used for the support of pottery-ware during kiln firing; since the use of clean fuels and electricity has made possible the

open-setting of ware, a multiplicity of refractory shapes has been introduced for this purpose. For individual items see COVER, CRANK, DOT, DUMP, PILLAR, PIN, PIP, POST (or PROP), PRINTER'S BIT, RING, SADDLE, SAGGAR, SETTER, SPUR, STILT, THIMBLE; see also Fig. 4. As well as fireclay refractory, recrystallized silicon carbide and nickel metal have been used.

Kiln Scum or Kiln White. See SCUM.

Kimmeridge Clay. A clay of the Upper Jurassic system used as a raw material for building bricks in Dorset and Oxford; the lower beds often contain gypsum and lime, the upper beds are shaly and bituminous.

King Closer. A brick cut diagonally, so that one end is full-width, the other half width (50cm).

King's Blue. A blue pigment, being mixed cobalt oxide and alumina.

Kink. A type of WAVINESS (q.v.) which intersects the edge of the wavy surface. (ASTM F109)

Kiss Marks. Local term for discoloured areas on the faces of bricks that have been in contact during the kiln firing.

Klebe Hammer. A device for the compaction of test-pieces of cement or mortar prior to the determination of mechanical strength; a weight falling from a fixed height on the test-piece mould ensures compaction under standardized conditions.

Klein Turbidimeter. Apparatus designed by A. Klein *(Proc. A.S.T.M.,* **34,** Pt. 2, 303, 1934) for the determination of the specific surface of portland cement. The sample is suspended in castor oil in a dish and the turbidity is measured photoelectrically. Because of the high viscosity of the suspending liquid the particles do not settle significantly. The specific surface is deduced from the turbidity by means of a calibration curve.

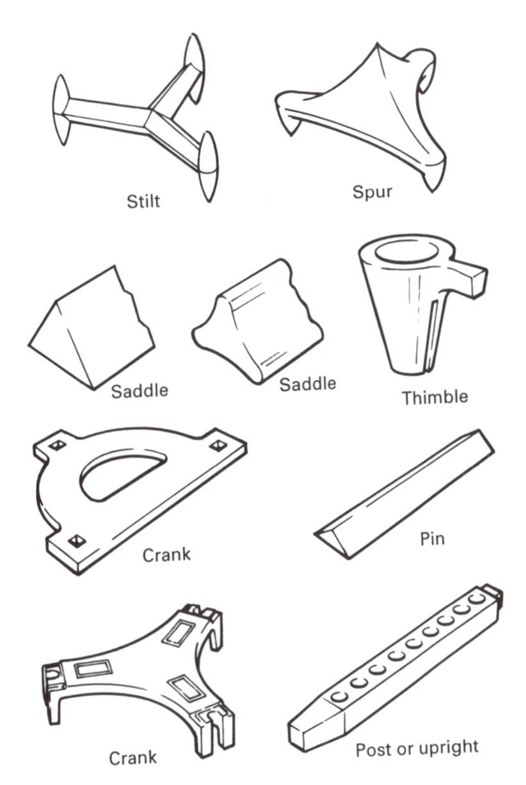

Fig. 4 See KILN FURNITURE
(The relative sizes are not to scale)

Kling. A type of TRANSFER LADLE (q.v.) for the conveyance of molten iron.

Klingenberg Clay. A refractory clay containing (unfired) 32–37% Al_2O_3; PCE. 32–34. From Klingenberg-am-Main S.W. Germany.

Klinker Brick. A type of building brick, which may be either of engineering-brick or facing-brick quality, made in Germany and Holland from clays that generally have a long vitrification range. The bricks have low water absorption and high crushing strength The colour may be yellow, red or variegated.

Klompje Brick. A miniature facing brick of a type used in S. Africa for decorative construction.
Knock-in Mould. A mould with a loose top die, for hand or light machine forming.
Knocking. The accidental removal, during the period between glaze application and the glost firing, of a patch of glaze from the surface of ceramic ware.
Knockings. US term for the oversize material remaining after ceramic slip has been screened; the UK word is KNOTTINGS.
Knock-out. See BUTTON.
Knoop Hardness. A pyramidal-diamond indentation test. The result is expressed in kg/mm^2, the applied load being indicated, eg. Knoop 100 or K100; a US abbreviation is KHN(Knoop Hardness Number) The test was introduced by F. Knoop, C G Peters and W. B. Emerson (*J. Res. Nat. Bur. Stand.*, **23**, 39, 1939). ASTM C849 describes its application to ceramic whitewares; C730 to glass.
Knot. A fault, in glass, consisting of a small inclusion of glass having a different composition (and so revealed by its different refractive index) from that of the surrounding glass.
Knottings. UK term for the oversize material remaining after a ceramic slip has been screened; the US word is KNOCKINGS.
Knotts. Term sometimes applied to the Lower Oxford Clay used for brickmaking in the Peterborough district, England.
Knuckling. Shaping a thrown pot with the knuckles of the hand.
Köhn Pipette. Apparatus for particle-size analysis by sedimentation; it has been used more particularly in the study of clay soils. (M. Kohn, *Z. Pflanz. Dung.*, **A11,** 50, 1928.)
Kolene Process. Trade-name: an electrolytic salt-bath treatment of sheet-steel before it is enamelled; (Kolene Corp., Detroit, USA).
Kopecky Elutriator. An elutriator consisting of three cylinders of different diameters, it has been used for the particle-size analysis of clays. *(Tonind. Ztg., 42,* 629, 1918.)
Kopp's Rule. The SPECIFIC HEAT of a compound is approximately equal to the sum of the heat capacities of its constituent elements.
Koranna Stone. Pyrophyllite (q.v.).
Kovar. A FeNiCo alloy, whose thermal expansion curve against temperature intersects that of alumina at 620°C. If the differential stresses during cooling can be withstood, a glass-metal seal made at 620°C will be stress-free at room temperature. Kovar is thus valuable for glass-metal and alumina metal seals.
Kozeny Equation. An equation relating the rate of flow (q) of a fluid of viscosity μ through a packed bed of particles of depth L and area A, under a pressure difference ΔP, the specific surface of the particles being S and the voids per unit weight being V:

$$q = K.\ AV^3/\mu S^2.\ \Delta P/L$$

This equation, due to J. Kozeny (Sitz. *Akad. Wiss. Wien., Math., Naturw. Kl.,* ***136***, 271, 1927) forms a basis for the determination of the specific surface of a powdered ceramic material by the gas permeability method.
Kramers-Kronig Dispersion Relations. Describe the scattering of light by small particles, based on fundamental considerations of causality.
Kreuger's Ratio. A ratio claimed by H. Kreuger *(Trans. Roy. Swedish Inst., Sci. Res.* No. **24**, 1923) to be a criterion of the frost resistance of clay building bricks; it is the ratio of the percentage water absorption after 4 days' immersion in

cold water to the total water absorption calculated from the specific gravity (cf. SATURATION COEFFICIENT).

Kreutzer Roof. A design for a furnace roof, particularly for open-hearth steel furnaces. Its feature is the system of transverse and longitudinal ribs, which divide the exterior of the roof into box-like compartments. The design was patented by C Kreutzer in Germany in 1948.

Kronig-Penny Model. A simplified, one-dimensional model of the crystal lattice, giving the essentials of the behaviour of electrons in a periodic potential. See also CRYSTAL STRUCTURE, ELECTRICAL CONDUCTIVITY.

Krupp-Renn Process. A rotary-kiln process, developed at Essen, Germany, in 1932–33, for the direct extraction of nodules *(luppen)* of iron from low-grade ores. Refractories in the 'luppen zone', nearest to the discharge end of the kiln are exposed to the most severe conditions: attack by a ferruginous slag at about 1300°C. Dense siliceous firebricks and high-alumina refractories have been used. The term derives from the German word Renn, meaning 'direct'.

Kryptol Furnace. The term used on the Continent and in Russia for a granular-carbon resistance furnace; the name derives from the German company, Kryptol Gesselschaft, that originally supplied the carbon granules.

K-S Test. The Kolmogorov-Smirnov goodness of fit test assesses the ability of a probability distribution calculated from WEIBULL STATISTICS (q.v.) to fit the experimental data. It and the *A-D (Anderson-Darling) test* (which is more sensitive to discrepancies at low and high probabilities of failure) are used as part of the CARES (q.v.) computer program for failure prediction. For a discussion of the K-S and A-D tests, see R.B. D'Agostino and M.A. Stephens, *Goodness of Fit Techniques*, Dekker 1986, p97–193. An additional test of goodness of fit is given by the Kanofsky-Srinivasan 90% confidence bands. (*Biometrika* **59** (3) 1972 p623.).

Kubelka-Munk Theory. This theory describes the diffusion of light through a translucent medium in terms of a scattering coefficient S and an absorption coefficient K. Equations relate these coefficients to the measured values of reflectance and transmission, suitably corrected for internal and external surface reflections when applied to ceramics. (P. Kubelka and F. Munk, *Z. Techn. Physik* **12**, 593, 1931. applied to ceramics by A. Dinsdale, *Am.Ceram. Soc. Bull* **55,** 993, 1976).

Kuhl Cement. A hydraulic cement introduced by H. Kuhl (Brit. Pat., 231,535; 31/3/25). It contains less SiO_2 but more Al_2O_3 and Fe_2O (about 7% of each) than does portland cement; its strength properties are similar to those of rapid-hardening portland cement. Also known as BAUXITLAND CEMENT.

Kurlbaum Method. A method for the determination of flame temperature by means of an optical pyrometer (F. Kurlbaum, *Phys. Z.,* **3**, 187, 1902).

Kyanite. *A* mineral having the same composition (Al_2SiO_5) as sillimanite and andalusite, but with different physical properties. The chief sources are Virginia and S Carolina (USA), and India. When fired, kyanite breaks down at 1300°C into mullite and cristobalite with a volume expansion of about 10%; it is therefore calcined before use. Calcined kyanite is used in making aluminous refractories.

Laboratory Glassware. Chemical resistance tests are specified in B.S. 3473. There are other standards for particular products (e.g. B.S. 2586 for pH electrodes; B.S. 572 and 2761 for conical and spherical joints.

Laboratory Porcelain. Quality requirements and testing methods are laid down in B.S 914.

Labradorite. A lime-soda FELDSPAR (q.v.). A typical composition is (%): SiO_2, 55; Al_2O_3, 30; CaO, 12; Na_2O, 0.5; H_2O etc, 2.5.

Labyrinth Factor. A correction to be applied in calculating the gas permeability of a porous refractory, to allow for the diminished air flow because pores are neither circular nor straight. The labyrinth factor is the ratio of that part of the total porosity which contributes to the air-flow, to the total porosity.

Laced Valley. A form of roof tiling designed to cover a re-entrant corner of a roof. The courses of tiling meet and interlace to provide efficicnt drainage.

Lacustrine Clay. A clay that was formed by deposition on the bed of a lake; the lacustrine clays of the Vale of York, England for example, are used in brickmaking

Ladd Circular Tunnel Kiln. A ROTARY-HEARTH KILN (q.v.) of a type introduced in 1941 by the Ladd Engineering Co., USA (see *Brick Clay Record,* ***99***, No 2, 42, 1941).

Ladder. A horizontal support for glass tubing that is to be cut into lengths (see also WASHBOARD).

Ladle Brick. A refractory brick of suitable shape and properties for use in the lining of a ladle; unless otherwise stated it is assumed that a ladle for the casting of steel is implied. Ladle bricks are almost invariably made of grogged fireclay; the features required are good shape, uniform size, low porosity and, if possible, a permanent expansion when exposed to high temperature. In the USA three qualities of ladle brick are specified in ASTM-C435.

Ladle Shroud. A refractory tube attached to the teeming system of a ladle, extending below the liquid surface in the tundish, to protect the steel stream from the atmosphere.

Ladling. An obsolescent term, describing the hot repair of small diameter rotary kilns by pouring a clay-based sloppy mortar onto the repair site from a ladle introduced through the kiln hood.

Lagoon, Lagoon Settling. A large shallow pond into which suspended solids are allowed to settle after a process, for recovery of the waste, or prevention of discharge of harmful effluent.

Laitance. The layer of water sometimes formed on the upper surface of freshly placed concrete as a result of the aggregate settling by sedimentation (see also BLEEDING).

Lambert's Law. For a TRANSLUCENT (q.v.) body, $I=I_0e^{-\mu t}$ where I is the intensity of the transmitted light; I_0 the intensity of the incident light; μ is a constant, the linear absorption coefficient of the material and *t* is its thickness.

Laminate. A material in sheet form consisting of several different layers united by a ceramic bond. Ceramic laminates have been made to provide lightweight heat insulation. Refractory laminates have been used as bats on which to fire electroceramics and ferrites; these laminates consist of a silicon carbide core with outer layers of alumina or mullite.

Laminated Glass. Safety glass made from alternating and adherent layers of glass and organic plastics, the glass layers being outermost; if broken, the fragments of glass are held in position by the plastics interlayer. For some purposes the interlayer may be made of glass fibre. Other laminated glassware is made by simultaneously forming two glasses with similar viscosities at the forming temperature. Thermometers are made by laminating a white OPAL GLASS (q.v.) stripe into a clear glass tube. Corning Glass developed a special vertical turret machine which forms glass tableware with an outer skin of lower thermal expansion, laminated to an inner core glass. When the ware cools, the surface is placed in compression, improving strength and thermal shock resistance.

Lamination. (1) Textural inhomogeneity in clayware resulting from the shaping process and particularly common in products that have been extruded. In structural clay products, lamination is a source of mechanical weakness and a cause of poor frost-resistance; in refractory products lamination may reduce resistance to slag attack and to thermal shock.
(2) An elongated excrescence, or line of such excrescences, in vitreous enamelware; the fault is usually a manifestation of swelling in the metal.

Lamotte Comparator. A pH meter of a type recommended by the US Porcelain Enamel Institute for use in the determination of the acidity or alkalinity of PICKLING (q.v.) solutions. (Lamotte Chemical Products Co., Baltimore, USA.)

Lamp-blown. Glass-ware shaped by means of an oxy-gas or air-gas burner; glass tubing or rod is the usual starting material.

Lancashire Kiln. A type of HOFFMANN KILN (q.v.) large wickets facilitate setting and drawing.

Lancaster Mixer. Trade-name: A counter-current pan-type mixer. (Posey Iron Works Inc., Lancaster, Pa., USA.)

Land-drain Pipe. See FIELD-DRAIN PIPE.

Lander. See LAUNDER.

Lance. (1) A refractory tube for the injection of gases into a melt.
(2) = STRIATION (q.v.) in fractography.

Langmuir's Adsorption Isotherm. See ADSORPTION.

Lanthanum Boride. LaB_6; m.p. 2100°C; sp. gr. 4.7; thermal expansion (25–1000°C), 5.7×10^{-6}

Lanthanum Carbide. LaC_2 exists in various forms: the tetragonal room-temperature form changes to hexagonal at intermediate temperatures; at 1750°C, through reaction with excess carbon, a cubic phase is produced.

Lanthanum Oxide. La_2O_3; m.p. approx. 2200°C; sp. gr. 6.51; thermal expansion (25–1000°C) 12×10^{-6}. A rare earth sometimes used in special optical glasses.

Lanthanum Titanates. Two compounds have been reported: $LaTi_2O_3$ and $La_2Ti_2O_9$; the former has a PEROVSKITE (q.v.) structure and can be synthesized by heating a mixture of La_2O_3 and Ti_2O_3 at 1200°C in a vacuum.

Lanxide Process. The directed oxidation of molten metal precursors, injected into a mass of filler. (Lanxide, Newark, Delaware, USA. See *Ceram. Eng. Sci. Proc.* **8** (7/8), 879, 1987).

Lap. (1) A faulty surface on glass-ware caused by a fold in the glass.
(2) A rotating disk, normally horizontal, carrying abrasive grain or powder and used for the finishing of work-pieces.
(3) The amount by which a roofing tile overlaps the course next but one below it.

Laporite. Trade-name for synthetic hectorite (q.v.) (originally made by the Fuller's Earth Union, which became part of the Laporte Group.)

Large Nine-inch Brick;. US term for a brick 9 × 6-¾ × 3 in. or 9 × 6-¾ × 2½ in.

Larnite. One of the crystalline forms of CALCIUM ORTHOSILICATE (q.v.).

Laser. A device for producing **l**ight **a**mplification by **s**timulated **e**mission of **r**adiation. The atoms of various materials will absorb light and remain for quite long periods in a metastable 'excited state'. The emission of a few photons will 'stimulate' or induce the emission of others so that all the excited atoms return together to their ground state energy. The light may be emitted as a very short pulse (ruby laser) or as a continuous beam (helium/neon or organic excimer ultraviolet laser). This radiation is of high intensity, of very specific wavelength (monochromatic) and unlike ordinary light, very coherent – having a constant phase relation. These properties allow lasers to be used to transmit information along OPTICAL FIBRES (q.v.) to be used in measuring equipment based on interference and diffraction effects, such as rangefinders and PARTICLE SIZE ANALYSIS; to heat materials to high temperatures in LASER ABLATION, powder preparation or in LASER MACHINING.

Laser Ablation. The formation of samples of fine powders by burning them from a solid surface using a laser.

Laser Beam Machining. A machining process in which material is removed by melting or vaporization with high energy focussed LASER beams. Very precise localised heating is possible, by which means, for example, fine holes may be drilled in highly refractory materials.

Laser Decoration. Glass surfaces may be etched or engraved using a focussed LASER (q.v.) beam. The appropriate pattern may be produced by covering areas with a metal mask; by a vector method, with linear movement of the beam controlled by a computer program; by raster scanning. The beam must be modulated to produce a series of points rather than a continuous melted line, which would lead to cracking.

Laser Particle Size Analysis. The precise and constant wavelength and coherent phase relationships in LASER (q.v.) beams are valuable in particle-size anaylsis. Techniques include those based on diffraction, and on laser Doppler effect measurement of the velocities of falling particles, in which a system of 'beats' is established between the unreflected beam and the beam reflected from the falling particles, to measure the frequency difference and hence the SEDIMENTATION (q.v.) velocity.

Lateral Crack. A crack beneath and parallel to a glass surface, produced during the unloading of contact with a hard, sharp object.

Laterite. Material rich in alumina and/or iron oxide; it is formed by the tropical weathering, *in situ* of appropriate silicate rocks. The most aluminous laterites are known as BAUXITE (q.v.).

Lath. See CLAY LATH.

Lathe. A machine for turning unfired hollow-ware, e.g. cups or vases.

Latorex. A building material made of LATERITE (q.v.) and a bond, without firing.

Lattice Brick. A hollow building brick, or block, in which the cells form a pattern of open lozenges; such bricks are claimed to have a high heat-insulating value because of the extended path of any heat flow through the solid material.

Lattice Colour. Ceramic colours in which the CHROMOPHORE (q.v.) or colouring atom forms part of the crystal lattice. (Cf. ENCAPSULATED COLOURS).

Lattice Constant, Lattice Vector. See CRYSTAL STRUCTURE.

Launder or Lander. An inclined channel, lined with refractory material, for the conveyance of molten steel from the furnace tap-hole to a ladle.

Lavaflame. A flame gunning system for the repair of coke-oven walls (Nippon Steel Corp., Japan.

Lawn. Term used in the British pottery industry for a fine-mesh screen or sieve. In the past, makers of these 'lawns' gave them somewhat arbitrary numbers; they are now largely replaced by meshes conforming to British Standards.

Lazy Slurry. A dilatant body which remains fluid when still or slowly stirred, but solidifies if stirred rapidly. It comprises about 100 parts of precipitated calcium carbonate (depending on the particle size); 10 parts of 5% $ZnNO_3$; 20 parts 5% Calgon; 40 parts water.

L-D Process. A process for the production of steel by the oxygen-blowing of molten iron held in a static vertical vessel. The latter is lined with tarred or fired dolomite refractories; dead-burned magnesite is also sometimes added to the batch. The process was first used in 1952, at Linz, Austria, and 'L-D' stands for Linzer Düsenverfahren, i.e. Linz Nozzle Process (not Linz-Donawitz, as is commonly believed).

Lea and Nurse Permeability Apparatus. A device for the determination of the specific surface of a powder by measurement of the permeability to air of a prepared bed of the sample; the calculation is based on the CARMAN EQUATION (q.v.). (F. M. Lea and R. W. Nurse, *J. Soc. Chem. Ind.*, **58**, 277, 1939.)

Leaching. In particular the process of removing soluble constituents from heat treated glasses, using hot dilute acids, to produce POROUS GLASS (q.v.).

Lead Antimonate (Naples Yellow). This compound, generally used in conjunction with tin oxide or zinc oxide, gives a good onglaze yellow for pottery decoration; it is also used, with a lead borosilicate flux, as a vitreous enamel colour.

Lead Bisilicate. See LEAD SILICATE.

Lead Chromate. The basic chromate, PbO. $PbCrO_4$, is normally used in ceramics. If the firing temperature is low, it can be used to produce coral reds in pottery glazes; in glassmaking it produces a deep green.

Lead Crystal Glass. A CRYSTAL GLASS (q.v.) specified in B.S. 3828 as containing ≥24% PbO; cf. FULL LEAD CRYSTAL GLASS.

Lead Niobate. $Pb(NbO_3)_2$ a ferroelectric compound having properties that make it useful in high-temperature transducers and in sensing devices. The Curie temperature is 570°C.

Lead Oxide. The more common lead oxides are litharge (PbO, m.p. 886°C) and red lead or minium (Pb_3O_4). These oxides are important constituents of heavy glassware, enamels, and some pottery glazes; because the oxides are poisonous, it is normal (in the UK compulsory) to form an insoluble lead silicate frit and to employ this non-toxic material as a source of lead for ceramic glazes.

Lead Poisoning. In the UK, lead poisoning has been virtually eliminated from the pottery, glass, and vitreous-enamel industries by successively more stringent factory regulations. The current regulations for the UK pottery industry are the 'Control of Lead at Work Regulations 1980' and the 'Potteries etc. (Modifications) Regulations'.

Lead Silicate. A material obtained by fritting lead oxide with silica in various ratios. The usual ratio approximates to $PbO.2SiO_2$ and a frit of this composition is known as lead bisilicate; a specification (*Trans. Brit. Ceram. Soc.*, **50**, 255, 1951) is 63–66% PbO, 2–3.5% Al_2O_3, and PbO + SiO_2 + Al_2O_3 to exceed 98%. A LEAD SOLUBILITY TEST (q.v.) is also specified. The bisilicate is used in pottery glazes, enamels, etc. Tribasic lead silicate, $3PbO.SiO_2$, finds some use in making lead glasses. There is also a so-called lead monosilicate with the approximate formula $PbO.0.7SiO_2$.

Lead Solubility Test. The (British) 'Pottery (Health and Welfare) Special Regulations, 1947 and 1950' defined this test, as applied to the specification of a Low Solubility Glaze as follows: 'A weighed quantity of the material which has been dried at 100°C and thoroughly mixed is continuously shaken for 1 h, at the common temperature, with 1000 times its weight of an aqueous solution of HCl containing 0.25% by wt of HCl. This solution is thereafter allowed to stand for 1 h and then filtered. The lead salt contained in the clear filtrate is then precipitated as PbS and weighed as $PbSO_4$.' The American test (ASTM – C555) exposes the glazed ware to acetic acid (1 : 20) at 60°C for 30 min. (See also LOW-SOLUBILITY GLAZE).

Lead Stannate. $PbSnO_3$: up to 5% of this material is sometimes added to barium titanate ceramics to reduce their tendency to depolarize when used as piezoelectrics.

Lead Tantalate. $PbTa_2O_6$; a ferroelectric compound of interest as an electro-ceramic. The Curie temperature is 260°C.

Lead Titanate. $PbTiO_3$: added in small amounts to barium titanate ceramics to improve their piezoelectric behaviour; a complex lead titanate-zirconate body (P.Z.T.) finds use as a ceramic component in piezoelectric transducers. The Curie temperature is 490°C.

Lead Zirconate. $PbZrO_3$; a ferroelectric material. It is also used in lead titanate-zirconate (P.Z.T.) piezoelectric ceramics.

Leaded Glass. Windows made from small pieces of glass which are held in position by strips of lead, H- or U-shaped in section.

Leadless Glaze. In the UK this is defined in the 'Potteries etc (Modifications) Regulations 1990 (S.I. 305)'. A glaze which does not contain more than 1% of its dry weight of a lead compound calculated as PbO. Under current environmental pressures to reduce the use of lead, bismuth glaze and colour flux systems are being actively studied.

Lean Clay. A clay of low plasticity; the adjective 'lean' is also applied to a body of low plasticity.

Lean Mix. (1) Fuel with a high proportion of combustion air.
(2) Concrete with low cement content.

Leather-hard. A term for clayware that has been partially dried; at this stage of drying, the shrinkage has been largely completed.

LECA. Abbreviation for LIGHTWEIGHT EXPANDED CLAY AGGREGATE (q.v.).

Le Chatelier Soundness Test. A procedure for assessing any expansion of hydraulic cement caused by the presence of excess lime, magnesia or sulphates. The gauged cement is put into a split brass cylinder 30 mm i.d., to which are attached two needles 156 mm long from the centre-line of the mould, one needle on each side of the split in the mould. The cement is allowed to set for 24 h immersed in water at 15–18°C; the distance between the ends of the needles is then measured. The mould is immersed in water and boiled for 1 h. When it has cooled, the distance between the ends of the needles is again measured. The difference between the first and second readings should not exceed 10 mm. The test is included in B.S. 12.

Ledge Rock. US term for GANISTER (q.v.) occurring in solid rock formations, as distinct from FLOE ROCK (q.v.).

Lee Board. See under HACK.

Leer. The original form of the word LEHR (q.v.).

Lehr. A heated chamber for annealing glass-ware; usually tunnel-shaped, the ware passing through on a conveyor. The spelling *lehr* originated in USA at the end of the 19th century; the older form – *leer* – first appeared in an English text of 1662. The origin of the word is obscure.

Leighton Buzzard Sand. An important source of sand from the Lower Greensand deposits of Bedfordshire, England. The sand is high in silica and well graded; it is used as a refractory foundry sand and as a standard sand for mixing with portland cement for testing according to B.S. 12.

Lens-fronted Tubing. Tubing for liquid-in-glass thermometers made in such a way as to magnify the width of the column of liquid.

Lepidolite. A lithium mineral, approximating to $LiKAlF_2Si_3O_4$; sp. gr. 2.9; m.p. 1170°C. The largest deposits are in Rhodesia but it also occurs in USA; the average Li_2O content is about 4%. Lepidolite is a source of Li for special glasses, pottery bodies and glazes, and vitreous enamels; account must be taken of its F content.

Lepol Kiln. See ACL KILN.

Leptize. To disintegrate a material, in the dry state, into very fine particles by impact, at least 50% of the product is <50μm and at least 1% is 2μm. The term was introduced by J. G. Bennett, M. Pirani, and M. W. Thring (Brit. Pat. 549,142, 9/11/42); it has not been generally adopted.

Lessing Rings. A particular shape for chemical stoneware tower fillings.

Leucite. $K_2O.Al_2O_3.4SiO_2$; there are two forms stable above and below 620°C respectively; sometimes formed when alkali attacks fireclay refractories, as in a blast furnace stack.

Lew Board. See under HACK.

LG Process. The large grain process for seawater magnesia. Greater density and grain size is achieved by using tube filter presses to dewater the slurry to the optimum 75% solids content. The product can be directly dead-burned. (*Refract J.* May/June, 1976, p13).

Lias Clay. Clay of variable composition, usually calcareous occurring in the Jurassic System. In Britain it is used to a limited extent in brickmaking, e.g. in Northamptonshire.

Libbey-Owens Process. See COLBURN PROCESS.

Libo. Trade-name; a lightweight building material made from very fine sand and lime (Nederlands-Belgische Maatschappij voor Lichte Bouwproducten, Gorinchem, Holland).

Liesegang Rings. The stratification of precipitated material formed when one reactant diffuses into another. Liesegang rings can be seen in minerals such as agate.

Lift. A fault in vitreous enamelware, a relatively large area of enamel coming from the metal; the fault is also known as PEELING. The cause may be inadequate cleaning of thc base metal or a defective ground-coat.

Ligand Field Theory. A theory concerned with the changes in electronic energy levels of ions, especially of transition element ions, which occur when other ions or polar groups (ligands) are brought into their immediate neighbourhood. Variations in the environment of the ion, e.g. in the nature of the ligands or the symmetry of their arrangement, cause variations in the spacings of the energy levels. This forms the basis of the modern explanation of the colours of crystals and of glasses.

Light. (1) As applied to optical glass the term means that the glass has a relatively low refractive index.

(2) When applied to part of a glass container this word means that that part is too thin; e.g. Light Shoulder.

Light-extinction Method. See TURBIDIMETER.

Light Magnesia. A form of magnesium oxide normally made by extraction from sea-water (see SEA-WATER MAGNESIA); the precipitated hydrate is subsequently purified by conversion to the carbonate which is then calcined at approx. 800°C. The bulk density is 0.1. On exposure to the air it takes up water and CO_2 to form a basic carbonate. Light magnesia finds use as a heat insulator and as a source of magnesia in various chemical processes.

Lightweight Expanded Clay Aggregate. A bloated clay aggregate made by the sudden heating of suitable clays either in a rotary kiln (the original method used in Denmark in 1939) or on a sinter-hearth. It is used as an aggregate for making lightweight concrete. B.S. 3797 specifies appropriate materials and properties for lightweight aggregates, structural concrete and concrete blocks. The US specification is ASTM C331.

Lignosulphonate. See SULPHITE LYE.

Lime. CaO; m.p. 2615°C; sp. gr. 3.3; thermal expansion (0–1000°C), 13.5×10^{-6}. Added as WHITING (q.v.) to some pottery bodies and glazes, and as crushed limestone ($CaCO_3$) or magnesian limestone to most glass batches. Limestone is a major constituent of the batch used to make portland cement clinker.

Lime Blowing. The falling away of small pieces from the face of a clay building brick as a result of the expansion (following hydration and carbonation by the atmosphere) of nodules of lime present in the fired brick. Some brick-clays contain nodules of calcite ($CaCO_3$) which are converted into CaO during the kiln firing; these nodules can be rendered innocuous by fine grinding and harder firing. A cure that is more of an expedient is DOCKING (q.v.).

Lime Refractory. Because of its abundance and high m.p.(2615°C), lime would be an attractive basic refractory but for its ready hydration and carbonation when exposed to the air. There has been much research on methods of stabilization, but the only accepted use of lime as a refractory has been as a container material for the melting of the platinum metals.

Lime Saturation Value. The ratio of the actual lime content of a hydraulic cement to that calculated from an equation deduced as representing the amount of lime combined as silicate, aluminate and ferrite. Several such equations have been proposed; all are empirical.

Lime-slag Cement. Alternative name for SLAG CEMENT (q.v.).

Limestone. A hard rock consisting of calcium carbonate, $CaCO_3$. It is used as a source of lime in cement manufacture and in glass-making; for the latter use see B.S. 3108 'Limestone for colourless glasses'.

Limit State Design. Structures reach their *limit state* when they become unfit for use, either by *collapse* or by reaching their *serviceability limit.* Attempts to calculate load-bearing capacity and maximum loads as precisely as possible, and then choose a safety factor in design so that the former is appreciably greater than the latter, are replaced in limit state design by a design philosophy in which the arbitrary safety factor is replaced by an estimate of the strength for which the probability of failure does not exceed some acceptably low value. Limit state design is now applicable to brick masonry structures.

Linde Flame Plating. See FLAME PLATING.
Lindemann Glass. A lithium beryllium borate glass that is highly transparent to X-rays. It is made from a batch consisting of 10 parts $Li_2B_2O_7$, 2 parts BeO and 3 parts B_2O. This glass is difficult to shape and is of low chemical durability. (C. L. Lindemann and F. A. Lindemann, *Z. Roentgenkunde,* **13,** 141, 1911.)
Lindemann-Danielson Test. See DANIELSON-LINDEMANN TEST.
Line Defect. See CRYSTAL STRUCTURE.
Line Reversal Method. See SODIUM-LINE REVERSAL METHOD.
Liner Plates. Vitrified clay units shaped in the arc of a circle, used as a protective internal lining for pipes, culverts and other structures. They usually have tenon ribs on their backs to key them firmly to the structure to be protected. ASTM C479 refers.
Lining. See BANDING.
Linhay. Cornish name for a storage shed for china clay.
Linseis Plastometer. A device designed by M. Linseis for the evaluation of the plasticity of clay on the basis of two parameters: cohesion (measured as tensile strength) and the capacity for relative movement of the clay particles without rupture. The apparatus is made by Netzsch Bros., Selb, Germany.
Lintel or Mantel. The metal supporting structure that takes the weight of the refractories and casing of the stack of a blast furnace.
Liparite. A term that has been used for rocks of various kinds but, in Europe, it has usually referred to a volcanic rock of the rhyolite type. The name derives from the Lipari Islands, off the N. coast of Sicily, where the rock in question consists chiefly of feldspars and quartz. Italian and Japanese liparites have been used as fluxes in porcelain manufacture.
Liquid Gold. See BRIGHT GOLD.
Liquid-liquid Opals. See OPAL GLASS.
Liquid-phase Sintering. A form of SINTERING in which a liquid phase is formed, and viscous flow and particle sliding leads to initial shrinkage, followed by a stage in which capillary forces draw the liquid to solid grain contact points which dissolve. Some material diffuses to free surfaces where it precipitates, reducing the distances between particle centres. If there is insufficient liquid phase to fill the pores (cf VITRIFICATION), final shrinkage must be by solid-state processes. Additives are used to produce liquid phases in covalently bonded ceramics which are otherwise nearly impossible to sinter. In many electronic ceramics the controlled boundary phases liquefy during firing.
Liquid Pressure Forming. Metal matrix composites are made by filling a cavity with a fibre pre-form, evacuating and refilling with molten metal under pressure. (Cray Advanced Materials, Yeovil. *Engineering* **222** (11) 643, 1987).
Liquid Shear Plane. See ZETA POTENTIAL.
Liquidus. The line on a binary equilibrium diagram (e.g. Fig 5. p230) showing the temperature at which any mixture of the two components, when under conditions of chemical equilibrium, becomes completely liquid: conversely, the liquidus shows the temperature at which a liquid of a given composition begins to precipitate solid material. In a ternary, i.e. three-component, system, the liquidus is represented by a curved surface.
Lisicon. The compound $Li_{14}Zn(GeO_4)_4$, a fast ion conductor (q.v.).
Litharge. Lead monoxide, PbO. See LEAD OXIDE.
Lithium Carbonate. Li_2CO_3; used as a source of Li_2O in vitreous enamel frits and in glazes.

Lithium Minerals. See AMBLYGONITE, EUCRYPTITE, LEPIDOLITE, PETALITE, SPODUMENE.

Lithium Niobate. $LiNbO_3$; a ferroelectric electroceramic having the ilmenite structure.

Lithium Silicate Glasses. Li_2SiO_3 can be precipitated from lithium silicate photosensitive glasses under the action of light and heat. The precipitated crystals are much more soluble in 10% HCl than is the glass. After dissolving out the Li_2SiO_3, X-ray or ultraviolet irradiation, followed by heat treatment, produces a non-porous glass-ceramic without distortion, with 1% shrinkage, and with higher strength and electrical resistivity than the original glass, useful for high temperature (500°C) printed circuit substrates and dielectrics.

Lithium Tantalate. $LiTaO_3$; a ferroelectric compound of potential value as a special electroceramic. The Curie temperature is above 350°C.

Lithium Titanate. A compound sometimes used as a flux in titanate bodies and in white cover-coats in vitreous enamelling. It is also characterized by very high thermal expansion: 19.5×10^{-6} (25–700°C).

Lithium Zirconate. Sometimes used as a flux in more refractory zirconate bodies of the type used as electroceramics.

Lithographic Decal. A transfer whose design is applied by dusting pigments on to varnished areas of the backing paper. See DECAL.

Lithography (UK); Decalcomania (USA). A method used for the decoration of pottery by means of transfers. A special paper is used but the ceramic colours cannot be printed directly and the actual printing is done in varnish and the colour then dusted on. The Litho or Decal (to use the common contractions of these words) is placed coloured-side down on the sized ware, rubbed firmly on. and the paper then sponged off.

Lithophane. A decorative type of translucent pottery which when viewed by transmitted light, exhibits a scene or portrait that is present in low relief on the ware. The process originated at Meissen, Germany, in 1828; examples of the ware were also produced in England during the 19th century.

Litre-weight Test. A works' test for the routine control of the firing of portland cement clinker; it was introduced by W. Anselm *(Zement,* **25**, 633, 1936).

Littler's Blue. A royal blue first produced in about 1750 by William Littler of Longton Hall, Stoke-on-Trent, by staining white slip with cobalt oxide, applying the slip to pottery-ware, firing, and salt-glazing.

Littleton Softening Point. See SOFTENING POINT.

Live-hole. A flue left in a CLAMP (q.v.) and filled with brushwood to start the firing process.

Liver. A vitreous enamel fault defined in ASTM-C286 as: 'In dry-process enamelling, a defect characterized by a wave-like form of abnormally thick enamel'.

Liver Spotting. See MOTTLING (of silica refractories).

Livering. A general term covering curious faults in casting slip which may develop if the slip is allowed to stand; the cause is lack of control in the DEFLOCCULATION (q.v.)

Livesite. A disordered kaolinite, particularly as found in some micaceous fireclays. The term is now little used; it was proposed by K. Carr, R. W. Grimshaw and A. L. Roberts *(Trans. Brit. Ceram. Soc.,* **51,** 339, 1952).

Load. The weight of glass produced by a glass-making furnace in a given time, usually 24 h. Known in USA (and also less commonly in UK) as PULL; other alternative terms are DRAW and OUTPUT.

Load-bearing Brickwork. Brick masonry of controlled construction and calculated strength, for high-rise buildings or engineered structures such as bridges.

Load-bearing Tile. US term for a hollow fired-clay building block designed to carry superimposed loads.

Loading. Choking of the pores in the face of an abrasive wheel with debris from the workpiece that is being ground.

LOF-Colburn Process. (LOF = Libbey-Owens-Ford Glass Co. See under COLBURN PROCESS.

L.O.I. Abbreviation sometimes used for LOSS ON IGNITION (q.v.).

London Clay. A tertiary clay used in making building bricks in Surrey, Berkshire, Essex, Suffolk, Dorset, Wiltshire, and Hampshire.

London Sink. A domestic or industrial sink with no overflow; the top edge is plain (cf. BELFAST SINK and EDINBURGH SINK).

London Stock Brick. See STOCK BRICK.

Long Glass. Glass that is slow-setting.

Longitudinal-arch Kiln. An ANNULAR KILN (q.v.) in which the axis of the arched roof, on both sides of the centre-line of the kiln. is parallel to the length of the kiln (cf. TRANSVERSE-ARCH KILN). Kilns of this type were much used in the firing of building bricks.

Loo Board. See under HACK.

Looping or Loops. See CURTAINS.

Los Angeles Test. The HARDNESS and ABRASION RESISTANCE of concrete aggregate is assessed by tumbling a standard sample in a standard ball-mill for a specified time.

Loss on Ignition. The loss in weight of a sample of material expressed as a percentage of its dry weight, when the sample is heated under specified conditions. In the testing of clays and similar materials, the temperature employed is 1050 ± 50°C, in the testing of frits, the temperature must be below the fusion point and below the temperature at which there is loss by volatilization.

Lost-wax Process. See INVESTMENT CASTING.

Low Cement Castable. Alumina and aluminosilicate castable refractories containing hydraulic setting cement and which have a total lime (CaO) content of greater than 1.0 to 2.5% on a calcined basis. Ultra-low cement castables have a lime content from 0.2 to 1.0% (ASTM C401–91).

Low-duty Fireclay Brick. Defined in ASTM – C27 as a fireclay refractory having a P.C.E. 15 and a modulus of rupture 600 p.s.i. (4.2 MNm^{-2})

Low-frequency Induction Furnace. See under ELECTRIC FURNACES FOR MELTING AND REFINING METALS.

Low-heat Cement. Hydraulic cement compounded so as to evolve less heat during setting than does normal portland cement. This feature is important when large volumes of concrete are placed, as in dams. It is achieved by reducing the proportions of $3CaO.SiO_2$ and $3CaO.Al_2O_3$. For specification see B.S. 1370.

Low-heat rejection diesel. See ADIABATIC DIESEL.

Low-solubility Glaze. The Potteries etc (Modifications) Regulations 1990 (S.I. 305) define this as a glaze which does not yield to dilute HCl more than 5% of its dry weight of a soluble lead compound, when determined in accordance with an approved method; see LEAD SOLUBILITY TEST (q.v.). for the original approval method.

Low Thermal Mass Kiln. A design of kiln requiring relatively little heat input per degree of temperature increase. The design was made practical by replacing conventional refractories with ceramic fibre hot-face lining; one-high firing which reduced kiln dimensions, and general lightweight (often prefabricated) construction. As well as general fuel economy, such kilns are suited to rapid firing and flexible production schedules.

LO-X. Ceramics with negative coefficients of thermal expansion, produced by Lockheed, California. (1970).

L.P.G. Abbreviation for *Liquefied Petroleum Gas*, (propane or butane).

LSP. Liquid Stabilised Plasma torch for thermally sprayed coatings.

LTM Kiln. See Low Thermal Mass Kiln.

Lubber's Process. This early process for the manufacture of sheet glass mechanized the hand cylinder method for BROAD GLASS (q.v.) A blow pipe with a large flanged metal disc, or 'bait', was lowered into a pool of molten glass. The bait was then raised and air blown through the pipe to blow a glass cylinder of the required length. This was severed at the bottom, turned sideways, split in half and flattened. Sheets 3ft × 30ft (1 × 10m) long were acheived (The American Window Glass Co. 1903).

Lucalox. Trade-name: a translucent, pure, polycrystalline alumina made by General Electric Co., Cleveland, Ohio, USA. The translucency results from the absence of micropores. Because the crystals are directly bonded to one another, without either matrix or pores between the crystal boundaries, the mechanical strength is very high: transverse strength, 350 MNm^{-2}; modulus of elasticity, 400 GNm^{-2}.

Ludwig's Chart. A diagram proposed by T. Ludwig *(Tonind. Ztg.,* **28**, 773, 1904) to relate the refractoriness of a fireclay to its composition, which is first recalculated to a molecular formula in which Al_2O_3 is unity, i.e. xRO.Al_2O_3.$y$$SiO_2$. In the chart, x is plotted as ordinate and y as abscissa; a series of diagonal lines indicates compositions of equal refractoriness. The diagram is inevitably a rough approximation only.

Luerkens Equations. Equations developed by J.K. Beddow (*Advances in Ceramics* Vol 21, 1987) to describe the shapes of particles by Fourier analysis. Application of the calculus of variations leads to a Fourier series in the 1-dimensional case; and to Fourier series multiplied by a Bessel function or by an Associated Legendre Polynomial in the 2- and 3- dimensional cases respectively. See also MORPHOLOGY; SHAPE FACTOR.

Lug. See SPACER LUG.

Lühr Filter. Dust-removal equipment in which the dusty air is drawn through a continuously-moving belt of filter cloth (H. Lühr, Stadthagen, Germany).

Lumenized. Lenses coated with MgF_2. (Trade-name – Kodak Ltd., Kingsway, London).

Luminous Wall Firing. Term sometimes used in USA for kiln or furnace-firing by SURFACE COMBUSTION (q.v.)

Lumnite Cement. Trade-name: a HIGH-ALUMINA CEMENT (q.v.) made by Universal Atlas Cement Division of U.S. Steel Corp., New York.

Lump. (1) A heap of unmelted batch floating on the molten glass in a tank furnace.

(2) The most defective saleable pottery-ware remaining after the sorting process; cf. FIRSTS, SECONDS.

(3) A fault in vitreous enamelware in the form of a small excrescence, ½–⅛ in. (12–3 mm) diam.; it is often caused by dripping of the spray gun.

Lump Man. A workman in a glassworks whose job is to observe and control any lumps of floating batch in a glass-tank furnace.

Lunden Conductive Tile Flooring. A method of using ceramic tiles in an ANTI-STATIC (q.v.) floor: electrically-conducting 'dots' of metal, or of special ceramic material that has been made conducting by loading the body with carbon, are interspersed among the normal tiles. The method was devised by S. E. Lunden, a Californian architect, in 1950.

Lurgi Process (for Cement Manufacture). A method for the production of hydraulic cement by sintering the charge on a grate.

Lustre. A form of metallic decoration on pottery or glass introduced in the Middle East before AD 900. Lustres for both materials can be made from metallic resinates, e.g. of Cu, Mn, or Co, in a solvent. Lustres are formed under reducing conditions achieved by the presence of reducing agents in the lustre composition and/or by adjustment of the kiln atmosphere. Lustre finishes can also be produced on vitreous enamelware, particularly cast iron ware for solid-fuel appliances; a metallic etching solution is sprayed on the enamelled surface while it is still red hot. See also BRIANCHON, BURGOS, CANTHARIDES and FRANCHET LUSTRES.

Luting. Joining two leatherhard ceramic surfaces with slip, as in sticking handles on cups.

Luting Cement. Refractory mortar for jointing spouts in glass forming, sealing coke oven doors etc.

L.W.F. Luminous Wall Firing (q.v.).

Lynch Machine. A machine for the manufacture of glass bottles; it is based on the original design introduced by J. Lynch in 1917 and operated on the 'blow-and-blow' principle. 'Press-and-blow' Lynch machines are also widely used.

Lytag. Trade-name: a lightweight aggregate for concrete made by sintering pulverized fuel ash (Lytag Ltd, Boreham Wood, England).

MA. See IL/MA.

MacAdam System. A method of colour notation in which nonspherical regions of equal perceptual differences are inscribed in the CIE (q.v.) space. (W. R. J. Brown and D. L. MacAdam, *J. Opt. Soc. Amer.*, **33**, 18, 1943.)

Machineable Ceramics. All ceramics can be machined in the green state, and in the fired state with diamond tooling. Machineable ceramics are those which can be shaped when fired, to engineering tolerances with ordinary high speed metal cutting tools. See MACHINING, and PYROPHYLLITE, DICOR, MACOR, PHOTON.

Machining. Shaping (usually to precise tolerances) by mechanical removal of material. There are three basic groups of processes: *grinding methods,* in which ABRASIVES (q.v.) bonded into the grinding wheels, burrs or other tools are the usual and most versatile methods; *free abrasive methods* use abrasive powders for lapping, honing, polishing, ULTRASONIC MACHINING and ABRASIVE JET MACHINING (q.v.); *non abrasive machining* techniques include LASER-BEAM MACHINING, ELECTRICAL DISCHARGE MACHINING, ION-BEAM and ELECTRON BEAM MACHINING, and CHEMICAL and ELECTROCHEMICAL MACHINING (q.v.) See also TOOL (CERAMIC) for machining metals.

Machlan Diagram. A diagram giving in general terms the liquidus lines in a ternary system of acidic oxide, alkaline oxide, and an alkaline earth oxide.

Mackensen Blower. An apparatus to determine the variation of hardness and inhomogeneity of structure across ceramic surfaces. A probe is zeroed against a scale, lifted from the surface and abrasive powder blown on to the surface at a pressure of 1.5 kg/cm². The displacement of the probe thereafter is a measure of the hardness of the surface.

Macor. Tradename. A machineable glass-ceramic based on fluormica crystals dispersed in a glass matrix. The tiny mica flakes break away during machining processes. (Corning Glass Co, USA. Br Pat 1345294 and 1365435, 1974)

Mack's Cement. A quick-setting plaster made by adding Na_2SO_4 or K_2SO_4 to plaster of Paris.

Mackenzie-Shuttleworth Equation. Relates to the progress of shrinkage, or densification, during the sintering of a compact, the latter being considered as a continuous matrix containing uniformly distributed spherical pores:

$$\frac{d\sigma}{dt} = (1 - \sigma)^{2/3} (4.5\pi\gamma n\sigma)^{1/3}/\eta$$

σ is the relative density, t the time in seconds, γ the surface tension, η the viscosity in poises and n the number of pores per cm^3 of real material. (J. K. Mackenzie and R. Shuttleworth, *Proc. Phys. Soc.. B*, **62,** 833, 1949.)

Mackler's Glaze. A type of AVENTURINE GLAZE (q.v.); (H. Mäckler, *Tonind. Zeit.,* **20,** 207, 219, 1896).

MacMichael Viscometer. A rotation-type viscometer designed for the testing of clay slips. (R. F. MacMichael, *Trans. Amer. Ceram. Soc.* **17**, 639, 1915.)

Macrodefect–free cement The flexural strength of cements can be increased fourfold to c. 70 MPa, if all defects >100mm can be removed. This is achieved by rheological control and careful mixing of fine powder. (J. D. Birchall et al, *Nature* **289**, 388, 1981)

Macro Process. Tungsten carbide powder is made direct from the one concentrate by an exothermic reaction causing simultaneous reduction and carburization.

Madelung Constant. An important constant for calculating the lattice energy of an ionic crystalline solid. If the distance between nearest adjacent ions is R, and the distance of the jth ion from a (negative) reference ion is r_j (always taken as positive), then the Madelung constant

$$\alpha = \sum_j \pm \frac{R}{r_j}$$

Maerz-Boelens Furnace. A type of open-hearth steel furnace with back and front walls sloping inwards, thus reducing the span of the roof; the design was introduced by E. Boelens, in Belgium, in 1952. (Also known as a PORK-PIE Furnace.)

Mafic Minerals. *M*agnesium, iron (*Fe*) and *c*alcium silicates, cheap minerals sometimes substituted for feldspars.

Magdolo. Trade-name: a half-way product in the extraction of magnesia from sea-water, used in Japan as a refractory material for the L-D PROCESS (q.v.). It contains 55–62%, MgO and 30–35%, CaO, together with minor amounts of SiO_2, Al_2O_3 and Fe_2O_3.

Magneli Phase. A series of structurally related compounds with regularly incremented oxygen contents. The titanium suboxides Ti_nO_{2n-1} with n=4 to 8 form a series of materials whose electrical conductivity falls rapidly from $3x10^5$ ohm.cm to a very small value.

Magnesia; Magnesium Oxide. A highly refractory oxide, MgO; MgO can be sintered to a dense engineering ceramic, and is also incorporated into other engineering ceramics as a sintering aid. For properties see under HEAVY MAGNESIA, LIGHT MAGNESIA and PERICLASE.

Magnesio-chromite. See PICROCHROMITE.

Magnesio-ferrite. $MgFe_2O_4$; m.p. 1750°C; sp. gr. 4.20; thermal expansion

(0–1200°C) 13.5 × 10^{-6}. This spinel is formed in some basic refractories by reaction between the periclase (MgO) and the iron oxide present.
Magnesite. Magnesium carbonate, $MgCO_3$; the natural mineral often contains some $FeCO_3$ (if more than 5%, the mineral is known as breunnerite). The main sources are Russia, Austria, Manchuria, USA, Greece and India. After it has been dead-burned at 1600-1700°C, magnesite is used as raw material for the manufacture of basic refractories (see also SEA-WATER MAGNESIA).
Magnesite Refractory. A refractory material, fired or chemically-bonded, consisting essentially of dead-burned magnesite so that the MgO content exceeds 80%. ASTM C-455 specifies a nominal MgO contents of 90, 95 and 98%, with the first having a minimum MgO content of 86%, below which the refractory is classified as MAGNESITE-CHROME (q.v.). Such refractories are used in the hearths and walls of basic steel furnaces, mixer furnaces and cement kilns, for example.
Magnesite-chrome Refractory. A refractory material made from dead-burned magnesite and chrome ore, the magnesite being present in the greater proportion so that the fired product contains 55–80% MgO. Such refractories may either be fired or chemically bonded and are used principally in the steel industry. A typical chemical analysis is (per cent): SiO_2, 3–4; Fe_2O_3, 4–6; Al_2O_3, 4–6: Cr_2O_3, 8–10; CaO, l-2; MgO, 70–75. ASTM C455 specifies 6 grades of chrome-magnesite and magnesite chrome refractories; with nominal MgO contents of 30 to 80% , with the minimum allowable being 5% less in each case.
Magnesium Aluminate. $MgAl_2O_4$; m.p. 2135°C; sp. gr. 3.6: thermal expansion (0–1200°C), 9.3 × 10^{-6}. This compound is the type mineral of the SPINEL (q.v.) group.
Magnesium Chromite. See PICROCHROMITE.
Magnesium Fluoride. MgF_2; m.p. 1255°C. Hot-pressed (200 MNm^{-2} at 650°C) compacts had the following properties: sp. gr. 3.185; hardness, Knoop 585, Rockwell C52; Youngs Modulus, 75 GNm^{-2}; thermal expansion, 11.5 × 10^{-6} (20-500°C). The hot-pressed material is suitable for use in infra-red components operating under severe conditions.
Magnesium Oxide. See MAGNESIA.
Magnesium Oxychloride. See SOREL CEMENT.
Magnesium Stannate. $MgSnO_3$; when added to titanate electroceramic bodies, the dielectric constant is usually somewhat increased up to the Curie point.
Magnesium Sulphate. $MgSO_4.7H_2O$; Epsom salt; used in the vitreous-enamel industry to improve the suspension of the enamel slip.
Magnesium Titanates. Three compounds exist. Mg_2TiO_4, m.p. 1730°C; this has a spinel structure and has been proposed for use as a resistor. $MgTiO_3$, m.p. 1630°C; ilmenite structure. $MgTi_2O_5$, m.p. 1650°C; has a low dielectric constant and has been added to other titanate bodies to reduce their dielectric constants.
Magnesium Zirconate. $MgZrO_3$; m.p. 2150°C. This compound is sometimes added in small amounts (up to 5%) to other electroceramic bodies to lower their dielectric constant at the Curie point.
Magnet Thickness Gauge. A non-destructive device for determining the thickness of a coating of vitreous enamel; it depends on the

principle that the force required to pull a permanent magnet off the surface of enamelware is inversely proportional to the thickness of the non-magnetic enamel layer on the magnetic base metal.

Magnetic Ceramics. See FERRITES.

Magnetic Separator. Equipment for removing iron particles from clay or other ceramic material; in the design commonly used in the pottery industry, the prepared body is allowed to flow as a fluid slip over a series of permanent magnets, which are periodically removed and cleaned.

Magnetite. Fe_3O_4; a spinel; m.p. 1590°C; sp. gr. 5 .14; thermal expansion (0–1000°C) 15×10^{-6}, This magnetic iron oxide is present in many ferruginous slags and in their reaction products with refractory materials.

Magnetoplumbite Structure. Ceramics with the general formula $M0.6Fe_2O_3$,, where M is usually Ba, Sr or Pb, have a hexagonal crystal structure with a unique 'c' axis. These are FERRITES (q.v.) and more complex structures with interlayered spinels are possible.

Magnit. Trade-name: A tarred magnesitic-dolomite refractory made by Vereinigte Österreichische Eisen und Stahlwerke, Linz, Austria. and used by them in the L-D PROCESS (q.v.) of steelmaking. The composition of the calcined raw material varies as follows: 65–80° MgO, 10–25% CaO, 2–5% SiO_2, 1% Al_2O_3, 4–6% Fe_2O_3, Loss on ignition, 0.5–2%.

Magnum. A 2-quart wine bottle.

Main Arch. (1) The refractory blocks forming the part of a horizontal gas-retort comprising the DIVISION WALLS (q.v.) and the roof that covers the retorts and the recuperators.
(2) General term for the central part of a furnace roof, particularly used as a synonym of the crown of a glass-tank furnace.

Maiolica; Majolica. Originally a porous type of pottery, with a glaze opacified with tin oxide, from the island of Majorca; this pottery was first made in the 16th century under the combined influence of Hispano-Moresque and Near Eastern wares; it was essentially similar, technically, to delft-ware and the original faience. As applied to present-day pottery, the term signifies a decorated type of earthenware having an opaque glaze, usually fired at a comparatively low temperature (900–1050°C). The term has also been applied to vitreous enamels that are transparent, but that contain colouring oxides; they are used over ground-coats and allow the latter to show where there is any relief.

Malachite Green. $CuO\text{-}CO_3\text{-}Cu(OH)_2$. A dye, used to test absorption of bodies, and to colour stoneware.

Malm. (1) A synonym of Marl in its original sense of a calcareous clay.
(2) The best quality of clamp-fired London Stock Bricks.

Manchester Kiln. A type of longitudinal-arch kiln that was introduced in the Manchester district of England for the firing of building bricks. It is top-fired. A distinctive feature is the flue system, with horizontal damper-plates, in the outside wall. The Manchester kiln usually has a hot-air system.

Mandrel. (1) A former used in the production of blown glassware from tubing or of precision tubing.
(2) A refractory tube used in the production of glass tubing or rod.
(3) A central core in ISOSTATIC PRESS TOOLING (q.v.) to enable hollow articles to be formed.

Mangalore Tile. A clay roofing tile of the interlocking type as made in the

Mangalore district of India. See Indian Standard 654.

Manganese-Alumina Pink. A ceramic stain, particularly for the colouring of pottery bodies; it is produced by calcination of a mixture of $MnCO_3$, hydrated alumina and borax. This stain will produce a strong, clean colour over a wide firing range but the glaze must be rich in Al_2O_3; this requirement makes it difficult to produce a smooth glaze.

Manganese Oxides. There are several manganese oxides, the commonest being MnO_2 (pyrolusite); other oxides are MnO (m.p. approx. 1780°C) and Mn_3O_4 (m.p. 1570°C). MnO_2 is used as a colouring oxide (red or purple); mixed with the oxides of Co, Cr, Fe, it produces a black. This oxide is also used to colour facing bricks, and to promote adherence of ground-coat vitreous enamels to the base-metal. When added in small amounts to glasses (particularly lead glasses) during pot melting, MnO_2 acts as a decolorizer.

Mangle. A type of dryer used in the tableware section of the pottery industry. The ware, while still in the plaster mould, is placed on trays suspended between two endless chains that pass over wheels above and below the working opening in the dryer; by means of these chains the trays, when filled, are moved vertically through the hot air current in the dryer until they arrive at a second opening where the moulds and dried ware are removed.

Mansard Roof. A tiled roof having a steeper pitch towards the eaves than towards the ridge; the term is also applied to a flat-topped roof with steeply-pitched tiling towards the eaves. (F. Mansard, 17th century French architect.)

Manson Effect. The BOILING (q.v.) of vitreous enamel that occurs when heated cast-iron is DREDGED (q.v.) with only a thin coating of enamel, which is allowed to fuse before a further coating of enamel has been applied. The effect was first observed by M. E. Manson *(Bull. Amer. Ceram. Soc.,* **11**, 204, 1932).

Mantel. See LINTEL.

Mantle Block. (1) A refractory block for use in the gable wall of a glass-tank furnace above the DOG-HOUSE (q.v.). (2) The upper course of refractory blocks, in a sheet-glass furnace, enclosing the Fourcault drawing pit.

Manx Stone. See CHINA STONE.

Marbleized. A vitreous-enamel finish resembling marble; it can be obtained by an offset process from natural marble.

Marbling. A procedure sometimes used by the studio potter: the marble effect is obtained by covering the piece of previously dried ware with coloured slips and then shaking the ware (while the slips are still wet) to make the colours run into one another.

Map Cracking. Random surface cracks in concrete, due to surface shrinkage or interior expansion.

Marcasite. A form of iron sulphide, FeS_2, sometimes found as particles or nodules in clays, forming dark spots when the clay is fired.

Margin. See GAUGE.

Mariotte Tube. An orifice-type viscometer designed by E. Mariotte (Paris, 1700). One such instrument with a tube diameter of 54 mm and a 3 mm orifice has been used for the assessment of the fluidity of clay slips (*J. Amer. Ceram. Soc.,* **27**, 99, 1944).

Marl. Strictly, an impure, calcareous clay; the word is used in N. Staffordshire, England, to denote a low-grade fireclay, e.g. saggar-marl. See also ETRURIA MARL, KEUPER MARL.

Marlow Kiln. A tunnel kiln fired with producer gas and preheated air; it was first used, in Stoke-on-Trent, England, for the biscuit- and glost-firing of wall tiles (J. H. Marlow, *Trans. Brit. Ceram. Soc.,* **21**, 153, 1922).

Marquardt Porcelain. A mullitic (65% Al_2O_3) porcelain introduced early in the present century by the State Porcelain Factory, Berlin, chiefly for pyrometer sheaths and furnace tubes.

Mars Pigments. PIGMENTS (q.v.) produced by calcining the precipitate resulting from mixing solutions of ferrous sulphate and calcium hydroxide. The calcination temperature determines the colour: yellow, orange, brown, red or violet.

Marseilles Tile. A clay roofing tile of the interlocking type, particularly of the pattern made in the Marseilles district of France. These tiles are made in a REVOLVER PRESS (q.v.).

Martensitic Transformation. A solid-state transformation between two phases with the same chemical composition, with a characteristic shape change in the transforming region. Any originally flat surface in the transformed region remains flat, but is tilted about its line of intersection with a definite *habit plane* (the interface common to the parent and product phases). Planes and straight lines in the parent phase are transformed into planes and straight lines in the product phase. Elastic and plastic deformation in the surrounding matrix accommodates the distortion due to the tilting. Martensitic transformations are usually athermal and diffusionless. They are usually associated with rapid cooling, but can lead to both equilibrium or metastable final phases. They are important in ceramics, contributing to TRANSFORMATION TOUGHENING. Their name derives from the transformation which forms *martensite* – a hard metastable phase developed in Fe -C alloys when they are quenched to strengthen them. See also BAIN DEFORMATION.

Martin's Cement. A quick-setting plaster made by adding K_2CO_3 to plaster of Paris.

Marumerizer. A machine for producing spherical granulate from cylindrical extrudate. The granules, 0.5 to 15mm diameter, are made by allowing short sections of extruded material to fall on to a rotating plate at the bottom of a cylindrical bowl. This breaks the extrudate into pellets about 1 diameter long. Collisions with the wall, the friction plate and other particles gradually produce spherical granules in a fluidized ring of particles slowly rotating against the wall of the Marumerizer. The machine is Japanese, licensed to Russell Finex, London)

Marver. A small metal or stone table on which a gather of glass is rolled or shaped by hand.

Mason's Ironstone China. A vitrified type of earthenware introduced by C. J. Mason, Stoke-on-Trent, England. According to his patent (Brit. Pat. 3724, 1813) the batch composition was: 4 pts. china clay, 4 pts. china stone, 4 pts. flint, 3 pts. prepared ironstone, and a trace of cobalt oxide. It is now known that the body did not contain ironstone, the name merely being a highly successful method of indicating to the public that the ware was very strong.

Masonry. Bricks (or other structural units of a similar type) bonded together and depending for their structural stability mainly on the strength of the bricks and of the mortar, but also on the bonding and weight. See also PREFABRICATED MASONRY; REINFORCED BRICKWORK. B.S. 8000 Pt 3 is a code of practice for site workmanship for masonry. B.S. 5628 is a code of practice for the structural use of masonry, including reinforced and prestressed masonry. B.S. 5080 and ASTM E488 give tests for tensile loading and shear for anchors in masonry. B.S. 6270 specifies techniques for cleaning and repairing the surfaces of stone, calcium silicate and clay brick masonry. ASTM E518 specifies two test methods (uniform and 3-point loading) for the flexural bond strength of masonry. ASTM E519 specifies a shear strength test in which a 1.2m masonry square is loaded across its diagonal. C1196 and C1197 are test methods for *in situ* compressive strength and deformability of masonry. The construction of dry-stacked surface bonded walls, in which walls of concrete blocks, laid close together without mortar are coated with mortar on both sides, is specified in ASTM C946.

Master Mould. See under MOULD.

Mat. See GLASS FIBRE.

Match Mark. See JOINT LINE.

Matching Piece. A short refractory channel between the spout of a glass-tank furnace and the pot spout for a revolving pot; also known as an INTERMEDIATE PIECE.

Matrix. That part of a ceramic raw material or product in which the larger crystals or aggregates are embedded. A fired silica refractory, for example, consists of crystalline silica (quartz, cristobalite and tridymite) set in a glassy matrix; a fireclay refractory may consist of quartz and mullitic GROG (q.v.) set in a largely amorphous matrix. (Also sometimes known as the GROUND-MASS.)

Matt Blue. A colour for pottery decoration depending on the formation of cobalt aluminate; zinc oxide is usually added, a quoted recipe being 60% Al_2O_3, 20% CoO, 20% ZnO.

Matt Glaze. A ceramic glaze that has partially devitrified; the effect is deliberately achieved, for example, on some types of glazed wall tile. The glazes used for the purpose are usually leadless and devitrification is encouraged by the introduction, into the glaze batch, of such oxides as TiO_2, CaO or ZnO.

Maturing Temperature or Maturing Range. (1) The firing temperature at which a ceramic body develops a required degree of vitrification.
(2) The firing temperature at which the constituents of a glaze have reacted to form a glass that, when the ware has been cooled, appears to the eye to be homogeneous and free from bubbles.

Maximum Exposure Limit. The maximum concentration of an airborne substance, (averaged over a reference period) to which employees may be exposed by inhalation under any circumstances. Cf OCCUPATIONAL EXPOSURE STANDARD.

Maxitherm. A brick with vertical perforations, and slots in the side into which expanded polystyrene cladding can be dovetailed.

Mazarine Blue or Royal Blue. A ceramic colour for on-glaze or under-glaze, based on the use of cobalt oxide (40–60%) together with a flux.

MBI Methylene Blue Index

MCC Test. (Metodo Centro Ceramica) is a test for the surface abrasion resistance of glazed tiles, specified in BS 6431 Pt 20. It uses a dry abrasive, in contrast to the PEI TEST (q.v.) which is also specified.

MCC – 1P. Static Leach Test. Speciments of known geometric surface are immersed in selected leachants, without agitation, for defined time intervals at defined temperatures. The test has been applied to dental ceramics, with leach periods of 1, 7, 14 and 28 days at 90°C, with leachants chosen to simulate saliva. (MCC = *M*aterials *C*haracterization *C*enter).

MCVD (Modified Chemical Vapour Deposition). See OPTICAL FIBRES.

MDF Cement – Macrodefect-free Cement, (q.v.)

MDHS *Methods for the Determination of Hazardous Substances* are published by the (UK) Health and Safety Executive.
MDHS 6,7,8 relate to lead and lead compounds;
MDHS 10,11 to cadmium and its compounds;
MDHS 14 to respirable dust;
MDHS 29 to beryllium and its compounds;
MDHS 37,38,51 to quartz in respirable dust;
MDHS 39 to asbestos fibres
MDHS 59 to man-made mineral fibres;
MDHS 61 to hexavalent chromium compounds;

Measles. Unbroken bubbles that have penetrated the groundcoat into the cover-coat of vitreous enamels.

Mechanical Analysis. A term sometimes used as a synonym of PARTICLE-SIZE ANALYSIS (q.v.).

Mechanical Boy. A device used for controlling the mould during the hand-blowing of glass-ware.

Mechanical Press. A press in which bricks (building or refractory) are shaped in a mould by pressure applied mechanically, e.g. by a toggle mechanism, as distinct from the procedure in a hydraulic press.

Median Crack. A crack in glass perpendicular to the surface, caused by a hard, sharp object.

Mediceram. Biocompatible calcium phosphate glass-ceramics developed in East Germany. (*Interceram* **37**, (3), 34, 1988)

Mechanical Shovel. The original form of excavator, still much used in the getting of brick-clays. It consists of a bucket (or 'dipper') on the end of an arm pivoted to a boom; the bucket scoops upwards, the excavator operating below the clay face.

Mechanical Spalling. SPALLING (q.v.) of a refractory brick or block caused by stresses resulting from pressure or impact.

Mechanical Water or Mechanically-held Water. The water that is present in moist clay and is not chemically combined. When the clay is heated, the mechanically-held water is virtually all removed by the time that the temperature has reached 150°C.

Medina Quartzite. A quartzite from Medina, Pennsylvania, USA, used as a raw material for silica refractories. A quoted chemical analysis is (per cent): SiO_2, 97.8; Al_2O_3, 0.9; Fe_2O_3,0.85; alkalis, 0.4.

Medium-duty Fireclay Brick. Defined in ASTM – C27 as a fireclay refractory having a P.C.E. 29 and a modulus of rupture 500 p.s.i. (3.5 MNm^{-2}).

Meissner Effect. The expulsion of a magnetic field from a SUPERCONDUCTOR (q.v.) when the temperature is lowered below the critical temperature T_c at which the material becomes superconducting.

MEL. Maximum Exposure Limit

Meldon Stone. A CHINA STONE (q.v.) of low quality from Cornwall, England. A quoted analysis is (per cent): SiO_2, 70; Al_2O_3, 18; Fe_2O_3, 0.4; CaO, 0.6; Na_2O, 4; K_2O, 4.5; loss on ignition, 2.5.

Melting. See CONGRUENT MELTING; INCONGRUENT MELTING. For melting points of various compounds and minerals, see under name of compound or mineral.

Melt Blowing. The production of refractory fibre by blowing a melt with steam or compressed air.

Melting End. The part of a glass-tank furnace where the batch is melted and the glass is refined.
Melt Infiltration. Production of a metal-ceramic composite by a controlled reaction between a ceramic preform and a liquid metal which pentrates its pores.
Melt Spinning. Ceramic fibres are produced from low viscosity metal melts, from which a molten jet is drawn and chemically stablilized to form the fibre.
Melting Temperature. For a glass, the ASTM definition is: 'The range of furnace temperatures within which melting takes place at a commercially desirable rate, and at which the resulting glass generally has a viscosity of $10^{1.5}$ to $10^{2.5}$ poises; for purposes of comparing glasses, it is assumed that the glass at melting temperature has a viscosity of 10^2 poises'.
Meltspinning. The production of ductile ribbons of active metal brazing materials by rapid quenching from the liquid state to form a glassy metal.
Membrane Curing. Spraying a bituminous seal on the surface of fresh concrete, to prevent moisture loss.
Membrane Filter. See DUST SAMPLING, FILTERS.
Membrane Theory of Plasticity. Attributes the plasticity of clay to the compressive action of a postulated surface envelope of water around the clay particles. (F. H. Norton, *J. Amer. Ceram. Soc.*, **31**, 236, 1948).
Mendheim Kiln. A gas-fired chamber kiln designed by G. Mendheim, of Munich, in about 1910 for the firing of refractories at high temperature. The gas enters at the four corners of each chamber, and burns within BAG WALLS (q.v.) which direct the hot products of combustion towards the roof; they then pass downward through the setting and are exhausted through the chamberfloor.
Menhaden Fish Oil. A dispersant used in TAPE CASTING (q.v.)
Menstruum Process. An alternative name for the Auxiliary Metal Bath Process (q.v.)
Merapon. Metal-complex solutions absorbed by unglazed ware, which then develop colour on firing. (Benckiser Kanpsack GmbH. *Sprechsaal* (10) 922, 1986)
Merch Bricks. Term sometimes used in USA for building bricks that come from the kiln discoloured, warped or off-sized.
Mercury Penetration Method. A procedure for the determination of the range of pore sizes in a ceramic material. It depends on the fact that the volume of mercury that will enter a porous body at a pressure of P dynes /cm^2 is a measure of the volume of pores larger than a radius r cm where $r = -2\sigma \cos\theta/P$, σ being the surface tension of mercury in dynes/cm and θ being the contact angle between mercury and the ceramic. A development of the method has been described by R. D. Hill (*Trans. Brit. Ceram. Soc.*, **59**, 198, 1960).
Mereses. The ribs found on the stems of some wine-glasses.
Merriman Test or Sugar Test. A quality test (now discarded) for hydraulic cement. The sample is shaken with a solution of cane sugar and the amount of cement dissolved is determined by titration with HCI (T. Merriman, *J. Boston Soc. Civil Eng.*, **26**, (1), 1939, p.1).
Merwinite. 3CaO. MgO. SiO_2; melts incongruently at 1575C; thermal expansion (0–1200°C) 13.4×10^{-6}. It may occur in, or be formed during service in, dolomite refractories.
Metahalloysite. A clay mineral of the kaolinite group and of the same approximate composition as kaolinite – $Al_2O_3.2SiO_2.2H_2O$. It is formed by the partial dehydration of HALLOYSITE (q.v.).

Metakaolin. An intermediate product formed when kaolinite is heated at temperatures between about 500° and 850°C; the layer structure of the parent kaolinite persists in modified form but collapse of the layers destroys any periodicity normal to the layers. At higher temperatures (925°C) metakaolin transforms to a cubic phase with a spinel-type structure; at 1050–1100°C mullite is formed.

Metal. In the glass industry, the molten glass is called 'metal'.

Metal Blister. A defect that may occur in vitreous enamelware if there is any gas trapped in a fault in the base metal.

Metal-cased Refractory. A basic refractory with a thin sheetmetal casing on three or four sides, leaving the ends of the brick exposed; the refractory material itself (generally magnesite or chrome-magnesite) is usually chemically bonded. Such bricks are used chiefly in steel furnaces; during use, the metal case at the hot end of the brick oxidizes and 'knits' each brick to its neighbours. There are many variants of this type of basic refractory but all derive from the steel tubes packed with magnesite as patented in 1916 by N. E. MacCallum of the Phoenix Iron Co., USA.

Metal Line. See FLUX LINE.

Metal-marking. See SILVER MARKING.

Metal Oxide Chemical Vapour Deposition. A process in which metal oxides are deposited from the vapours of organic compounds of the relevant material.

Metal Protection (at high temperature). Metals can be protected from oxidation at high temperature by various types of ceramic coating, the commonest being flame-sprayed alumina and refractory enamels. Such coatings have found particular use on the exhaust systems of aircraft.

Metal Release. The action of foods, or of test solutions of acids, may leach toxic metals from the glaze or decoration of ceramic whitewares. Propensity to metal release may depend on complex interactions between glaze, decoration and kiln atmosphere. The usual tests are for Pb, often Cd, and sometimes for other metals. The usual test solution is 4% acetic acid, sometimes hot if cooking ware is being tested. There are varied but strict limits in the amounts (in the p.p.m. range) of metal leached into the test solutions. These limits vary from country to country, and are enforced by legislation. British Ceramic Research Ltd Special Publication 136, 1993 gives a detailed summary of tests and limits in over 40 countries. The most important are US FDA Compliance Policy Guide 7117.06 and 7117.07 for cadmium and lead; European EC Directive 84/500/EEC 1984 and GB Ceramic Ware (Safety) Regulations SI 1647, 1988. The test methods for ceramic tableware and cooking ware and vitreous enamelware have been unified in Britain in BS6748; 1986. In USA, ASTM C738, C895, C1035 give test methods for glazed ceramics, ceramic tile and cookware. C927 is a test for the tip and rim areas of glass tumblers.

Metallic Glasses. Electrically conducting glasses can be prepared from combinations of predominantly metallic elements by rapid cooling or vapour deposition. Their magnetic properties make them suitable for low-loss transformer cures, and their mechanical strengths approach theoretical fracture strength, as local dislocation fracture mechanisms do not operate.

Metallizing. Electroceramics are metallized when it is required to join the ceramic to a metal or form a seal. An alumina ceramic, for example, can be metallized by painting it with a powdered mixture of Mo and Fe and firing it in a protective atmosphere to bond the metals to the surface; the metallized area is then plated with Cu or Ni. This two-stage procedure is known as the TELEFUNKEN PROCESS (Telefunken G.m.b.H., Berlin). Single stage metallizing is possible if the hydride of Ti or Zr is used, together with a hard solder: this is sometimes known as the BONDLEY PROCESS (R. J. Bondley, *Electronics,* **20**, (7), 97, 1947).

Metalloceramic. One of the old, now obsolete, names of CERMET (q.v.).

Metal Strain. Lines, often parallel, of small blisters in vitreous enamelware; they may be the result of severe stress in the metal or they may be caused by damage to the fired ground-coat before the cover-coat is added.

Metameric Colours. Colours that appear to be the same under one type of illuminant but, because they have different spectral reflectivity curves, will not match under a different illuminant. Two white vitreous enamels may be metameric, for example, if one is opacified with titania in the anatase form whereas the other contains titania in the form of rutile.

Methuselah. A 9-quart wine bottle.

Methylene Blue Index A simple approximate method for assessing the surface area of a clay. 0.01M methylene blue dye is added to aqueous suspension of clay, in successive small amounts, to determine the least amount of dye required to impart colour to the water suspending the clay. Dye up to this amount is absorbed by the clay, so that this is a measure of the clay's cation exchange capacity (see IONIC EXCHANGE). For most pottery clays this is proportional to the surface area to a good approximation. ASTM C837 specifies a test for clays.

Methyltrichlorosilane. A gaseous PRECURSOR, whose breakdown leads to vapour-deposited silicon nitride.

Mettlach Tile. A vitreous floor tile (especially of the multicoloured type) as first made by Villeroy and Boch at Mettlach, in the Saar. The term (often mis-spelt 'Metlach') is now commonly used on the Continent and in Russia for any vitreous floor tile.

MEXE Method. A calculation method for determining the permissible loading of brick and masonry arches. For a critique and development of this method originally due to Pippard, see R.P. Walklate and J.W. Mann, *Proc. Inst. Civil Engrs.* Pt. 2 **75** (Dec) 585, 1983, and discussion Pt. 2 **77** (Sept). 401, 1984. MEXE is the Military Engineering Experimental Establishment.

Mexim. A glass-ceramic joining compound for metals (English Electric, Stafford) cf MEXIMS.

Mexims. Matched Expansion Insulator/ Metal Systems. These are ceramic-to-metal seals. (Trade-name: English Electric Co., Stafford, England).

MG–Block, MG–plank. Perforated building units, 9in × 4½ × 9in (228 × 114 × 228 mm) (block) or storey-height (plank), made from fired clay. Developed by the *M*ellor *G*reen Laboratories of British Ceramic Research Ltd.

Mg Point. See TRANSFORMATION POINTS.

MHD Magnetohydrodynamics. The study of the motion of an electrically-conducting fluid (which may be a plasma) in a magnetic field. Ceramics are used to contain the fluid in direct electricity generation by this process.

Mica Convention. See under RATIONAL ANALYSIS.

Mica (Glass-bonded). An electroceramic made by the bonding of mica (natural or synthetic) with a glass of high softening point. Some products of this type can be used up to 500°C. They have good dielectric properties. Transverse strength, 90–105 MNm^{-2} Thermal expansion 11×10^{-6} . Sp. gr. 3.0.

Mica (Natural). Group name for a series of silicates, the most important of which, to the ceramist, is MUSCOVITE (q.v.). a common impurity in clays.

Micas Hydrated aluminium silicates with sheet-type crystal structures which allow cleavage into thin plates. Their general formula is $M[Al_2(OH)_2][Si_3Al)O_{10}]$ where M is K (MUSCOVITE q.v) or Na. The Al may be substituted by Mg or Ca. Micas crystallize in the monoclinic system as $Al_2(OH)_2$ sandwiched between two $Si_{1.5}$ $Al_{0.5}O_5$ sheets. They are excellent electrical insulators.

Michigan Slip Clay. A clay occurring in Ontonagon County Michigan; it is similar to ALBANY CLAY (q.v.).

Micoquille. Thin glass similar to COQUILLE (q.v.) but with a radius of curvature of 180 mm.

Microballoons. Hollow glass spheres 20–300μm diameter, with wall thicknesses of a few microns, made by introducing particles comprising a mixture of glass-forming components and a blowing agent (e.g.urea) into the fuel supply of a burner. The particles melt to form glassy spheres which are blown into microballoons by gaseous decomposition products of the blowing agent.

Microcharacter Hardness. A scratch hardness test in which a loaded diamond is used; it has been applied to the testing of glazes. The 'Microcharacter' instrument was developed by C. H. Bierbaum *(Trans. Amer. Soc. Mech. Engrs.,* p. 1099, 1920; p. 1273, 1921).

Microcline. See FELDSPAR.

Microcracking See FRACTURE, CRACK NUCLEATION; CRACK PROPAGATION; TOUGHNESS.

Microcracks (sub-micron fractures) are the origin of weaknesses in BRITTLE ceramics. However, regions containing extensive pre-existing microcracking can inhibit the growth of a larger crack by blunting the crack tip and reducing the stress concentration. See STRAIN HARDENING; CRACK BRANCHING.

Micro-glass. Thin glass for making cover-slips for use in microscopy.

Microlite (1) An expanded volcanic mineral, white spheroidal granules of a metastable amorphous aluminium silicate with a honeycomb, microcellular structure.

(2) Trade-name: a dense corundum, consisting of uniform micro-crystals (1–3μm), introduced in 1951 by the Moscow Hard-Alloy Combine. It contains 99% Al_2O_3 and a small amount of MgO to inhibit crystal growth; the primary use is in tool tips.

Micromeritics. The science of small particles. This term, which has not achieved popularity, was proposed by M. J. DallaValle in his book *Micromeritics* (Pitman, 1943).

Micromerograph. Trade-name: an instrument for particle-size analysis depending on the rate of fall of particles in air; it is suitable for use in the range l-250μm. (Sharples Corp. Research Labs., Bridgeport, Pa., USA.)

Micron. One-thousandth of a millimetre; symbol μm.

Micronizer. A type of fluid-energy mill used for very fine grinding, e.g. talc to 5μm. (This term was originally a trade-name registered by The Buell Combustion Co. Ltd., London).

Microscopy. The classic books on the use of the microscope in ceramics are: *Thin-Section Mineralogy of Ceramic Materials* by G. R. Rigby. (British Ceramic Research Assoc., 2nd Ed., 1953) and

Microscopy of Ceramics and Cements by H. Insley and D. Frechette (Academic Press, New York, 1955).
Microstructure. The arrangement of grains, pores and cracks and mineral species on the scale of approximately 1 micron. The microstructure of ceramics, and its wideranging importance, was extensively reviewed by R. Rice (in *Treatise on Materials Science and Technology,* Vol II, Academic Press 1977, pp 199–381.
Microwave Heating. Materials absorb energy from radiation when the frequency of that radiation corresponds to the natural or resonant frequency of an interatomic bond in the material. Typically microwave energy is absorbed by the OH bond stretching frequency and so can be used for drying or firing of ceramics. As heating is uniform throughout the body, there are technical advantages, but also economic limitations to the processes.
Midfeather. A dividing wall between two flues, in a gas retort or glass-tank furnace for example; in the latter, the wall may also be called a TONGUE.
Midland Impact Test. A test for the ABRASION RESISTANCE (q.v.) of PAVERS (q.v.), which is a modified version of B.S. 368, 1965. The surface is abraded by ball bearings, and the volume removed is calculated from the measured loss in mass.
Midrex Process. In this process for the direct reduction of iron, iron ore is reduced in a shaft furnace by H_2 and CO at up to 760°C. (*J. Inst. Refract. Engrs.* Winter 1980, p17)
Mie Scattering. Detailed calculations based on the full electromagnetic theory of light are required to explain light scattering by particles of sizes comparable to the wavelength of light. (G. Mie *Ann Phys* **25**, 1908, 377). Such scattering by colloidal particles is responsible for the red colour of gold-ruby glasses.
Migeon Kiln. A zig-zag brick and tile kiln, but with a single gallery. (Br. Pat 701337, 23/12/53. QRE Migeon, France.
Mil. One-thousandth of an inch.
Milford-Astor Machine. A simple printing machine that has been used for the small-scale production of transfers for the decoration of pottery. (Brittains Ltd., Stoke-on-Trent, England.)
Milk Glass. Alumina and fluorspar is added to soda-lime glass to produce an opaque white glass. (cf OPAL GLASS; ALABASTER GLASS).
Mill Addition. Any of the materials, other than the frit, charged to the ball mill during the preparation of a vitreous enamel slip. Such materials include clay, opacifier, colouring oxide and settingup agent.
Millefiori. Decorative glassware, particularly paperweights, made by setting multi-coloured glass cane to form a design within a clear glass matrix. This type of glass-ware originated in Italy and derives its name from Italian words meaning 'A Thousand Flowers'.
Mineral Dressing. A term formerly used in UK instead of the now more common U.S. term BENEFICIATION (q.v.)
Mineral Wool or Rock Wool. There are three main types of mineral wool: *Rock Wool,* made from natural rocks or a mixture of these; *Slag Wool* made from metallurgical slag; *Glass Wool,* made from scrap glass or from a glass batch. All three varieties are used as heat-insulating materials. B.S. 3958 deals with the properties of mineral fibre thermal insulation. US Dept. Commerce Standard CS 117–49 defines 5 classes of mineral wool having the following max. temperature limits of use: Class A – 600°F; Class B – 1000°F; Class C –

1200°F; Class D – 1600°F; Class E – 1800°F.

Mineralizer. A substance that, even though present in only a small amount, assists the formation and/or crystallization of other compounds during firing. A small amount of alkali, for example, mineralizes the conversion of quartz to cristobalite; boric oxide acts as a mineralizer in the synthesis of spinel ($MgAl_2O_4$); the presence of iron compounds facilitates the growth of mullite crystals.

Minium. Red lead, Pb_3O_4. See LEAD OXIDE.

Minter System. A method of interconnecting periodic brick kilns by a flue system that permits the withdrawal of hot gases from one kiln and their transfer to another kiln in the system. (M. M. Minter, *Clayworker,* **70**, 332, 1918).

Minton Oven. A down-draught type of pottery BOTTLE OVEN (q.v.) patented by T. W. Minton (Brit. Pat. 1709, 10th May 1873).

Mirac Process. Trade-name: a process for the treatment of steel claimed to permit one-coat enamelling; (Pemco Corp., Baltimore, Md., USA).

Miraceram Barium titanate ceramics with positive temperature coefficient of resistance, used to make heating elements for domestic appliances, which then need no separate overheating cutout. The heaters are made as honeycomb with high specific surface.

Miraware Disposable 'china' containing 50% vermiculite developed by the American Can Co. U.S. Pat. 3689611, 5/9/72; *Ceram Ind* **97** (1) 28 1971

MISIA An automatic loading/unloading system for bricks and blocks. The blocks are loaded on trays, which are transported on special cars (in tunnel dryers) or moved by fork-lift truck (onto chamber dryers). (Ennepi Nosenzo Poggi SpA, Italy).

Mismatch. If the properties, particularly the thermal expansion, of the components of a heterogeneous solid are mutually incompatible, there is said to be 'mismatch'. A typical example is the case of a ceramic material consisting of crystals set in a glassy matrix; it may be important that there is no mismatch between these two components.

Mistron Vapour. Talc, marketed by United Sierra, USA.

Mitographic Decal A transfer whose design is applied by silk-screen printing.

Mitre Seam. See JOINT LINE.

Mix. A BATCH (q.v.) after it has been mixed.

Mixer. In the clay industries the usual types of mixer are:

(1) *Batch-type Mixer*; operates by rotating arms.

(2) *Shaft Mixer*; a continuous mixer for wet or plastic material which is fed into an open trough along which it is propelled and mixed by one or two rotating shafts carrying blades.

(3) *Pug Mill;* a shaft mixer with a closed barrel instead of an open trough; the term PUG MILL should not be confused with PUG (q.v.)

Mixer Furnace. See HOT-METAL MIXER.

MMC Metal matrix composite – see COMPOSITES.

Mobilometer. See GARDNER MOBILOMETER.

Mocca or Mocha Ware. A type of pottery having a brown moss-like or dendritic decoration similar in appearance to the Mocha, or Mecca, stones (brown agates) used in Arabia for brooches. The effect is produced by allowing differently coloured slips to flow over the surface of the ware.

Mock Acid Gold. See ACID GOLD.

MOCVD The *M*etal *O*xide *C*hemical *V*apour *D*eposition process (q.v.).
Modified Chemical Vapour Deposition (MCVD). See OPTICAL FIBRES.
Modular Co-ordination; Module. A system for the standardization of the dimensions of building components on the basis of multiples of one or more modules, i.e. basic units of length. The British Standards Institution has issued B.S. 6649 specifying a 200 × 100 × 75 mm format for clay and calcium silicate bricks, giving a work size of 190 × 90 × 65 mm. ISO 1006/1 specifies a 100 mm module for tiles and other building products. ASTM C909 specifies sizes for refractory bricks and shapes based on a 38 mm module. E835 specifies a metric 100 mm module for clay and concrete building units. See also PREFERRED INCREMENT.
Modulus of Elasticity. The ratio of stress to strain within the elastic range. See YOUNG'S MODULUS.
Modulus of Rigidity. The MODULUS OF ELASTICITY (q.v.) in shear.
Modulus of Rupture. The maximum transverse breaking stress applied under specified conditions, that a material will withstand before fracture. The modulus of rupture, *M*, of a test piece of rectangular cross-section is given by: $M = 3Wl/2bd^2$, where *W* is the breaking load, *l* the distance between the knife-edges in the transverse-strength test, *b* the breadth of the test piece and *d* its depth. For a cylindical test-piece of diameter *d*, the equation is : $M = 8Wl/\pi d^3$. B.S. 1902 Pts 4.4 and 4.5 specify the test methods for refractories at ambient and at elevated temperatures. Pt 7.6 specifies the test for castable refractories. ASTM C133 and C93 specify tests for refractories and for insulating firebricks; C583 specifies a test for refractories at high temperatures; C491 a test for air-setting plastic refractories. C-1099 for carbon containing refractories. ASTM C1161 specifies a test for advanced ceramics at ambient temperature, C1211 at elevated temperature, F417 a test for electronic grade ceramics. C689 is a test for unfired clays; for C648 test for tiles; C674 for whitewares; C158 for glass.
Mogensen Sizer. A classifer consisting of a system of cascading screens (Fredrik Mogensen AB, Sweden).
Mohammedan Blue. See CHINESE BLUE.
Mohr-Coulomb Failure Law. Predicts the shear stress which a powder compact will support before particle rearrangement or flow begins.
$\tau_f = C + \sigma_f \tan \phi$
where τ_f is the shear stress required to produce flow; C is a constant; σ_f is the applied axial load, ϕ is the angle of internal friction, a material constant, depending on the mechanical and morphological properties and surface chemistry of the powder. ϕ gives a measure of the fluidity of the powder; low values indicate that the powder is easy to shear.
Mohs' Hardness. A scale of scratch hardness based on a series of minerals in ascending order of hardness. The original scale, due to Mohs, in the early 19th century, is still used, with minor changes, but an extended scale takes account of the need for greater differentation between hard materials. See APPENDIX C.
Moil. The excess glass remaining on a shaping tool or in a glass-making machine; it is re-used as CULLET (q.v.).
Moistorer. Japanese (Shinegawa Refractories) equipment for preventing the slaking of dolomite bricks.
Moisture Content. In chemical analysis the moisture content is determined by drying the sample at 110°C until the

weight is constant. For works control of the moisture content of a powder or of clay, instruments are available based on measurement of the pressure developed by the evolution of acetylene from a mixture of the sample with calcium carbide. Methods depending on the variations of the electrical properties of a material with changes in its moisture content are not applicable to clays owing to the great influence of exchangeable bases and soluble salts. The neutron absorption method is more promising.

Moisture Expansion. See ADSORPTION.

Molecular Sieve. The crystal structures of some compounds (e.g. ZEOLITES, q.v.) contain holes or tunnels which permit the passage of certain other molecules if small enough.

Moler. A deposit of DIATOMITE (q.v.) of marine origin occurring on the island of Mors, Denmark; it has been worked since 1912 for use as a heat-insulating material, as a constituent of special cements and for other purposes.

Moller and Pfeiffer Dryer. A multi-track tunnel dryer with concurrent air flow and heat recuperation from the hot air at the exit end. This design of dryer is well suited to heavy clay products requiring gentle initial drying.

Molochite. Trade-name: a pre-fired china clay. Composition (per cent): SiO_2 54–55; Al_2O_3 42–43; Fe_2O_3 0.7; Alkalis 1.5–2.0. P.C.E. 1770°C. Thermal expanson 4.7×10^{-6} (0°-1000°C). It is used as a CHAMOTTE (q.v.) in blast-furnace stack refractories, kiln furniture, refractory cements, etc. (English Clays Lovering Pochin & Co. Ltd, St. Austell, England).

Molten-cast Refractory. A refractory made by FUSION CASTING (q.v.); also, and in the UK more commonly, known as a FUSION-CAST REFRACTORY.

Molybdenum Borides. Five compounds have been reported: Mo_2B; m.p. 2120°C; sp. gr. 9.3. Mo_3B_2, dissociates at 1900°C. MoB exists in two crystalline forms – α-MoB, m.p. 2350°C sp. gr. 8.8; β-MoB, m.p. 2180°C, sp. gr. 8.4. Mo_2B_5 dissociates at approx. 1600°C; sp. gr.. 7.5. MoB_2; m.p. 2100°C; sp. gr., 7.8; thermal expansion, 7.7×10^{-6}.

Molybdenum Carbides. MoC; m.p. 2700°C; sp. gr. 8 .; thermal expansion (20–1000°C) 5.7×10^{-6}, (20–2000°C) 11.5×10^{-6}. Mo_2C; m.p. 2685C; sp. gr. 8.9; thermal expansion (20–1000°C) 6.2×10^{-6}.

Molybdenum Disilicide. See MOLYBEDENUM SILICIDES.

Molybdenum Enamel ('Moly' Enamel). A vitreous enamel containing about 7.5% molybdic oxide, which improves the adherence and acid resistance.

Molybdenum Nitrides. MoN and Mo_2N; both compounds are unstable above about 700°C.

Molybdenum Silicides. There are at least three compounds: $MoSi_2$, Mo_5Si_3 and Mo_3Si. The most important is the disilicide $MoSi_2$ m.p. 1980°C (dissociates into Mo_5Si_3); hardness 80–87 Rockwell A; thermal expansion (20–1000°C) 8×10^{-6}, This special refractory material resists oxidation up to 1700°C. Because of its relatively high electrical conductivity, it has found some use as a material for furnace heating elements.

Moly-manganese Process. The most widely-used BRAZING (q.v.) process for alumina ceramics. A slurry of powders of Mn, Mo or their oxides, with glass-forming compounds, is painted on the ceramic and fired in wet hydrogen at 1500°C. Glassy material from the ceramic bonds the Mo constituent to its surface. Braze metals may then be used, which wet this metallized interface, though not the original ceramic.

Monazite. Strictly, the mineral $(La,Ce)PO_4$; the term is commonly used for the naturally occurring sands rich in rare earths.

Monel. Trade name: Henry Wiggin & Co. Ltd., Birmingham, England. A nickel-copper alloy having good resistance to acids and therefore used for equipment in the PICKLING (q.v.) process in the vitreous-enamel industry.

Monitor. Equipment for directing a high-pressure (2 MPa) jet of water against a clay face in the hydraulic process of winning china clay.

Monk-and-Nun Tile. German name for SPANISH TILE (q.v.).

Monk Bond. A method of laying bricks in a wall; each course is laid as two stretchers followed by one header; the headers in alternate courses cover vertical joints, those in the intermediate courses are set in the middle of a stretcher. Also known as *flying bond* it is a modified FLEMISH BOND (q.v.).

Monnier Kiln. A tunnel kiln designed for the firing of building bricks; it is mechanically fired from the top with coal, which burns among the bricks as in a HOFFMANN KILN (q.v.). Although the design is of French origin (Patented by J. B. Monnier) the first such kiln was built in Kent, England, in 1979.

Monocottura. The process of single firing of glazed tiles, and the resulting product, of particular properties.

Monofire. Tradename. A tile setter for single-fired glazed tiles. Ferro Corp. *Interceram* (2) p147, 1978.

Monolithic. Literally, 'single stone'. The older meaning was attached to furnace linings made by casting, ramming or gunning rather than by bonding together bricks or shapes. The newer meaning relates to ceramics made from a single material, as opposed to composites.

Monometer Furnace. Trade-name: Monometer Mfg. Co. Ltd, London. An oil-fired rotary furnace, particularly for the melting of cast iron; a rammed refractory lining is used, generally similar to that in a CUPOLA (q.v.).

Monticellite. The complex orthosilicate $CaO.MgO.SiO_2$; dissociates at 1485°C with the formation of a liquid phase, sp. gr. 3.2; thermal expansion (0–1200°C) 12.8×10^{-6}. Monticellite is formed in basic refractories when appreciable amounts of lime and silica are present; it is for this reason that specifications for chrome ore and dead-burned magnesite usually stipulate low maxima for the SiO_2 and CaO contents.

Montmorillonite. A magnesian clay mineral approximating in composition to $5Al_2O_3.2MgO.24SiO_2.6H_2O$; some of the H^+ groups are usually replaced by Na^+ or by Ca^{2+}. Montmorillonite is the principal constituent of bentonite; Na-bentonite is characterized by considerable swelling when it takes up water, whereas little, if any, swelling occurs with Ca-bentonite. The mineral was originally found (1847) at Montmorillon, France.

Moonstone Glass. A type of OPAL GLASS (q.v.) resembling moonstone (a whitish, pearly feldspar).

Moore Bin Discharger. Trade-name: a system of vibrating fins arranged to assist the discharge of clay, or other 'sticky' material, from a storage bin. (Conveyors (Ready Built) Ltd., Stroud, Glos., England.)

Moore-Campbell Kiln. An early type of electric tunnel kiln; its principal feature was the free suspension of the heating elements on refractory knife-edges. (B. J. Moore and A. J. Campbell, Brit. Pat. 270 035, 5/5/27.)

Morehouse Mill. A mill of the attrition type consisting essentially of a vertically-driven shaft, with horizontal milling elements, made of sintered alumina. It has been used for the preparation of vitreous-enamel slips and ceramic glazes.

Morgan-Marshall Test. A sand-blast test developed by Morgan Refractories Ltd., and Thos. Marshall & Co. Ltd., England, for the evaluation of the abrasion resistance of refractory bricks. In the UK the original apparatus *(Trans. Brit. Ceram. Soc.,* **54,** 239, 1955), modified in a few details, has been generally adopted by the refractories industry. See B.S. 1902 Pt 4.7, which test is obsolete, though still listed. Some typical ranges of results of this test are given in the following table:

Material Abradability Index
Mild Steel 5
Silicon carbide refractory 15–120
Floor quarry 30–50
Blue engineering brick 50–80
Plate glass 65–70
Fireclay refractory 70–300
High-alumina refractory 40–300
Carbon refractory 130–160
Magnesite refractory 160–300
Silica refractory 200–300

Morphology. The study of size, shape and texture of particles and particle compacts. B.S. 2455, 1958 defines a variety of particle shapes:

Spherical – globule shaped, approximately spherical
Nodular – rounded irregular
Granular – approximately equi-dimensional, but irregular
Irregular – lacking all symmetry
Crystalline – geometric, as freely developed crystals in a fluid
Dendritic – branched crystalline
Acicular – needle shaped
Angular – sharp-edged, polyhedral
Flaky – plate-like
Fibrous – threadlike

Mortar. Material for the joining of structural brickwork. B.S. 4551 includes tests for sampling, chemical analysis, consistency, flow properties, stiffening rate and strength of mortars. It gives compositions, methods for their caluculation, and tables of mix proportions.. See also REFRACTORY MORTAR, and B.S. 1902 Pt 11. The usual types of building mortars, and the proportions (by vol.) recommended in B.S. 5628 Pt 3, are:

Type of Mortar	*Cement*	*Lime*	*Fine Aggregate*
Cement Mortar	1	0–¼	3
Cement-Lime Mortar	1	1	5–6
Cement-Lime Mortar	1	2	8–9
Cement-Lime Mortar	1	3	10–12
Lime Mortar	–	1	2–3

ASTM standards for mortar include C270; C1142 for ready-mixed mortar; C780 for evaluation of mortars. Nine other ASTM standards specify aggregates, grouts and special tests. See also ACID RESISTANT CEMENTS.

Morted Ware. A fault in pottery-ware resulting from a small local concentration of soluble salts forming on the surface of the green ware or the biscuit ware. In the green state the fault may result from a drop of water falling on the ware, during fettling for example. In the biscuit state, the fault arises from moisture condensing on the ware during the early stages of firing.

Mosaic. Small pieces of coloured glass or clay tile arranged to form a pattern or picture and cemented to a wall or floor. (The word is cognate with Muse, a Greek goddess of the arts.) In the USA the largest facial area for a ceramic mosaic tile is 6 in^2 (3750 mm^2).

Mottling. (1) A form of decoration applied to vitreous enamelware, to glazed ceramic tiles, and to some studio pottery. For tiles the effect is produced by hand, with a sponge, or by a rubber roller.
(2) The staining that sometimes develops during the firing of silica refractories; the stains are irregular in shape and vary from red to brown. The cause is precipitation of Fe_2O_3 from solution in the matrix of the brick. These stains are also sometimes referred to as LIVER SPOTTING or FLOWERS.
Mould (for the Shaping of Pottery). The plaster moulds used for the casting, jiggering and jolleying of pottery-ware are made as follows. *Stage 1, Model:* the prototype of the piece of ware that is to be made is usually shaped of clay, occasionally of wood or plaster. *Stage 2, Master Mould:* plaster is cast round the model: depending on the shape of the latter, the mould, when set, may need to be cut into several pieces. *Stage 3, Case Mould:* a replica of the original model, usually made by casting plaster in the master mould, which is first coated with a separating agent, e.g. soft soap. More recently, organic plastics have been used in the making of case moulds; they are costly but, owing to their hardness, have a long life. *Stage 4, Working Mould:* as many moulds as may be required are made by casting plaster around the case mould.
Mouldable. A mixture, containing graded refractory aggregate and other plasticizing and/or bonding materials, which will harden when heated, by the formation of a chemical and/or a ceramic bond. Such a mixture is supplied mixed with an appropriate addition of water or other liquids, and in a workable condition ready for installation by hand or pneumatic ramming. (Mouldables are distinguised from RAMMING MIXES (q.v.) by having a WORKABILITY (BS 1902 Pt 7.2) greater than 15%.
Mould Brick. A fireclay brick, or a heat-insulating brick shaped to fit in the top of an ingot mould to help to maintain the steel at the top of an ingot molten until the main part of the ingot has solidified.
Mould Mark. A fault on glass-ware caused by an uneven mould surface, or, in the case of curved glass sheets, by the dies or frame used in its shaping (see also JOINT LINE).
Mould Plug. A piece of refractory material, or metal, for sealing the bottom of an ingot mould in the casting of steel. (See also PLUG.)
Mould Release. After casting or pressing, the ceramic shape must be removed from the mould without damage or distortion. Such *mould release* depends partly on good mould design, but can be aided by lubricating the mould surface prior to use with a *mould release agent*. Mould release from metal moulds can be achieved electrically. An applied potential causes water in the ceramic compact to migrate to the mould surface, there providing lubrication.
Mould Runner. A man or youth who, in old pottery factories, carried filled moulds from the making department to the drying room and empty moulds back again to the maker; this work has now been eliminated by the introduction of dryers of various types placed adjacent to the making machines.
Mouldable Refractory. A mixture of graded refractory aggregate (or aggregates) and a plasticizer, usually clay, supplied mixed with water so that it is already workable; chemical bonding agents may also be incorporated; the workability is such that the material can readily be placed by hand malleting. Also known as a PLASTIC REFRACTORY.

Mounted Point. A small abrasive unit, variously shaped, permanently mounted on a spindle.
Mouth (of a Converter). See NOSE.
Mouthpiece. There is a difference in usage of this term as applied to a PUG (q.v.) or AUGER (q.v.): more commonly it refers to the die only, but sometimes to the die-plus-spacer.
Move. In the hand-made glass industry, a stipulated output per CHAIR (q.v.) for an agreed rate of pay.
Moving-fire Kiln. See ANNULAR KILN.
Mud Hog. A US machine for the disintegration of dry or moist plastic clay. It consists of a rotating swing hammer operating close to a series of anvils linked together to form a steeply inclined slat conveyor.
Muff. A blown cylinder of glass, cut and flattened while plastic, to make small pieces of flat glass.
Muffle. A refractory internal shell, forming a permanent part of some fuel-fired kilns. Ware placed in a muffle is protected against contact with the combustion gases. In small laboratory kilns the muffle is usually a single refractory shape; in industrial kilns the muffle is built up with refractory brickwork. In a SEMI-MUFFLE the refractory protection acts only as a shield against direct flame impingement, the combustion gases subsequently being free to pass through the setting.
Mulberry. A cobalt stain modified to a purple colour by the addition of manganese dioxide.
Muller Mixer. A WET PAN (q.v.) in which the mullers are suspended out of contact with the pan bottom. Thus the material charged to the pan is mixed without further grinding taking place.
Mullet. A knife-shaped piece of iron used in severing a handmade glass bottle from the blow-pipe.
Mullite. The only compound of alumina and silica that is stable at high temperatures (m.p. 1850°C). The composition is $3Al_2O_3.2SiO_2$ but it can take Al_2O_3 into solid solution. Mullite is formed when any refractory clay is fired and is an important constituent of aluminosilicate refractories, aluminous porcelains, etc. In USA, mullite refractories are defined (ASTM – C467) as containing 56–79% Al_2O_3.
Multi-bucket Excavator. A type of mechanical excavator sometimes used in the getting of brick-clays. Buckets are carried by endless chains round a steel boom (known as a 'ladder') and as they travel they scoop up clay. This type of excavator normally gets clay from below the level on which the truck of the machine is standing.
Mul-T-Cote. Tradename. A coating machine for glass containers particularly bottles. A single polyurethane-based coating can be applied for decoration, or a double coating with a second 'high modulus' polyurethane to provide resistance to shock, scratches chemical attack, and to ultraviolet light. Coating is by dipping and drying. (Moon-Star, Japan and Smit Oven, Neths).
Multiflies. A tiling system in which ten tiles may be fixed simulaneously on to a reinforced screed covered with a contact slip. Various joint patterns can be made, with exact bond widths. (*Fliessen Platten* **37** (3) 64 1987).
Multiform Process. Dry-milled glass powder is mixed with an organic binder and a fluid vehicle, spray-dried and the agglomerates pressed, to form small articles of complex shape such as thread guides and television tube electron gun mounts.
Multijet Kiln. Outside fresh air is fed through the roof to the top of the kiln, so that convection automatically

establishes an even temperature distribution. (Solaquaire Corp. USA).

Multi-passage Kiln. A kiln consisting of a number of adjacent tunnels, each of small cross-section, through which the ware is pushed on refractory bats. These kilns may be electric or gas-fired. The typical electric multi-passage kiln has 24 passages arranged 4 across and 6 high; the ware is pushed in opposite directions in adjacent passages. Gas-fired multi-passage kilns usually have only 4 passages and the ware is pushed in the same direction in each passage.

Multiple-hearth Furnace. A furnace designed primarily for the roasting of ores in the non-ferrous metal industry, but which has more recently been adapted to the calcination of fireclay to produce CHAMOTTE (q.v.). The furnace is in the form of a large, squat, vertical cylinder and consists of several hearths around a central shaft on which rabble arms are mounted. The material to be calcined is charged to the top hearth and slowly descends, hearth by hearth, to the bottom, where it is discharged.

Munsell System. A method of colour notation based on evaluation of three parameters: lightness, hue and saturation. (A. H. Munsell, *A Colour Notation*, Munsell Color Co., Inc., Baltimore, USA.)

Munter Wheel. A rotary air/air heat exchanger, fitted between the exhaust and supply air ductwork of kilns and dryers. The device was originally developed in Scandinavia for the printing trade, in which additional economies were achieved by the recovery of condensed solvents.

Murgatroyd Belt. That part of the side wall of a glass container in which stresses are concentrated following thermal shock; the belt normally extends from the base to about three-quarters of the height of the wall. (J. B. Murgatroyd, *J. Soc. Glass Tech.*, **27**, 77, 1943)

Murray-Curvex Machine. A device for off-set printing from an engraved copper plate on to pottery flatware by means of a solid convex pad of gelatine. Cold-printing colours are used, the medium being formulated from synthetic resins and oils. Output is about 1200 pieces/day. (G. L. Murray, Brit. Pat., 735 637, 24/8/55; 736 312, 7/9/55.) In modern embodiments of this machine, silicone rubber replaces the gelatine pad, to obtain greater dimensional stability when temperature and humidity vary; inks need to be compatible with the silicone; the engraved copper plate may be replaced by a PHOTOPOLYMER PLATE (q.v.) for quicker and cheaper production of designs. See also TOTAL TRANSFER PRINTING.

Muscovite. A mica approximating in composition to $(Na, K)_2O.3Al_2O_3.6SiO_2.2H_2O$; it is in this form that most of the alkali is present in many clays. Thermal expansion (20–800°C) 9.6×10^{-6}. When heated, mica begins to lose water at about 250°C but, at normal rates of heating, dehydration is not complete until about 1100°C; the product is a glass containing mullite and some free Al_2O_3.

Musgrave-Harner Turbidimeter. An instrument developed in 1947 by J. R. Musgrave and H. R. Harner of the Eagle-Picher Co., Missouri, USA: it has been used for the particle-size analysis of clays and other ceramic raw materials in the range 0.25–60μm.

Mussel Gold. An old form of prepared gold for use in the decoration of pottery. It was made by rubbing together gold leaf, sugar (or honey) and salt; the paste was then washed free from soluble material and, traditionally, stored in mussel shells.

Mycalex. Trade-name: a particular type of MICA (GLASSBONDED) (q.v.). Mycalex and T.I.M. Ltd., Cirencester, England).
Mylius Test. See IODOEOSIN TEST.
***n*, n_D**, etc. Symbol for refractive index, the subscript indicates the spectral line to which the quoted value refers, e.g. n_D = the refractive index for a wavelength equivalent to the sodium D line.
***ν*-value.** Alternative name for ABBE VALUE (q.v.).
Nabarro-Herring Creep. CREEP (q.v.) controlled by the diffusion of vacancies through the lattice (bulk diffusion). The creep rate is proportional to the square of the grain size. (F.R. Nabarro, Conference on the strength of solids, Bristol University 1947, and C. Herring *J. Appl. Phys* **21,** 1950, p437).
Nacrite. $Al_2O_3.2SiO_2.2H_2O$. A rare, exceptionally well crystallized, clay mineral of the kaolinite group.
Naftoflex. Trade-name. A range of fluid polymers, which can be cured with a hardener to form rubber-like or hard plastics at room temperature. They are used to make case moulds. (Metalgesellschaft AG, Frankfurt/Main.
Nalfloc. Trade-name. Colloidal silica. (Nalfloc, Northwich, 1970).
Nanoperm. See PERM.
Nanostructure. Particle size, porosity and other structural features of ceramics on the scale of approximately 1nm (10^{-9} metre) of microstructure.
Naples Yellow. See LEAD ANTIMONATE.
Napoleon Green. A ceramic colour that can be fired-on at temperatures up to 1000°C. It consists of (%): CoO, 30; Cr_2O_3. 45; hydrated Al_2O_3, 15; ZnO, 10.
Nasicon. The compound $Na_3Zr_2PSi_2O_{12}$ is a superionic conductor with application as a solid electrolyte.
Natch. The key (a hemispherical hollow and complementary protrusion) in a two- or multi-part plaster mould for pottery making; the purpose is to ensure that the parts fit correctly. Occasionally known as a JOGGLE.
Natural Clay Tile. US term defined (ASTM – C242) as a tile made by either the dust-pressed method or the plastic method from clays that produce a dense body having a distinctive, slightly textured, appearance.
Natural Gas. Gas from large natural underground deposits (in Britain, chiefly below the North Sea). The gas is chiefly methane, and has provided a readily controllable, clean burning, sulphur free fuel for the ceramics industry since the 1970's. Cf. LPG.
Navvy. Old name for a MECHANICAL SHOVEL (q.v.) – from its original use in canal ('Navigation') excavation.
NDE. Non-Destructive Evaluation (q.v.).
NDT. Non-Destructive Testing (q.v.)
Neapolitan Yellow. See LEAD ANTIMONATE.
Neat Work. In a brick structure, the brickwork set out at the base of a wall.
Near-net shaping. Shaping methods for green ceramics which provide a close approximation to the final fired shape, minimising the need for expensive diamond grinding of the fired shape to achieve the required shapes and size tolerances. This is particularly important when making complex shapes to close tolerances from hard, wear resistant engineering ceramics.
Nebuchadnezzar. A 20-quart wine bottle (the largest size).
Neck. (1) That part of a bottle between the FINISH (q.v.) and the shoulder.
(2) A stack of unfired bricks as set in a CLAMP (q.v.).
Neck Brick. US term for FEATHER-END (q.v.).
Neck Mould or Neck Ring. The mould that shapes the neck of a glass bottle.

(Also known as a FINISH MOULD or RING MOULD.)

Needle or Plunger. A refractory part of a feeder in a glassmaking machine; it forces glass through the orifice and then pulls it upwards after the shearing operation.

Needling. A process for bonding refractory fibres into a blanket.

Negative thixotropy. A phenomenon, displayed by suspensions of some clay minerals, in which fluidity increases with resting time, and the hysteresis loops are clockwise.

Nelson Stud. A proprietory form of stud, fusion welded in place.

Neo-Ceramic Glass. A term proposed by B. V. Jana Kiramarao *Glass Tech.*, **5**, (2). 67, 1964) for DEVITRIFIED GLASS (q.v.). In defining the term, absence of nucleation after 4 h heating at the deformation temperature is taken as an arbitrary criterion for distinguishing between neo-ceramic glasses and stable glasses.

Neodymium Oxide. Nd_2O_3; m.p. 2270°C; thermal expansion (25–1000°C) 11.4×10^{-6}. This oxide imparts a violet colour to glass.

Neophane Glass. A yellow glass for automobile sunroofs, sunglasses, etc. tinted with neodymium oxide to reduce glare.

Nepheline Syenite. An igneous rock composed principally of nepheline ($K_2O.3Na_2O.4Al_2O_3.8SiO_2$) and feldspar. There are large deposits in Ontario (Canada), and Stjernoy (Norway). It is used as a flux in whiteware bodies, and as a constituent of some glasses and enamels.

Nernst Body. A sintered mixture of thoria, zirconia and yttria together with small amounts of other rare-earth oxides. After it has been preheated to about 2000°C this body becomes sufficiently electrically conducting for use as a resistor in high temperature laboratory furnaces. (W. Nernst, Brit. Pat. 6135, 12/3/1898.)

Network-forming Ion. One of the ions in a glass that form the network in the glass structure as postulated by Zachariasen (see under GLASS). The ratio of the ionic radius of the network-forming ion to that of the oxygen ion must lie between 0.155 and 0.225 for triangular co-ordination, or between 0.225 and 0.414 for tetrahedral co-ordination; such ions include B^{3+}, Al^{3+}, Si^{4+} and P^{5+}.

Network-modifying Ion. One of the ions in a glass which, according to Zachariasen's theory (see under GLASS), do not participate in the network. They must have a rather large radius and a low valency, e.g. the alkali metals and the alkaline earths.

Neuberg Blue. A pigment comprising copper carbonate, and iron ferrocyanide and sulphate.

Neumann – Kopp Rule. The heat capacity of a compound is equal to the sum of those of its components in the same standard state at the same temperature. This implies that there is no change in heat capacity on formation of the compound, and so the entropy of formation is zero – an assumption more likely to be valid for non-ionic compounds.

Neutral Glass. A name sometimes applied to glass that is resistant to chemical attack; such glasses commonly contain 8–10% B_2O_3. (cf. NEUTRAL-TINTED GLASS).

Neutral Refractory. A refractory material such as chrome ore that is chemically neutral at high temperatures and so does not react with either silica or basic refractories.

Neutral-tinted Glass. A grey glass, usually of the borosilicate type; these glasses are used in light filters to reduce the transmission without (as far as possible) selective absorption of particular wavelengths (cf. NEUTRAL GLASS).

Neutralizer. A dilute solution of alkali or of sodium cyanide used in the treatment of the base-metal for vitreous enamelling after the PICKLING (q.v.) process. *Alkali Neutralizer* consists of a warm (65–70°C) solution of soda ash, borax or trisodium phosphate, the strength being equivalent to 0.3–0.4% Na_2O. *Cyanide Neutralizer* is a 0.10–0.15% solution of sodium cyanide: other ingredients may be present to neutralize the hardness of the water.

Neutron-absorbing Glass. For thermal neutron-absorption a glass must contain cadmium, which is the only common glass constituent having a high neutron-capture cross-section; it is used to form a cadmium borate glass, which is made more chemically durable by the addition of TiO_2 and ZrO_2.

New Mine Fireclay. A fireclay (there are five seams) occurring below the OLD MINE FIRECLAY (q.v.) in the Stourbridge district. England.

Newcastle Kiln. A type of intermittent kiln formerly popular in the Newcastle-on-Tyne area. In its original form it is a rectangular kiln with two or three fireboxes at one end and openings for the exhaustion of waste gases at the base of the other end-wall, which incorporates a chimney. In a later design the kiln is of double length, there are fireboxes at each end, and the waste gases are removed from the centre of the kiln. Such kilns found particular use in the firing of refractories and salt-glazed ware.

Newtonian Fluid. A fluid for which the ratio of the shear stress to the shear rate is constant; this constant is the viscosity. Ceramic slips approach Newtonian behaviour when completely deflocculated. (cf. BINGHAM BODY; DILATANCY; RHEOPEXY; THIXOTROPY)

Nextel. Trade-name. 3M Corp. Mullite fibres.

Nib. (1) A protrusion at one end of a roofing tile serving to hook the tile on the laths in the roof.
(2) A fault in flat glass, in the form of a protrusion at one corner, caused during cutting.

Nibbed Saggar. A SAGGAR (q.v.) with internal protrusions to support a bat (thus permitting the placing of two separate layers of ware) or to allow a cover to be placed inside the top of the saggar.

Nibber. A US term for the squeegee blade in SCREEN PRINTING (q.v.)

Nicalon. Silicon carbide yarn (i.e. a continuous thread, not simply discrete fibres) made by Nippon Carbon Co.

Nickel Dip. Also sometimes known as NICKEL FLASHING or NICKEL PICKLING. A process for the treatment (after PICKLING (q.v.)) of base-metal for vitreous enamelling to improve adherence and reduce the likelihood of COPPERHEADS (q.v.) and FISH-SCALING (q.v.). A solution of a nickel salt, e.g. the sulphate or double ammonium sulphate, is used; boric acid and ammonium carbonate are also added in small amounts to give the correct pH. In an alternative process the nickel is deposited on the metal by the addition of a hypophosphite solution

Nickel Oxide. There are two common oxides: grey, NiO, m.p. 1980°C; black, Ni_2O_3. They are used (about 0.5% of either type) in ground-coat enamels for sheet steel to promote adherence, and in ferromagnetic ferrites.

Niobates. Compounds of metals with Nb_2O_3, usually made by firing the oxides or their precursors together. Best known are lithium niobate ($LiNbO_3$) and lead niobate ($PbNb_2O_6$). They have a structure similar to, but of lower symmetry than the PEROVSKITES, (see TUNGSTEN BRONZE) and have high polarizability, and so are piezoelectric, pyroelectric or ferroelectric compounds. Lithium niobate is used in laser and microwave electronics applications. Lead niobate is used for high-temperature piezoelectric transducers. Some niobates are RELAXOR (q.v.) ferroelectrics. PNN, PZN, PFN, PMN are lead niobates with nickel, zinc, iron and magnesium respectively.

Niobium Borides. Several compounds have been reported, including NbB_2; decomposes at 2900°C; sp. gr. 7.0: thermal expansion, 5.9×10^{-6} parallel to *a* and 8.4×10^{-6} parallel to *c*. NbB; m.p. 2300°C; sp. gr., 7.6. Nb_3B_4; melts incongruently at 2700°C, sp. gr. 7.3.

Niobium Carbide. NbC; m.p. 3500°C; sp. gr. (theoretical), 7.8 g/ml; modulus of rupture (25°C), 25 MNm^{-2}; thermal expansion (25–1500°C) 7.5×10^{-6}.

Niobium Nitrides. Three nitrides have been reported: NbN, Nb_2N and Nb_4N_3. During reaction between Nb and N_2 at 800–1500°C the product generally consists of more than one compound. Most of the phases are stable at least to 1500°C.

Niobium Oxide. The commonest oxide is Nb_2O_5; m.p. 1485°C; sp.gr.4.6.

Niopside. The nickel analogue, $CaO.NiO.2SiO_2$, of diopside.

Nip. A small glass bottle for mineral water.

Nitre. See POTASSIUM NITRATE.

Nitrides. A group of special ceramic materials: see under the nitrides of the following elements: Al, Be, Cr, Ga, Hf, Mo, Nb, Si, Ta, Th, Ti, U, V, W, Zr.

Nitriding. The production of nitrides from a porous metal shape, by direct reaction with nitrogen gas infiltrating the pores.

Nitrogen Ceramics. The nitrides, oxynitrides, sialons and related compounds containing nitrogen.

Nitty Enamel. Vitreous enamel containing almost imperceptible small surface pits.

Nobel Elutriator. An early type of multiple-vessel elutriator for particle-size analysis; it is described in *Z. Anal. Chem.*, **3**, 85, 1864.

Noborigama. A step kiln invented in China c. 400 B.C. and introduced into Japan c. 1300 A.D. It comprises successive chambers built into the side of a hill, and will fire pottery to 1300°C. (*Keram Z.* **35** (3) 149 1983).

No-Cement Castable. Alumina and aluminosilicate castable refractories which do not contain hydraulic setting cement and in which the bonding agents contribute no significant amount of lime (CaO). The product might contain up to 0–2% total lime (CaO) on a calcined basis as contributed by the aggregate. (ASTM C401–91).

Nodular. see MORPHOLOGY.

Nodular Fireclay. An argillaceous rock containing aluminous and/or ferruginous nodules.

No-fines Concrete. Concrete made with aggregate from which all the fine material, i.e. less than 9 mm, has been removed; the proportion of cement to aggregate varies from 1:8 to 1:12 (vol.). The strength is low (7–8.5 MNm^{-2}) but so too are heat and moisture transfer.

Nominal Dimension. As applied to hollow clay building blocks this term is defined in the USA (ASTM – C43) as: A dimension that may be greater than the specified masonry dimension by the thickness of a mortar joint.

Nomogram or Nomograph. Two or more scales drawn as graphs and arranged so that a calculation involving quantities represented by the scales can be reduced to a reading from the diagram. Examples of the use of nomograms in ceramics include the determination of porosity, the solids concentration in slips, the drying and firing shrinkage of clay pipes.

Non-Destructive Evaluation, Non-Destructive Testing. Methods of measuring the properties of materials which preserve the items tested; e.g. strength measurements relying on measuring elastic properties ultrasonically, rather than by breaking rods. Such techniques are of greater importance in the testing of expensive engineering ceramics, especially when it is desirable to evaluate the properties of shaped and fired components. ASTM C1175 specifies test methods and standards for the non-destructive testing of advanced ceramics, largely ultrasonic and radiographic techniques.

Non-load Bearing Tile. US term for a hollow fired-clay building block for use in masonry carrying no superimposed load.

Non-metallic Inclusion. A non-metallic particle that has become embedded in steel during its processing. Tests, particularly those employing radioactive tracers, have shown that (contrary to earlier belief) these inclusions rarely originate from the refractory materials of the furnace, ladle or casting-pit.

Non-Oxide Ceramics. A useful general term for ceramics (such as borides, carbides, nitrides) in whose structure oxygen plays no rôle.

Non-Plastics. (1) Those components of a TRIAXIAL WHITEWARE BODY (q.v.) having no inherent plasticity of their own. (2) Ceramic materials (e.g. alumina) with no inherent plasticity, requiring the addition of a PLASTICIZER (q.v.) to allow them to be shaped satisfactorily.

Non-stoichiometric. A chemical compound is said to be nonstoichiometric if the ratio of its constituents differs from that demanded by the chemical formula. This may happen with oxides that are readily reducible, or with compounds containing an element of variable valency, or when interstitial atoms are present in the lattice. Some non-stoichiometric ceramics are of interest as being semi-conducting.

Non-vitreous or Non-vitrified. These synonymous terms are defined in USA (ASTM-C242) in terms of water-absorption: the term non-vitreous generally implies that the ceramic material has a water absorption above 10% except for floor tiles and wall tiles which are considered to be non-vitreous if the water absorption exceeds 7%.

Norfloat, Norflot. A Norwegian feldspar (the latter is the German spelling).

Normal Spinel. See SPINEL.

Norman Brick. A long brick, 2⅔rds × 4 × 12 in. (68 × 102 × 305 mm).

Norman Slabs. A type of glass for stained windows; it is made by blowing bottles of square section and cutting slabs of glass from the four sides.

Normative Analysis. The CALCULATED MINERALOGY (q.v.)

Nose or Mouth. The constricted circular opening at the top of the refractory lining of a CONVERTER (q.v.); through this opening molten pig-iron is charged and steel is subsequently poured.

Nose-ring Block. A block of refractory material specially shaped for building into the discharge end of a rotary cement kiln; also known as a DISCHARGE-END BLOCK.

Nose-ring Dam. A deep dam, or (band of thicker refractory) at the discharge

end of a rotary kiln, to ensure an extended soaking period for the charge.
Nostril. The term used for the refractory gas- or air-port in a gas retort.
Nostril Blocks. See RIDER BRICKS.
Notch. See SLAG NOTCH.
Notch Test. A test proposed by R. Rose and R. S. Bradley (*J. Amer. Ceram. Soc.*, **32**, 360, 1949) for the assessment of the low-temperature spalling tendency of fireclay refractories. A notch is made in a transverse-strength test-piece and the effect of this on the strength at 800°C is determined; if the transverse strength is but little reduced by the notch, the low-temperature spalling resistance of the fireclay refractory will be good.
Novamur. See HELTON.
Novorotor. A double-rotor hammer mill for clay preparation. (Hazemag, Germany cf UNIROTOR).
NO_x. A general (punning) term for the oxides of nitrogen, now subject to emission limitation rules for noxious gases.
Nozzle. A cylindrical fireclay shape traversed by a central hole of uniform diameter; the top of the nozzle is contoured to form a seating for a STOPPER (q.v.). Nozzles are fitted in the bottom of ladles used in the teeming of steel. In the USA, three qualities are specified in ASTM – C435.
NSC Clay. A blended clay with controlled casting concentration, tailored to casting sanitaryware (English Clays, St Austell, Cornwall).
NTC. New Tundish Coating. The automatic spraying of a tundish with refractory, 75–90% MgO. (Foseco, Japan).
Nucerite. Trade-name. A glass-ceramic coating for metal (Pfaudler Permutit Inc, New York, 1966; Pfaudler Balfour, Leven, Scotland, 1990).
Numerical Aperture (NA). The sine of the maximum launch angle (the *acceptance angle*) that allows total internal reflection and hence transmission of a signal down an optical fibre. Optical fibres usually have a core glass of refractive index n_1, surrounded by a cladding glass of refractive index n_2. $(NA)^2 = n_1^2 - n_2^2$.
Nu-Value. See ABBE VALUE.
Nuclear Reactor Ceramics. These special ceramics include: fuel elements – UO_2, UC and UC_2; control materials – rare-earth oxides and B_4C; moderators – BeO and C.
Nucleation; Nucleating Agent. Nucleation is the process of initiation of a phase change by the provision of points of thermodynamic non-equilibrium; the points are provided by 'seeding', i.e. the introduction of finely-dispersed nucleating agents. The process has become of great interest to the ceramist since DEVITRIFIED GLASS (q.v.) became a commercial product.
NZP. Sodium Zirconium Phosphate. Materials of this structural group have very low and highly anistropic coefficients of thermal expansion. The structure consists of PO_4 tetrahedra and ZrO_6 octohedra linked by shared oxygen atoms at the corners. Each member of the alkali zirconium phosphate family can be derived from this structure by rotating the polyhedra in the network.
Obsidian. A natural glass. A dark transparent, siliceous igneous material whose composition is similar to granite.
Obscure Glass. Glass designed or surface treated to diffuse the light it transmits, to prevent objects behind it being seen clearly.
Occlusion. Trapping of gas or liquid within a solid.
Occupational Exposure Standard. The concentration of an airborne substance, (averaged over a reference period) at which according to current knowledge,

there is no evidence that it is likely to be injurious to employees if they are exposed by inhalation day after day to that concentration. Control of exposure is considered adequate if the OES is not exceeded, or if it is exceeded, the employer has identified the reason and is taking appropriate steps to comply with the OES as soon as is reasonably practicable. See *Occupational Exposure Limits, Criteria Document Summaries.* HSE, London HMSO 1993.

Ochre. Impure hydrated iron oxide used to a limited extent as a colouring material for coarse pottery and structural clayware.

Ocrate Process. The treatment of concrete with gaseous SiF_4 to transform any free CaO into CaF_2. The treated concrete has improved resistance to chemicals and to abrasion. (Ocriet-fabriek NV, Baarn, Holland.)

Octa-comb System. This system for installing ceramic fibre panels requires no pre-anchoring plan, but uses stud-welding of octagonal packs.

Octahedrite. See ANATASE.

Octopus. A system of firing intermittent kilns with gas or pulverised fuel, so-called from the spoke-like arrangement of burners across the kiln roof. The system was developed by National Coal Board and Gibbons Bros (see e.g. *Clayworker* **83**, (983), 16, 1974)

Oddments or Fittings. Special glazed clayware shapes used in conjunction with glazed pipes. These shapes include Bends, Junctions, Tapers, Channels, Street Gullies, Syphons, Interceptors and Yard Gullies; the first five of these are made partly by machine and partly by hand, the last three items have to be entirely hand-moulded. For dimensions see B.S. 539.

Oden Balance. Apparatus for particle-size analysis; one of the pans of a delicate balance is immersed in the settling suspension and the change in weight as particles settle on the pan is measured. (S. Oden, *Internat. Mitt. Bodenk.*, **5**, 257, 1915)

OES Occupational Exposure Standard.

Off Gases. The exhaust or stack gases from a kiln firing ceramics, particularly those that originate from the body being fired or the burnout of binders.

Off-hand. Hand-made glass-ware produced without the aid of a mould.

Off-hand Grinding. See FREEHAND GRINDING.

Offset. A recessional set-back of brickwork, e.g. in a domestic chimney.

Offset Finish. See FINISH.

Offset Printing. A process of decoration as applied, for example, in the MURRAY-CURVEX MACHINE (q.v.).

Offset Punt. Term applied to the bottom of a bottle if it is asymmetric to the axis.

OH Furnace. See OPEN-HEARTH FURNACE.

Oil Spot. A surface fault, seen as a mottled circle, on electric lamp bulbs or valves; it is caused by carbonization of a contaminating drop of oil.

Old Mine Fireclay. A fireclay occurring in the Brierley Hill district, near Stourbridge (England); it usually contains 56–64% SiO_2, 25–30% Al_2O_3 and > 2.5% Fe_2O_3. A smooth plastic fireclay formerly much used for making glass-pots, but now largely worked out.

Old Roman Tile. Roofing tiles designed for use in pairs: a flat under-tile (the Tegula) is laid adjacent to a rounded over-tile (the Imbrex). cf. ROMAN TILE; SPANISH TILE.

Olivine. A mineral consisting principally of a solid solution of FORSTERITE (q.v.) and FAYALITE (q.v.). Rocks rich in olivine find some use as a raw material for the manufacture of forsterite refractories and, when crushed, as a foundry sand.

Omnicon. A linescan computerised image analyser for measuring particle size and shape. (Trade-name, Bausch & Lomb Corp.).

Once-fired Ware (UK) or Single-fired Ware (USA). Ceramic whiteware to which a glaze is applied before the ware is fired, the biscuit firing and glost firing then being combined in a single operation. Because the glaze must mature at a relatively high temperature, it is usually of the leadless type. Sanitary-ware is the principal type of ware made in this way.

One-Coat ware. Articles with a single coat of vitreous enamel, though the term is sometimes used as a contraction of *one-cover coat* ware, in which case a single cover coat is applied over a ground coat.

One-fire Finish. US term for vitreous enamelware produced by a single firing.

One-high firing. The American term for SINGLE LAYER SETTING (q.v.).

O'Neill Machine. A machine for making glass bottles originally designed by F. O'Neill in 1915; it was based on the use of blank moulds and blow-moulds. The principle in modern O'Neill machines is the same; they are electrically operated and are capable of high outputs.

On-glaze Decoration. Decoration applied to pottery after it has been glazed; the ware is again fired and the colours fuse into the glaze, the decoration thus becoming durable. Because the decorating fire can be at a lower temperature with on-glaze decoration, a more varied palette of colours is available than with under-glaze decoration. A test for the resistance to detergents is provided in ASTM – C556.

Ooms-Ittner Kiln. An annular, longitudinal-arch kiln divided into chambers by permanent walls and with a flue system designed for the salt-glazing of clay pipes; the fuel can be coal, gas or oil. Ooms-Ittner Co., Braunsfeld, Cologne, Germany.)

Opacifier. A material added to a batch to produce opal glass, or to a glaze or enamel to render it opaque. Glass is usually opacified by the addition of a fluoride, e.g. fluorspar or cryolite; for glazes and enamels the oxides of tin, zirconium or titanium are commonly used.

Opacity. Defined in ASTM – C286 as: The property of reflecting light diffusely and non-selectively. For a definition relating specifically to vitreous enamel see CONTRAST RATIO.

Opal. Non-crystalline hydrated silica.

Opal Glass. A translucent to opaque glass with a 'fiery' appearance. (Cf. ALABASTER GLASS). Opal glasses are made by phase-separation processes which produce dispersed crystals in the glass. In *crystalline opals* these crystals form spontaneously during cooling from the melt or later heat treatment during manufacture. NaF or CaF_2 crystals are commonest. *Liquid – liquid opals* are formed by separation of the glass into two distinct glassy phases while it is still soft. *Reheat* or *restrike* opals are clear after forming, and are reheated to produce the phase separation. Their opacity is due to light scattering, and depends on the difference in refractive index between the phases, the degree of phase separation and the size distribution of the dispersed second phase. Opal glasses are usually made creamy white, and used for dinnerware. Liquid-liquid opal is more expensive to make, but process control is better.

Opalizer. A substance (usually a fluoride) which will produce opalescence (iridescent reflection) in glass.

Opaque Ceramic Glazed Tile. US term (ASTM – C43) for a hollow clay facing block the surface faces of which are covered by an opaque ceramic glaze which is coloured and which has a bright satin or gloss finish.

Opatowski Process. A process for the extraction of BeO from siliceous ores. The ground ore is mixed with NH_4HSO_4 and water and heated to 95°C to dissolve metal salts. Mn is oxidised using $NH_4S_2O_8$, and ammonia treatment precipitates the metals as hydroxides. After separation, treatment with NaOH dissolves the beryllia as Na_2BeO_2, which is hydrolized to Be(OH), which on heating to 250°C, decomposes to beryllia, BeO. The process is effective for concentrations of BeO in the ore of 51%, and the $(NH_4)_2SO_4$ formed is recycleable. (J. Opatowski, Canadian Pat. 663 776 1962).

Open Flame kiln. A kiln in which the ware being fired is exposed directly to the burning gases of the flame.

Open Handle. A cup handle of the type that is attached to the cup at the top and bottom only, the side of the cup itself forming part of the finger-opening (cf. BLOCK HANDLE).

Open-hearth Furnace. A large rectangular furnace in which steel, covered with a layer of slag, is refined on a refractory hearth; it is heated by gas or oil, and operates on the regenerative principle which was first applied to this type of furnace by Sir Wm. Siemens in 1867 (hence the earlier name SIEMENS FURNACE). The type of refractory lining depends on the particular steelmaking process used: see ACID OPEN-HEARTH FURNACE, BASIC OPEN-HEARTH FURNACE and ALL-BASIC FURNACE. The process is now obsolete.

Opening Material. Term occasionally used for a non-plastic that is added to a clay to decrease the shrinkage and increase the porosity; such materials include grog, chamotte, sand and pitchers.

Op-po Mill. Tradename. A mill for the vibratory grinding of fine powders. Podmore & Sons, Stoke-on-Trent.

Optical Blank. A piece of optical glass that has been pressed to approximate to the shape finally required; also called a PRESSING.

Optical Crown Glass. Any glass of low dispersion used for optical equipment (cf. flint glass). There are many varieties, their names indicating their characteristic composition, e.g. barium crown, borosilicate crown, fluor-crown, phosphate crown, zinc crown.

Optical Fibres. Fine glass fibres along which information can be transmitted by total internal reflection of coherent light signals. One fibre may be several km long and carry as much information as can several thousand telephone wires. Glasses in the GeO_2-SiO_2 system are usually used, with additions of P_2O_5 to ease fabrication and of chlorides to adjust the resultant refractive index to that of silica. The fibres are made by one of several vapour-phase techniques. Volatile halides ($SiCl_4$, $GeCl_4$ and $POCl_4$) and high purity gases (Cl_2, O_2, $C_2Cl_2F_2$ etc) deposit glass onto a suitable substrate after a high temperature oxidation or hydrolysis reaction. The resulting preform is drawn into fibres with appropriate variations in compostion across their cross-section. In ***MCVD (modified chemical vapour deposition)*** the vapour stream reacts inside a rotating silica tube heated by an oxyhydrogen torch. A white soot is deposited downstream of the torch, and sintered into an internal glassy layer by moving the torch downstream. The tube is collapsed into a solid preform rod by increasing the flame temperature. In ***OVPO (outside vapour-phase oxidation)*** and ***VAD (vapour-phase axial deposition)*** the hydrolyzed reactants are deposited on the side of a rotating starting rod (OVPO) or on the end of a rotating rod which is gradually raised from the flame as the deposit grows (VAD). Heating may also be by plasma oxidation at RF- or microwave frequencies, with higher reaction efficiencies. Optical fibres are characterized by high strength (several GPa) due to careful surface treatment eliminating flaws from which microcracks can develop.

Optical Flint Glass. See under FLINT GLASS.

Optical Glass Classification. A system by which an optical glass is classified according to its refractive index, n_D and its Abbe Value, υ. Standard borosilicate crown glass, for example, has $n_D = 1.510$ and $\upsilon = 64.4$; its classification by this system is 510644 or 510/644. Further identiiication is often provided by letters, preceding the number; e.g. BSC = Boro-Silicate Crown; LF = Light Flint, etc.

Optical Pyrometer. A device for the measurement of high temperatures depending ultimately on visual matching. The type most used in the ceramic industry is the DISAPPEARING FILAMENT PYROMETER (q.v.); a less accurate type is the WEDGE PYROMETER (q.v.).

Optimat. A de-airing extruder designed on the basis of model experiments and theoretical studies of the plastic flow of clays. (Rieterwerke, Germany *ZI Int* **32,** (5) 1980, p267).

Optoelectronics. The study and application of the variation of the optical properties of materials with applied electric fields. See FERPICS.

Orange Peel. A surface blemish, adequately described by its name, sometimes occurring on vitreous enamelware, glass-ware and glazed ceramics. In vitreous enamelling, the fault can usually be prevented by an alteration in the spraying process and/or in the specific gravity of the enamel slip.

Orbit Kiln. A continuous kiln for firing tableware. Each item is fired separately, and is mounted on an alloy support, coated with alumina, which rotates the ware on its own axis as it passes through the kiln, thus ensuring uniform temperature distribution. There is no other kiln furniture, and the ware can be loaded and unloaded automatically. (Kerabedarf, Germany).

Organic Bond. (1) A material such as gum or starch paste that can be incorporated in a ceramic batch to give it dry-strength; the organic bond burns away during the firing process, which develops a ceramic bond or causes sintering.
(2) Some abrasive wheels have an organic bond such as synthetic resin, rubber or shellac, and are not fired.

Organotin Compounds. Chemical compounds in which the element Sn is combined with organic radicals. Organotin compounds can catalyse the hydrolysis of ETHYL SILICATE (q.v.) to a gel, or mineralise MULLITE formation.

Orifice Ring. See under BUSHING.

Ormosil. Organically-modified silicates.

Ornamental Ware. Decorative items such as statuettes, figures, flowers and ornamental plates. The COMBINED NOMENCLATURE (q.v.) regards a plate as ornamental if it has on its face an obviously decorative design; is not part of a table service; has a pierced rim for hanging, or other integral stand, and/or is generally unsuitable for culinary purposes because of shape, weight, material, decoration or difficulty in cleaning its non-smooth surface. The

distinction is significant because ornamental plates do not usually need to meet METAL RELEASE (q.v.) regulations.
Orthoclase. See FELDSPAR.
Orthosilicates. See SILICATE STRUCTURES.
Orton Cones. PYROMETRIC CONES (q.v.) made by the Edward Orton Jr. Ceramic Foundation, Columbus, Ohio. They are made in two sizes: 2½ in. (63.5 mm) high for industrial kiln control, and 1 1/8th in. (29 mm) high for P.C.E. testing. For nominal equivalent softening temperatures see Appendix 2.
Osborn-Shaw Process. See SHAW PROCESS.
Osmosis. See ELECTRO-OSMOSIS.
Ostwald Ripening. Grain growth due to differential dissolution between particles of different sizes and shapes. Solute dissolves at sharp corners, or from small particles, and tends to reprecipitate on coarser, larger or rounder particles, which grow at the expense of fine particles.
Osumilite – see BARIUM OSUMILITE.
Outside Vapour - phase Oxidation (MCVD, OVPO and VAD). See OPTICAL FIBRES.
Oven. Obsolete term for a kiln, more particularly the old type of BOTTLE-OVEN (q.v.).
Ovenware. Ceramic whiteware or glassware (casseroles etc) of good thermal shock resistance for use in cooking.
Overalls. Workers overalls in the pottery industry are designed without lapels, pockets and other dust traps. They are made from terylene, a smooth fibre which does not retain DUST (q.v.).
Overarching. A progressive departure of circumferential joints in arches or rings from the required radial alignment. Sometimes due to the use of too many sharply tapered bricks.
Overburden. The soil and other unwanted material that has to be removed before the underlying clay or rock can be worked by the open-cast method.
Overflow Process. A DOWNDRAW PROCESS (q.v.) for special glasses, ranging from soft photochromic glasses to hard chemically strengthened aluminosilicate automobile glass. Molten glass flows into a trough, then overflows and runs down either side of a pipe, fuses together at the root, and is drawn downward until cool. The thickness of the sheet depends on the velocity of pulling. Sheets 0.4mm up to 1cm can be drawn, with both surfaces fire-polished and comparable to float glass. The process is higher viscosity than the FLOAT GLASS PROCESS (q.v.) with rather lower production rates. (Corning Glass Works, USA).
Overflush. A fault in glass-ware caused by the flow of too much glass along the line of a joint (cf. FIN).
Over-glaze Decoration. See ON-GLAZE DECORATION.
Over-glazed. Pottery-ware having too thick a glaze layer, particularly on the bottom; this thick glaze is likely to be crazed. Causes of this fault are incorrect dipping, the use of slop glaze of too high a density, or biscuit ware that is too porous, i.e. underfired.
Over-pickling. PICKLING (q.v.) for too long a period or in too strong a solution; this causes blisters in vitreous enamelware.
Overpress. A fault, in glass-ware, in the form of an inside FIN (q.v.).
Oversailing. A projection in brickwork, e.g. in a domestic chimney.
Overspray. (1) In the application, by spraying, of enamel slip to base-metal or of slop glaze to ceramic whiteware, that proportion of slip or glaze that is not deposited on the ware; it is normally collected for re-use.
(2) In vitreous enamelling, the application of a second (usually thinner) coat of enamel slip over a previous layer of enamel that has not yet been fired.

Overswing. See TORSION VISCOMETER.
OVPO (Outside Vapour - Phase Oxidation). See OPTICAL FIBRES.
Owens Machine. A suction-type machine for making glass bottles; the original machine was designed by M. J. Owens between 1897 and 1904. Since then, there have been many improvements but the basic principle remains the same.
Ox Gall. This material (an indefinite mixture of fats, glycocholates and taurocholates) has been added to glaze suspensions to prevent crawling.
Oxford Clay. A clay of the Upper Jurassic system providing raw material for 30% of the building bricks made in the UK; the Fletton brick industry of the Peterborough area is based on this clay, which contains so much carbonaceous material that the drypressed bricks can be fired with very little additional fuel.
Oxford Feldspar. A ceramic grade of feldspar occurring in Maine, USA. A quoted analysis is (%) SiO_2, 69; Al_2O_3, 17; Fe_2O_3, 0.1; CaO, 0.4; K_2O, 7.9; Na_2O, 3.2; Loss, 0.3.
Oxidation Period. The stage in the firing of clayware during which any carbonaceous matter is burned out, i.e. the temperature range 400–850°C. It is important that all the carbon is removed before the next stage of the firing process (the vitrification period) begins, otherwise a black core may result.
Oxide Ceramics. Special ceramics made from substantially pure oxides, usually by dry-pressing or slip-casting followed by sintering at high temperature. The most common oxide ceramics are Al_2O_3, BeO, MgO, ThO_2, and ZrO_2.
Oxygen Sensor. An application for ceramics whose electrical conductivity is sensitive to changes in the oxygen partial pressure of the surrounding atmosphere. Zirconia is a *potentiometric* sensor material, whose conductivity is based on a solid-state ionic transfer process. Titania is a *conductimetric* sensor dependent upon semiconduction processes.
Oxynitride Glasses. GLASS FORMING SYSTEMS (q.v.) are found amoung the silicon and aluminium oxynitrides, as well as among the silicates. Such oxynitride glasses have increased resistance to devitrification, higher refractive index, higher dielectric constant and greater viscosity and glass-transition temperatures. They are more difficult to prepare than silicate glasses, needing higher temperatures and an atmosphere with a low oxygen content. Crystallisation processes in them yield polycrystalline nitride ceramics with good high-temperature properties. (The strength and creep-resistance of nitrogen ceramics depend markedly on the amount and properties of the intergranular glasses). Silicon oxynitride surface coatings on silica optical fibres greatly enhance their mechanical strength.
Packing Density. The BULK DENSITY (q.v.) of a granular material, e.g. grog or crushed quartzite, when packed under specified conditions. A common method of test, particularly for foundry sands, involves the use of a SAND RAMMER (q.v.).
Packing Factor. The ratio of the TRUE VOLUME (q.v.) to the BULK VOLUME (q.v.).
Pad. The refractory brickwork below the molten iron at the base of a blast furnace.
Pad Printing. Printing ceramic decoration using silicone pads. See MURRAY-CURVEX process.
Paddling. The process of preliminary shaping of a piece of glass, while it is in a furnace and soft, prior to the pressing of blanks of optical glass.
Padmos Method. A comparative method for the determination of the coefficient of thermal expansion of a glass; the glass being tested is fused to a glass of known

expansion and similar transformation temperature. From the birefringence resulting from the consequent stress at the junction of the two glasses, the difference in expansion can be calculated. (A. A. Padmos, *Philips Res. Repts.*, **1**, 321, 1946.)

Paillons. Small pieces of vitreous enamelled metal foil, for jewelry or artware.

Pale Glass. Glass of a pale green colour.

Palissy Ware. Fine earthenware with a brightly coloured tin glaze.

Pall Ring. A type of ceramic filling for towers in the chemical industry – the 'ring' is a small hollow cylinder with slots in the sides and projections in the core. The design was introduced by Badische Anilin und Soda Fabrik – see *Industr. Chemist,* **35**, 36, 1959.

Pallet. A board, small platform or packaging unit sometimes used, for example, in the transport of refractories or building bricks; cf. STILLAGE.

Pallette. See BATTLEDORE.

Palmqvist Cracks. Short cracks, semi-circular in a section perpendicular to the surface, which form at the corners of the indentation made by the diamond pyramid in a VICKERS HARDNESS (q.v.) test. S. Palmqvist (*Jernkontorets Ann.* **141,** 1957, p300; *Arch. Eisenhüttenwesen* **33**, (6), 1962, p.629) first suggested, by empirical reasoning, that the indentation hardness test could also be used as a FRACTURE TOUGHNESS TEST (q.v.).

Palygorskite. A fibrous, hydrated magnesium aluminium silicate. Structurally, this mineral is the same as ATTAPULGITE (q v.). There is a deposit of economic size at Cherkassy, Russia.

PAM. Pneumatically Applied Mortar.

PAMM. British Ceramic Plant and Machinery Manufacturer's Association. PO Box 28, Biddulph, Stoke-on-Trent ST8 7AZ.

Pan Mill. Term sometimes applied to an EDGE-RUNNER MILL (q.v.) but more properly reserved for the old type of mill (also known as a BLOCK MILL) for grinding flint for the pottery industry. It consists of a circular metal pan paved with chert stones over which are moved heavy blocks of chert (RUNNERS) chained to paddle arms.

Pandermite. A hydrated calcium borate that has been used as a component of glazes.

Panel Brick. A special, long, rectangular tongued and grooved refractory brick, laid stretcher-fashion in the wall of a coke-oven.

Panel Spalling Test. A test for the spalling resistance of refractories that was first standardized in 1936, in USA. A panel of the bricks to be tested is subjected to a sequence of heating and cooling cycles and the loss in weight due to the spalling away of fragments is reported. General details of the apparatus and test are given in ASTM – C38. ASTM C107 gives procedures for High-duty Fireclay Refractory, ASTM – C122 for Super-duty Fireclay Refractory. B.S. 1902 Pt 5.11 specifies a panel test for refractories. See also THERMAL SHOCK TESTS.

Paper Resist. See FRISKET.

Pannetier's Reds. Brilliant on-glaze ceramic colours, based on iron oxide, developed in the early 19th century by Pannetier in Paris.

Pantile. A single-lap roofing tile having a flat S-shape in horizontal section. Details of clay pantiles are specified in B.S.1424; the minimum pitch recommended is 35°.

Parallel-plate Plastometer. See WILLIAMS' PLASTOMETER.

Parge, Parget, Pargeting. The coating of the surface of brickwork, as with plaster. The term 'pargeting' is used in the building trade, e.g. for the internal coating of chimney flues with lime

mortar. In the refractories industry the term 'parging' is more common, e.g. for coating or patching a furnace lining.
Parian. A white, vitreous, type of pottery made from china clay and feldspar (in the proportion of about 1:2); some recipes specify the addition of CULLET (q.v.). Parian ware is generally in the form of figures – hence the name, from Paros, the Aegian island from which white marble was quarried for sculpture in Classical Greece. Parian was created by the firm of Spode, England: it was subsequently made by several other firms and was popular between about 1840 and 1860. (PARIAN CEMENT or PLASTER is an alternative name for KEENE'S CEMENT (q.v.).)
Paris Green. A ceramic colour suitable for firing up to about 1050°C and consisting of (%) Cr_2O_3, 25; CoO, 4; ZnO, 8; whiting, 12; borax, 8; flint, 43.
Paris White. See WHITING.
Parison. A piece of glass that has been given an approximate shape in a preliminary forming process ready for its final shaping. (From the French *paraison*, derived from *parer* – to prepare.)
Parker Pre-namel. A process for the treatment of steel prior to enamelling (US Pat., 2 809 907, 15/10/57).
Partially Stabilized Zirconia. See ZIRCONIA.
Particle Shape. See MORPHOLOGY.
Particle Size or (preferably) Particle Mean Size. A concept used in the study of powders and defined as: 'The dimensions of a hypothetical particle such that, if a material were wholly composed of such particles, it would have the same value as the actual material in respect of some stated property' (cf. EQUIVALENT PARTICLE DIAMETER).
Particle-size Analysis. The process of determining the proportions of particles of defined size fractions in a granular or powdered sample; the term also refers to the result of the analysis. The methods of determination available include: ADSORPTION, in which the particle size is assessed on the basis of surface area; the DIVER METHOD; the use of a CENTRIFUGE; COULTER COUNTER; ELUTRIATION; TURBIDIMETER; the ANDREASEN PIPETTE; and, for coarser particles, a SIEVE. (For further details see under each heading.) Automatic operation, electronic analysis of results, image analysis of direct microscopic observation, and laser techniques have been added to the range of techniques available. ASTM C678, 690, 721, 775, 958, 1070 and 1182 specify various techniques for measuring particle size distributions of clays, alumina and quartz.
Particle Size Distribution. The graph of the number of particles in a given size range, against the size of the particles, measured in some consistent fashion (e.g. by diameter, mass, volume, etc.).
Parting Agent. A MOULD RELEASE (q.v.) agent.
Parting Line. See JOINT LINE.
Parting Powder. Refractory powder spread between brick courses to prevent them sticking together when fired.
Parting Strength. The tensile strenth of a ceramic fibre product. Test methods are specified by B.S. 1902 Pt. 6, Section 9.
Parting Wheel. See CUTTING-OFF WHEEL.
Partition Tile. US term for a hollow fired-clay building block for use in the construction of interior partitions but not carrying any superimposed load.
Paste. A literal translation of the French word *Pâte* in the sense of ceramic body; see HARD PASTE and SOFT PASTE.
Paste Mould. A metal mould lined with carbon and used wet in the blowing of glass-ware.

Pat Test. A qualitative method for assessing the soundness of hydraulic cement. Pats of cement are made about 3 in. (75 mm) dia., ½ in. (12.5 mm) thick at the centre but with a thin circumference. They are immersed in cold water for 28 days or in boiling water or steam for 3–5 hours. Unsoundness is revealed by distortion or cracking.
Patch. A repair to a brick lining, especially to one in which the refractories are bricked in rings, and no whole ring is replaced.
Pâte Dure, Pâte Tendre. HARD FIRED and SOFT FIRED (i.e. too high or low temperature) whitewares respectively.
Pâte-sur-Pâte. A method of pottery decoration in which a bas-relief is built up by hand-painting with clay slip as the paint. The method is more particularly suited to the decoration of vases in the classical style; fine examples were made by M. L. Solon at Mintons Ltd., England, in the late 19th century.
Pattern Cracking. See MAP CRACKING.
Paul Floc Test or Paul Water Test. See FLOC TEST.
Paver. Dense, well vitrified bricks of normal size or of lesser thickness for use as a material for walkways and paths. B.S. 6677 Pt. 1 specifies materials, sizes, strengths and skid resistance of clay and calcium silicate pavers. (Pts 2 and 3 specify design practice and method of construction of pavements). In the USA a paver may be a dust-pressed, unglazed relatively thick floor tile with a superficial area of at least $6in^2$ ($3871mm^2$). ASTM C-902 is the US standard for paving bricks for light traffic areas.
Paving Brick. ASTM C410 specifies four grades of industrial floor bricks, type 'T' resistant to thermal and mechanical shock; type 'H' resistant to chemicals and thermal shock; type 'M' with low absorption (usually highly abrasion resistant) and type 'L' with low absorption, high chemical resistance and high abrasion resistance, though normally low resistance to thermal and mechanical shocks.
Paviour. A term applied to clamp-fired STOCK BRICKS (q.v.) that are not of first quality but are nevertheless hard, well-shaped and of good colour.
PBNT. Lead Barium Neodymium Titanate has very low dielectric loss and its high frequency permittivity is stable over a wide temperature range, and so it is used in microwave resonators and filters.
PCE. Abbreviation for PYROMETRIC CONE EQUIVALENT (q.v.).
PDZ. Plasma Dissociated Zircon.
Peach. A term used by Cornish miners, originally for chlorite, but latterly applied to veins of quartz and schorl forming part of the waste from china-clay pits.
Peach Bloom. A glaze effect on pottery produced by the Chinese and characterized by its soft pink colour with patches of deeper red. It is achieved by the addition of copper oxide to a high-alkali glaze but requires very careful control of the kiln atmosphere; the 'bloom' results from incipient devitrification of the glaze surface.
Peacock Blue. A ceramic colour made from a batch such as 33% Cobalt oxide, 7% STANDARD BLACK (q.v.), 45% China Stone, 15% Flint.
Peacocking. An iridescent discolouration (predominantly purple and blue) on blue bricks exposed to rain and the prevailing winds. A film of silica forms on the surface of the brick, thin enough to show light interference effects.
Peak Lopping. Electricity from the national grid supply is supplemented by on-site generation at periods of peak demand, so that the load on the grid is

kept to lower values which attract preferential tariffs.

Pearl-ash. Potassium carbonate, K_2CO_3; sometimes used in glass and glaze batches.

Pearlite. A lamellar aggregate of iron and cementite (Fe_3C) found in carbon steels.

Pearson Air Elutriator. A down-blast type of ELUTRIATOR (q.v.) designed by J. C. Pearson (*US Bur. Stand. Tech. Paper No. 48*, 1915) and used for determining the fineness of portland cement.

Pearson Distribution. A skew statistical distribution applied to find the appropriate sample size in metal release testing. (F. Moore. *Trans J. Brit Ceram. Soc.* **76** (3) 52 1977 where the equation and properties of the distribution are discussed).

Pebble Heater. A heat exchanger in which refractory 'pebbles' (which may be made of mullite, alumina, zircon or zirconia) are used as heat carriers. One type of pebble heater consists of two refractory-lined chambers joined vertically by a throat; both chambers are filled with 'pebbles', which descend at a steady rate, being discharged from the bottom of the lower chamber and returned to the top of the upper chamber. In the latter they are heated by a countercurrent of hot gases; in the lower chamber they give up this heat to a second stream of gas or air.

Pebble Mill. A BALL-MILL (q.v.) in which flint pebbles are used as grinding media.

Peck. A unit of volume (16 pints) sometimes used in the pottery industry in the wet process of body preparation. See also SLOP PECK and STANDARD SLOP PECK.

Pedersen Process. A process devised in 1944 by H. Pedersen, a Norwegian, for the extraction of alumina from siliceous bauxite; the bauxite is first melted in an electric furnace with limestone and coke, the reaction product then being leached with NaOH.

Peeling or Shivering. (1) The breaking away of glaze from ceramic ware in consequence of too high a compression in the glaze layer; this is caused by the glaze being of such a composition that its expansion coefficient is too low to match that of the body (a certain degree of compression in the glaze is desirable. however).
(2) A similar effect sometimes occurs on the slagged face of a refractory.
(3) A fault in vitreous enamelling that is also known as LIFT (q.v.).

Pegasus. A precision grinder, with the workpiece fed normal to a metal or resin bonded diamond cupwheel, for superfinishing ultrahard surfaces.

Pegmatite. A coarsely crystalline rock occurring as veins, which may be several hundred feet thick. Feldspars are a common constituent of pegmatites, which are of ceramic interest for this reason.

PEI. Porcelain Enamel Institute, Washington D.C., USA.

PEI Test. An abrasive charge consisting of steel balls, alumina grit and deionized or distilled water is rotated on the surface whose ABRASION RESISTANCE is to be assessed. An abraded circle 80mm diameter is produced. Glazed tiles thus tested lose only little material, and their abrasion resistance is assessed by a visual estimate of gloss and colour under standard lighting conditions after 150, 600 and 1600 revolutions. (The test is specified in ASTM C1027 and B.S. 6431 Pt. 20, which also specifies a dry method, the MCC TEST).

Pelletized Clay. Ball clays, especially for sanitaryware production, which have been extruded and the rods cut into short lengths for easier handling and dispersion in water.

Pencil Edging. The process of rounding the edges of flat glass.

Pencil Ganister. A GANISTER (q.v.) with black markings, as though pencilled, formed by decomposition of the rootlets that were present in the original seat-earth from which the ganister was formed.

Pencil Stone. Term sometimes applied to PYROPHYLLITE (q.v.).

Pendulum Test. See TRRL PENDULUM TEST (for slip resistance of floors) and IMPACT RESISTANCE.

Penlee Stone. Stone from Penlee Quarries, Penzance, comprising 53% silica, and used as a road stone.

Pennvernon Process. See PITTSBURGH PROCESS.

Pentacalcium Trialuminate. $5CaO.3Al_2O_3$; orthorhombic; sp. gr. 3.03–3.06.

Peptize. Term used in colloid chemistry for the process in ceramic technology known as DEFLOCCULATION (q.v.).

Percussion Cone. A circular or semi-circular crack on a surface subjected to impact damage, which propagates into the body of the material as a spreading cone.

Perfluent Sintering. Lightweight clay blocks are made by a process in which combustion gases stream through the pores of a partially expanded pelletized clay contained in a sagger.

Perforated Brick. A building brick made lighter in weight by its being pierced with numerous, relatively small (6–12 mm dia.) holes, usually in the direction of one of the two short axes. In the UK a perforated brick is defined (B.S. 3921) as one in which has holes not exceeding 25% of the volume of the brick with the holes so disposed that solid material is never less than 30% of the width of the brick; in USA it is normally also over 25% (cf. HOLLOW CLAY BLOCKS).

Periclase. Crystalline magnesium oxide, MgO; m.p. 2850°C: sp. gr. 3.6; thermal expansion (20–1000°C), 13.8×10^{-6}. Periclase, generally with a little FeO in solid solution, is the main constituent of magnesite refractories. SEA-WATER MAGNESIA (q.v.) can be produced as a nearly-pure periclase. Large single crystals of periclase have measurable ductility which may reveal the factors determining ductility and how this property might be achieved in ceramic products.

Periodic Kiln. See INTERMITTENT KILN.

Peripheral Speed. The rate of movement of a point on the edge of a rotating disk or cylinder: it is the product of the circumference and the rate of revolution. The peripheral speed is of importance, for example, in the operation of abrasive wheels and ball mills.

Perish. To disintegrate as a result of slow hydration on exposure to moist air; calcined dolomite disintegrates in this manner if stored for more than a short period.

Perkiewicz Method for Preventing Kiln Scum. A process in which clay bricks, prior to their being set in the kiln, are coated with a combustible, e.g. tar or a mixture of gelatine and flour; should sulphur compounds condense on the bricks during the early stages of firing, the deposit will fall away when the combustible coating subsequently burns off. (M. Perkiewicz, Brit. Pat. 3760, 15/2/04).

Perlite. A rock of the rhyolite type that is of interest on account of its intumescence when suddenly heated to a temperature of about 1000°C. Composition (%): SiO_2, 70–75; Al_2O_3, 12–14; Alkalis, 6–8; Loss-on-ignition, 3.5–6. Perlite occurs in economic quantity in USA, Hungary, Turkey and New Zealand. The expanded product has a bulk density of 0.1–0.2 g/ml and is used as a heat-insulating material.

Perm, (Pm). A c.g.s. unit of gas permeability expressed in [$(cm^3/sec.)$ cm. Poise]/[$cm.^2(Dyne/cm.^2)$]. As this unit is very large, the Nanoperm is used when c.g.s. units are applied to the permeability of ceramic materials; 1 Nanoperm (nPm) = 10^{-9} Perm; 1 Perm = 0.9871×10^{-18} DARCY.

Permanent Linear Change. A preferred term including the two terms AFTER CONTRACTION (q.v.) and AFTER-EXPANSION (q.v.). ASTM C113, C210 and C179 are the American tests for refractory bricks, insulating firebrick and refractory plastic and ramming mixes, respectively. ASTM C605 describes the determination of permanent volume change for fireclay nozzles and sleeves, from which the PLC can be calculated. BS 1902 specifies PLC for shaped insulating products (Pt. 5.9); for dense refractories (Pt. 5.10); for unshaped refractories (Pt. 7.6); while Pt.6 Section 5 specifies a SHRINKAGE test for refractory fibre products.

Permeability. The rate of flow of a fluid (usually air) through a porous ceramic material per unit area and unit pressure gradient. This property gives some idea of the size of the pores in a body – whereas the measurement of POROSITY (q.v.) evaluates only the total pore volume. From the permeability of a compacted powder the SPECIFIC SURFACE (q.v.) of the powder can be deduced. B.S. 1902 Pt. 3.9 describes a gas permeability test applicable to refractory materials, as does ASTM C577. The ASTM C866 test for the filtration rate of whiteware clays depends on their water permeability. See BLAINE TEST; CARMAN EQUATION; LEA AND NURSE PERMEABILITY APPARATUS; PERM; RIGDEN'S APPARATUS.

Pernetti. (1) Small iron PINS or spurs for setting ware in the kiln.
(2) marks on the ware caused by sticking to the pins.

Perovskite. $CaTiO_3$; m.p. 1915°C; sp. gr. 4.10. This mineral gives its name to a group of compounds of similar structure, these forming the basis of the titanate, stannate and zirconate dielectrics. Extensive solid solution is possible so that the potential range of compositions and ferroelectric properties is very great. R. S. Roth (*J. Res. Nat. Bur. Stand.*, **58**, 75, 1957) has classified the perovskites on the basis of ionic radii of the constituent ions; a graph of this type can be divided into orthorhombic, pseudocubic and cubic fields with an area of ferroelectric and antiferroelectric compounds superimposed on the cubic field. In a development of this classification a three dimensional graph includes the polarizability of the ions as the third dimension.

Perpend. An alignment of crossjoints in brickwork.

Perrit. A support, made of heat-resisting alloy, designed to carry vitreous enamelware through an enamelling furnace.

Persian Blue. See EGYPTIAN BLUE.

Persian Red. Pigments of ferric oxide and basic lead chromate.

Perthite. A mineral consisting of a lamellar intergrowth of feldspars, especially of albite and orthoclase.

PESTS. Porcelain Enamel Steel Substrates for screen printed or hybrid circuitry.

Petalite. A lithium mineral, $Li_2O.Al_2O_3.8SiO_2$; sp. gr. 2.45; m.p. 1350°C. The chief sources are Rhodesia and S. W. Africa. Normally it is a less economical source of Li_2O than is SPODUMENE (q.v.) but it is usually free from iron compounds. A major use is in low-expansion bodies resistant to thermal shock.

Petersen Air Elutriator. An up-blast ELUTRIATOR (q.v.) that has found some use in determining the fineness of portland cement.

Petrography, Petrology. The description, classification and study of minerals, their compositions and structure.

Petuntse, Petunze. The Chinese name of CHINA STONE (q.v.).

PFA. Abbreviation for PULVERIZED FUEL ASH (q.v.).

Pfefferkorn Test. K. Pfefferkorn (*Sprechsaal*, **57**, 297, 1924) suggested that the plasticity of a clay could be assessed by determining the amount of water that must be added to produce a standard deformation when a test-piece of the clay is subjected to a standard impact. The test-piece used was a cylinder 40 mm high and 33 mm diam.; the impact was that given by a disk weighing 1.2 kg falling through 14.6 cm; the standard deformation was 33%. The test involves plotting a graph relating water content to deformation; from this graph the water content for 66% deformation is read.

PFN. Lead iron niobate.

PFW. Lead iron tungstate (from the chemical symbols Pb, Fe, W).

pH. Symbol for the acidity or alkalinity of a solution; it is a number equivalent to the logarithm, to the base 10, of the reciprocal of the concentration of hydrogen ions in an aqueous solution. The point of neutrality is pH7; a solution with a pH below 7 is acid, above 7 is alkaline. The pH of a casting slip is of practical importance in determining its rheological properties.

Phase Diagram. Equilibrium diagrams of interest to the ceramist represent crystallographic changes and the melting (or conversely the solidification) behaviour of compounds, mixtures, and solid solutions) under conditions of chemical equilibrium. As an example, the AlO-SiO equilibrium diagram, one of the most important for the ceramist, is shown in Fig. 5. Items of interest are:
(1) The melting points of three compounds are shown, namely those of silica (1723°C), mullite (1850°C) and alumina (2050°C).
(2) There are two EUTECTICS (q.v.) that between silica and mullite melting at 1595°C, and that between alumina and mullite melting at 1840°C.
(3) The ability of mullite to take a small amount of alumina into SOLID SOLUTION (q.v.). For a useful discussion of equilibrium diagrams from the standpoint of the ceramist, and a very comprehensive collection of such diagrams, see E. M. Levin, C. R. Robbins and H. F. McMurdie, *'Phase Diagrams for Ceramists' (Amer. Ceram. Soc.* 1964). This publication has now reached Vol. VIII; Vols VI to VIII being published by the ACS jointly with the US National Institute of Standards and Technology. There are regular bibliographical updates, and prototype computer databases for an on-line service.

Phase Rule. The number of degrees of freedom of a system at equilibrium plus

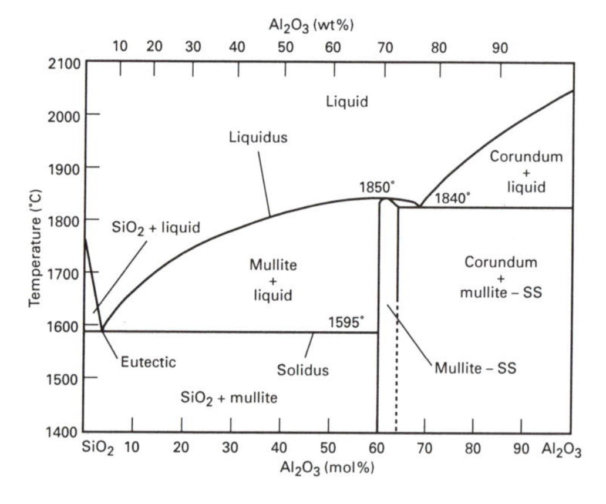

Fig. 5 The Alumina-Silica Phase Diagram.

the number of phases, is equal to the number of components plus 2.

Phase Separation. If a homogeneous liquid mixture is cooled below a critical immiscibility temperature, two liquid (or glass) phases may form, initiated by homogeneous nucleation. Usually one phase has a higher surface tension and forms droplets in the other. Glass ceramics may be formed in this way.

Phlogopite. A magnesium MICA (q.v.) containing little iron, but otherwise similar to BIOTITE (q.v.) Phlogopite micas are excellent electrical insulators, and form the basis of some machineable glass-ceramics.

Phosphate Bond. Cold-setting bonds for refractories can be formed by reactions between oxides and phosphoric acid, phosphate-phosphoric acid reactions or the direct use of liquid phosphate bonds. The bond is due to the formation of an acid phosphate, with weakly basic or amphoteric oxides.

Phosphate Glasses. These are low durability glasses with structures based on an oxygen-phosphorus tetrahedron. Iron-containing phosphate glasses are transparent to visible light, but absorb heat in the infra-red range. They are more resistant to fluorides than silicate glasses. Fluorphosphate glasses have very low optical dispersion (ABBÉ NUMBERS (q.v.) of 70 to 80).

Phosphides. Phosphide ceramics include AlP, BP, Be_3P_2, InP and TiP. Some of these compounds have potential use as semiconductors.

Phosphor. A fluorescent material of the type used, for example, in the coating of the screen of a cathode-ray tube. Ceramic phosphors have been made from Ca silicate, stannate or phosphate activated with Sn, Mn or Bi; some of the Ca can be replaced by Sr, Zn or Cd.

Phosphoric Acid. A syrupy liquid of variable composition sometimes used as a binder for refractories (see ALUMINIUM PHOSPHATE).

Photochromic Glass. Glass whose transmission of visible light decreases when exposed to ultraviolet or visible light, and increases again when that exposure ceases. Commercial photochromic glasses contain silver halide crystallites about 10 nm diameter, dispersed about 100 nm apart. These are precipitated thermally without exposure to radiation (cf. PHOTOSENSITIVE GLASS). When light shines on the glass, small amounts of metallic silver form on these silver halide crystals, darkening the glass much as a photographic image is formed. The glass traps the reaction products close to each other. When the glass is no longer exposed to the light, the reaction reverses and the silver is re-absorbed. Darkening times range from seconds upwards, while some glasses are stable for days in their dark state. The glasses darken less and fade more quickly at higher temperatures. Applications are to sun glasses, opthalmic lenses, automobile sunroofs and architectural sheet glass.

Photoelectric Pyrometer. A device for measuring high temperatures on the basis of the electric current generated by a photoelectric cell when it receives focused radiation from a furnace or other hot body. Instruments have been designed for the measurement of temperatures exceeding 2000°C, but others are available for use at lower temperatures, e.g. for checking the temperature of the molten glass used in lamp-bulb manufacture.

Photon. Trade-name. A machineable ceramic made by Photon Ceramics Ltd, Tokyo.

Photopolymer Plate. A substitute for the engraved metal plate in the MURRAY-CURVEX process. A complex design is developed photographically on a polymer sheet.

Photosensitive Glass. A glass containing small amounts (< 0.1 %) of gold, silver or copper, which render the glass sensitive to light in a manner similar to photographic films; the image is developed by heat treatment. Such glass undergoes a change in some property, usually optical absorption, on exposure to light and simultaneous or subsequent heat treatment. There are three types: *direct* in which absorption develops at the time of the exposure to light; *developed* in which later heat treatment changes the property; *nucleated* in which the heat treatment produces nucleation and growth of a second phase. In the first type, absorption is by photoelectrons and holes trapped in the lattice. In the others, metal particles are produced by chemical reduction, and these absorb the light. See also PHOTOCHROMIC GLASS; POLYCHROMATIC GLASS.

Photovoltaic Materials. Materials which generate an electric potential when exposed to light. CuS, CdS, GaAs and Si are such materials, which find applications in photocells for the generation of solar power.

Photozone Stream Counter. Particle size characteristics are deduced from the magnitude and spatial distribution of reflected light signals from the particles to be measured.

Piano-wire Screen. A screen formed by piano wires stretched tightly, lengthwise, on a frame 2–3 ft (0.6–1 m) wide and 4–8 ft (1.3–2.6 m) high. The screen is set up at an angle of about 45° and crushed material is fed to it from above. The mesh size varies from about 4 to 16. Because there are no cross-wires, and because the taut wires can vibrate, there is less tendency for 'blinding', but some elongated particles inevitably pass the screen.

Pick-up. The amount of vitreous enamel slip (expressed in terms of dry weight per unit area) after the dipping and draining process. For ground-coats, the pick-up is usually 350–450 $g.m^{-2}$.

Pickings. Term applied to clamp-fired STOCK BRICKS (q.v.) that are soft though of good shape.

Pickle Basket. A container for vitreous enamelware during PICKLING (q.v.); the basket is made from a corrosion-resistant metal, e.g. Monel.

Pickle Pills. Capsules, containing Na_2CO_3 and an indicator such as methyl orange, for use in works tests of the strength of the PICKLING (q.v.) bath employed in vitreous enamelling.

Pickling. The chemical treatment of the base metal used in vitreous enamelling to ensure a satisfactory bond between the enamel and metal. The various methods that have been used may be classified as follows: (1) the use of reducing acids, e.g. H_2O_4. HCl and acid salts; (2) the sulphur compounds, H_2S and $Na_2S_2O_3$; (3) metal coatings, e.g. Ni, As and Sb; (4) oxidizing acids and salts, e.g. H_3PO_4, HNO_3, solutions of ferric salts in acids, or $NaNO_3+H_2SO_4$. Method 2 is no longer used; metal coating with Ni (Method 3) is not regarded as pickling in the UK.

Picotite. A complex spinel of general formula $(Fe,Mg)O.(Al,Cr)_2O_3$; the proportion of Al exceeds that of Cr and the Fe:Mg ratio is from 3 to 1. Picotite may occur in slagged basic refractories.

Picrochromite. Magnesium chromite, $MgO.Cr_2O_3$; m.p. 2250°C: sp. gr. 4.41; thermal expansion (0–1200°C) 8.2×10^{-6}, This spinel can be synthesized by heating a mixture of the two oxides at 1600°C; it is formed (usually with other spinels in solid solution) in fired chrome-magnesite refractories. Picrochromite is highly refractory but when heated at 2000°C the Cr_2O_3 slowly volatilizes.

Piezoelectric. A material is stated to be 'piezoelectric' when an applied stress results in the setting up of an electric charge on its surface; conversely, if the material is subjected to an electric field, it will expand in one direction and contract in another direction. Typical piezoelectric materials are the titanate and zirconate ceramics.

Pigeonholes. basket-weave checkers (q.v.)

Pigment. Pigments are solid particles of colour, which are suspended in glass to form ceramic colours. Ceramic pigments, as well as having good colouring properties, must be able to withstand firing temperatures of at least 750°C, and possibly up to 1400°C, in contact with fluxes and glazes. They must be chemically inert and colour stable, non-toxic (meeting METAL RELEASE (q.v.) regulations) and alkali-stable to withstand machine dishwashing detergents. See also ENCAPSULATED COLOURS; FRIT; DRY COLOUR MANUFACTURERS' ASSOCIATION (A brief list of ceramic colours is to be found (by colour) in P.Rado. *Introduction to the Technology of Pottery*, Inst. of Ceramics, 1988 p 154.

Pigskin. A defect in the surface texture of vitreous enamelware and of glazed sewer-pipes; it is caused by variations in the thickness of the enamel or glaze.

Pilkington Twin Process. A continuous process for the production of polished plate glass; a continuous ribbon of glass is rolled, annealed and then simultaneously ground on both faces. The process was introduced by Pilkington Bros. Ltd., England, in 1937. (For a more recent process introduced by this firm see FLOAT GLASS PROCESS.)

Pill Test. See SLAG ATTACK TESTS.

Pillar. (1) A column of brickwork; for example the refractory brickwork between the doors of an open-hearth steel furnace.
(2) An item of KILN FURNITURE (q.v.) forming one of the upright parts of a CRANK (q.v.); cf. POST.

Pillared Clays. Monmorillonite clays, in which the interlayer spacing is widened by intercalation of complex aluminium hydroxide ions.

PIM. Powder INJECTION MOULDING (q.v.).

Pin. An item of KILN FURNITURE (q.v.); it is a small refractory SADDLE (q.v.) for use in conjunction with a CRANK (q.v.). See Fig. 4, p177.

Pin Beater Mill. A type of HAMMER MILL (q.v.).

Pin Disk Mill. A type of rotary disintegrator sometimes used (more particularly on the Continent and in USA) for the size reduction of clay.

Pin-hole. (1) A fault in vitreous enamelware. It is commonly the result of a blister that has burst and partially healed; the usual sources of the gas that gave rise to the blister are a hole in the base-metal or a speck of combustible foreign matter in the cover-coat. Pin-holes can also occur on castings at points where there is extra metal thickness (lugs, etc). At such points the enamel is apt to be underfired and unmatured.
(2) Pin-holes in glazes also result from burst bubbles; here, most of the gas originates from air trapped between the particles of powdered glaze as the glaze begins to mature.
(3) A frequent source of pin-holes in pottery biscuit-ware, and in subsequent

stages of processing, is air occluded in the clay during its preparation.
(4) Pin-holes in plaster moulds originate in air attached to the particles of plaster during blending; this can be eliminated by blending the plaster in a vacuum.
Pin Mark or Point Mark. A fault in vitreous enamelware caused by the imprint of the supports used during the firing process.
Pin-on-Disc Test. A method for measuring the coefficient of friction between two materials in which on material forms the pin, and is pressed against the other material, which forms a rotating disc.
Pinch Effect. Term applied by J. W. Mellor (*Trans. Brit. Ceram. Soc.*, **31**, 129, 1932) to the crazing of wall tiles as a result of contraction of the cement used to fix the tiles on a wall, while it is setting.
Pinch Spalling. Mechanical damage to the hot face of a refractory arch or ring, indicating excessive compression. It may develop radially or axially.
Pining Rod. A metal rod used to measure the contraction of a setting of bricks in a kiln; from colloquial use of the word 'pine' in the sense of 'shrink.'
Pinite. A rock consisting largely of sericite. The composition is (per cent): SiO_2, 46–50; Al_2O_3, 35–37; K_2O, 8–10; loss on ignition, 4–6. A deposit in Nevada, USA, was worked from 1933–1948 as a raw material for making dense refractories for rotary frit kilns and for the cooler parts of cement kilns where high abrasion-resistance is required
Pinning. The process of fixing PINS (q.v.) in kiln furniture.
Pint Weight. The weight in oz of 1 pint of a suspension of clay, flint etc., in water. (One of the last quantities to hold out against metrication, where *slip density* is expressed in kg/litre.
Pip. An item of KILN FURNITURE (q.v.). A pip is a small refractory button with a point on its top surface; ware is supported on the point.
Pipe. (1) See FIELD-DRAIN PIPE; SEWER PIPE, CONCRETE PIPES, PIPECLAY.
(2) A cavity at the top of a fusion-cast refractory resulting from the contraction of the molten material as it cools.
Pipe Blister. A large bubble sometimes produced on the inside of hand-made glass-ware by impurities or scale on the blow-pipe.
Pipeclay. A white-firing, siliceous, clay of a type originally used for making tobacco pipes.
Pipette Method. A method for the determination of particle size: see ANDREASEN PIPETTE.
Pistol Brick. Brick slips cut into L-shapes, to achieve decorative patterns.
Piston Extruder. See STUPID.
Pit. A fault in vitreous enamelware similar to, but smaller than, a DIMPLE (q.v.).
Pitch. The slope of a tiled roof. For clay roofing tiles the common pitch, measured internally at the ridge between the two sides of the roof, is 105°. Less common are the SQUARE PITCH (90°), GOTHIC PITCH (75°) and SHARP PITCH (60°).
Pitch-bonded Refractory. See TAR-BONDED REFRACTORY.
Pitch-impregnated Refractory. See TAR IMPREGNATED REFRACTORY.
Pitch Polishing. The polishing of glass with a polishing agent supported on pitch instead of the more usual felt.
Pitchers. Pottery that has been broken in the course of manufacture. Biscuit pitchers are crushed, ground and re-used, either in the same factory or elsewhere; the crushed material is also used in other industries as an inert filler. Because of the adhering glaze, glost pitchers find less use.

Pittsburgh Process. A process for the vertical drawing of sheet glass invented by the Pittsburgh Plate Glass Co. in 1921. The sheet is drawn from the free surface of the molten glass, the drawing slot being completely submerged; the edges of the sheet are formed by rollers. (This process has also been referred to as the PENNVERNON PROCESS.)
Place Brick. A clamp-fired STOCK BRICK (q.v.) of very poor quality – of use only for temporary erections. The term went out of use before 1940.
Placing. For the purposes of COSHH (q.v.), the approved code of practice defines Glost Placing to include placing ware coated with unfired glaze onto kiln cars; removal of such items from kiln cars except in tunnel kilns.
Placer. A man who sets pottery-ware in SAGGARS (q.v.) or with KILN FURNITURE (q.v.) ready for firing in a kiln. The man in charge of a team of placers was known in the N. Staffordshire potteries as a COD PLACER.
Placing Sand. Fine, clean silica sand used in the placing of earthenware in SAGGARS (q.v.); calcined alumina is used for this purpose in the bedding of bone china. With the increased use of open setting in tunnel kilns and top-hat kilns, placing sand is less used than formerly.
Plagioclase. See FELDSPAR.
Plain Tile. See under ROOFING TILE, WALL TILE (Fig. 7, p350).
Plaining (of Glass). See REFINING.
Planches. Supports for firing enamel artware.
Planiceram. Large thin tongue and groove blocks with mortar-free internal walls.
Plaque. A flat refractory slab, often with triangular indentations, to support PYROMETRIC CONES (q.v.)
Plascast. A clay-bonded castable refractory which is claimed to combine the advantages of mouldables and castables, being more resilient and stable at high temperatures than castables, but more difficult to install. (Morgan Refractories. *Steel Times* **216**, (3) 1988 p155).
Plasma Activated Sintering. A plasma is generated among powder particles by applying a high power electric pulse. This cleans the particles of surface oxides and entrapped gases. The powder is then resistance heated and uniaxial pressure applied. A brief application (minutes) causes sufficient material flow for rapid consolidation with minimum microstructural change.
Plasma-dissociated Zircon. Zircon ($ZrSiO_4$) can be dissociated into zirconia and silica in a plasma at > 1800°C, with rapid cooling to prevent reassociation. The free silica is removed by dissolving in sodium hydroxide, leaving pure zirconia powder.
Plasma Spraying. The process of coating a surface (of metal or of a refractory) by spraying it with particles of oxides, carbides, silicides or nitrides that have been made molten by passage through the constricted electric arc of a PLASMA GUN; the temperature of the plasma arc 'flame' can be as high as 30 000°C. An arc is struck between a thoriated tungsten cone (the cathode) and an anode nozzle. As the plasma leaves the nozzle, it entrains a large quantity of external air (*APS – air plasma spraying*) at velocities of 1000 m/s. The powder is injected into the plasma orthogonally to the jet axis. The outside gas may be argon (*IPS – inert plasma spraying*) or with a suitably redesigned gun, *VPS, vacuum plasma spraying* may be used to spray carbides or cermets without oxidation. The usual purpose of a refractory coating applied in this way is to protect a material, e.g. Mo or C, from

oxidation when used at high temperature.

Plaster of Paris. Calcium sulphate hemihydrate, $CaSO_4.½H_2O$; prepared by heating GYPSUM (q.v.) at 150–160°C. There are two forms: α, produced by dehydrating gypsum in water or saturated steam; β, produced in an unsaturated atmosphere. Plaster usually contains both forms. It is used for making MOULDS (q.v.) in the pottery industry.

Plaster-base Finish Tile. US term (ASTM – C43) for a hollow clay building block the surfaces of which are intended for the direct application of plaster; the surface may be smooth, scored, combed or roughened.

Plastic Cracking. Cracking in concrete before it has set and hardened. Its cause may be differential settlement, or shrinkage.

Plastic Deformation. Deformation due to the movement of dislocation in the crystal lattice. In ceramics it is the mechanism for the initial formation of new cracks. See CRYSTAL STRUCTURE; CRACK NUCLEATION.

Plastic Forming. All processes of shaping ceramics in the plastic condition: extrusion, plastic pressing, injection moulding, roller-head shaping, jiggering, jolleying, throwing.

Plastic Making. PLASTIC FORMING (q.v.)

Plastic Pressing. The shaping of ceramic ware in a die from a plastic body by direct pressure.

Plastic Refractory. ASTM C71 defines this as a refractory material, tempered with water, that can be extruded and that has suitable workability to be pounded into place to form a monolithic structure. See FIRECLAY PLASTIC REFRACTORY, MOULDABLE REFRACTORY.

Plastic Shaping. See PLASTIC MAKING.

Plasticity. The characteristic property of moist clay that permits it to be deformed without cracking and to retain its new shape when the deforming stress is removed. Plasticity is associated with the sheet structure of the clay minerals and with the manner in which water films are held by the clay particles.

Plasticizer. A material, usually organic, which is added to non-plastic ceramic materials such as oxide powders, to form a dough-like plastic mass more readily shaped by pressing or moulding.

Plastometer. An instrument for the evaluation of plasticity, see, for example, BINGHAM PLASTOMETER, LINSEIS PLASTOMETER, PFEFFERCORN TEST; WILLIAMS PLASTOMETER.

Plat. Cornish term for the overburden of a china-clay mine.

Plate Glass. Flat, transparent, glass both surfaces of which have been ground and polished so that they give undistorted vision. Plate glass is used, for example, in shop windows.

Platting. See under SCOVE.

Plauson Mill. The original 'colloid mill' designed by H. Plauson in Germany, and covcred by a serics of patents beginning with Brit. Pat. 155 836, 24/12/20. The novel feature was the high speed of operation (up to 30 m/s) and the concentration of the grinding pressure at a small number of points.

P.L.C. Abbreviation for PERMANENT LINEAR CHANGE (q.v.).

Plique-a-Jour. One type of vitreous-enamel artware: a pattern is first made of metal strips, and coloured enamels are fused into the partitions thus formed; after the object has been polished on both sides, it can be held to the light to give a stained-glass window effect. (French words: 'open-work plait'.)

Pluck, Plucking. (1) A surface blemish sometimes found on pottery where the

glaze has been removed by the pointed supports for the ware used in the firing process.
(2) A surface fault on rolled glass arising if the glass sticks to the rollers.
Plug. Part of a glass-blowing machine for hollow-ware; it moves in a blank-mould with a reciprocating action, forming a cavity for blowing; this part is sometimes referred to as a PLUNGER. (See also MOULD PLUG.)
Plugging Compound. See FILLER.
Plumbago. A refractory material composed of a mixture of fireclay and graphite; some silicon carbide may also be included. It finds considerable use as a crucible material for foundries. In USA refractories of this type are termed CARBON CERAMIC REFRACTORIES.
Plunger. See NEEDLE; PLUG.
Ply Glass. Glass-ware, particularly for lamp shades and globes, made by covering opal glass (usually on both sides) with transparent glass of matched thermal expansion (cf. CASED GLASS).
PLZT. Lead lanthanum zirconate titanate. PLZT is an OPTOELECTRONIC material (q.v.).
PMN. Lead Magnesium Niobate is a useful capacitor material, with high permittivity and high breakdown voltage.
Pneumatolysis. The breakdown of the minerals in a rock by the action of the volatile constituents of a magma, the main body of the magma having already become solid. The china clays of Cornwall, England, were formed by the pneumatolysis of the feldspar present in the parent granite.
Pneumoconiosis. Disability caused by the inhalation, over a long period, of various dusts; the form of the disease encountered most commonly in the ceramic industry is SILICOSIS (q.v.).
PNN. Lead nickel niobate.
PNZT. Lead niobium zirconate titanate.
Pocket Clay. A highly siliceous clay sometimes found in large pockets in Carboniferous Limestone. In England it is found, typically, in Derbyshire and is worked as a refractory raw material at Friden, about 10 miles S.W. of Bakewell; a quoted composition is (per cent): SiO_2 78; Al_2O_3 , 13.5; Fe_2O_3, 2.0; TiO_2, 0.7; CaO, 0.2; MgO, 0.3; K_2O, 1.6; Na_2O, 0.1; loss on ignition, 3.9.
Pocket Setting. See BOXING-IN.
Podmore Factor. A factor proposed by H. L. Podmore (*Pottery Gazette*, **73**, 130, 1948) to indicate the intrinsic solubility of a frit:
Podmore factor = (solubility × 100) ÷ (specific surface). The solubility of any given frit is approximately proportional to the surface area of frit exposed to the solvent, i.e. to the fineness of grinding; the Podmore factor is independent of this fineness.
Poge. A tool formerly used for lifting the 'balls' of ball-clay.
Point. (1) Fill and neaten masonry joints with mortar.
(2) See MOUNTED POINT.
Point Bar. One type of support for vitreous enamelware used during the firing process.
Point Defect. See CRYSTAL STRUCTURE.
Point Mark. See PIN MARK.
Poise. A unit of viscosity, of 10^{-1} $Nm^{-2}s$, named after Poiseuille. The viscosity of water is about 1 cP.
Poisson's Ratio. The ratio of the lateral contraction per unit breadth to the longitudinal extension per unit length, when material is stretched. Its value is normally between 0.25 and 0.4. Young's modulus (E) and the modulus of rigidity G are related to Poisson's ratio ς by $E = 2G(1 + \varsigma)$

ASTM C1198 specifies a sonic resonance test for E and hence ς, for advanced ceramics.

Poke Hole. An opening in a refractory lining for compressed air lances or steel rods to free blockages.

Polcal Process. This system for the continuous calcining of powders has three pre-heating cyclones (or one preheating cyclone and a flash dryer); a calciner with a separating cyclone; a cooling stage with one or more cyclones. It has been applied to calcining natural magnesite to caustic MgO. (Krupp Polysius AG, Germany).

Poling. (1) A method of removing bubbles from molten glass in a pot; the glass is stirred with a wooden pole, gases sweep through the molten glass and remove the smaller bubbles, which would otherwise rise to the surface only sluggishly. If a block of wood is used, the process is called BLOCKING.
(2) The alignment with an external electric field of the electrostatic charge domains in ferroelectric ceramics.

Polished Section. A section of material that has been ground and plane-polished on one face for examination, under a microscope, by reflected light (cf. THIN SECTION).

Polishing. (1) A finishing process for plate glass and optical glass. Plate glass is polished by a series of rotating felt disks with rouge (very fine hydrated iron oxide) as the polishing medium. For the polishing of optical glass cerium oxide, rouge or (less commonly) zircon are used.
(2) The grinding away of small surface blemishes from the face of glazed pottery-ware.

Polychromatic Glass. A special PHOTOSENSITIVE GLASS (q.v.) which when activated by ultraviolet light and heat, can produce within the same glass, in 2 and 3 dimensions, patterns of white or coloured opacity, or transparent colours of all hues. Permanent full-colour 3-d photographic images can be produced. The mechanism is the formation of metallic silver on a precipitated sodium halide microcrystalline phase, each formed by separate photosensitive processes.

Polyphant Stone. An impure SOAPSTONE (q.v.) from the village of Polyphant, near Launceston, Cornwall, England. Composition (per cent): SiO_2, 78–36; Al_2O_3, 6–9; $FeO+Fe_2O_3$, 10–12; CaO 4–5, MgO, 23–27; loss on ignition, 10–12. The m.p. is 1300–1400°C. Blocks of this stone are used as a refractory material in alkali furnaces.

Polytypes. Crystal structures, modified by stacking faults, but too near the ideal to classify as separate structures.

Polytypoids. Mixed POLYTYPES (q.v.) Two phase solid solutions with minor variations in crystal structure. Aluminium nitride polytypoids are phases in the Si-Al-O-N system with constanct (Si+Al):(O+N) ratio, whose structures can be classified by RAMSDELL NUMBERS (q.v.).

Pontesa Ironstone. Earthenware made by Alfares de Pontesampayo, Vigo, Spain.

Pontil. See PUNTY.

Pooles Tile. An interlocking clay roofing tile of a design specified in B.S. 1424.

Pop off or Popper. A fault sometimes occurring during vitreous enamelling on sheet steel, small disks of ground-coat becoming detached and rising into the first cover-coat. The fault may be caused by poor metal surface.

Popout. A fault in concrete. A conical piece is pushed out from the surface by the expansion of an aggregate particle at its apex.

Popping. Term sometimes used for (1) LIME BLOWING (q.v.)

(2) the heat treatment of PERLITE (q.v.) to cause expansion.

Porcelain. One type of vitreous ceramic whiteware. The COMBINED NOMENCLATURE (q.v.) defines this as: completely vitrified, hard, impermeable (even before glazing), white or artificially coloured, translucent (except when of considerable thickness) and resonant. In the UK the term is defined on the basis of composition: a vitreous white-ware made from a feldspathic body (typified by the porcelain tableware made in Western Europe and containing 40–50% kaolin, 15–25% quartz and 20–30% feldspar). In the USA the term is defined on the basis of use: a glazed or unglazed vitreous ceramic white-ware used for technical purposes, e.g. electrical porcelain, chemical porcelain, etc. Note, however, that the term electrical porcelain is also used in the UK. The firing of porcelain differs from that of earthenware in that the first firing is at a low temperature (900–1000°C), the body and the feldspathic glaze being subsequently matured together in a second firing at about 1350–1400°C. Tensile strength 140 MPa; Young's modulus 200 GPA; Poisson's Ratio c. 0.25. See also HARD PORCELAIN, SOFT PORCELAIN, SEMI-PORCELAIN, CHINA, BONE CHINA, STONEWARE.

Porcelain Enamel. The term used in USA for VITREOUS ENAMEL (q.v.) and defined in ASTM – C286 as: A substantially vitreous or glassy inorganic coating bonded to metal by fusion at a temperature above 800°F (427°C).

Porcelain Tile. US term defined (ASTM – C242) as a ceramic mosaic tile or PAVER (q.v.) that is generally made by dust-pressing and of a composition yielding a tile that is dense, fine-grained, and smooth, with sharply-formed face, usually impervious. The colours of such tiles are generally clear and bright.

Porcelainite. MULLITE, (q.v.) The usage is obsolete.

Porcelit. A non-vitreous porcelain-type body fired at c.1250°C to produce a type of ware between earthenware and stoneware. Made in Poland at Pruszkow. (*Tableware Int.* **4**, (10) 84, 1974).

Pore-size Distribution. The range of sizes of pores in a ceramic product and the relative abundance of these sizes. This property is usually expressed as a graph relating the percentage volume porosity to the pore diameter in microns. Pore-size distribution is difficult to determine, the most generally satisfactory methods are the MERCURY PENETRATION METHOD (q.v.) and the BUBBLE PRESSURE METHOD (q.v.). B.S. 1902 Pt. 3.16 specifies the mercury penetration method for refractories.

Poriso. A fine-pored brick made from colliery washings, shale, sawdust and fly-ash, requiring no fuel for firing other than the combustible body compenents. (Hendrik Steenfabrick, Netherlands; *ZI Int.* **33**, (1), 30 1981).

Pork-pie Furnace. See MAERZ-BOELENS FURNACE.

Porodur. A plastics material for moulds for plastic shaping of tableware. (*Sklar Keram* **28** (3), 76, 1978).

Porosimeter. A term variously applied to apparatus for the measurement of POROSITY (q.v.) or of PORE-SIZE DISTRIBUTION (q.v.)

Porosity. A measure of the proportion of pores in a ceramic material, defined as:

Apparent Porosity. The ratio of the open pores to the BULK VOLUME (q.v.), expressed as a percentage. ASTM C830 specifies a vacuum pressure test for refractories; ASTM C20 a boiling water test.

Fractional Porosity. A porosity value expressed as a decimal instead of as a percentage, e.g. 0.25 rather than 25%.
Sealed Porosity or Closed Porosity. The ratio of the volume of the sealed pores to the bulk volume, expressed as a percentage.
True Porosity. The ratio of the total volume of the open and sealed pores to the bulk volume, expressed as a percentage.
B.S. 1902 Pt. 3.8/ISO 5017 specifies an immersion test. Closed pores are then those which are not penetrated in this test: open pores are those that are. ASTM C373 specifies tests for porosity and related products of whitewares. C949 specifies a dye penetration test.

Poroton. Insulating bricks in which foamed polystyrene beads are mixed with the clay as a combustible component, leaving a porous structure. (*J Brit. Ceram Soc.* **7**, (1) 3, 1970.)

Porous Glass. An alkali borosilicate glass is made, which solidifies in two phases, one soluble in dilute acid. This is leached out, leaving a porous glass with a nearly pure silica skeleton. Such porous glasses can be consolidated by careful heat treatment at c. 1200°C to sinter them into an impervious clear high silica glass. Volume shrinkage is then c. 35%. The affinity of porous glass for moisture makes it a useful non-dusting drying agent, used in various shapes in scientific instruments.

Porous Plug. See GAS BUBBLING BRICK.

Port. An opening through which fuel gas or oil, or air for combustion, passes into a furnace; ports are usually formed of refractory blocks or monolithic material.

Portakiln. Trade-name. A low thermal mass prefabricated brick kiln. (Bricesco, *Euroclay* (3), 12, 1979).

Port Walker. A workman who observes the sheet of glass issuing from a FOURCAULT (q.v.) tank-furnace and gives warning of faults.

Portland Blast-furnace Cement. A hydraulic cement made by mixing portland cement clinker with up to 65% blast-furnace slag and grinding them together (see B.S. 146 and cf. SLAG CEMENT).

Portland Cement. A hydraulic cement produced by firing (usually in a rotary kiln) a mixture of limestone, or chalk, and clay at a temperature sufficiently high to cause reaction and the formation of calcium silicates and aluminates; the proportion of $3CaO.SiO_2$ is about 45%, and that of $2CaO.SiO_2$ about 25%, the other major constituents being $3CaO.Al_2O_3$ and $4CaO.Al_2O_3.Fe_2O_3$. The chemical composition is (per cent): SiO_2, 17–24; Al_2O_3, 3–7; Fe_2O_3, 1–5; CaO, 60–65; MgO, 1–5; alkalis, 1; SO_3. 1–3. Properties are specified in B.S. 12 and ASTM-C150. The name was given to this cement by its inventor, Joseph Aspdin (Brit Pat. 5022, 1824) on account of the resemblance of the colour of the set cement to that of the well-known building stone from Portland, England.

Post. (1) The gather of glass after it has received preliminary shaping and is ready to be hand-drawn into tube or rod. (2) An item of KILN FURNITURE (q.v.). Posts, also known as PROPS or UPRIGHTS, support the horizontal bats on which ware is set on a tunnel-kiln car. (See Fig. 4, p177.)
(3) A discrete portion of bond between abrasive grains in a grinding wheel or other abrasive article. When the abrasive grain held by a post has become worn, the post should break to release the worn grain so that a fresh abrasive grain will became cxposed.

Post Clay. A term sometimes used in N.E. England for an impure siliceous fireclay.

Post-sintering. To produce dense engineering ceramics, a forming method which leads to a porous fired component is followed by sintering in a controlled atmosphere to densify the compact with minimum shrinkage. Complex near net shapes can be produced. For example, injection moulded silicon nitride components were sintered in nitrogen after forming, with shrinkage of only 6%. (UK Pat. 2010913A and 2010915A, FIAT SpA).

Post-tensioned Brickwork. Steel bars pass through the brickwork in aligned perforations or pockets in the bonding. The bars are anchored top and bottom and tightened mechanically after construction, to place the brickwork in compression. See REINFORCED BRICKWORK.

Pot. A large fireclay crucible used for melting and refining special types of glass. A pot may either be open (the surface of the glass being exposed to the furnace gases) or closed by an integrally moulded roof, a mouth being left for charging and gathering; a CLOSED POT is also known as a COVERED or HOODED POT.

Pot Arch. A furnace for preheating the POTS (q.v.) used in one method of glass-making.

Pot Bank. Obsolete term in N. Staffordshire for a pottery factory.

Pot Clay. A somewhat siliceous fireclay that is used in the manufacture of glass-pots or crucibles for use in melting steel; the term is also applied to a mixture of such a clay with grog, so that it is ready for use.

Pot Furnace. A furnace in which glass is melted and refined in POTS (q.v.), or in which a frit is melted for use in a glaze or in a vitreous enamel.

Pot Ring. See RING.

Pot Spout. A refractory block used in the glass industry to connect the working end of a glass-tank furnace to a revolving pot.

Potassium cyanide. KCN; used as a neutralizer in vitreous enamelling.

Potassium Niobate. $KNbO_3$; a ferroelectric compound having a perovskite structure at room temperature. The Curie temperature is 420°C.

Potassium Nitrate. KNO_3; used to some extent as an oxidizing agent in glass-making. The old name was NITRE.

Potassium Tantalate. $KTaO_3$; a ferroelectric material having a dielectric constant exceeding 4000 at the Curie temperature (–260°C).

Potassium Titanate. This compound, which approximates in composition to $K_2Ti_6O_{13}$ and melts at 1370°C, can be made into fibres for use as a heat-insulating material.

Potette. A refractory shape used in the glass industry; it is partly immersed in the molten glass and protects the gathering point from the furnace gases and from any scum that is floating on the glass.

Potsherd. A piece of broken pottery. The word is not now used in the industry; the N. Staffs. dialect word is 'shord'.

Potter's Horn or Kidney. A thin kidney-shaped piece of horn or metal used, until the early 20th century, by pottery pressers. To make dishes, a bat of prepared body was placed on a plaster mould and hand-pressed to shape with a piece of fired ware; the horn was used for final smoothing of the surface.

Potter's Red Cement. A POZZOLANA (q.v.) type of cement consisting of crushed fired clay mixed with portland cement; (C. J. Potter, *J. Soc. Chem. Ind.*, **28**, 6, 1909).

Potter's Shop. 'COSHH in Production of Pottery, Approved Code of Practice'

1990 defines a potter's shop to include all places where pottery is formed by casting, pressing or any other process and all places where shaping, fettling or other treatment of pottery prior to placing for biscuit fire is carried on.

Pottery. This term is generally understood to mean domestic ceramic ware, i.e. tableware, kitchenware and sanitaryware, but the pottery industry also embraces the manufacture of wall- and floor-tiles, electroceramics and chemical stoneware. There are subdivisions within these main groups of pottery-ware, e.g. tableware may be earthenware, bone china or porcelain; similarly, sanitaryware may be sanitary fireclay, sanitary earthenware or vitreous china sanitaryware. 'COSHH in the Production of Pottery, Approved Code of Practice' 1990 defines POTTERY to include china, earthenware and any article made from clay or from a mixture containing clay and other materials. See also COMMON POTTERY, FINE POTTERY. In USA the term WHITEWARE (q.v.) is used. A *pottery* is also general usage for the factory where pottery is made.

Potting Material. A material for the protection of electrical components, e.g. transformers. The use for this purpose of sintered alumina powder, within hermetically sealed cans, has proved successful.

Pour. To place concrete. A *pour* is a single such placement.

Pour Point. The optimum temperature for pouring a molten batch.

Pourbaix Diagrams. Plots of equilibrium pH vs electrochemical potential E describe the effects of aqueous corrosion on borosilicate and silicate glasses. They are applicable to weathering studies and to ground water attack on nuclear waste glasses. The diagrams display any immune zone between active corrosion and passivation due to the formation of a surface layer. M. Pourbaix, *Atlas of Electrochemical Equilibria in Aqueous Solutions (translation) Nat. Assoc. Corrosion Engs. Houston, Texas*, 1974.

Pouring-pit Refractories. Alternative term (particularly in USA) for CASTING PIT REFRACTORIES (q.v.).

Powder Blue. See SMALT.

Powder Density. See TRUE DENSITY.

Powdering. A process sometimes used for the on-glaze decoration of the more expensive types of pottery-ware. Colour mixed with an oil medium is first brushed on the ware, dry powdered colour then being applied to achieve a stippled effect.

Pozzolana. A material that, when ground and mixed with lime and water, will react with the former to produce compounds having hydraulic properties. There are both natural and artificial pozzolanas. The original Natural Pozzolana was a volcanic tuff worked at Pozzoli (or Pozzuoli), Italy, in Roman times. One of the principal types of Artificial Pozzolanas is produced by firing clay or shale at 600–1000°C.

PPG Delivery System. See FLOAT GLASS PROCESS.

PPG Ring-roll Process. PPG = Pittsburgh Plate Glass Co.; see RING-ROLL PROCESS.

Prague Red. Red iron oxide pigment.

Prall Mill. An impact mill consisting of an impeller rotating clockwise at 1000 rev/min, a baffle plate moving anti-clockwise at 1500 rev/min, and a second baffle plate that is stationary.

Praseodymium Oxides. There are at least two oxides: Pr_2O_3, and PrO_2.

Praseodymium Yellow. A ceramic colour made by calcining a stoichiometric mixture of ZrO_2 and SiO_2 with about 5% PrO_2 or Pr_2O_3. This clean, bright yellow can be used in glazes firing from about 1100–1300°C; the zirconium silicate acts as a stabilizer.

PRE. (Produits Réfractaires Européen) Fédération Européanne des Fabricants de Produits Réfractaires. (European Federation of Refractories Producers) Via Corridoni 3, Milan, Italy. The *PRE Refractory Materials Recommendations* 1978 comprised over 40 recommended test methods and classifications of various types of refractory materials. The *PRE Glossary* listed refractories terminology in French, German, Italian and English. (See Appendix B) SIPRE (Siderugie et PRE) was a body whose members were drawn from PRE and the European iron and steel industry.

Precipitation Hardening. See DISPERSION STRENGTHENING.

Precision-bore Tubing. Special glass tubing made by heating ordinary glass tubing until it is soft and then shrinking it on to a steel mandrel.

Precursor. Precursors are materials which can be converted into ceramic materials by suitable heat or chemical treatment. The properties of the precursor may make it possible to achieve e.g. finer powders with greater uniformity by precipitation from solution, the ceramic being insoluble.

Prefabricated Masonry. Panels or beams of specially designed hollow clay blocks formed by the insertion of metal reinforcing wires through a number of the blocks, laid side by side, followed by infilling with cement. Many designs of such prefabricated panels and sections have been developed to accelerate building construction. ASTM C901 is a specification for prefabricated masonry panels.

Preferred Increment. In the building industry this term is defined as: 'the smallest interval between two consecutive sizes in any range of preferred dimensions for spaces or components.' The (UK) Ministry of Public Buildings and Works, in its *Dimensional Co-ordination for Industrialized Building* (Feb. 1963), recommended preferred increments of 1, 4 and 12 in.; these are nominal dimensions. See also MODULAR CO-ORDINATION.

Preform. (1) Term used in the glass industry, and particularly in the making of glass-to-metal seals, for a compact of powdered glass that has been fired to a non-porous state.
(2) A shaped powder compact which is to be further processed (e.g. by post-sintering, reaction bonding, isostatic pressing, machining) to form the final shape, particularly of an engineering ceramic component.

Prepared Body. Instead of individual whiteware producers themselves mixing the raw materials, the body is prepared centrally by a specialist supplier. At the cost of variety, body may be prepared under better control, and using processes (e.g. spray drying) which to be economically viable require a scale of production greater than that needed by an individual manufacturer, who may also avoid waste by buying the required amount of body appropriately ground and packaged.

President Press. Trade-name: a dry-press brickmaking machine of a type that gives two pressings on both the top and bottom of the mould, a short pause between the pressings allowing any trapped air to escape. (W. Johnson & Son (Leeds) Ltd., England.)

Press-and-Blow Process. A method of shaping glass-ware, the PARISON (q.v.) is pressed and then blown to the final shape of the ware.

Press Cloth. See FILTER CLOTH.

Pressing. In the pottery industry, pressing was an old method of shaping ware by hand-pressing a prepared piece of body between two halves of a mould. See also DRY PRESSING; DUST PRESSING; ISOSTATIC PRESSING; HOT PRESSING; PLASTIC PRESSING; RAM PRESSING. Glassware, such as dinnerware, lenses and other flat shapes, is pressed between a plunger and a mould, usually in an automatic rotary press. See OPTICAL BLANK.

Pressure Casting. Hydraulic pressure is applied to the slip in SLIP CASTING, to increase the casting rate. The technique was first applied particularly to sanitaryware, as a high pressure process, or latterly to speed up BATTERY CASTING. More recently it has been applied to tableware shaping, where it is becoming more widespread. The higher stresses on the moulds led to studies of sand-resin, plastics or even porous metal mould systems, but these require long production runs to make them economic, and they have by no means ousted strong plaster moulds.

Pressure Check. A crack in a glass article caused by too high a forming pressure.

Pressure Cracking. Cracking of the compacted semi-dry powder immediately after it has been shaped in a dry-press; the cause is sudden expansion of air that has been trapped and compressed in the pores of the compact. The fault has been largely eliminated by designing presses so that the plunger descends twice to its lowest level, the trapped air having time to escape between the first and second pressings.

Pressure Sintering. This term has the same meaning as HOT PRESSING (q.v.).

Pressureless Sintering. Sintering at atmospheric pressure.

Prestocal. A machine for applying HEAT ACTIVATED DECALS (q.v.). (Commercial Decal Inc, USA).

Preston Density Comparator. An instrument designed in 1950 at the Preston Laboratories, USA, for use in the routine quality control of glass on the basis of an observed relationship, for any specific glass, between any change in composition and the associated change in density; the sink-float method is employed.

Preton. A brick panel system (Keller AG, Germany).

Primary Air. The air that mingles with the fuel and effects the initial stages of its combustion. In a ceramic kiln of a type fired by solid fuel on a grate, the primary air passes through the grate and the fuel bed (cf. SECONDARY AIR and TERTIARY AIR).

Primary Boiling. Gas evolution during the initial firing of vitreous enamel; this may result in faults in the ware.

Primary Clay or Residual Clay. A clay still remaining in the geographical location where it was formed; in the UK such a clay is typified by the CHINA CLAY (q.v.) of Cornwall (cf. SEDIMENTARY CLAY).

Primary Colours. See TRI-STIMULUS VALUES.

Primary Crusher. A heavy crusher suitable for the first stage in a process of size reduction, the product being about 25–75 mm in size. Gyratory and jaw crushers, and Kibbler rolls, fall into this class.

Primary Phase. The first crystalline phase to appear when a liquid is cooled. For example, in the Al_2O_3-SiO_2 system (see Fig. 5, p230), if a liquid is formed containing 40% Al_2O_3, and if this liquid is then cooled. the primary phase, when solidification begins, will be mullite; if the liquid contains 80% Al_2O_3, however, it can be seen that the primary phase will be corundum.

Prince Rupert's Drops. Drops of glass that have been highly stressed by quenching; when the 'tail' of one of these drops is broken the glass explodes to dust, but the drop itself is immensely strong. These drops were first made by Prince Rupert, nephew of King Charles I of England.

Printer's Bit. An item of KILN FURNITURE (q.v.). It is a small piece of refractory material for use as a distance-piece in the stacking of decorated pottery-ware before and during firing.

Prism Test. See THERMAL SHOCK TESTS.

Prismatic Glass. Glass that has been pressed or rolled to produce a pattern of prisms; these refract light passing through the glass. (The term is also sometimes, erroneously, applied to LENS FRONTED TUBING (q.v.)

Prismo. Trade-name. Ceramic fibre modules for lining and lids of ladles and reheating furnaces (Lafarge Aluminous Cement Co).

Proarcus. A range of partly curved wall tiles, to provide smooth transition between wall and floor surfaces in wet rooms, shower trays, etc (Korzilius GmbH, *Fliessen Platten* **37**, (6), 52, 1987).

Process Fish-scaling. A fault in vitreous enamelling, FISHSCALING (q.v.) occurring during the drying or firing of the covercoat.

Proctor Dryer. The original 'Proctor Dryer' of the early 1920s was a tunnel dryer for heavy-clay and refractory bricks; drying was achieved by air recirculating over heated steam coils, or over pipes carrying hot waste gases. Proctor dryers operating on the same principle were subsequently made in a variety of types suitable for all kinds of ceramic product. The name derives from the manufacturers: Proctor and Schwartz, Philadelphia, USA.

Profile Setter. A SETTER (q.v.) whose shape matches that of the article to be supported for firing.

Progression China. Tableware with a cordierite body, made by Noritaki, Japan.

Prokaolin. A term that has been applied to an amorphous intermediate product in the process of KAOLINIZATION (q.v.).

Prop. See POST.

Proppants. Millimetre-sized ceramic particles, forced with hydraulic fluids into fractures in oil and gas wells, to keep them open. The productivity of the wells is thus enhanced.

Prosper Stone. A form of CORNISH STONE (q.v.).

Prostheses. Artificial functional replacements for body parts, especially bones and joints. Ceramic materials for the latter are usually alumina, apatite or hydroxyapatite, and some glass-ceramics. Carbon composites have been used for heart valves. B.S. 7253 Pt. 2 specifies alumina bone substitutes. See BIOCERAMICS.

Proto-Enstatite. See ENSTATITE.

Prouty Kiln. A tunnel kiln of small cross-section suitable for rapid firing of pottery-ware, which is carried through the kiln on bats. (T. C. and W. O. Prouty, US Pat., 1 676 799, 10/7/28.)

Proximate Analysis. See RATIONAL ANALYSIS.

Prunt. A crest, or other device, that has been fused on glassware subsequent to the general shaping process.

Prussian Blue. An iron-ferrocyanide blue pigment.

Prussian Red. Potassium ferrocyanide red pigments.

PSD. Abbreviation for PORE SIZE DISTRIBUTION (q.v.).

Pseudo-wollastonite. See under WOLLASTONITE.

p.s.i. Abbreviation for pounds per square inch. (Obsolete. 1 p.s.i. $\cong$ 7 kNm^{-2}).

Psychrometric Chart. A graphical representation of the relationship between the relative humidity, specific volume, weight ratio of moisture to air, dry-bulb temperature, vapour pressure, total heat, and dew-point of moist air. The chart finds use in the ceramic industry, particularly in the control of dryers.

PSZ. Partially Stabilized Zirconia. See Zirconia.

PT. Lead Titanate.

PTCR. Positive temperature coefficient of resistance. Ceramics with PTCR's include barium titanates.

Pucella. A tool for widening the top of a wine glass in the handmaking process; from Italian word meaning a virgin.

Puddling Clay. Clay used to produce an impervious layer to line a pool or reservoir. The term relates to the application, not to the composition of the clay.

Puesta. An Italian vitreous enamelling process using electrostatic powder coating.

Puff-and-blow Process. A modified BLOW AND BLOW (q.v.) technique used to make thin-walled hollow glassware. The first use of compressed air (puff) produces a hollow PARISON (q.v.) which is shaped in a rotating paste mould, by the second blow. Light bulbs, flash bulbs and Christmas tree ornaments are made from a continuous ribbon of glass (Corning *ribbon machine*) while larger articles are made by the *turret-chain machine* which uses individual glass gobs and produces less cullet.

Pug. A machine for consolidating plastic clay or body into a firm column. It consists of a steel cylinder ('barrel') which tapers at one end to a die, through which the clay or body is forced by knives mounted on a shaft which rotates centrally to the barrel (cf. AUGER and MIXER).

Pug Mill. See MIXER.

Pull. See LOAD.

Pulled Stem. See STEMWARE.

Pullout. See FIBRE PULLOUT; TOUGHNESS.

Pulpstone. Sandstone used as a grinding wheel.

Pulsed Laser Deposition. A pulsed laser (e.g. a Kr-F excimer laser) is used to vaporise material from a target, and deposit it to form a coating on a substrate. The technique is particularly applicable to the production of electronic devices using high temperature superconductors.

Pulverized Fuel Ash. Finely divided ash carried over from coalfired power-station boilers. It has found some use in the manufacture of building materials, e.g. THERMALITE YTONG (q.v.), LYTAG (q.v.), and to a less extent in clay building bricks. The composition of the ash is (per cent): SiO_2 43–50; Al_2O_3, 24–28; Fe_2O_3, 6–12; CaO, 2–4; MgO, 2–3; alkalis, 3–5; loss on ignition, 2–10. In the UK, the Central Electricity Generating Board Specification is: $> 35\%$ SiO_2; $< 4\%$ MgO; $< 3\%$ SO_3; $> 12\%$ loss on ignition.

Pulverizer. See FINE GRINDER.

Pump Casting. A process in which castable refractories are pumped through pipes to their point of application.

Punch Test. A simple test to determine whether the glaze on a piece of fired pottery is in tension or compression. A steel centrepunch with a blunt end 0.8–1.6 mm dia. is placed on the glazed surface and hit sufficiently hard with a hammer to break the glaze. If the latter was in tension, one or more cracks will be found to have radiated from the point of impact; if the glaze was in compression a circular crack will have formed round the punch mark or a conical piece of glaze will have become detached.

Punch Ware. US term for thin hand-blown glass tumblers, etc.
Punt. The bottom of a glass container.
Punt Code. A mark of identification on the base of a glass bottle.
Punty. An iron rod used in the hand-making of glass-ware to hold the base of the ware during its manipulation; from Latin puntellum, the word used in 16th century manuscripts on glassmaking.
Pup. A brick (particularly a refractory brick) of a shape and size based on that of a standard square but with the end faces square (or nearly square). This shape of brick is also known as a SOAP or CLOSER. See Fig. 1, p39.
Purcell Method. Name sometimes given to the MERCURY PENETRATION METHOD (q.v.) for determining pore-size distribution (W. R. Purcell, *J. Petroleum Tech.*, **1**, (2), 39, 1949).
Purple of Cassius. Colouring material produced by adding a mixture of tin chlorides to a dilute solution of gold chloride; the latter is reduced to the metal and hydrated stannic oxide is precipitated, the colour resulting from the finely-divided gold on the SnO_2 particles.
Purpling. A fault liable to occur with CHROME-TIN PINK (q.v.) ceramic colour if the amount of alkali and borax is too high and the amount of lime too low.
Push-down Cullet. A fault occasionally found in sheet glass as a result of the presence of CULLET (q.v.) in the zone of the furnace from which the glass was drawn.
Push-up. Alternative name for PUSHED PUNT (q.v.).
Pushed-bat Kiln or Sliding-bat Kiln. A tunnel kiln, of small cross-section, through which the ware is conveyed on sliding bats instead of on the usual cars; when there are a number of such tunnels in a single kiln, it is known as a MULTI-PASSAGE KILN (q.v.).
Pushed Punt. The concave bottom of a glass wine-bottle or other container.
Putnam Clay. A fine-grained, plastic, Florida kaolin; it fires to a good white colour.
Puzzuolana. Obsolete spelling of POZZOLANA (q.v.).
PVA. Polyvinyl alcohol, frequently used as a binder and plasticizer.
PVD. Physical Vapour Deposition. Vapour is created by evaporation or sputtering and transported along a line-of-sight path to the substrate to be coated (normally in vacuum) see SPUTTERING, ION-PLATING.
Pycnometer. A small bottle having an accurately-ground glass stopper pierced with a capillary. The bottle is of known volume and is used for the accurate determination of the specific gravity of a material.
Pyrex. Trade name. A borosilicate glass introduced by the Corning Glass Co., USA in 1915 and made in England by J. A. Jobling & Co. Ltd., since 1922. This glass has high thermal endurance and is much used for making chemical ware.
Pyrite, Pyrites. Iron sulphide, FeS_2: sometimes present in clays causing slaggy spots on the fired product.
Pyroceram. Trade-name (Corning Glass Works, USA) for the original commercially available GLASS-CERAMIC (q.v.); the equivalent material made in the UK under licence is known as PYROSIL (q.v.)
Pyro-clad. Trade-name. A process for depositing multilayer coatings of metals and ceramics. (Aremco Products Inc. USA).
Pyroelectric Materials. Pyroelectrics develop an observable spontaneous electric moment only when heated. Cf FERROELECTRICS. Pyroelectric ceramics include some niobate-zirconate systems, barium titanate modified with zirconia and hafnia, and lanthanum-modified lead

titanate. Applications are to infra-red detectors and other electronic devices.
Pyrofix. Trade-name. (1) A highly siliceous sleeve brick. Harima Refractories, Japan.
(2) A refractory concrete anchoring system based on a ribbed steel bar. The ribs eliminate the need to bend the bar into complex shapes. (George Clark, (Sheffield) Ltd. *Refract. J.* May/June 1976, p18)
Pyroflam. Trade-name. Glass-ceramic cooking ware marketed in the UK as PYROSIL (q.v.) (Corning Nederlande Fabricken).
Pyrogel. Trade-name. Mixed oxide compositions made by SOL-GEL processing (UKAEA, Harwell).
Pyrolite. Trade-name. A lithium-aluminium-silicon oxide ceramic. (Krohn Ceramic Corp, USA).
Pyrolusite. MnO_2; see MANGANESE OXIDE.
Pyrolytic Coatings. Pyrolysis is the decomposition of a material by heat. A pyrolytic coating is a thin coating produced by the breakdown of a volatile compound on a hot surface. Some types of resistor are made by the pyrolytic coating of an electroceramic rod with carbon. Pyrolytic coatings of BN, SiC, and Si_3N_4 have been applied to components for their protection during exposure to high temperatures.
Pyron. Trade-name. A range of colours for articles, including glazed ceramics, glass as well as wood or plastics, which are not to be subsequently fired. (CMS Colours Ltd, UK).
Pyrometer. A device for the measurement of high temperature, e.g. a thermocouple or a radiation pyrometer. B.S. 1041 deals with temperature measurement and various pyrometers.
Pyrometric Cone. A pyramid with a triangular base and of a defined shape and size; the 'cone' is shaped from a carefully proportioned and uniformly mixed batch of ceramic materials so that, when it is heated under stated conditions, it will bend due to softening, the tip of the cone becoming level with the base at a definite temperature. Pyrometric cones are made in series, the temperature interval between successive cones usually being 20°C. The best known series are Seger Cones (Germany), Orton Cones (USA), Staffordshire Cones (UK); for nominal temperature equivalents see Appendix D.
Pyrometric Cone Equivalent. A measure of refractoriness: the identification number of the standard pyrometric cone that bends at the temperature nearest to that at which the test-cone bends under the standardized conditions of test. These conditions are specified in the following national standards: Britain, B.S. 1902 Pt. 5.1 specifies pyrometric cones, and Pt. 5.2 the determination of PCE (refractoriness); USA, ASTM – C24 specifies the determination of PCE for firebrick and high alumina refractories; France, B49–102; Germany, DIN 51 063.
Pyrophyllite. A natural hydrated aluminium silicate, $Al_2O_3.4SiO_2.H_2O$. It occurs in N. Carolina, Newfoundland, Japan, and S. Africa, and finds some use in whiteware bodies. The massive material can be turned on a lathe and then fired at 1000–1200°C with no appreciable change in dimensions.
Pyroplastic Deformation. The irreversible deformation suffered by many ceramic materials when heavily stressed at high temperatures. The term has been applied more particularly to the slow deformation of fireclay refractories when loaded at high temperatures.
Pyroscope. A device that, by a change in shape or size, indicates the temperature

or, more correctly, the combined effect of time and temperature (which has been called HEAT-WORK). The best-known pyroscopes are PYROMETRIC CONES (q.v.), BULLERS RINGS (q.v.) HOLDCROFT BARS (q.v.); and WATKIN HEAT RECORDERS (q.v.).

Pyrosil. (1) Glass-ceramic cooking ware. See PYROFLAM, PYROCERAM.
(2) Kiln car tops containing PYROPHYLLITE (q.v.) (South Yarra Fire Brick Co Pty, Australia).

Pyrotenax. Mineral insulated cables or thermocouple probes, made by BICC Pyrotenax Ltd, Hebburn. Used in electrical power distribution in ships, and where there is a high fire risk.

Pyroxenes. See SILICATE STRUCTURES.

Pythagoras Ware. Trade-name: W. Haldenwanger, Berlin Spandau. A gas-tight mullitic porcelain that can be used at temperatures above 1400°C. P.C.E. 1825°C. Thermal expansion (20–1000°C), 5.7×10^{-6}, Tensile strength, 50 MNm^{-2}; crushing strength, 700 MNm^{-2}.

PZN. Lead Zinc Niobate.

PZP. A zircon refractory ramming mix made by Savoie Refractaire, France.

PZT. Lead zirconate-titanate, $Pb(Zr,Ti)O_3$; used as a ceramic component in piezoelectric transducers. Trade-name of Vernitron Ltd.

Quality Assurance. An organised programme which specifies tests, and responsibilities for testing and the recording of results. It is designed to ensure that a claimed quality of product or service is consistently delivered, and the origin of any shortfall identified. The standards for such quality systems are B.S. 5750 and ISO 9000.

Quality Control. A system of regular tests to ensure that products attain the required standards of performance. See also ACTION LIMIT.

Quantimet Image Analyser. A method of analysing the image in microscopy, by displaying it on a television screen scanned at 720 lines, which allows areas of interest to be scanned at 625 lines with 800 image points, or 500,000 observable dots. Grain size and shape can be studied, and phases identified by distinguishing black, grey and white areas.

Quarl Block. A refractory shape forming the whole, or a segment, of a gas- or oil-fired burner, particularly in a boiler furnace or glass-tank furnace.

Quarl Throat. An opening in the wall of a boiler furnace, into which a burner is set.

Quarry. (1) An open-cast working of rock.
(2) See FLOOR QUARRY.

Quartz. The common form of silica, SiO_2. Quartz is the principal constituent of silica sand, quartzite and ganister, and it occurs as an impurity in most clays. The room-temperature form, α-quartz (sp. gr. 2.65) undergoes a reversible crystalline change to β-quartz at 573°C; this inversion is accompanied by a linear expansion of 0.45 %. At 870°C quartz ceases to be stable but, in the absence of fluxes, does not alter until a much higher temperature is reached, when it is converted into cristobalite and/or tridymite, depending on the temperature and nature of the fluxes present.

Quartz Glass. This term is derived from the German Quarzglas and as such refers to Transparent Vitreous Silica; the term is not recommended. For definitions of the types of VITREOUS SILICA see under that term.

Quartzel. 99.99% SiO_2 silica yarn which can be woven into textiles.

Quartzite. A silica rock of a type in which the quartz grains as originally deposited, have grown at the expense of the siliceous cement. Its principal use in the ceramics industry is as raw material for the manufacture of silica refractories. For this purpose the quartzite should contain as little alkali and alumina compounds as possible; the porosity should be low (< 3%) and the rock should consist of small quartz grains set in a chalcedonic or opaline matrix. Quartzite for silica refractories is worked in Britain in N. and S. Wales, the Sheffield area and the N. Pennines (cf. GANISTER; SILCRETE).

Quasi-brittle fracture. See FRACTURE. Quasi-brittle fracture is characterized by some toughening mechanism which increases the fracture energy to a range from 0.1 kJm^{-2} (in concrete) to several kJm^{-2} in fibre-reinforced materials. There is usually a CRACK BRIDGING ZONE just behind the crack tip, with a small (sub-millimetre) microcrack region at the crack tip, and the cracking process zone extends over several mm. This is the cracking mechanism in some large-grained ceramics, and in whisker reinforced ceramics. See CRACK BRIDGING; TOUGHNESS.

Quasi-isostatic Pressing. A variant of ISOSTATIC PRESSING, in which rubber or synthetic elastomer, rather than a fluid, is used as the pressure transmitting medium. It is claimed to be of value for large products. In the Dorst press for making tableware plates, the lower plunger is a hollow membrane pressurized from within with oil. (Dorst-Keramikmachinenbau, Br.Pat 1589666 and UK Pat 2060470A, 1981).

Quebracho Extract. An extract of tannin from the S. American tree Quebracho Colorado. It has found some use, particularly as sodium tannate, as a deflocculant for pottery slips.

Queen Closer. A brick cut longitudinally to half width (50cm), used to maintain the bond.

Queen's Ware. The name given by Queen Charlotte, wife of George III, to the fine white earthenware introduced by Josiah Wedgwood in 1763.

Quench Tests. See THERMAL SHOCK TESTS.

Quenching of Frit. Molten glaze-frit or enamel-frit is quenched to break it up, thus making it easier to grind. The simplest method is to allow the stream of molten frit to fall into water, but this does not give uniform quenching and fracture. Better methods are to expose the stream of molten frit to a blast of air and water, or to pass the stream between water-cooled rolls; the latter process gives a flaky product.

Quetta Bond. A brickwork bond which provides vertical channels the height of the wall, into which reinforcement can be placed. The walls are usually 1½ bricks thick. Named after the town of Quetta, in Pakistan, where the bond was introduced after an earthquake, to provide reinforced structures capable of withstanding seismic damage. See REINFORCED BRICKWORK.

Quick Clay. A term in soil mechanics for a clay that, when undisturbed, has a certain shear strength and can be regarded as a solid body but which, when remoulded, can be regarded as a liquid (cf. THIXOTROPY).

Quickfire Kiln. An experimental low cross-section low thermal mass tunnel kiln for tableware, designed for short firing cycles of ware set one-high, enabling flexible operation, with the firing schedule matched to daily shifts, or readily changed between small batches of different ware. The kiln was designed by British Ceramic Research Association, and a very similar kiln marketed as the FIREPOWER (q.v.).

Quintus Press. Trade-name. An isostatic press made by Asea, Sweden.

Quoin. An external corner of brickwork.

R-Curve. R is the crack resistance force, which for many ceramics increases as the crack propagates. This is termed 'R-curve behaviour,' but the shape of the curve R vs a/W is not a characteristic of the material, but depends also on the test conditions. a/W is the crack length a, normalized for a specimen height W. The crack resistance force R is given by: $R = (P^2/2b).\ d/da\ (d/P)$ where P is the load, d the deflection and b is the specimen width. See CRACK BRIDGING.

R-value. The partial dispersion ratio of a glass expressed as $(n_D - n_C)(n_F - n_C)^{-1}$ where n_C, n_D and n_F are the refractive indices at wavelengths equivalent to the spectral lines C, D and F. (cf. ABBE NUMBER.)

Rack Mark. A surface blemish on rolled glass resulting from a mechanical defect in the drive actuating the forming roller.

Radial Crack. A semi-elliptical crack perpendicular to a surface damaged by pressing on to it a hard, sharp object.

Radial Brick. A brick with the two end faces curved to form parts of concentric cylinders (cf. CIRCLE BRICK and see Fig. 1. p39)

Radio-frequency Heating. See under INDUCTION HEATING.

Radome. A protective cone for the radar equipment in the nose of an aircraft, rocket or space vehicle. At high speeds the skin temperature may exceed 500°C and ceramic radomes become necessary; alumina and glass-ceramics have been used.

Rain Penetration. The degree to which rain penetrates brickwork greatly affects the thermal insulation value and the resistance to frost damage. It depends on the integrity of the structure, the workmanship, as well as the porosity of the bricks and mortar. It is influenced by wind conditions and site exposure to driving rain. Test cabinets with water sprays to simulate real conditions have been devised. B.S. 8104 recommends two methods for assessing the exposure of walls to driving rain. ASTM E514 is a panel test to assess water penetration of brickwork. In Britain, the British Ceramic Research panel test is used. Other ASTM tests (E331, E5467, E1105) relate to windows, curtain walls and doors.

Rain Spalling. Severe mechanical damage to the refractory lining of a rotary kiln, indistinguishable from CHEVRONING (q.v.), but sometimes due to sudden contraction of exposed parts of the kiln in heavy rain.

Rain Spot. A dark circular mark, often surrounded by a halo, on vitreous enamelware; the fault arises if a dirty drop of water falls on the biscuit ware.

Rake. CULLET CUT (q.v.).

Rake-out Slots. Gaps left in a kiln car deck for the removal of debris.

Raking Stretcher Bond. A method of laying bricks in a wall, all the courses being stretchers and all vertical joints being broken by the length of a quarter of a brick.

Raku Ware. A type of pottery used in Japan for the Tea Ceremony. The ware is thick-walled, is covered with a very soft lead borosilicate glaze, and is once-fired at about 750°C.

RAL Colour Register. (*Reichs Ausschusz für Lieferbedingungen* – (German) National Board for Conditions of Delivery). This register is a collection of samples of those colours predominantly used in industry. (Cf. B.S. 5252 1976). Every named colour in the register is placed in the RAL Colour List, a set of cards with the original colour shade on their left-hand side. The colour register makes possible rationalisation and standardization in dealing with colours in all parts of industry. It is relevant to ceramic tiles and sanitaryware. It is not a colour system based on a strict scientific colour measurement, but has been developed empirically to classify colours used in practice.

Ram Pressing. A process for the plastic shaping of tableware and sanitaryware by pressing a bat of the prepared body between two porous plates or mould units; after the pressing operation, air is blown through the porous mould parts to release the shaped ware. The process was patented in 1952 by Ram Incorporated (US Pats, 2 584 109 & 10).

Ramming. The forming of refractory shapes or monolithic linings by means of a portable device (usually pneumatic) by which the material is compressed by repeated impact from a blunt ended tool (cf. TAMPING).

Ramming Material. Graded refractory aggregate, or a mixture of aggregates, with or without the addition of a plasticizer; ramming materials are usually supplied at a consistency that requires a mechanical method of application; chemical bond(s) may be incorporated. The rammed mass becomes strong and monolithic by vitrification or sintering *in situ.*

Ramming Mix. A mixture containing graded refractory aggregate and other bonding materials which will harden when heated, by the formation of a chemical and/or ceramic bond. Such a mixture may be supplied either dry or mixed with water or other liquids and, in either case, requires to be installed by the use of mechanical or pneumatic tools or vibration. (B.S. 1902 Pt. 7). ASTM C71 defines a ramming mix as a refractory material, usually tempered with water, that cannot be extruded, but can be rammed in place to form a monolithic structure.

Ramonite. A material developed for investment casting and lost-wax casting, from a mixture of enriched raw materials containing up to 52% alumina.

Ramp. That part of the roof of an open-hearth steel furnace that rises from the end of the main roof to the top of the end wall; as the furnace is symmetrical, there is a ramp at each end.

Ramsdell Numbers. A shorthand notation for labelling POLYTYPES (q.v.) of crystal structures The numbers are of the form nR or nH were n is the number of layers in the UNIT CELL (q.v.) necessary for the arrangement of the layers to repeat itself. R, H stand for rhombohedral or hexagonal crystal form respectively. Originally devised to classify silicon carbide polytypes, the system has been applied to other carbides and nitrides. (L.S. Ramsdell, *Amer. Miner* **32**, 147, p64).

Randalusite. ANDALUSITE (q.v.) from the Transvaal.

Rankine Temperature Scale. The Fahrenheit scale displaced downwards so that Absolute Zero (-492°F) is 0° Rankine: 32°F = 492° Rankine; etc.

Rankinite. $3CaO.2SiO_2$; melts incongruently at 1464°C to form dicalcium silicate and a liquid; thermal expansion (20–1000°C) 13×10^{-6}. It is *not* found in portland cement.

Rapid Firing. Reducing the conventional firing schedules for traditional ceramics from days to hours, or less, by methods such as SINGLE LAYER SETTING, radiant heating, LOW THERMAL MASS KILNS and improved body and glaze formulations.

Rapid-hardening Portland Cement. The UK equivalent of the US term HIGH EARLY-STRENGTH CEMENT. It is more finely ground than ordinary portland cement; B.S. 12 stipulates a specific surface 3250 cm^2/g compared with 2250 cm^2/g for ordinary cement. This extra fineness results in the 1-day strength being as high as the 3-day strength of ordinary cement.

'Rapid' Method of Silicate Analysis. Analysis of clays and ceramics, using selected reagents and spectrophotometry to complete an analysis in about 8 hours, instead of several days by the 'standard' wet chemical technique. Modern instrumental techniques such as X-ray fluorescence can produce complete analysis in minutes.

Rapid Omnidirectional Compaction. This method for shaping mixed ceramic and metal powders combines high pressure and short compaction time with densification by plastic deformation to achieve near net shape forming (*Ceram Eng Sci. Proc.* **9,** (7/8), 965, 1988).

Rapidor Mill. A ball mill with a rapidly-rotating vertical barrel. Tradename Wm Boulton Ltd., Stoke-on-Trent.

Rare Earths. Oxides of the rare earth elements of Group III in the Periodic System. Those used in ceramics are the oxides of LANTHANUM, CERIUM, PRASEODYMIUM, NEODYMIUM, and YTTRIUM (q.v.).

Raschig Ring. A thin-walled hollow cylinder made of chemical stoneware, glass, carbon or metal, for the packing of absorption towers. These rings are made in various sizes from about 12.5 × 12.5 mm (300000/m^3) to 50x50 mm (5000/m^3) and even up to 200x200 mm. (F. Raschig, *German Pat.,* 286 122.)

Rasorite. Natural BORAX (q.v.).

Rat Hole. A hole eroded through the refractory structure, especially the crown, of a glass tank.

Rate-controlled Sintering. Densification rate (the slope of the density/time curve) is used as the independent variable in the control of the sintering process (instead of the temperature, as is usual). Densification is made to proceed at the lowest temperatures which will maintain the desired density-time curve. The power input to the furnace is varied by feedback from continuous dilatometry of the sintering sample. (The resulting temperature-time curve will be quite non-linear, and may include periods of cooling).

Rational Analysis. The mineralogical composition of a material as deduced from the chemical analysis. With materials whose general mineralogical composition is not already known, the calculation is made only after micro- and/or X-ray examination has shown what minerals are present and their approximate proportions. With pottery clays the calculation is made without such guidance, either on the FELDSPAR CONVENTION or the MICA CONVENTION. The former is the older procedure and assumes that the minerals present are kaolinite, feldspar and quartz. It is now known, however, that the alkali impurity in English pottery clays is present as mica and according to the mica convention the rational analysis is calculated on the basis of kaolinite, mica and quartz. (The term PROXIMATE ANALYSIS is sometimes applied to the rational analysis of clays.)

Rattler Test. A test introduced in 1913 by ASTM, as a method for evaluating the resistance of paving bricks to impact and abrasion; it is now a standard test (ASTM – C7). A sample of 10 bricks is subjected to the action of 10 cast-iron balls 3.75 in. (47.5 mm) dia. and 245–260 balls 1.875 in. (95 mm) dia. in a drum (28 in. dia; 20 in. long (0.7 m × 0.5 m)) rotating at 30 rev/min for 1 h. The severity of abrasion and impact is reported as a percentage loss in weight.

Raw Glaze. A glaze in which none of the constituents has been previously fritted.

RBAO. Reaction Bonded Aluminium Oxide.

RBM. REINFORCED BRICK MASONRY (q.v.)

RBSC. Reaction-bonded Silicon Carbide. See SILICON CARBIDE.

RBSN. Reaction bonded silicon nitride. See SILICON NITRIDE.

RCVD. Reactive Chemical Vapour Deposition (q.v).

RE, REL, REO, REOL, RER, RES, REX, REXH, REXR. Symbols for various shapes of ceramic wall tiles and fittings (see Fig. 7, p350).

Reaction-bonding. The formation of a polycrystalline ceramic by a chemical reaction between the powder and a gas or a liquid. Cf. REACTION SINTERING and see SILICON NITRIDE, SILICON CARBIDE.

Reaction-sintering. SINTERING (q.v.) of a two-component powder, in which a chemical reaction also takes place to produce a dense single phase or a multiphase material. Ceramic-ceramic composites may be thus produced – e.g. dense mullite zirconia composites from zircon and alumina. See also REACTIVE HOT-PRESSING.

Reaction-sintered Silicon Carbide. A misnomer, for reaction bonded silicon carbide. See SILICON CARBIDE.

Reactive Chemical Vapour Deposition. (RCVD). A variant of CVD, in which more than one material is deposited at the same time, to optimise the desired properties of the composite coating.

RESA (Reactive Electrode Submerged Arc Technique). Fine (10–1000nm anhydrous oxide powders are made as sols by striking an arc between two metal electrodes submerged in a dielectric liquid, typically water. (A. Kumar and R. Roy *J Mater Res* **3** (6), 1988 p. 1373).

Reactive Hot-Pressing. A term proposed by A. C. D. Chaklader *(Nature,* **206**, 392, 1965) for the special case in which the material being pressed decomposes, e.g. by losing H_2O or CO_2, at the temperature employed; increased strength and decreased porosity can be achieved in this way.

Reactive Spraying. A mixture of metal, metal oxide and a peroxide is sprayed on to a surface pre-heated to 1200–1500°C (e.g. a hot refractory lining in need of repair). A monolithic ceramic layer is formed *in situ.*

Reactolite. Trade-name. PHOTOCHROMIC GLASS (q.v.) used particularly for spectacles. (Chance-Pillington).

Ready-mixed Concrete. See SHRINK-MIXED and TRANSFER-MIXED.

Real Brick. Thin brick slips are cut from green bricks and fired alongside the normal products. They are intended as do-it-yourself facings. (Michegan Brick Co. *Brick Clay Rec* **168,** (2) 23, 1976).

Ream. A fault in flat glass resulting from the presence of a heterogeneous layer, or layers, within the glass. The fault is seen as a series of parallel lines when the glass is viewed through an edge; the appearance is similar to a pile of paper, hence the name.

Rearing. The setting of pottery flatware on edge for the firing process (cf. DOTTLING).

Réaumur Porcelain. A fritted porcelain ('soft paste') made from 75% frit, 17% chalk and 8% clay; the frit contained 60% SiO_2 the remainder being K_2O, Na_2O, Al_2O_3 and CaO. It was first made by a Frenchman, R. A. Réaumur (1683–1757).

Réaumur Temperature Scale. A little-used scale in which the temperature interval between the freezing point and boiling point of water is divided into 80 degrees.

Reboil. The appearance of bubbles in glass or vitreous enamel during manufacture. Molten glass may reboil after it has apparently freed itself from gases; in vitreous enamelling, bubbles may appear on the surface of sheet-steel ground-coats during the second or third firing process.

Recessed Wheel. An abrasive wheel having a contoured central recess in one or both sides.

Recesso. Ceramic bathroom fittings, e.g. recessed soap-trays. (Trade-mark, Richards Tiles Ltd., England.)

Reciprocal Lattice. A mathematical construct to express elegantly the conditions for an X-ray beam to be diffracted by a crystal lattice. If the directions of the crystal lattice are given by vectors **a**, **b** and **c**, then the reciprocal lattice vectors **a***, **b*** and **c*** are such that **a*** is perpendicular to the plane of **b** and **c**, and **a***.**a** = 1, etc. Then a vector **r*** from the origin to a point (h, k, l) of the reciprocal lattice is perpendicular to the plane (h, k, l) of the crystal lattice, and its length is the reciprocal of the spacing of the (h, k, l) planes of the crystal lattice. Every point in the reciprocal lattice corresponds to a possible X-ray reflection from the crystal lattice. See also X-RAY CRYSTALLOGRAPHY.

Reckna Clay Beams. Prefabricated beams made of specially designed hollow clay building blocks reinforced with steel rods; the load-bearing capacity is about $40MNm^{-2}$. The design originated in Germany *(Ziegelindustrie,* **3,** 60, 1950).

Reclaim. Overspray (enamel or glaze) that, after it has been reconditioned, is suitable for re-use. In vitreous enamelling, reclaim is generally used only in first-coats.

Reconstructed Glass. POROUS GLASS (q.v.) can be impregnated with aqueous solutions of nitrates or chlorides of colouring ions. The impregnated glass is washed and CONSOLIDATED in air or oxygen, to form heat-resistant coloured glasses. Black glasses of various electrical resistivities can be made by impregnating with furfuryl alcohol, which polymerises to a resin. This is pyrolyzed to an amorphous carbon phase in the pores, and the glass consolidated.

Recrystallization. The process of producing strain-free grains in a plastically deformed matrix. One crystal species grows at the expense of other(s) of the same substance which are smaller. *Primary recrystallization* is by nucleation and growth of new strain-free grains in the plastically deformed matrix. In *secondary recrystallization* large grains grow at the expense of a fine-grained strain-free matrix. There is a threshold temperature below which recrystallization does not take place. ALUMINA and SILICON CARBIDE are the most common materials available in recrystallized form.

Rectangular Kiln. An intermittent kiln, rectangular in plan, with fireboxes at intervals along the two long sides. Such kilns find limited use in the heavy-clay and refractories industries.

Recuperator. A continuous heat exchanger in which heat from furnace gases is transferred to incoming air or gas through walls (usually tubes) of metal or refractory material.

Red Clay. A term applied in the pottery industry to any of the ferruginous clays used for making tea-pots, floor tiles and plant pots; in England these clays are typified by the ETRURIA MARLS (q.v.)

Red Edge. A fault in plate glass characterized by numerous small cavities, containing rouge from the polishing, around the edges.

Red Lead. Pb_3O_4; see LEAD OXIDE.

Red Mud. The residue from the manufacture of alumina by the Bayer process. Red mud has been used to make building bricks.

Red Ware. See ROCKINGHAM WARE.

Redwood Cup. An orifice-type VISCOMETER. See FLOW CUP.

Redston-Stanworth Annealing Schedule. A procedure for determining the optimum conditions for annealing small glass articles; it is based on the Maxwell model of stress release, modified to take account of the variation of viscosity with time. (G. D. Redston and J. E. Stanworth, *J. Soc. Glass Tech.,* **32,** 32, 1948.)

Reel Cutter. A type of CUTTING-OFF TABLE (q.v.) in which the wires are stretched within a large circular frame, the axis of which is slightly above the line of the extruding clay column. This type of cutter can operate on a stiff clay and, if powered and automatic, has a high output.

Rees-Hugill Flask. A 250 ml flask, with a neck graduated directly in specific gravity units, designed expressly for the determination of the specific gravity of silica and other refractories. It was introduced by W. Hugill and W. J. Rees *(Trans. Brit. Ceram. Soc.*, **24,** 70, 1925) and has since become the subject of a British Standard (B.S. 2701).

REFEL. Registered tradename (Reactor Fuel Element Laboratory) of the UKAEA Springfields laboratory, for REACTION BONDED SILICON CARBIDE developed there on the basis of P. Popper's original work at British Ceramic R.A. (Power Jets (R&D) Ltd, Br. Pat. 866,813, 3/5/61).

Refining. The process in glass-making during which the molten glass becomes virtually free from bubbles; this is effected in the REFINING ZONE of a glass-tank furnace, or in a pot.

Refire Opal. See OPAL GLASS.

Refiring. A second firing given to articles of ceramic sanitaryware or vitreous enamel, which fires a partial glaze or enamel coating applied to repair or further decorate the piece.

Reflectance. This term is defined in ASTM-C286 as: 'The fraction of incident light that is diffusely reflected, measured relative to magnesium oxide under standard conditions.'

Reflectivity. As applied to vitreous enamel, this term is defined in ASTM-C286 as: 'The REFLECTANCE (q.v.) of a coating so thick that additional thickness does not change the reflectance. The method of test is given in ASTM – C347.' See also GLOSS.

Reflex Glass. Glass strip, for use in vessels containing liquids, made with prismatic grooves to facilitate the reading of the liquid level.

Reformulation. If one raw material is substituted for another nominally similar in a ceramic body, significant changes in behaviour can result, which may be overcome by the stepwise replacement of other ingredients. Reformulation technology replaces this experimentation by numerical calculation techniques based on measurements of a small number of characterizing features of the body. (G.W. Phelps,, *Am. Ceram. Soc. Bull.* **55**, (5) 1976 p.528 – R. L. Lehman, *Ceram Eng. Sci. Proc.* **13** (1/2) 1992 p. 321).

Refractoriness. The ability of a material to withstand high temperatures; it is evaluated in terms of the PYROMETRIC CONE EQUIVALENT (q.v.).

Refractoriness-under-Load. The ability of a material to withstand specified conditions of load, temperature and time. Details of variations in this test will be found in the following national standards: Britain-B.S. 1902; USA-ASTM-C16; Germany-DIN 51064; France-AFNOR B49–105. B.S. 1902 specifies a dilatometer method (Pt. 4.8)

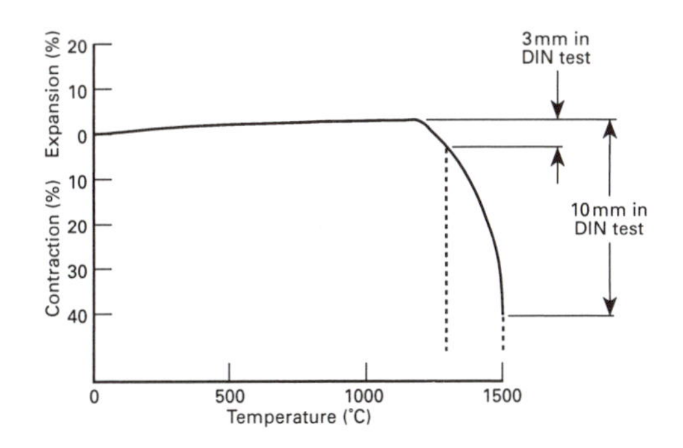

Fig. 6 REFRACTORINESS-UNDER-LOAD

and a differential method (Pt. 4.9). It is usual to show the result of this test as a curve (Fig. 5) relating the expansion/ contraction to the temperature. In the ASTM method the test-piece is two whole bricks; unless the brick fails by shear, the subsidence caused by the test is determined by direct measurement of the brick after it has cooled to room temperature. The German procedure reports two temperatures: t_a at which the curve has fallen 3 mm from its highest point, and either t_e, (10mm subsidence from its highest point) or t_b (the temperature at which the test-piece completely collapses).

Refractory Cement. A finely-ground refractory material, or mixture, used for the jointing of furnace brickwork. The term is often used to denote all types of such cement but should more properly be reserved for materials that harden only as a result of ceramic bonding at high temperature.

Refractory Coating. A refractory material for the protection of the surface of refractory brickwork or of metals (e.g. pyrometer sheaths or aircraft exhaust systems). Examples of such coatings include Al_2O_3, $ZrSiO_4$, $MoSi_2$ and (for the protection of metals) refractory enamels. Refractory coatings for furnace brickwork are sometimes known as WASHES.

Refractory Concrete. A concrete made from a graded aggregate of crushed refractory material together with high-alumina hydraulic cement, e.g. Ciment Fondu or Lumnite. The concrete may be cast in position in a furnace lining or it can be made into blocks and special shapes. With an aggregate of crushed 40%-Al_2O_3 firebrick, such a concrete can be used up to about 1300°C. Insulating concrete can be made for use up to 900°C; its thermal conductivity is about 0.25–0.35 W/mK.

Refractory Material. Also known as REFRACTORY PRODUCT, or simply REFRACTORY. Typical refractory materials are fireclay, silica, magnesite, chrome-magnesite, sillimanite; dolomite and carbon. They are materials used in large quantities to withstand high temperatures, in furnace linings and similar industrial applications. B.S. 3446 Pt. 1, 1990 defines a REFRACTORY MATERIAL as a non-metallic material or product (but not excluding those containing a proportion of metal) having heat resisting properties, usually measured by classes in PYROMETRIC CONE EQUIVALENT and selected according to the operating temperature and the level of protection required.

Refractory Mortar. Jointing material for furnace brickwork. (B.S. 1902 Pt.11 which specifies properties and tests) ASTM C71 defines refractory mortar as a finely ground preparation which becomes trowelable and plastic when tempered with water, and is suitable for laying and bonding refractory brickwork. A *heat-setting mortar* is defined as a refractory mortar whose potential strength is dependent on use at furnace or process temperature. ASTM C198 specifies a cold bonding strength test; C199 a test for their refractoriness, in which a brick pier laid with the test mortar is heated in prescribed manner to see if the mortar flows from the joints.

Refractory Putty. A MOULDABLE with maximum grain size of 0.5mm, for small-scale sealing, patching and coating. See B.S. 3446 and B.S. 1902 Pt. 11, which specifies properties and tests.

Refrasil. Fused silica fibre. (Trade-name of Chemical & Insulating Co., Darlington).

Refrax. Trade-name. A silicon carbide/silicon nitride composite, used for special refractory shapes and tiles. The silicon nitride bond is formed in-situ between the grains of silicon carbide, imparting a finer crystal structure to the refractory, so improving strength and refractoriness, to 1800°C. (Refrax S.1 is silicon nitride).

Regenerator. A periodic heat exchanger in which refractory checker bricks alternately receive heat from furnace gases and give up heat to incoming air or gas; this is achieved by reversing the direction of gas flow.

Reheat Test. The prescribed heat treatment of a fired refractory to determine its dimensional stability when reheated to a second temperature higher than the temperature to which it was first subjected. ASTM C113 specifies eight heat treatment schedules which can be used for various types of refractories. See PERMANENT LINEAR CHANGE.

Rehoboam. A 6-quart wine bottle.

Reinforced Brickwork or Reinforced Masonry. Brickwork that is reinforced by steel rods or wires inserted into continuous cavities through the brickwork, the space between steel and brick then being filled with poured concrete. Such brickwork can withstand tensile, as well as the usual compressive, stresses. Rules for the design of such brickwork are included in B.S.5628 Pt2.

Reintjes Bricking Rig. A scaffolding system for lining cylindrical furnaces, cement kilns in particular. When the lower part of the lining has been laid, the rig is assembled and further bricks are laid over the top of the rig by bricklayers working from a scaffold integral with and moving with the rig as it advances upwards course by course. The rig allows individual rings of bricks to be tightened using hydraulic jacks, with shims from the side, rather than by applying forces to the hot-face.

Relaxors. FERROELECTRIC (q.v.) ceramics in which the ferroelectric-paraelectric phase transition takes place over a range, rather than at a well-defined CURIE TEMPERATURE (q.v.). The *relaxation* (or exponential decay) of the dielectric constant takes place over a range of temperature. One explanation of this DIFFUSE PHASE TRANSITION is that the materials have a structure comprising an assembly of microregions, each with its own Curie temperature, but there is a distribution of Curie temperatures from different individual microvolumes. Thus the transition from ferroelectric to paraelectric behaviour takes place over a wide temperature range, within which some of the microregions are paraelectric, while the others are still ferroelectric. Ceramic relaxors include some PLZT ceramics and some niobates. Their properties are also frequency dependent.

Relict. A porous AGGLOMERATE of oxide particles, formed by calcination of salt particles with the same size and grouping as the resulting agglomerate. Relicts have ultrafine porosity and are good thermal insulators and barriers to gas diffusion.

Relieving Arch. An arch built into structural or refractory brickwork to take from that part of the wall below the arch the weight of brickwork above the arch.

Rendering. A coating of mortar or plaster on brickwork or the process of forming such a coating.

Renn Press. See KRUPP-RENN PROCESS.

Repressing. A second pressing sometimes given to bricks (both building and refractory) to improve their final shape; with building bricks the purpose can also be to provide some special surface effect.

Repetitive Strain Injury (RSI). Injury due to overstressing by constant repetition of a movement, or prolonged holding of a limb in an unnatural position. Chronic pain or loss of mobility may result.

Replicast CS Process. In this metal casting process, a refractory coating is applied to high quality expanded polystyrene, which replaces the wax in manufacture by ceramic shell moulding. Relatively large components with good surface finish can be produced. The process was developed by SCRATA, Sheffield, using coatings developed by Foseco International.

Replication. The technique of reproducing the topology of a polished section or a fracture surface by applying to it a thin film such as a softened plastic type or a rapidly polymerizing liquid silicone rubber. The topology may then be examined by microscopy, using the replica, while the original is available for other tests.

Re-setting (of Portland Cement). The property of portland cement, after it has been mixed with water and allowed to set, to set for a second time (although giving a great reduced strength) if it is again finely ground. The effect is due to the fact that the hydration causing setting is confined to the surface (10mm) of the cement particles, some of which are as large as 100mm.

Residual Clay. See PRIMARY CLAY.

Resinous Cement. A term used in chemical engineering for an acid-proof cement, based on synthetic resin, for jointing chemical stoneware or acid-resisting bricks. The cement, when set, is impervious and very hard.

Resist. (1) A fluoride-resisting layer used during acid etching to protect those parts of glass or ceramic ware not required to be etched. Wax, asphalt, resin, etc., dissolved in turpentine are suitable for this purpose.

(2) A method for the on-glaze decoration of pottery, particularly for use with lustres. The decoration is painted on the ware with glycerine, or other resist; the ware is then coated with the lustre solution, dried, the painted area washed clear with water, re-dried and fired. Alternatively, an infusible resist can be used, e.g. china clay, and the ware can be fired before the resist is gently rubbed away with a soft abrasive powder.

Respirable Dust. For dust to penetrate to the lungs, it must be less than 10mm diameter. Maximum deposition in the lungs occurs for particles below 5mm, with a peak at 1 to 2mm. For the purposes of estimating airborne dust in relation to pneumoconiosis, samples should represent only the respirable fraction. This is defined in terms of the free-falling speed of the particles. $C/C_o = 1-f/f_c$ where C and C_o are the concentrations of particles of falling speed f in the respirable fraction and the whole sample, while f_c is a constant: twice the falling speed in air of a sphere of unit density and 5mm diameter.

Residual Carbon. The amount of carbon retained in test-pieces of pitch-bonded, resin-bonded or -impregnated refractory after carbonization. The carbonized test-piece is ignited in an oxidising atmosphere at a specified temperature for a specified time. The weight loss expressed as a percentage of the weight of the original uncarbonized test-piece, is the *carbon yield,* or amount of residual carbon present. ASTM C831 specifies a test for residual carbon and carbon yield in coked bricks and shapes.

Resilience. The resilience of ceramic fibre is the percentage increase in thickness of a test-piece after 5 min, following removal of the pressure required to compress the fibre to 50% of its original thickness. Cf. COMPRESSIBILITY. B.S.1902 Pt.6 Section 8 specifies tests for refractory fibres.

Resistazone Stream Counter. Particle size analyzers based on the change in electrical resistance of a well-defined cylinder of conducting fluid, due to the presence of fine particles. The particles are suspended in an electrolyte, and the voltage pulses generated when they pass between measurement electrodes are recorded and analysed electronically.

RESS – Rapid Expansion of Supercritical Solutions. A jet of dense gas is expanded through a jet. Solid material dissolved in the gas at high temperature and high pressure is then precipitated as a powder from a concentrated solution, or deposited as a film from a dilute solution using a short nozzle. (D.W. Matson *et al, J. Mater Sci* **22,** (6) 1987, p.1919).

Restrike Opal. See OPAL GLASS.

Retarder. (1) A material, such as sodium citrate or keratin, that can be added to plaster to retard setting.
(2) Retarders for the setting of portland cement generally consist of cellulose derivatives (e.g. carboxymethyl-cellulose), starches or sugars; these are additional to the 1–3% of gypsum added to all ordinary portland cements to retard the setting time to within convenient limits (see INITIAL SET and FINAL SET).

Retort. A structure, generally built of refractory material, into which raw materials are charged to be decomposed by heat, e.g. a zinc retort or gas retort.

Retouch Enamel. To cover or protect imperfections in vitreous enamel, using a fine overspray or brushed-on coating.

Retroflex Curves. Solubility curves for substances whose solubility decreases with increasing temperature.

Revergen Kiln. Trade-mark: See DAVIS REVERGEN KILN.

Reverse Cam Spalling. PINCH SPALLING (q.v.) of refractory brick CAMS (q.v.) and their loss by SLABBING (q.v.). The intervening thinner refractory is usually undamaged, with the recessed sites of the lost cams between.

Revolver Press. A type of press for the shaping of clay roofing tiles. It consists of a pentagonal or hexagonal drum, mounted on a horizontal shaft and carrying bottom press-moulds on each flat surface. As the drum rotates, discontinuously, each mould in turn is brought beneath a vertical plunger which consolidates a clot of clay in the mould. While the drum is stationary (during pressing) a clot of clay is fed to the mould next due to arrive beneath the plunger.

Revolving Pot. A shallow refractory pot, which is slowly rotated and from which molten glass is gathered by a suction machine.

Revolving Screen or Trommel. A cylindrical screen, usually of the perforated-plate type, set at an inclined angle and made to rotate about its axis.

RF. Radio-frequency, as in RF-heating, RF sputtering.

RH-degassing. The Rheinstahl-Heraeus Process for SECONDARY STEELMAKING. By introducing a gas stream (argon) into the steelmaking vessel through a snorkel, the steel is sucked into a vacuum chamber. After degassing, the steel (now in a column of greater than barometric height) flows down again, continuously. As the steel enters the vacuum chamber, the release of argon and the evolution of CO produces a large surface area in the form of droplets, ensuring rapid degassing.

Rheology. The science of the deformation and flow of materials, e.g. the study of the viscosity of a glass, glaze or enamel; the study of the plasticity of clay.

Rheopexy. The complex behaviour of some materials, including a few ceramic bodies, which show DILATANCY (q.v.) under a small shearing stress followed by THIXOTROPY (q.v) under higher stress.

Rhythmical Firing. The original name for the SWEEP-FIRE system (q.v.)

Rib. A wooden or metal tool for smoothing the outside of a vase or bowl while it is being thrown.

Rib Marks. The marks found on the surface of broken glass; they are in the form of raised arcs perpendicular to the direction in which fracture occurred, usually convex in the direction of crack propagation. See HACKLE MARKS.

Ribbed Roof. A furnace roof (particularly of an open-hearth steel furnace) in which some of the refractory bricks, while conforming with the smooth internal surface of the roof, project outwards to form continuous ribs across the furnace; these ribs, because they remain cool, confer strength on the roof even when it has worn thin.

Ribbon. An ornamental course of tiles on a roof (cf. RIBBON COURSES).

Ribbon Courses. Successive courses of roofing tiles laid to alternately greater and lesser exposures.

Ribbon Machine. See PUFF-AND-BLOW PROCESS.

Ribbon Process. A glass-making process in which the glass, after it has been melted and refined, is delivered as a continuous ribbon.

Ribbon Test. See THERMAL SHOCK TESTS.

Rice Hulls or Rice Husks. The waste husks from rice consist when calcined, of about 96·5% SiO_2 2·2% CaO and minor amounts of other oxides. Because this composition is highly refractory, and because the porosity of the calcined material is nearly 80%, this waste product has been used, particularly in Italy where rice is an important crop, as a raw material for the manufacture of insulating refractories.

Riddle-tile Hearth. Passages beneath a kiln for the removal of waste gases.

Rider Bricks. Refractory bricks, which may be solid, perforated or arched, used in the base of a regenerator chamber to form a support for the checker bricks; in a coke-oven regenerator, also known as SOLE-FLUE PORT BRICKS or NOSTRIL BLOCKS; in a glass tank regenerator, also known as the RIDER ARCH, BEARER ARCH or SADDLE ARCH.

Ridge Tile. A special fired-clay or concrete shape for use along the ridge of a pitched roof. There are a number of varieties, e.g. SEGMENTAL, HOG'S BACK, WIND RIDGE, etc.

Riffling. Dividing a stream of particles into representative samples, using a mechanical device. E.g. a *spin-riffler* is an annular set of sieves which is rotated under a stream of powder, so that each in turn receives a random sample of the stream.

Rigden's Apparatus. An air-permeability apparatus for the determination of the specific surface of a powder; air is forced through a bed of the powder by the pressure of oil displaced from equilibrium in a U-tube. (P. J. Rigden, *J. Soc. Chem. Ind.*, **62**, 1, 1943.) The apparatus is essentially the same as that used in the BLAINE TEST (q.v.).

Rigidizer. A liquid binder for rendering ceramic fibre products hard and erosion resistant.

RILEM. Réunion Internationale des Laboratoires d'Essais et de Recherches sur les Materiaux et les Constructions. This body is the author of various test methods and standards relevant to constructional ceramics.

Ring. (1) A free-standing, self supporting annular structure formed of tapered refractory brick, used for the construction of cylindrical and conical sections of furnaces, vessels and ducts. Design is flexible. The sizes and numbers of bricks to make different sizes of rings are variable. Axial and radial tightness depends on the area of application, the properties of the refractory and the severity of the process. Various BRICKING RIGS have been designed for easier and/or faster construction of rings.
(2) A refractory ring that floats on the molten glass in a pot to keep any scum from the gathering area within the ring (cf. FLOATER).
(3) The part of the mould that forms the rim of pressed glassware.
(4) A SAGGAR (q.v.) without a bottom, also sometimes known as a RINGER.

Ring Mould. See NECK MOULD.

Ring-roll Crusher. A grinding unit consisting, typically, of a steel casing in which is housed a free-running grinding ring, within which revolve three crushing rolls; the top roll is mounted on the driving shaft while the two lower rolls are carried on floating shafts supported by springs. The three rolls are pressed against the inner surface of the ring by these springs and material is ground as it passes between the ring and the rolls. This type of crusher has been used successfully with ceramic raw materials such as bauxite.

Ring-roll Process. A method for the production of blanks for plate-glass manufacture. Molten glass passes between a heated 'ring-roll' casting table, of large diameter, and a smaller forming roll.

Ring Setter. A SETTER designed to prevent or correct warping of tableware during firing. Cf STRAIGHTENING RING.

Ring Test. (1) GLAZE FIT: a test first used by H. G. Schurecht and G. R. Pole (*J. Amer. Ceram. Soc.*, **13**, 369, 1930). The testpieces are hollow cylindrical rings 2 in. (50 mm) dia. glazed on the outside only using the glaze to be tested. The glazed ring is fired and two grooves or holes are then cut in one edge of the ring, approx. 2 in. (50 mm) apart and large enough to hold glass capillary tubes 1/6th in. (4 mm) long and 1/32 in. (0.8 mm) dia.; these capillaries provide sharp reference marks, the distance between which is measured with a micrometer microscope. The ring is then cut open with a diamond saw and the distance between the reference marks is again measured. Similar measurements are made on unglazed rings so that the true expansion or contraction caused by partially releasing the stress between glaze

and body can be determined. If the glaze is in tension the ring will expand when cut open; if in compression it will contract.
(2) A THERMAL-SHOCK test proposed by W. R. Buessem and E. A. Bush *(J. Amer. Ceram. Soc.,* **38**, 27, 1955). A stack of ceramic rings, each 2 in. (50 mm) o.d. and 1 in. (25 mm) i.d., and ½ in. (12.5 mm) long, are heated from the inside by a heating element and cooled from the outside by a calorimetric chamber. Both thermal conductivity and thermal-shock resistance can be evaluated (cf. BRITTLE-RING TEST).
Ring Wall. See SHELL WALL.
Ringed Roof. A furnace roof consisting of arches of bricks unbonded with adjacent arches (cf. BONDED ROOF).
Ringelmann Chart. A chart divided into five (Nos. 0–4) shades of darkness introduced in the late l9th century by Professor Ringelmann, of Paris, as a means of designating the blackness of smoke emitted from industrial chimneys. The charts have been standardized (B.S. 2742) and are used in the ceramic industry in compliance with the Clean Air Act in which dark smoke is defined as equal to, or denser than, Shade 2 on the Ringelmann Chart.
Ringer. See RING (4).
Rinman's Green. A colouring material consisting of a solid solution of CoO in ZnO; it finds limited use as a ceramic colour. (S. Rinman, *Kong. Vet. Akad. Handl.*, p. 163, 1780.)
Ripple. See ORANGE PEEL.
Ripple Mark. = WALLNER LINE (q.v.).
Rippled. See DRAGGED.
Rise. (1) In a sprung arch, the vertical distance between the level of the springer and the highest point of the arch at its inner surface.
(2) The vertical distance though which the centre of the sprung roof of a furnace rises as a result of expansion when the furnace is heated to its operating temperature.
Riser Brick or End Runner. A RUNNER BRICK (q.v.) with a hole near one end of its upper face and (generally) a short tubular projection from this hole to lead molten steel into the bottom of an ingot mould.
Rittinger's Law. A law relating to the theoretical amount of energy required in crushing or grinding: the energy necessary for the size-reduction of a material is directly proportional to the increase in total surface area. (P.R. von Rittinger, Berlin, 1867).
River marks. Cleavage steps on individual grains of a polycrystalline material, or on a single crystal, spreading from the point of origin. A special case of TWIST HACKLE (q.v.).
RO Fusion-cast Refractory. A fusion-cast refractory made in an inclined mould designed to concentrate the shrinkage cavity in one corner of the block: such blocks are made in France (RO = Retassure Orientée, i.e. oriented cavity; cf. DCL, RT and SR).
Robertson Kiln. Several types of tunnel kiln were designed by H.M. Robertson but that most commonly associated with his name was a tunnel kiln for salt-glazing; the salt was introduced via fireboxes in the side walls and the fumes extracted in the cooling zone. (Brit Pats., 331 224 and 331 225, 9/5/28.).
Robinson Trolley. A circular trolley, weighing 68.1 ± 2.3 kg with three equally-spaced wheels, which can be fitted with various types of rim or tyre. It is rotated over a 4ft × 4ft (1.23 × 1.23 m) test panel of floor tiles to assess resistance to wear or damage by small-wheeled vehicles such as fork-lift trucks, (ASTM C627–76).
Robey Oven. A down-draught type of pottery BOTTLE OVEN (q.v.) patented by

C. Robey (Brit. Pat. 970, 17th March, 1873).

Rocaille Flue. An alternative (now less common) name for STRASS (q.v.).

ROC Technique. Rapid Omnidirectional Compaction (q.v.).

Rock Cement. US term for ROMAN CEMENT (q.v.).

Rock Crystal. Large, naturally occurring, transparent crystals of quartz. In USA the term is also used for CRYSTAL GLASS (q.v.)

Rock Wool. Trade-name. A form of MINERAL WOOL (q.v.).

Rocker. A glass bottle that has a faulty, convex, bottom.

Rocket Nozzle. The flame temperature in a rocket nozzle exceeds 2500°C; in addition, thermal shock is very severe. Special refractory materials that have been used in rocket nozzles include silicon carbide (usually with graphite additions), silicon nitride, boron nitride, beryllia and various refractory carbides.

Rockingham Ware. The term originally referred to the ornate porcelain made at a pottery at Swinton, Yorkshire, England, on the estate of the Marquis of Rockingham during the years 1826–1842. Ware with a brown manganese glaze was also produced and it is this type of glaze, which in the UK is usually applied to teapots made from red clay, to which the term Rockingham Ware now refers. In the USA, Rockingham Ware was made at Bennington.

Rockwell Hardness. Symbol HR. A criterion of hardness based on indentatiion, either by a steel ball or by a diamond cone. Details are given in B.S. 891. It has been but little used in the ceramic industry, except in relation to some special ceramics.

Rod Cover or Sleeve. A cylindrical fireclay shape having an axial hole and terminating in a spigot at one end and a socket at the other. These refractory sleeves are used to protect the metal stopper-rod in a steel casting ladle. In the UK six sizes of rod cover are standardized in B.S. 2496. In the USA three qualities are specified in ASTM-C435.

Rod Mill. A steel cylinder into which is charged material that is to be finely ground together with metal rods as grinding media. This type of mill will reduce material from 1 in. (25 mm) to 10 mesh in one stage and the product tends to be granular. Rod mills are employed for the preparation of the materials used in the manufacture of sand-lime bricks.

Rokide Process. Trade-name: (Norton Company, USA): a process for the production of a refractory coating, on metal or on ceramic, by atomizing directly from the fused end of a rod of the coating material, e.g. Al_2O_3 or ZrO_2; the molten particles are blown against the cool surface that is to be coated. The original patent was granted to W.M. Wheildon (US Pat. 2 707 691, 3/5/55).

Rolands' Cement. Trade-name: a HIGH-ALUMINA CEMENT (q.v.) made by Rolandshütte A.G., Lubeck, Germany.

Roll Back. CRAWLING (q.v.) of vitreous enamel. The coating pulls away or rolls up at the edges of the metal, or over areas of dirt.

Roll Quenching. See under QUENCHING OF FRIT.

Rolled Edge. The edge of a plate or saucer is said to be 'rolled' if its diameter is greater than the general thickness of the rim of the ware.

Rolled Glass. (1) Flat glass that has been rolled so that one surface is patterned or textured; also known as Rough Cast. (2) In USA the term is also applied to optical glass that has been rolled into plates at the time of manufacture, as distinct from TRANSFER GLASS (q.v.).

Roller-bat Machine. A machine for making, from stiff-plastic clay, bats for a final pressing process in one method of roofing-tile manufacture.
Roller Elutriator. An air-type elutriator designed for the determination of fineness of portland cement. (P. S. Roller, *Proc. ASTM,* **32**, Pt. 2, 607, 1932.)
Roller's Equation. A relationship between the percentage weight and the size of powders: $w = a\sqrt{d}. exp(-b/d)$, where w is the weight percent of all material having diameters less than d, and a and b are constants. Other equations were deduced relating to specific surface and to the number of particles per gram or powder. (P. S. Roller, *J. Franklin Inst.,* **223**, 609, 1937.)
Roller's Plasticity Test. A method for the assessment of plasticity on the basis of the stress/deformation relationship when clay cylinders are loaded. (P. S. Roller, *Chem Industries,* **43**, No. 4, 398, 1938.)
Roller-head Machine. A machine for the shaping of pottery flatware on a rotating mould, as in a JIGGER (q.v.), but with a rotary shaping tool instead of a fixed profile. The rotary tool is in the form of a shallow cone of the same diameter as the ware and shaped to produce the back of the article being made. The ware is completely shaped, by relatively unskilled labour, in one operation at a rate (depending on the size of the ware) of about 12 pieces per minute. The machine, developed from earlier attempts to improve on the use of a fixed tool, was patented by T. G. Green & Co. and H. J. Smith (Brit. Pat., 621 712, 14/4/49) with subsequent improvements by J. A. Johnson (Brit. Pat., 765 097, 2/1/57; 895 988, 9/5/62). It is made by Service (Engineers) Ltd., Stoke-on-Trent, England.
Roller-hearth Kiln. A tunnel kiln through which the ware, placed on bats, is carried on rollers. Tiles may be carried directly, without bats.
Roller Mark or Roller Scratch. A surface blemish on vertically drawn sheet glass caused by contact with the rollers.
Rolling (of Glaze). A term sometimes used instead of CRAWLING (q.v.).
Rolling-out Limit. The lower limiting water content in the ATTERBERG TEST (q.v.).
Roll Mark. A machine for applying transfers, especially to tiles. (Meyercord Co. USA).
Rollrod Kiln. A roller-hearth kiln for the rapid firing of tiles. The tiles rest on supporting rods which themselves glide on the kiln rollers at right angles to them. (Agrob Ab).
Rolls. See CRUSHING ROLLS.
Roman Brick. US term for a building brick of nominal size 12 × 4 × 2 in. (305 × 102 × 50 mm).
Roman Cement. The misleading name given in the early 19th century to a naturally-occurring mixture of clay and limestone after its calcination; the product had hydraulic properties and was the forerunner of PORTLAND CEMENT (q.v.). A small quantity of Roman Cement is still made. In USA it is known as ROCK CEMENT.
Roman Tile. A type of roofing tile made in the Bridgwater area of England and developed from the OLD ROMAN TILE (q.v.). There are two varieties: Single Roman and Double Roman. Single Roman is a flat tile with an upturned edge along one side and a roll at the other which, in use, covers the upturned edge of the adjacent tile; Double Roman is larger and has an additional roll moulded on the centre line of the tile.
Roofing Tile. A tile for the covering of a pitched roof. The terminology of the

many designs of roofing tile is confused: it is particularly difficult to equate the terms used in different languages. For definitions of British terms see: DOUBLE ROLL VERGE TILE; INTERLOCKING TILE; ITALIAN TILE; MARSEILLES TILE; OLD ROMAN TILE; PANTILE; POOLES TILE; ROMAN TILE; SINGLE-LAP ROOFING TILE; SPANISH TILE. Most of the roofing tiles now being made in England are concrete; these must meet the requirements of B.S.473, 550 Concrete Roofing Tiles and Fittings. Clay roofing tiles are dealt with in B.S. 402. B.S. 5534 is a code of practice for slating and tiling with cross-reference to B.S. 8000 Pt. 6. Pt.2 provides design charts for fixing roof tiles against wind uplift. The ASTM specification for clay roof tiles is C1167.

Roseki. A Japanese rock, which is mainly pyrophyllite. With small additions of kaolin and diaspore it is commonly used to make glass-melting crucibles, and has also been used experimentally in pottery bodies. Roseki is also the name of an 80% silica ramming material, made by Hiroshimaken, Japan.

Rosin-Rammler Equation. An equation relating to fine grinding: for most powders that have been prepared by grinding, the relationship between *R*, the residue remaining on any particular sieve, and the grain-size in microns (x) is exponential:

$$R = 100 \exp(-bx^n)$$

where *b* and n are constants. (P. Rosin, E. Rammler and K. Sperling, *Ber. C52 Reichs-Kohlenstaubs,* Berlin, 1933.)

Rosso Antico. A red unglazed stoneware made in the 18th century by Josiah Wedgwood; it was a refinement of the red ware previously made in North Staffordshire by the Elers.

Rotablock. A system of furnace construction in which the refractory blocks are suspended on metal hangers, which hook on to an external suspension bar. Individual blocks can be pulled back from the lining by rotating the hanger about the bar. (Foseco Trading AG Br. Pat 1534914, 1978).

Rotap. Testing sieve shaking equipment (C-E Tyler Co, Mentor, Ohio).

Rotocast System. A system for installing castable refractories in cylindrical kilns, particularly cement kilns. Anchors and ties are welded to the shell, and formwork fitted around the circumference. Concrete is placed and vibrated behind the formwork. As work progresses the kiln is progressively turned, and the formwork moved on.

Rotary Disk Feeder. See DISK FEEDER.

Rotary Drum Test. See SLAG ATTACK TESTS.

Rotary-hearth Kiln. A circular tunnel with a slowly rotating platform for conveyance of the ware through thc kiln. An early example was the WOODHALL-DUCKHAM KILN (Brit. Pat. 212 585 23/2/23). The principle has since been adapted in the CLARK CIRCLE SYSTEM of brickmaking introduced in Australia in 1953 and in some modern pottery-decorating kilns. (This type of kiln is sometimes also known as a ROTATING-PLATFORM KILN.)

Rotary Kiln. A kiln in the form of a long cylinder, usually inclined, and slowly rotated about its axis; the kiln is fired by a burner set axially at its lower end. Such kilns are used in the manufacture of portland cement and in the dead-burning of magnesite, calcination of fireclay, etc.

Rotary Smelter. A batch-type cylindrical furnace which can be rotated about its horizontal axis while frit is being melted in it; when the process of melting is complete the furnace is tilted so that the molten frit runs out. Such furnaces are fired by gas or oil and their capacity varies from about 0·05 – 0·5 tonne.

Rotary Vane Feeder. See VANE FEEDER.

Rotating-platform Kiln. See ROTARY-HEARTH KILN.

Rotolec. Trade-name: a circular electric decorating kiln for pottery-ware. (Gibbons Bros. Ltd, Dudley, England.)

Rotomixair. Trade-name: a reciprocating airjet system of drying bricks and tiles; the system originated in Italy in 1958. (W. G. Cannon & Sons Ltd., London S.W.19.)

Rotor Process. A process for the production of steel by the oxygen-blowing of molten iron held in a horizontal, rotating vessel. This is usually lined with tarred dolomite refractories, but tarred magnesite refractories have also been used.

Rouge. Very fine (<1μm) ferric oxide used for polishing glass. *Rouge pits* are residual imperfections in the glass surface, containing traces of rouge.

Rouge Flambé. (1) As applied to old Chinese porcelain this term means a SANG DE BOEUF (q.v.) glaze that has streaks of purple or blue to give a flame-like (French, *flambé)* appearance. (2) As applied to modern pottery the term refers to a reduced copper glaze produced by an on-glaze process (cf. SANG DE BOEUF); although the colour is produced by on-glaze treatment, micro-examination shows that the layer of colloidal copper occurs within the glaze layer.

Rough-cast. See ROLLED GLASS.

Round Edge Tile. For the various types of wall tiles with round edges see Fig. 7, p350.

Round Kiln or Beehive Kiln. An intermittent kiln, circular in plan, with fireboxes arranged around the circumference. Such kilns find use in the firing of blue engineering bricks, pipes, some refractory bricks, etc.

Rouse-Shearer Plastometer. See R & S PLASTOMETER.

Rowlock. A form of brickwork which includes courses of HEADERS (q.v.) laid on edge.

Royal Blue. See MAZARINE BLUE.

RPS. Reinforced Porcelain System. Porcelain dental crowns are reinforced with thimble couplings with metal extensions in the form of shelves, belts or blades. *(J Prosthetic Dentistry* **50,** (4), 489, 1983)

RSBN. Reaction sintered boron nitride.

RSI. Repetitive Strain Injury (q.v.).

R & S Plastometer. A device for the assessment of the rheological properties of a clay slip in terms of the time taken for a given volume of the slip to flow through a tube of known diameter; it was designed by Rouse and Shearer Inc., a firm of ceramic consultants in Trenton, N.J., USA. (*J. Amer. Ceram. Soc.,* **15**, 622, 1932.)

RT Fusion-cast Refractory. A refractory made by a process devised by L'Electro-Réfractaire, Paris, for ensuring freedom from large cavities in fusion-cast blocks, RT = Remplissage Totale (cf. DCL, RO and SR).

RTS System. Radial Trocken Strangpresse. In this DRY BAG ISOSTATIC PRESS (q.v.) for the continuous production of ceramic tubes, a length of tube is isostatically pressed in a vertical cavity between pressure tube and mandrel. The top of the tube is widened into a short cone. The formed tube length is then extruded downwards, leaving the cone at the bottom of the mould. A second length of tube is then isostatically pressed above the first, and the cone closes over its lower end to form a continuous tube. The process may be repeated to make tubes of any desired length. (H. Ittner, Händle GmbH, Germany. *Ceram. Eng. Sci. Proc.* **7** (11/12) 1391, 1986).

Ruabon Kiln. A CHAMBER KILN (q.v.) in which the hot gases entering a chamber are deflected by a BAG-WALL (q.v.) to the roof before descending through the setting. The design originated at Ruabon, N. Wales.

Rubber. A building brick made from a sandy clay and lightly fired so that it can be readily rubbed to shape for use in gauged work. The crushing strength of such a brick is about $7 MNm^{-2}$

Rubbing Block. A shaped block of abrasive material used in the grinding of blocks of marble or other natural stones.

Rubbing Stone. A block of fine-grained abrasive, e.g. corundum, for the STONING (q.v.) of vitreous enamel.

Rubbing Up. When flatware is placed in a bung for biscuit firing, the spaces between the rims of the ware are filled with sand (if the ware is earthenware) or calcined alumina (if the ware is bone china); this is done by taking handfuls of sand, or alumina, from a heap around the bung and allowing it to fall between the rims of the ware. The process is known in N. Staffordshire as 'rubbing up'.

Ruby Glass. Glass having a characteristic red colour resulting from the presence of colloidal gold, copper or selenium. To produce GOLD RUBY a batch containing a small quantity of gold is first melted and cooled; at this stage it is colourless, but when gently heated it develops a red colour as colloidal gold is formed. For COPPER RUBY, produced in the same manner but with copper substituted for the gold, the batch must contain zinc and must be smelted in a non-oxidizing atmosphere. The most common ruby glass today is SELENIUM RUBY; a recommended batch contains 2% Se, 1% CdS, 1% As_2O_3 and 0.5% C; the furnace atmosphere must be reducing. The colour is due to MIE SCATTERING (q.v.) of light by the colloidal particles.

Ruby laser. (see also LASER) consists of single-crystal alumina with some Al replaced by Cr. The Cr electrons can be 'pumped' to higher energy levels; a beam of light making repeated passes along the crystal will trigger the Cr electrons back to their ground state and will itself gain in energy. The resultant laser beam offers a ready means of heating small samples to very high temperatures; it will drill holes in ceramics.

Ruckling. See CRAWLING.

RuL. Abbreviation for REFRACTORINESS UNDER LOAD (q.v.).

Rumbling. See SCOURING.

Run or Run-down. A fault in vitreous enamelling resulting from an excessive amount of cover-coat becoming concentrated in one area of the ware.

Runner. (1) The channel in which molten iron or slag flows from a blast furnace when it is tapped. The runner is usually lined with fireclay refractories which are then covered with a layer of refractory ramming material, e.g. a mixture of fireclay, grog and carbon. Refractory concrete has also been used to line runners.
(2) A large block of chert as used in a PAN MILL (q.v.).

Runner Brick. A fireclay shape, square in section and about 30 cm long, with a hole about 25 mm dia. along its length and terminating in a spigot at one end and a socket at the other end. A number of such refractory pieces, when placed together, form a passage through which, during the bottom-pouring of steel, the molten metal can pass from the CENTRE BRICK (q.v.) to the base of the ingot mould. Two standard sizes are specified in B.S. 2496.

Running-out Machine. Name sometimes applied to a batchtype extrusion machine of the type more commonly called a STUPID (q.v.).

Rust-spotting. A fault in vitreous enamelling: the appearance of rust-coloured spots in ground-coat enamels during drying. The fault is most frequent with ground-coats of high Na_2O:B_2O_3 ratio; it can often be cured by the addition of 0.05–0.10% of sodium nitrite to the enamel slip.

Rustic Brick. A fired clay facing brick with a rough textured surface; this effect can be obtained by stretching a wire across the top of the die so that it removes a thin slice of clay from the moving column.

Rusticating. A mechanical process for the roughening of the facing surface of a clay building brick, while it is still moist, to give it a 'rustic' appearance when fired; wire combs or brushes, or roughened rollers are used.

Rutile. The most common form of TITANIUM OXIDE (q.v.) sp. gr. 4.26. It occurs in the beach sands of Australia, Florida and elsewhere; used as an opacifying agent in enamels and glazes. Rutile is also used in the production of titania and titanate dielectrics; pure rutile has a dielectric constant of 89 perpendicular to the principal axis and 173 parallel to this axis; the value for a polycrystalline body is 85–95 at 20°C.

Rutile Break. A crystalline effect in pottery (particularly tile) glazes, obtained by the addition of rutile to a suitable basic glaze composition.

Ryolex. A volcanic aluminosilicate used as lightweight thermal insulation.

S-Brick. Bricks of S shape, into whose bends polystyrene sheet is placed as thermal insulation and to align the next course of bricks.

S-crack. Lamination, in the form of a letter 'S', in a clay column from a pug having a poorly designed mouthpiece. The crack develops from the central hole in the clay column formed at the end of the shaft in the pug, the hole being deformed to an 'S' in the rectangular mouthpiece.

S-Glass (Strength)**.** A low alkali glass composition for glass fibres, used for specialised applications in which its higher (than 'E' glass) strength and elastic modulus are important.

Sacrificial Red. Old Chinese name for the ceramic colours known as ROUGE FLAMBÉ and SANG DE BOEUF (q.v.).

Saddle. (1) An item of KILN FURNITURE (q.v.). It is a piece of refractory material in the form of a bar of triangular cross-section. See Fig. 4, p177.
(2) The upper part of a two-piece low-tension porcelain insulator of the Callender-Brown type.

Saddle Arch. See RIDER BRICKS.

Sadler Clay. An ALBANY CLAY (q.v.).

Safety. See HEALTH AND SAFETY.

Safety Glass. See LAMINATED GLASS; TOUGHENED GLASS; WIRED GLASS. B.S. 6206 is an impact resistance test for flat safety glass. B.S. AU178 and B.S. 857 specify safety glasses for road and railway vehicles.

Saffil. Trade-name. Alumina, and sometimes zirconia, fibres manufactured by ICI plc.

Saggar. A fireclay box, usually oval (0.6 × 0.45 m), in which pottery-ware can be set in a kiln. The object is to protect the ware from contamination by the kiln gases and the name is generally thought to be a corruption of the word 'safe-guard'. Since the bottle-oven has become obsolete as a kiln for the firing of pottery and since clean fuels and electricity have become increasingly used in the industry, the use of saggars has been largely displaced by the setting of ware on lighter pieces of KILN FURNITURE (q.v.) though special saggars are still used in very specialised applications in which protection from kiln atmosphere is vital (eg some electronic ceramics).

Saggar-maker's Bottom-knocker. The man whose job it was to beat out, by means of a heavy wooden tool, a wad of grogged fireclay to form a bottom for a SAGGAR (q.v.). To the regret of folklorists, the species is now virtually extinct.

Sagging. (1) The permanent distortion by downward bending of vitreous enamelware that is inadequately supported during firing
(2) The flow of enamel on vertical surfaces of vitreous enamelware while it is being fired; the fault is visible as roughly horizontal lines or waves and is caused by the enamel being too fluid, by overfiring, or by the enamel layer being too thick.
(3) A method of shaping glass-ware by heating the glass above a mould and allowing it to sag into the contours of the mould.
(4) The dropping away from the casing of a ring or series of rings in a rotary kiln lining.

Salamander. The mass of iron which, as a result of wear of the refractory brickwork or blocks in the hearth bottom of a blast furnace, slowly replaces much of the refractory materials in this location.

Salmanazar. A 12-quart wine bottle.

Salmon. Salmon Brick. US term for an underfired building brick (in allusion to its colour). See CHUFF BRICK.

Salt Cake. Commercial grade sodium sulphate, Na_2SO_4; it is added to glass batches, together with carbon, to prevent scumming – particularly in flat-glass manufacture; it is also added to facilitate melting and refining but to a less extent than formerly.

Salt Expansion. Irreversible expansion that may occur in a porous ceramic product containing SOLUBLE SALTS (q.v) if it becomes moist and is then dried; it is generally believed that the expansion is associated with crystallization of the salts.

Salt Glaze. In the 17th century and 18th century salt-glazing was used in the manufacture of domestic pottery. Now, except for its use by some studio potters, the process is obsolete. Before its demise in the face of environmental clean air restrictions, it was last used in the production of salt-glazed sewer-pipes. The pipes (made from a siliceous clay to ensure a good glaze) were fired in the usual way up to 1100–1200C and salt then thrown on the fire, where it volatilizes; the salt vapour passes through the setting of pipes and reacts with the clay to form a sodium aluminosilicate glaze. Borax was sometimes added with the salt in the ratio of about 1:4. (See also SEWER PIPE.)

Saltpetre. See SODIUM NITRATE.

Samel. An underfired brick from the outer part of a CLAMP (q.v.).

Samarium Oxide. Sm_2O_3; m.p. approx. 2300°C; sp. gr. 7.4; thermal expansion (20–1000°C) 10×10^{-6}.

Samian Ware. See TERRA SIGILLATA.

Sampling. The choice of procedure for the obtaining of a representative sample of a consignment of ceramic raw material or product prior to testing, is of great importance. In USA a procedure for the sampling of clays is laid down in ASTM – C322; in the UK, B.S. 812 deals with the sampling of mineral aggregates and sands, and B.S. 3406 (Pt. 1) deals with the sampling of powders for particle-size analysis. B.S. 6065 specifies sampling and acceptance testing of shaped refractories and B.S. 6434 is a guide to its use. Details of the number of bricks that should be taken as a sample for testing are given in B.S. 1902 (Pts. 3,4 and 5) (Refractories) and B.S. 3921 (Building Bricks). B.S. 1902 Pt.7.1 gives guidance on taking samples from unshaped refractories. In a slightly different sense, ASTM C1190 describes how test samples should be cut from tar-bonded and burned basic refractory bricks. In his book *'Particle Size Measurement.'* (3rd Ed 1981) T. Allen gave a 'golden rule' for powder sampling for size distribution measurements that a sample should be taken from a moving stream, and all of the stream caught for a short time.

Sand. Typically, sand consists of discrete particles of quartz: some beach sands contain a concentration of heavy minerals, e.g. rutile, ilmenite and zircon. Silica sand occurs widely and is used as a major constituent of glass, as a refractory material for furnace hearths and foundry moulds, and for mixing with portland cement to make cement mortar and concrete. The purest sands are used for glassmaking; B.S. 2975 deals with their sampling and analysis. For specifications for building sands see B.S. 812; B.S. 1199 and 1200; B.S. 4550 Pt. 5 and Pt. 6.

Sand-creased. A texture produced on the surface of clay facing bricks. In the hand-made process the clot of c!ay is first sprinkled with, or rolled in, sand; where the sand keeps the clay away from the mould a crease is produced. An imitation of this texture can be achieved in machine-made bricks by impressing the surface of the green bricks with a tooled steel plate.

Sand-faced. A clay building brick that has been blasted with sand while in the GREEN (q.v.) state; alternatively. with extruded bricks, the sand-facing can be applied by gravity to the column of clay. A wide range of surface textures can be obtained by sandfacing.

Sand-gritting. The process of roughening those parts of the glaze of an electrical porcelain insulator where cement will be applied in the final assembly. The 'sand' (a specially-prepared mixture of coarse-graded porcelain together with powdered glaze) is applied to areas of the glazed insulator before it is fired.

Sand Holes. Small fractures in glass surfaces produced during rough grinding and not completely removed by polishing.

Sand-lime Brick. See CALCIUM SILICATE BRICK.

Sand Rammer. The standard rammer for making test-pieces allows a 14 lb (6.25 kg) weight to fall 50 mm on sand contained in a 50 mm diam. mould; three blows from the rammer are used in making a standard test-piece. Its use in compaction and workability tests for unshaped refractories is specified in B.S. 1902 Pt. 7.2.

Sand resin moulds. Moulds for pressure-cast sanitaryware have been made from cured organic resins with a graded sand filler. The material has controlled porosity and is stronger and more abrasion resistant than plaster.

Sand Seal. A device for preventing hot gases from reaching the undercarriage of the cars in a tunnel kiln. It consists of two troughs running parallel to the internal walls of the kiln and filled with sand; into this sand dip vertical metal plates (APRONS) which are fixed to the sides of each kiln car.

Sand Slinger. A rotating head which projects refractory powder on to the inner wall of a circular ladle or furnace.

Sanding. (1) The sprinkling of sand or finely crushed brick between courses of bricks as they are being set in a kiln; the object is to make even any irregularity in level and to prevent sticking.
(2) The treatment with sand, during manufacture, of the surface of facing brick to give it a pleasing texture and colour.
(3) The BEDDING (q.v.) of earthenware in sand.
Sandstock Bricks. Made in Australia since 1788, these are handmade bricks with a distinctive appearance due to the use of sand as the mould release agent. This appearance is simulated in modern bricks by replacing the sand by coloured brick frit or flake.
Sandstone. A sedimentary rock which may be of several types, e.g. siliceous, calcareous, argillaceous, depending on whether the grains of silica in the rock are bonded with secondary silica, with lime or with clay. Siliceous and argillaceous sandstones both find some use in the refractories industry; crushed sandstone is used as a source of silica in glass manufacture.
Sandstorm Kiln. A system of coal-firing for kilns for earthenware tiles, in which an alternating arrangement of grates and secondary air flues was claimed to provide a more even distribution of hot gases over the fired ware. (G.H.Downing and H.R.Holding Br. Pat 391 175, 1933 and 464 128, 1937).
Sandstruck Brick. A brick (hand) moulded at 20 to 30% moisture, in a mould coated with sand to prevent sticking.
Sandwich Kiln. A tunnel kiln designed for rapid firing; the height of the setting is small compared with the width and the firing is from above and below the setting.
Sang de boeuf. A blood-red in-glaze effect (cf. ROUGE FLAMBÉ) achieved by the Chinese potters by adding a very small amount of copper oxide to the glaze batch and firing under reducing conditions to form colloidal copper; the glaze also usually contained some tin oxide.
Sangrind Pantograph. A device for the surface grinding of sanitaryware. (W.S. Industrial Designs Ltd.).
Sanitary Earthenware. A type of sanitaryware made from white-firing clays but often covered with a coloured glaze; the body itself has a water absorption of 6–8% (cf. VITREOUS-CHINA SANITARYWARE). The body is made from a batch containing 22–24% ball clay, 24–26% china clay, 15–18% china stone and 33–35% flint; it may be once- or twice fired, the firing temperature being about 1120°C.
Sanitary Fireclay. A type of sanitaryware made from a grogged fireclay body, which is covered with a white ENGOBE (q.v.), which in turn is covered with a glaze. A typical body composition is 60–80% fireclay, 20–40% grog. The engobe contains 5–15% ball clay, 30–50% china clay, 15–30% flint, 20–35% china stone, 0–10% feldspar; the proportions of china stone and feldspar vary inversely as one another.
Sanitaryware. The various types include: SANITARY EARTHENWARE (q.v.); VITREOUS CHINA SANITARYWARE (q.v.); SANITARY FIRECLAY (q.v.); VITREOUS-ENAMEL SANITARYWARE, e.g. baths. B.S. 1125 and B.S. 1188 specify W.C. flushing cisterns and ceramic wash basins and pedestals respectively. B.S. 5503 and B.S. 5504 specify materials, dimensions and quality of vitreous china washdown W.C.'s and wall hung W.C.'s respectively. B.S. 5627 specifies plastics connectors for vitreous china W.C. pans. B.S. 5505 is a specification for bidets; B.S. 5506 for pedestal and wall hung wash basins; B.S. 5520 for vitreous china bowl urinals;

B.S. 6340 specifies shower trays (Pt. 6 – vitreous enamelled cast iron; Pt. 7 vitreous enamelled sheet steel; Pt. 8 glazed ceramic). B.S. 6465 specifies the required scale of sanitary ware installation. B.S. 3402 specifies the quality of vitreous china sanitary ware. B.S. 8000 Pt. 13 specifies site workmanship.

Santorin Earth. A POZZOLANA (q.v.) from the Grecian island, Santorin. A quoted composition is (per cent): SiO_2, 64; Al_2O_3, 13; Fe_2O_3, 5.5; TiO_2, 1; CaO, 3.5; MgO, 2; alkalis, 6.5; loss on ignition, 4.

Saphiber. 99% alumina fibres, a trade-name of Alcoa, USA.

Saphikon. Trade-name. Single-crystal sapphire filaments. Tyco Laboratories, Waltham, Mass. USA.

Sapphire. Single-crystal alumina; sapphire boules can be made by the VERNEUIL PROCESS (q.v.) and find use as bearings and thread guides.

Sardamag. Trade-name. Sardinian seawater magnesia.

Sarking. A layer of insulating material under the main roof covering. In Scotland the term has the special meaning of sawn roof boarding usually 15 mm thick (cf. TORCHING).

Satin Glaze, Satin-vellum Glaze or Vellum Glaze. A semi-matt glaze, particularly for wall tiles, with a characteristic satin appearance. Such glazes are generally of the tin-zinc-titanium type.

Satin Gold. Bright gold applied over a finely etched or satin finish decoration on glassware.

Saturation Coefficient. The ratio of the amount of water taken up by a porous building material after it has been immersed in cold water for an arbitrary period (e.g. 24 h) to its water absorption, normally determined by boiling in water for 5 h; the ratio is usually expressed as a decimal fraction. e.g. 0.85. A low saturation coefficient generally indicates good frost resistance but the correlation is far from perfect.

Saucer Wheel. An abrasive wheel shaped like a saucer.

Saxon Slabs. A less common name for NORMAN SLABS (q.v.).

Saybolt Cup. An orifice type viscometer. See FLOW CUP.

SBPM. Sonon-based particle model, in which ultrasonic properties are modelled by particle interactions, and related to sintering and processing in space under zero gravity.

Scab. (1) A fault in the base-metal for vitreous enamelling; the 'scab' is a partially detached piece of metal (which may subsequently have been rolled into the metal surface) and is liable to cause faults in the applied enamel coating. (2) A fault in glass caused by an undissolved inclusion of sodium sulphate; also known as SULPHATE SCAB and as WHITEWASH.

SCACB. Strengthening by Chemical Activation of Chemical Bonds. Unfired compacts of fused silica and sand were strengthened by treatment at room temperature and above (including hydrothermal treatment) with sodium silicate solution.

Scalding. A term that has been used to describe the fault in the glost firing of pottery when glaze falls off the ware before it has fused; a cause is too great a difference between dimensional changes of body and applied glaze.

Scale. A fault, in glass or vitreous enamel-ware, in the form of an embedded particle of metal oxide or carbon.

Scaling. Preparation of the base-metal for vitreous enamelling by heating it (often in an atmosphere of acid fumes or of sulphur gases) to form a surface coating of oxide – 'scale' – which is subsequently removed by PICKLING (q.v.).

Scallop. (1) The rims and edges of pottery-ware are sometimes trimmed to give a scalloped effect, i.e. small segments are symmetrically removed from the edges before the ware is fired. Until 1955, when a scalloping machine was introduced, the process had always been carried out by hand.
(2) Term sometimes used in the Staffordshire potteries for a dish.
Scandium Oxide. Sc_2O_3; m.p. >2400°C; sp. gr. 3.86.
Scanning Acoustic Microscopy. The use of high frequency (100–200 MHz) ultrasound to detect surface defects by generating surface waves on the sample with a wide-angle acoustic lens.
Scanning Electron Microscopy. See ELECTRON MICROSCOPY.
Scanning Laser Acoustic Microscopy. SLAM. A transducer beneath the sample produces a collimated ultrasonic beam which is altered by voids, cracks or inclusions in its path. The resultant upper surface displacement pattern of ripples is scanned by a laser beam and converted into a magnified visual image.
SCARE. See CARES.
Scarp. See INTERSECTION SCARP, TRANSITION SCARP.
Scatter Coefficient. See COEFFICIENT OF SCATTER.
SCC Test. The test for the ABRASION RESISTANCE (q.v.) of PAVERS (q.v.) used by Sydney City Council. It involves ball bearings both rotating on and impacting the surface.
Scheidhauer and Giessing Process. See S.u.G. PROCESS.
Schellbach Tubing. A type of ENAMEL-BACK TUBING (q.v.) with a central blue line.
Schenck Porosimeter. Apparatus for the determination of PORE SIZE DISTRIBUTION (q.v.) by the MERCURY PENETRATION METHOD (q.v.); it has been applied to the study of refractories. (H. Schenck and J. Cloth, *Arch Eisenhüttenw.* **27**, 421 1956.)
Schlenkermann's Stone. A German FIRESTONE (q.v.); it contains about 90% SiO_2.
Schmidt Hammer. A device for the non-destructive testing of set concrete; it is based on the principle that the rebound of a steel hammer, after impact against the concrete, is proportional to the compressive strength of the concrete. (E. Schmidt. *Schweiz. Bauzeitung. 15/7/50).* It is also used to test for the adherence of tiles to their substrates. Adherence markedly affects the measured coefficient of restitution, and the impact resistance of tiles.
Schnitzler's Gold Purple. A tin-gold colour, produced by a wet method: it has been used for the decoration of porcelain.
Schöne's Apparatus. An elutriator consisting of a tall glass vessel tapering towards the bottom, where water enters at a constant rate. Schöne's formula is: $V = 104.7(S - 1)^{1.57}D^{1.57}$ where V is the velocity of water (mm/s) required to carry away particles of diameter D and sp. gr. S. (E. Schöne, *Uber Schlämmanalyse und einen neuen Schlämmapparat,* Berlin, 1867.)
Schottky Defect. See CRYSTAL STRUCTURE.
Schugi Mixer. A continuous mixer for powders or for powders and liquids (Schurmans & Van Ginneken).
Schuhmann Equation. An equation for the particle-size distribution resulting from a crushing process: $y = 100(x/K)^a$, where y is the cumulative percentage finer than x, a is the distribution modulus, and K is the size modulus: a and K are both constants. (R. Schuhmann, *Amer. Inst. Min. Engrs., Tech. Paper,* 1189, 1940) cf. GAUDIN'S EQUATION).

Schuller Process. See UP-DRAW PROCESS.

Schulze Elutriator. The original type of water elutriator, it has since been improved by H. Harkort *(Ber. Deut. Keram. Ges.,* **8,** 6, 1927).

Schurecht Ratio. A term that has been used for SATURATION COEFFICIENT (q.v.); named from H. G. Schurecht (USA) who introduced this coefficient in his research on frost-resistance of terra cotta carried out at the National Bureau of Standards in 1926 but never published; the term 'Schurecht Ratio' was first applied by J. W. McBurney.

Scleroscope. An instrument for determining the relative hardness of materials by a rebound method.

Scone. See SPLIT.

Scotch Block. One form of gas port in an open-hearth steel furnace, the distinguishing feature is that it is monolithic, being made by ramming suitably graded refactory material around a metal template.

Scotch Kiln. See SCOVE.

Scouring. The cleaning and smoothing of biscuit-fired ceramic ware by placing the ware in a revolving drum together with coarse abrasive material, e.g. PITCHERS (q.v.). This process is used in the making of bone china and of some other types of vitreous tableware (in USA); when applied in the manufacture of electrical porcelain, the process is known (in UK) as RUMBLING.

Scouring Block. An abrasive block of ceramically bonded SiC or Al_2O_3 used in the grinding of steel rolls. Such a block is 125–150 mm long and 50–75 mm 'square' (the actual section may be trapezoidal, roughly semicircular or of other special shape).

Scove or Scotch Kiln. An early type of up-draught intermittent kiln for the firing of bricks, etc. It was rectangular and consisted of side-walls and end-walls only, with fire holes in each side-wall and WICKETS (q.v.) in each end wall. The top of the setting was covered with a PLATTING consisting of a layer of fired bricks, with ashes or earth above.

SCR Brick. A perforated clay building brick introduced in 1952 by the Structural Clay Products Research Institute, USA, whence its name. The brick is 11½ × 5½ × 2¹⁄₁₆th in., with 10 holes 1⅜ths in. dia. and a ¾ in. square jamb-slot in one end. The weight is 8 lb: (292 × 140 × 52 mm, with 10 holes 34 mm dia, and a 19 mm slot. The weight is 3.6 kg)

Scrapings. OVERSPRAY (q.v.) removed from the sides and bottom of a spray-booth in vitreous enamelling; it is re-milled and used as a first coat.

Scrapping. The removal of excess body from a shaped piece of pottery-ware while the latter is still in or on the mould; (cf. FETTLING).

Scraps. Excess body removed during the shaping of potteryware, together with any broken, unfired, pieces. Scraps are usually returned to the blunger for re-use.

Screed. A layer of mortar or other material applied to a sub-floor and brought to a defined level. (B.S. 6100 Pt.6). A fine-grained mix is used to provide a level surface for TILE FIXING (q.v.).

Screeding Board. See TRAVERSING RULE.

Screen. An industrial-scale unit for size classification (a SIEVE (q.v.) is for laboratory use). A screen consists of a wire mesh or a perforated metal plate: the usual types are REVOLVING SCREENS (q.v.) and VIBRATING SCREENS (q.v.).

Screen Printing. A decorating method that can be applied to pottery, glassware or vitreous enamelware. The simplest silkscreen equipment consists of a frame over which is stretched silk bolting cloth, or fine wire gauze, of 50–60 meshes per cm. A stencil is then placed on the frame and varnish is applied to fill in those parts of the screen not covered by the design. Colour, dispersed in a suitable oil, is pressed through the open parts of the screen by means of a roller and the pattern of the stencil thus reproduced on the ware. The stencil can be made photographically, and the whole process mechanised. A further development was the screen printed collodion film transfer, which has itself been improved to give screen-printed cover-coat transfers. Screen-printing is particularly applied to the decoration of wall-tiles.

Screw Press. See FRICTION PRESS.

Scrub Marks. A surface blemish on glass bottles; see BRUSH MARKS.

Scuffing. The dull mark that sometimes results from abrasion of a glazed ceramic surface or of glass-ware.

Scull. See SKULL.

Scum; Scumming. (1) A surface deposit sometimes formed on clay building bricks. The deposit may be of soluble salts present in the clay and carried to the surface of the bricks by the water as it escapes during drying: it is then known as DRYER SCUM. The deposit may also be formed during kiln firing, either from soluble salts in the clay or by reaction between the sulphur gases in the kiln atmosphere and minerals in the clay bricks; it is then known as KILN SCUM (cf. EFFLORESCENCE).
(2) Undissolved batch constituents floating as a layer above the molten glass in a pot or tank furnace.
(3) Areas of poor gloss on a vitreous enamel: the fault may be due to the action of furnace gases, to a non-uniform firing temperature, or to a film of clay arising from faulty enamel suspension.

Scurf. A hard deposit, mainly of crystalline carbon, formed by the thermal cracking of crude coal gas on the refractory walls of gas retorts and coke ovens. When the scurf is periodically removed, by burning off, there is a danger that the refractory brickwork of the retort or oven may be damaged by overheating.

Scutch. US term for a hand tool for trimming bricks.

Sea-water Magnesia. Sea-water contains approx. 0.14% Mg. It can be extracted by treatment with slaked lime or with lightly calcined dolomite:

$MgCl_2 + Ca(OH)_2 \rightarrow Mg(OH)_2 + CaCl_2$

$MgSO_4 + Ca(OH)_2 \rightarrow Mg(OH)_2 + CaSO_4$

When calcined dolomite is used as precipitant, the yield is almost doubled because the MgO in the calcined dolomite is also largely recovered. The precipitated $Mg(OH)_2$ is settled in tanks, filtered. and then calcined or dead-burned to produce MgO. The first small-scale plant was put into operation in California in 1935. The first large-scale production was at West Hartlepool, England, in 1938; most of the magnesia needed for the production of basic refractories in Britain has since been derived from the sea; typical analysis %: SiO_2, 2; Al_2O_3 0.5; Fe_2O_3, 1.5; CaO, 2; MgO, 93.

Séailles Process. For the simultaneous production of alumina and portland cement from siliceous bauxite, or from an aluminous slag or clay. An appropriately proportioned batch is fired in a rotary kiln to give $5CaO.\,3Al_2O_3$ and $2CaO.SiO_2$; leaching of this product yields alumina and reburning of the residue with more lime yields a cement clinker. The process was introduced in 1928 by J. S. Séailles who later worked in collaboration with W. R. G. Dyckerhoff (whence the alternative name Séailles-

Dyckerhoff Process), see their joint British Patent 545 149, 2/8/38.

Seal. See CERAMIC TO METAL SEAL; GLASS-TO-METAL SEAL; METALLIZING; SEALING GLASS.

Sealed Porosity. See under POROSITY.

Sealing Glass. A glass that is suitable for use in sealing a glass envelope of an electronic valve, for example, to metal. The usual basis for the selection of such a glass is matching its thermal expansion and contraction with that of the metal over the range of temperature from that at which the seal is made to room temperature; however, a glass that is sufficiently soft (e.g. a lead glass) can accommodate considerable stress at a glass-metal seal by slowly yielding. Some sealing glasses are crystallized during processing to form glass-ceramics with improved properties. ASTM F79 specifies sealing glasses.

Seam. See JOINT LINE.

Seat. A place prepared on the SIEGE (q.v.) of a glass-pot furnace for the support of a pot.

Seat Earth or Underclay. The material immediately beneath a coal seam, normally a fireclay but in some cases a silica rock (GANISTER, q.v.).

Seating Block. A block of fireclay refractory, shaped to support a boiler.

Secar. Trade-name: a pure calcium aluminate cement, suitable for use in making special refractory castables or shapes, made by Lafarge Aluminous Cement Co. Ltd. It is available in two grades; the purer (Secar 250) contains 70–72% Al_2O_3, 26–29% CaO, 1% SiO_2 and 1% Fe_2O_3.

Secondary Air. Air that is admitted to a kiln (that is being fired under oxidizing conditions) in an amount and in a suitable location to complete the combustion of the fuel initiated by the PRIMARY AIR (q.v.). In a kiln fired by solid fuel on a grate, the secondary air passes over the fuel bed and burns the combustible gases arising therefrom.

Secondary Clay. See SEDIMENTARY CLAY.

Secondary Crusher. A machine for the size-reduction of a feed up to 25 mm to a product passing about 8 mesh. This group of machines includes the finer types of jaw crusher and gyratory crusher, and also crushing rolls, hammer mills and edge-runner mills.

Secondary Steelmaking. The treatment of steel in the ladle (e.g. by vacuum degassing) to improve cleaness, composition control and steelplant productivity. See Appendix B.1.

Seconds. Pottery-ware with small, not readily noticeable, blemishes (cf. FIRSTS, LUMP).

Sedigraph. Trade-name. A particle size analyzer based on the different SEDIMENTATION (q.v.) rates of different sizes of particles, which provides an automatic graphical printout of the PARTICLE SIZE DISTRIBUTION.

Sedimentary Clay or Secondary Clay. A clay that has been geologically transported from the site of its formation and redeposited elsewhere. The English ball clays, for example, are secondary kaolins (cf. PRIMARY CLAYS).

Sedimentation. The settling out of solid particles from a liquid. Some deposits of clay, sand etc, have been formed by sedimentation from lakes, estuaries and the sea. Sedimentation is used as a method of purifying clays, heavy impurities (e.g. iron compounds) settling out more quickly than the lighter particles of clay. Sedimentation is also a method for determining the particle size of clays and powders (cf. ELUTRIATION).

Sedimentation Volume. The volume occupied by solid particles after they have settled from suspension in a liquid. With most clays, the sedimentation volume depends on the degree of flocculation or deflocculation. Determination of the sedimentation volume of brick clays provides some indication of the fineness of the clay, its working moisture content, and drying shrinkage.

Seed. A fault, in the form of small bubbles, in glass. When near the surface of plate glass they sometimes become exposed, as minute depressions, during the polishing process; they are then known as BROKEN SEED.

Seekers. Mechanisms for the location of kiln cars.

Seger Cone. The first series of PYROMETRIC CONES (q.v.) was that of H. A. Seger *(Tonind. Zeit.,* **10**, 135, 168, 229, 1886); the standard cones of Germany are still known by his name. For table of equivalent temperatures see Appendix 2.

Seger Formula. A procedure introduced by H. A. Seger *(Collected Writings,* Vol. 2. p. 557), and still commonly used, for the representation of the composition of a ceramic glaze. The chemical composition is recalculated to molecular fractions and the constituent oxides are then arranged in three groups: the bases, which are made equal to unity; R_2O_3; RO_2. Example: (0.3 Na_2O. 0.2 CaO. 0.5 PbO). 0.2 Al_2O_3. (3.0 SiO_2.0.7 SnO_2).

Seger's Green. A ceramic colour that can be fired up to about 1050°C and that consists of (%): $K_2Cr_2O_7$, 36; $CaCl_2$, 12; whiting, 20; CaF_2, 10; flint, 20.

Seger's Porcelain. A German porcelain introduced in 1886 by H.A. Seger: 30% feldspar, 35–40% quartz, 30–35% kaolin. It is covered with a glaze prepared from 83.5 parts feldspar, 26 parts kaolin, 35 parts whiting, 54 parts flint. It is biscuit fired at a low temperature and glost fired at Cone 9.

Seger's Rules. A series of empirical rules put forward by the German ceramist H A. Seger *(Collected Writings,* Vol. 2. p. 577) for the prevention of crazing and peeling. To prevent crazing, the body should be adjusted as follows: decrease the clay, increase the free silica (e.g. flint; replace some of the plastic clay by kaolin; decrease the feldspar; grind the flint more finely; biscuit fire at higher temperature. Alternatively, the glaze can be adjusted: increase silica and/or decrease fluxes; replace some SiO_2 by B_2O_3; replace fluxes of high equivalent weight by fluxes of lower equivalent weight. To prevent peeling, the body or glaze should be adjusted in the reverse direction.

Seger's Solution. A solution for testing the acid resistance of ceramics (Portuguese Stand. P.144, 1956) 262 ml of sulphuric acid (s.g. 1.84) and 108 ml of nitric acid (s.g. 1.42) are added to 1 litre of distilled water.

Seggars. A N.E. England dialect term for mixed fireclay and coal, used as a blast-furnace tap hole clay.

Segmental Arch. A sprung arch having the contour of a segment of a circle.

Segmental Tile. See RIDGE TILE.

Segmental Wheel. An abrasive wheel that has been built up from specially made segments of bonded abrasive; wheels up to 1.5 m dia. can be made in this way.

Segregation. Partial re-separation of a previously mixed batch of material into its constituents; this can occur either as a result of differences in particle size or in density. Segregation is liable to occur in storage bins, on conveyors and in feeders during the dry or semi-dry processing of ceramic materials.

Seignette-electric. Seignette salt is the alternative name for Rochelle salt (Na-K tartrate). Crystals of this composition are markedly piezo-electric and were used, for this property, before titanate ceramics were introduced. The term 'seignette-electric' is still sometimes used in Western Europe and Russia to signify ferroelectric.

Selenium. Compounds of Se are used for decolourising glass in the tank-melting process; under other conditions this element will produce red coloured glass (see RUBY GLASS), pottery, or vitreous enamel.

Self-bonded Silicon Carbide. See REACTION BONDED SILICON CARBIDE.

Self-cleaning Enamel. Vitreous enamels which contain components designed to promote oxidation of grease and oven spills during use.

Self-glazing Tiles. Russian pressed floor tiles, based on volcanic glasses, cullet and ceramic scrap, are dried by radiation, which causes alkalis to migrate to the upper surface as the water evaporates. There they act as a flux during firing, increasing vitrification and forming a glazed surface.

Self-propagating High Temperature Synthesis. Self-sustaining highly exothermic solid state reactions result in a combustion wave which converts the reactants to the desired product as it propagates through a compacted mixture of the reactant powders.

Self-slip. A term introduced by archeologists to describe the fine surface layer on pottery vessels, produced by wet-surface smoothing.

Selfstak. A system of handling bricks for delivery to sites. The bricks are stacked in cubes, on pallets or shrink-wrapped, about 1m side. The cubes are lifted by a small crane mounted on the delivery lorry. The system originated in Canada, but was developed by the London Brick Co., whose fletton bricks are delivered nationwide.

Self-reinforced Ceramic. A ceramic reinforced by in-situ growth of whisker-like elongated grains of the same material as the bulk ceramic.

Selvedge. The formed edge of a ribbon of rolled glass.

SEM. Scanning Electron Microscopy.

Semi-conducting Glaze. Porcelain insulators that are covered with a normal glaze are liable, particularly if the surface gets dirty, to surface discharges which cause radio interference and may lead to complete flashover and interruption of the power supply. This can be largely prevented if the glaze is semi-conducting as a result of the incorporation of metal oxides such as Fe_2O_3, Fe_3O_4, MnO_2, Cr_2O_3, Co_3O_4, CuO or TiO_2; SiC and C have also been used as semi-conducting constituents.

Semi-conductor. A material with modest ELECTRICAL CONDUCTIVITY (q.v.) which increases with temperature. Semi-conducting ceramics include silicon carbide (used in heating elements), other carbides, oxides of variable valency metals (e.g. ZnO) and specialized electronic ceramics such as GaAs. (Most ceramics become semi-conducting at high temperatures.)

Semi-continuous Kiln. A TRANSVERSE-ARCH KILN (q.v.) having only a single line of chambers, so that when the firing zone reaches one end of the kiln the process of fire travel must be re-started at the other end. Kilns of this type were never common.

Semi-dry Mortar. MORTAR (q.v.) which is mixed using just sufficient water to make a mix which coheres in the hand (much less than in a conventional mix). This leads to lower shrinkage, an advantage in TILE-FIXING. (q.v.).

Semi-dry Pressing. See DRY PRESSING.

Semi-muffle. See under MUFFLE.

Semi-porcelain. A trade term sometimes applied, both in thc UK and the USA, to semi-vitreous tableware. The term is confusing but is nevertheless included in the COMBINED NOMENCLATURE (q.v.) and defined as ware 'prepared, decorated and glazed to give the commercial appearance of porcelain. Without being really opaque like earthenware, or truly translucent like porcelain, these products may be slightly translucent in the thinner parts such as thc bottoms of cups. These materials can, however, be distinguished from real porcelain because their fracture is rough-grained, dull and non-vitrified. They are therefore porous beneath the glaze and the fracture clings to the tongue. They are easily scratched with a steel knife, and are not considered as porcelain or china.'

Semi-silica Refractory; Semi-siliceous Refractory. See SILICEOUS REFRACTORY; note, however, that ASTM-C27 defines a semi-silica refractory as containing 72% SiO_2. A tentative ISO recommendation defines a semi-siliceous refractory as containing 85–93% SiO_2.

Semi-universal Bricking. See SPIRAL BRICKING (q.v.)

Semi-vitreous or Semi-vitrified. These synonymous terms are defined in the USA (ASTM – C242) as signifying a ceramic whiteware having a water absorption between 0.5% and 10%, except for floor tiles and wall tiles which are deemed to be semi-vitreous when the water absorption is between 3% and 7%.

S.E.N. Submerged Entry Nozzle.

S.E.N.B. Single Edge Notched Beam. See FRACTURE TOUGHNESS TESTS.

Sensitizing Compound. A substance which, in solution, can be applied to a ceramic surface to facilitate silvering or plating. Salts of Al, Ti, Fe or certain other metals can be used.

Sentinel Pyrometers. Small cylinders made from blended chemical compounds (non-ceramic) and so proportioned that they melt at stated temperatures within the range 220–1050°C, used principally in controlling the heat treatment of metals.

S.E.P.B. Single Edge Precracked Beam. See FRACTURE TOUGHNESS TESTS.

Sepiolite. $3MgO.4SiO_2.5H_2O$; the magnesian end-member of the series of clay minerals known as PALYGORSKITES (q.v.); it has been found in some of the Keuper Marls used for brickmaking in Central England.

Sepulchre Kiln. A shaft kiln used for the calcination of dolomite and other materials. (F. Sepulchre, Soc. Anon. de Marche-les-Dames, Namur, Belgium.)

Serpentine. A naturally occurring hydrated magnesium silicate, $3MgO.2SiO_2.2H_2O$; some of the Mg is commonly replaced by Fe. Serpentine is a common gangue mineral associated with chrome ore. Some types of serpentine are used as raw materials (together with dead-burned magnesite) for the manufacture of forsterite refractories.

Serrated Saddle. A refractory support for pottery-ware during kiln firing; this particular item is a rod of triangular section, its upper edge being serrated to facilitate the REARING (q.v.) of the ware.

S.E.S. Submerged Entry Shroud.

SESCI Furnace. A rotary furnace, designed to burn low-volatile coal, and used for the melting of cast iron. A rammed siliceous refractory is generally used as lining material. The name derives from the initials of the French makers: Sociéte des Entreprises Speciales de Chauffage Industriel.

Sessile Drop Test. A method for the measurement of SURFACE TENSION γ. If the drop, resting on a horizontal surface, is so large that the curvature at

the top is negligible $\gamma = \frac{1}{2}h^2gd$ where h is the height of the drop and d its density. If not, $\gamma = \frac{1}{2}h^2gd(1 + 0.609h/r)$ where r is the drop radius and h is the distance from the top of the drop to its greatest horizontal diameter.
Sessile drops can be measured at high temperature using a travelling telescope through a furnace window. ASTM C813 specifies a test for the 'Hydrophobic Contamination of Glass by Contact Angle Measurement'. The angle of contact is measured by a comparator microscope, or with a protractor on a photograph of the drop.

Set. The general flow behaviour of vitreous enamel slip determining the rate of draining, residual thickness, and uniformity of coating.

Setter. (1) A man who sets bricks or tiles in a kiln; (the man who sets pottery-ware is known as a PLACER).
(2) An item of KILN FURNITURE (q.v.); it is a piece of fired refractory material shaped so that its upper surface conforms to the lower surface of the piece of ware that it is designed to support during kiln firing.

Setting. (1) The arrangement of ware in a kiln. With pottery-ware, the setting generally consists of the individual pieces together with the kiln furniture that supports them; a setting of building bricks or refractories consists merely of the bricks themselves.
(2) A group of gas retorts or chambers within a bench, the group being heated independently of other groups of retorts in the same bench; also known as a BED.
(3) The process of hardening of a cement (see INITIAL SET and FINAL SET) or of a plaster.

Setting Density. The fired weight per unit volume of ceramic ware set in a kiln for the firing process; it is usually expressed in kgm^{-3} (or lb/ft^3. In British Standard Test Codes for ceramic kilns (BS 1081 and 1388, now withdrawn) two setting densities were defined: the *Overall Setting Density* is the fired weight divided by the overall setting volume; the *Actual Setting Density* is the fired weight divided by the actual setting volume. The *Overall Setting Density* indicates the effectiveness of filling the available kiln volume; the *Actual Setting Density* excludes the spaces in the kiln that are not utilized and thus gives a better indication of the closeness of packing of the ware within the setting itself.

Setting Rate. The time taken for a glass to cool over its WORKING RANGE (q.v.).

Setting-up Agent. An electrolyte, e.g. K_2CO_3, $MgSO_4$, etc., added to a vitreous enamel slip or to slop glaze to flocculate the clay particles and thus to hold the coarser and heavier frit particles in suspension.

Settle Blow. The stage in the BLOW-AND-BLOW (q.v.) glass-making process when glass is forced into a finish or ring mould by air pressure.

Settle Mark. Any slight variation in the wall thickness of a glass container; sometimes called a SETTLE WAVE.

Settlement. Movement of a kiln setting during firing, due to dimensional changes.

7.6 Temperature. See SOFTENING POINT.

Sewer Brick. A specification for clay bricks intended for this use is given in ASTM–C32.

Sewer Pipe. An impervious clay pipe (usually glazed) for sewerage and trade-effluent disposal. Relevant British Standards are: B.S. 65, and B.S. 8000 Pt. 14 (a code of practice for below-ground drainage) Relevant US specifications are: ASTM-C12, C-301, C-425, C700, C-828, C-1091, C-1208 which specify installing clay pipe lines, the pipes and joints and air tests on the lines. See also CONCRETE PIPES.

Sgraffito. A mode of decoration sometimes used by the studio potter: a coloured ENGOBE (q.v.) is applied to the dried ware and a pattern is then formed by scratching through the engobe to expose the differently coloured body. (From Italian *graffito*, to scratch.)

Shadow Wall. A refractory wall in a glass-tank furnace erected on the bridge cover; it may be solid or may have openings, its purpose being to screen the working end from excessive heat radiation. It is also sometimes known as a BAFFLE WALL or CURTAIN WALL but the latter term is by some authorities reserved for a suspended wall serving the same purpose.

Shaft. See STACK.

Shaft Kiln. A vertical kiln charged at the top and discharged at the bottom. If solid fuel is used it is fed in with the charge, but shaft kilns can also be fired with gas or oil by burners placed nearer to the bottom of the shaft. Such kilns are used for the calcination of flint, dolomite, fireclay, etc.

Shaft Mixer. See MIXER.

Shaker Table. A slightly tilted vibrating table, sometimes with a riffled surface. Particles poured on it are separated, the larger and denser reaching the bottom first.

Shaking Gold. BURNISH GOLD (q.v.) suspensions, which require vigorous shaking to disperse settled particles. (a US term.)

Shale. A hard, laminated, generally carbonaceous, clay. Colliery shales are frequently used as raw materials for building brick production.

Shale Cutter or Shale Planer. A mechanical excavator sometimes used for getting clay from deposits that are both hard and friable, and that will maintain a steep clay face. The main feature of the machine is an endless chain that carries a series of cutters which bear downwards on the clay, removing a layer about 18 mm thick. The machine makes semi-circular sweeps into the clay face before being moved forward or sideways.

Shaling. A fault in vitreous enamelware; also known as SPALLING (q.v.).

Shape Factors. Ratios of characteristic lengths or areas of particles (e.g. the ratio of the longest to the shortest dimensions) to give a dimensionless numerical assessment of the geometrical appearance of the particle. See also FRACTAL.

Shaping Block. A piece of wood for the preliminary shaping of glass on the blowing-iron before it is blown in a mould.

Shark's Teeth. A STRIATION (q.v.) of dagger-like step fractures from the scored edge to the compression edge (the region in which the brittle material is subject to compressive stress).

Sharon Quartzite. A quartzite occurring in Ohio, USA, and used in the manufacture of silica refractories. A quoted analysis is (per cent): SiO_2, 98.7; Al_2O_3 0.3; Fe_2O_3. 0.3; alkalis, 0.3.

Sharp. The bricks with larger radial taper in a two-taper system of arch or ring construction. The opposite of SLOW.

Sharp Finish. See FINISH.

Sharp Fire. Combustion with excess air and short, hot flame.

Sharp Pitch. See under PITCH.

Shatterbox. Trade-name. A laboratory grinder producing 1 to 300 mesh powder. (Glen Creston Co., Middx.).

Shaw Kiln. A gas-fired chamber kiln; one feature is that some of the hot gases pass beneath the kiln floor to diminish the temperature difference from top to bottom of the setting. The kiln was designed by Shaw's Gas Kiln Co. Ltd., in about 1925.

Shaw Process. A process for the precision casting of small metal components; its main feature is the use of silicon ester as the bond for the refractory powder, e.g. sillimanite. from which the mould is made. (N. Shaw, Brit. Pat., 716 394, 6/10/54 and later patents.)

S.H.C.C. Strain Hardening Cementitious Composite. See STRAIN HARDENING.

Shear-cake. A counterbalanced refractory slab acting as the door of a small furnace. (USA).

Shear Fire. A thin flame, used to sever a glass article from the tool after shaping.

Shear Hackle. A HACKLE MARK generated by interaction of a shear component with the principal tension driving the crack.

Shearicon. See TIMBRELL-COULTER SHEARICON.

Shear Marks on glassware are caused by the cooling action of the cutting off shears on the hot but solid glass.

Shearer Plastometer. See R & S PLASTOMETER.

Sheet Glass. Flat, transparent glass made by drawing or blowing.

Sheet Silicates. See SILICATE STRUCTURES.

Shell. (1) The falling away of a 25–50 mm internal layer of refractory from the roof of an all-basic open-hearth steel furnace; the probable cause is the combined effect of flux migration, temperature gradient and stress. This form of wear is also known as SLABBING.
(2) A flake of glass chipped from the edge of glass-ware, or the hollow left by such a flake.
(3) The 'shell' of a hollow clay building block refers to the outer walls of the block.

Shell Moulding. A method for the precision casting of metal in refractory moulds made from silica sand bonded with synthetic resin, sodium silicate or silicon ester; alumina and crushed high-silica (96% SiO_2) glass have also been used as the refractory component of these moulds.

Shell Wall or Ring Wall. The wall of fireclay refractories that protects the outer steel casing of a HOT-BLAST STOVE (q.v.).

Shellback, Shellbacked. An erroneous variant of SCHELLBACH.

Shelling. See SHELL.

Shell-stone. A CHINA STONE (q.v.) from Cornwall containing too much iron (as brown mica) for use as a flux in pottery glazes but of potential value in sewer-pipe glazes.

Shelving. The effect produced in the refractory lining of a glass-tank furnace by severe erosion of the horizontal joints between the tank blocks.

Sherd. POTSHERD (q.v.) Sometimes spelled SHARD.

Shetty's Criteria. The critical crack size relative to the size of the stress-concentrating defect is the key parameter determining the effect of stress-state on flaw severity. The relative critical crack size is a parameter of the material. In commercial aluminas it is 2 to 3 times the pore size. In RBSN it is much smaller. (D.K. Shetty *et al,* in *Fracture Mechanics of Ceramics* **5**, Bradt *et al,* Plenum 1983, p 531). See FRACTURE, FRACTURE MECHANICS.

Shetty's Mixed Mode Equation.
$K_I/K_{IC} + (K_S/CK_{IC})^2 = 1$
where K_I is the STRESS INTENSITY FACTOR (q.v.) for Mode 1 loading (stress perpendicular to the crack plane); K_{IC} is the critical stress intensity; K_S is the stress intensity for Mode II or Mode III loading (parallel to the crack plane)

whichever is dominant. C is a constant which is adjusted to best fit the data, and is usually in the range $0.8<C<2.0$ for large flaws. As C increases, the material becomes less sensitive to shear. See FRACTURE; FRACTURE MODES.

Shielding Glass. A protective glass for use in nuclear engineering; although it is transparent to visible light, it absorbs high-energy electromagnetic radiation. Such glasses contain a maximum proportion of oxides of heavy elements, e.g. PbO, Ta_2O_5, Nb_2O_3, WO_3.

Shillet. Local term for the Devonian slates used for brick-making in the Plymouth and Torquay areas of England; the word is a dialect form of 'shale'.

Shiners or Shiner Scale. Minute FISH SCALES (q.v.) that sparkle in reflected light; they are liable to occur in vitreous enamelware if the ground-coat is overfired.

Ship-and-Galley Tile. US term for a special FLOOR QUARRY (q.v.) with an indented anti-slip pattern on its face.

Shivering. See PEELING.

Shoe. A hollow refractory shape that is placed in the mouth of a glass pot and used for heating the BLOWING IRONS (q.v.).

Shop. (1) A department or room in a pottery, etc.
(2) A team of men operating a blowing or pressing process in a glass-works.

Shore Hardness Tester or Scleroscope. A procedure for the determination of the hardness of a surface by dropping a ball from a fixed height above the surface and noting the height of rebound. This technique was first proposed by A. F. Shore, an American, in 1906. Although primarily for the testing of the hardness of metals, it has also been applied to a limited extent in the testing of ceramics.

Shore-lines. A fault in vitreous enamelware similar in appearance to the marks left on a shore by receding water. The basic cause is the drying out of soluble salts. The cure is adjustment of the mill additions and of the drying process.

Short. Term applied to a clay body that has little workability or to a glass that quickly sets.

Short Glazed. See STARVED GLAZE.

Shot. Non-fiberized material, appearing as small spherical particles, in ceramic, glass or mineral wool fibres. It impairs thermal insulation. B.S. 1902 Pt. 6 section 7 describes a test for the shot content of refractory fibres. See SLUG.

Shotcrete. A mixture of sand, cement and water applied by a compressed air hose to surface concrete. Cf. GUNCRETE.

Shoulder Angle. A special shape of ceramic wall tile (see Fig. 7, p350).

Shouldering. The splay at the top right-hand and bottom lefthand corners of a single-lap roofing tile.

Shraff. N. Staffordshire word for the waste (e.g. broken fired ware, broken saggars, old plaster moulds) from a pottery.

Shredder. See CLAY SHREDDER.

Shrend. US term for the process of making CULLET (q.v.) by running molten glass into water (cf. DRAG-LADLE).

Shrinkage. Reduction in dimensions. Ceramic fibre products shrink when heated. A test is described in B.S. 1902 Pt. 6 section 5. See also DRYING SHRINKAGE; PERMANENT LINEAR CHANGE.

Shrinkage Cracking. Cracking due to excessively rapid evaporation of moisture from (e.g.) concrete, mortar or a drying ceramic. It can be reduced by AIR ENTRAINING (q.v.)

Shrink-mixed Concrete. Concrete whose mixing is completed in the mixer lorry on the way to the site.

Shrink-wrapping. A packaging technique, used for pottery and for bricks. The ware is covered with a thick polymer sheet, which is sealed and shrunk onto the ware by hot air.
SHS. Self-propagating High-temperature Synthesis (q.v.)
Shuff. A clamp-fired STOCK BRICK (q.v.) that is of too poor a quality for use. Specifically, it is a Stock Brick from the top of a CLAMP (q.v.) that has been subjected to heavy rain while under full fire, the thermal shock resulting in a shattered texture and no 'ring'.
Shuttle Kiln. An intermittent bogie kiln consisting of a box-like structure with doors at each end and accommodating kiln cars (usually two in number). Pottery-ware is set on the refractory decks of the cars which are then pushed along rails into the kiln, displacing two other cars of fired ware from the kiln. The fired ware is taken from the displaced cars which are then re-set with more ware to be fired. The shuttle movement of the kiln cars is is then repeated (cf. BOGIE KILN).
Shuttle Press. A press with two mould cavities to be used alternately.
S.I. (Système International). The internationally accepted system of units. ISO 1000 1992; B.S. 5555, 1993. For a description of this system, see *S.I. The International System of Units. Approved translation of the International Bureau of Weights and Measures publication 'Le Système International d'Unités (SI)'.* National Physical Laboratory, London, HMSO, 1986. ASTM C899 describes the reporting of properties of refractories in metric units.
Sial. A borosilicate glass, of high chemical resistance and thermal endurance, made by the Kavalier Glassworks, Sazava, Czechoslovakia.
Sialcor. A double layer nozzle for steel casting. A 42% alumina outer shell is lined with 90% alumina. (Savoie Refractaire, France.)
Sialons. Materials in the Si-Al-O-N and related systems, originally made by hot-pressing alumina and silicon nitride mixtures K.H. Jack and W.I. Wilson, *Nature Phys. Sci.* **238**, (80), 1972, p 28). Si is partly substituted by Al, and N by O, in an expanded β silicon nitride lattice, to form β^1 sialon. β^1-sialon is a silicon oxynitride solid solution containing Al, with a mullite-like structure. α^1-sialon (harder than β^1) is an expanded structure based on α-silicon nitride in with Si^{4+} is substituted by Al^{3+} with charges balanced by interstitial Y^{3+}, Ca^{2+} or Li^+ or rare earths. Aluminium nitride POLYTYPOIDS (q.v.) (Si-Al-O-N phases with constant (Si+Al):(O+N) ratio) with AlN structure are mixed with α-Si_3N_4 (with its SiO_2 surface film) and Y_2O_3 as a sintering aid, and pressed and sintered in nitrogen under a BN protective powder bed, to form β^1-sialon. Residual yttria forms a grain-boundary glass or yttrium aluminium garnet phase which controls the high temperature properties. Sialons are used as engine components, seals, bearings, wear parts and cutting tools. α^1,β^1 composites perform better in the last two applications. Similar systems are possible with oxides other than alumina. See SIMONS.
Sibeon. A solid solution between β-silicon nitride and beryllia, developed as a crucible material for molten silicon.
Side Arch. A brick with the two largest faces symmetrically inclined towards each other, see Fig. 1, p39.
Side-blown Converter. See CONVERTER.
Side-construction Tile. A US term for a hollow block designed to undergo compression at right angles to the axis of the perforations.

Side Feather. See FEATHER BRICK.

Side Lap. The distance by which the side of a roofing tile overlaps the joint in the course of tiles next below.

Side Pocket. Alternative name for SLAG POCKET (q.v.) as applied to glass-tank furnaces.

Side Skew. A brick with one of the side faces completely bevelled at an angle of 60°. An arch can be sprung from such a brick. See Fig. 1, p39.

Siderite. Natural ferrous carbonate. $FeCO_3$, found as an impurity in some clays.

Siege. The refractory floor of a glass furnace, particularly of a pot furnace: from French *siège*, a seat).

Siemens or Siemens-Martin Furnace. See OPEN-HEARTH FURNACE.

Sieve. This term is generally reserved for testing equipment: the corresponding industrial equipment is generally called a SCREEN (q.v.). Automated sieving equipment has been developed. (see AUTOSIEVE; SORSI). ASTM C285 specifies Methods of Test for Sieve Analysis of Wet Milled and Dry-Milled Porcelain. B.S. 1902 Pt. 3.3 specifies the sieve analysis of refractory materials, as does ASTM C92. ASTM C325 specifies wet sieve analysis of whiteware clays. C371 and C925 are methods for non-plastic ceramic powders, using wire-cloth sieves and wet sieving using precision electroformed sieves respectively. There are several standard series of test sieves; those most frequently met with in the ceramic industry are:
British Standard sieves (conforming to B.S. 410)
USA Standard sieves (conforming to National Bureau of Standards LC-584; or ASTM – E11)
French Standard sieves (AFNOR NF 11–501)
German Standard sieves (DIN 4188)
Tyler sieves, in which the ratio between the mesh sizes of successive sieves in the series is 2; thus the areas of the openings of each sieve are double those of the next finer sieve.
Institution of Mining and Metallurgy, London (I.M.M. Sieves).
(For a comparison of these series see Appendix E.) All now follow the ISO sieve sizes. (ISO 2591, 3310)

SIFCA. Slurry Infiltrated Fibrous Castable (q.v.)

Sigma Cement. Trade-name: a hydraulic cement made by mixing portland cement consisting only of particles <30μm with 16–50% of an inert extender, e.g. limestone, basalt or flint, the particle size of which is 30–200μm. (S. Gottlieb, Brit. Pat., 580 291, 26/6/44.)

Signal Glass. Coloured glass for light signals on railways, roads, airfields and at sea; the glass must conform to a close specification (B.S. 1376).

Silazanes. Organic compounds in which the $SiNH_2$ group is substituted for a hydrogen atom. They are pyrolysed to form silicon nitride films.

Silcomp. A composite made by infiltration of molten silicon into fibrous graphite above 1450C.

Silcrete. A type of quartzite that is either cryptocrystalline or chalcedonic. It occurs, typically, at Albertinia, Mossel Bay, S. Africa; the porosity is low and the Al_2O_3 content is less than 0.5%.

Silex. A name sometimes applied to CHERT (q.v.), particularly to the trimmed blocks of this material, from Belgium, used in the lining of ball mills. For the milling of vitreous-enamel frits, silex blocks are used more in mills for ground-coats than for covercoats.

Silica. Silicon dioxide, SiO_2. For the various forms in which silica occurs see CHALCEDONY, CHERT, COESITE, CRISTOBALITE, FLINT, KEATITE,

QUARTZ, SAND, TRIDYMITE and VITREOUS SILICA.

Silica Fireclay. A US term defined as: A refractory mortar consisting of a finely ground mixture of quartzite, silica brick and fireclay.

Silica Gel. A form of silica produced by treatment of a solution of sodium silicate with acid and/or other precipitant. The dried gel is highly porous. It has found little use in the ceramic industry.

Silica Glass. This term is not recommended; see VITREOUS SILICA.

Silica Modulus. The ratio of SiO_2: $(Al_2O_3 + Fe_2O_3)$ in a hydraulic cement. In portland cement this modulus usually lies between 2 and 3. A cement with a low silica modulus can be expected to have high early strength, but if this modulus is high the final strength will be the greater.

Silica Refractory. Defined in B.S. 1902 Pt. 2 as a refractory material that, in the fired state, contains not less than 92% SiO_2. Refractories of this type were first made in 1822, by W. W. Young, from the quartzite of Dinas, S. Wales, and are still often referred to as Dinas bricks in Germany and Russia. They are characterized by high RuL (q.v.) but are sensitive to thermal shock at temperatures up to 600C. Their principal use is in steel furnaces, coke-ovens, gas-retorts and glass-tank furnaces. B.S. 4966 specifies the properties required of silica refractories for coke-ovens; they must contain at least 93% SiO_2. In the USA silica refractories are defined (ASTM C416) as containing ≤ 1.50% Al_2O_3; ≤ 0.20% TiO_2; ≤ 2,5% Fe_2O_3; ≤ 4% CaO; they are classified on the basis of a 'flux factor' equal to the percentage Al_2O_3 plus twice the total percentage of alkalis; Type A has a flux factor of 0.50 or less, Type B includes all other silica refractories meeting the general part of the specification. (See also SUPER DUTY SILICA REFRACTORY.)

Silica Sol. Colloidal silica in the form of a dispersion in water. The modern method of manufacture involves the passage of sodium silicate solution through an ion-exchanger. As made, the sols contain about 3% SiO_2 but the concentration can be increased to 20% by evaporation; to give stability, a concentrated sol must have a larger particle size or a stabilizer, e.g. NaOH, must be added. In the ceramic industry, silica sol has been used to a small extent as a bond.

Silicate Bond. (1) As applied to ceramics other than abrasives, this term means that the fired material consists of larger grains set in a matrix of a complex, usually glassy, silicate.
(2) In the abrasives industry this term means a bond consisting essentially of sodium silicate, the abrasive wheels so bonded being baked at about 250°C.

Silicate Cement. As used in chemical engineering, this term denotes an acid-proof cement, consisting of an inert powder bonded with silica gel that has been precipitated in situ from Na- or K-silicate (usually in the presence of sodium silicofluoride), or from silicon ester.

Silicate Structures. Silicates have a variety of crystal structures, based on the *silica tetrahedron* comprising a silicon ion with four oxygen ions surrounding it symmetrically. In *island structures* the tetrahedra are linked to each other through additional non-silicon cations – the ORTHOSILICATES, which include zircon and the aluminosilicates. In *isolated group structures* two or more tetrahedra are linked by common corners. This results in $(Si_2O_7)^{6-}$ groups, or in ring structures, which include ZEOLITES. *Chain structures* form when two oxygen ions in every tetrahedron are shared, with general formula $n(SiO_3)^{2-}$. The charge balance is maintained by adjacent foreign cations which link the chains in three dimensions. Such PYROXENES include enstatite ($MgSiO_3$), wollastonite ($CaSiO_3$) and spodumene $LiAl(SiO_3)_2$. In the AMPHIBOLES, the chains are linked by shared oxygen ions. If three oxygen ions are shared, *sheet structures* are formed, the layers having the general formula $n^2(Si_2O_5)^{2-}$. Most clay minerals have this structure, which imparts plasticity and determines their behaviour in slips. In *framework silicates* all four oxygen atoms are shared. As well as the various crystal forms of SILICA (SiO_2) itself, some silicon atoms may be replaced by Al, with alkali or alkaline earth atoms then incorporated to maintain charge balance, as in the FELDSPARS. See R.W. Grimshaw. *The Chemistry and Physics of Clays and Allied Ceramic Materials, Ch III 4th Ed* London, Ernest Benn, 1971.

Siliceous Refractory. Defined in B.S. 3446 as a refractory material that, in the fired state, contains 85–93% SiO_2 the remainder being essentially Al_2O_3: an ISO standard stipulates 85–93% SiO_2. SEMI-SILICA and SEMI-SILICEOUS refractories also fall within this definition. The specification for refractory materials for the gas and coking industries laid down by the Gas Council in collaboration with the Society of British Gas Industries and the British Coking Industry Association subdivides siliceous refractories into three groups: Clay-bonded Siliceous 'B', with <88% SiO_2; Clay-bonded Siliceous 'C', with <85% SiO_2; Semi-siliceous Firebrick, with 78–85% SiO_2.

The PRE definition of a siliceous refractory is that the SiO_2 content shall not exceed 93%, a further quality specification being according to use.

Silicides. A group of special ceramic materials: see under the silicides of Cr, Mo, Ti, W, Zr.

Silicon Borides. Several compounds have been reported. SiB_4, oxidation-resistant to 1370°C; thermal expansion (20–1000°C) 58×10^{-6}; thermal conductivity (70°C); 0.023c.g.s.; microhardness, 2100 kg/mm^2 (KHN-100). SiB_6, m.p. 1950°C; thermal expansion (20–1000°C) 5.1×10^{-6} ; thermal conductivity (70°C), 0.023 c.g.s.; microhardness, 2350 kg/mm^2 (KHN-100). A special refractory has been made by reacting Si and B in air, the product containing SiB_4 and Si in a borosilicate matrix; it is stable in air to at least 1550°C and has good thermal-shock resistance. Other compounds reported are SiB_3 and SiB_{12}.

Silicon Carbide. SiC. A non-oxide ceramic with a wide range of types and uses which depend on the method of production, the resulting microstructure and purity, and the grain size.

SiC dissociates at approx. 2250°C. Density 3.2 kg; thermal expansion (25–1400°C) 5×10^{-8} ; thermal conductivity usually below 100 W/m.K. The value is greatly dependent on the type of bond, as is the electrical conductivity. SiC is a semiconductor.

The uses of SiC are very varied, and depend on the method of production, the resulting microstructure and purity, and the grain size. SiC resists oxidation better than most carbides, due to the formation of a protective surface film of silica. ASTM C863 evaluates the oxidation resistance of SiC. β-SiC is a cubic crystal, stable above 2000C, produced by vapour phase reactions. α-SiC is rhombohedral or (usually) hexagonal, and is the low temperature form. There are many POLYTYPES (q.v.)

of α-SiC, giving rise to a special classifying notation (see RAMSDELL NUMBERS. α-SiC is made by reducing silica sand with carbon in an arc furnace. The resulting grain is used as an ABRASIVE (q.v.) or to manufacture refractory shapes. The abrasive products may be resin or glass bonded.

Clay-bonded Silicon Carbide is made by mixing coarse SiC grain with up to 50% clay and shaping and firing to form a porous thermal-shock resistant refractory.

Recrystallized Silicon Carbide is a single-phase material with a fairly coarse grain size and a highly faceted surface structure. A bimodal α-SiC grain size distribution is shaped (usually slip-cast) and sintered without additives in an inert gas at 2200C. The product is used for tubes, other refractory shapes, kiln furniture and heating elements. There are strong intergranular bonds and high porosity. There is no shrinkage, and RECRYSTALLIZATION (q.v.) takes place by a complex evaporation/condensation mechanism. SiC in the fine-grained half of the bimodal distribution reacts with the silica surface layer always adhering to SiC particles. The reaction $SiC + SiO_2 \rightarrow Si + CO + \frac{1}{2} O_2$ proceeds in the opposite direction to produce coarse particles. In CRYSTAR (q.v.) electronic grade silicon is incorporated to promote this reverse reaction.

Silicon Carbide Heating Elements make use of SiC's semiconducting properties. They are made by proprietary methods, either from pressed or tamped α-SiC fired at 2200°C, or from β-SiC at a lower temperature. The objective is to form a central zone of high electrical resistance in a rod or tube, e.g. by using capillary suction to infiltrate the tube ends with molten silicon metal. The result is a coarse crystal structure cross-section with an α-SiC 'recrystallized' faceted surface.

Reaction-bonded silicon carbide. Finer silicon carbide grains are shaped with graphite and a binder into a porous compact. The binder carbonizes in the pores. The compact is heated further in contact with molten silicon, which is drawn into the pores by capillary action. In ***self-bonded silicon carbide*** the silicon reacts with the carbon, depositing further silicon carbide epitaxially on the original grains. All the carbon is converted, strengthening and densifying the compact, while excess silicon fills the remaining interconnected pores. This material, originally developed as a cladding for nuclear fuel elements (see REFEL) is used for many engineering applications including bearings, engine components and seals. It is sometimes (inappropriately) known as ***reaction sintered silicon carbide.***

Nitride bonded silicon carbide is formed similarly, with less graphite and more silicon. The latter may be incorporated as a powder in the original mix. It reacts with nitrogen in the atmosphere to form an intergranular bond with a high proportion of Si_2N_2O. Engineering ceramics of almost zero porosity can be made. ***Sintered SiC*** uses B_4C as a sintering aid. ***Hot-pressed SiC*** has about 2% alumina added as a densification aid. Sintered SiC is less oxidation resistant than reaction-bonded SiC.

SiC films can be made by pyrolysis of chlorinated silanes, as can ***SiC fibres*** some 0.1mm diameter. Finer continuous fibres are made from C-Si polymer fibres. ***SiC whiskers*** (q.v.) are much used in fibre-reinforced composites.

Typical values of Young's modulus for various forms of SiC illustrate the dependence of mechanical properties on microstructure and production method:

sublimed SiC 460–470 GPa; CVD 440–470 GPa; hot-pressed (2% Al_2O_3) 415–445 GPa; sintered α-SiC 410 GPa; reaction-bonded 380–390 GPa; recrystallized 145 GPa.

Silicon Monoxide. SiO deposited by CVD and PVD methods to form dielectrics and insulating layers in integrated circuits, and as a protective coating.

Silicon Nitride. Si_3N_4. There are two forms, both hexagonal. α ($Si_{12}N_{16}$) changes irreversibly to β (Si_6N_8) at approx. 1700°C, above which temperature decomposition sets in. Silicon nitride powder is mainly α form. It can be made by nitriding silicon powder, by the carbothermal reduction of silica in nitrogen, or by reacting silicon tetrachloride with ammonia. Dense engineering ceramics can be made by most of the advanced ceramic forming processes, including:

Reaction Bonded Silicon Nitride – a porous silicon compact is made (e.g. by isostatic pressing) and heated in nitrogen at 1100 to 1450°C. The silicon (solid, liquid or vapour depending on the temperature) reacts with the nitrogen to form sub-micron grains of α – and β – silicon nitride, which fill up the pore space of the original silicon compact, without shrinkage. The porosity is 15–30%, with some large pores (50μm) due to the melting of impurity particles from the silicon. Compacts can be partially nitrided to sufficient strength but low enough hardness for machining to final size and shape with normal machine tools. Nitriding is then continued. Engineering components of complex geometry can be made, as can pouring tubes and nozzles for molten aluminium.

Sintered Silicon Nitride. Dense engineering ceramics can be made by using MgO, Y_2O_3 or other oxides as SINTERING AIDS for α-silicon nitride powder (which usually has a surface layer of silica). The powder mixture is shaped and fired in nitrogen, protected by a bed of boron nitride powder. An oxynitride liquid is formed, which dissolves α-Si_3N_4 which then precipitates as acicular β-Si_3N_4, densifying and strengthening the compact. Depending on the oxide additive the liquid cools to an amorphous or crystalline grain boundary phase, which controls the high temperature properties of the ceramic.

Hot pressed Silicon Nitride is similar to sintered silicon nitride, but uniaxially pressed in an induction heated graphite die at 1700–1850°C. See also SIALONS.

Pyrolytic Silicon Nitride. Tubes and protective coatings for engineering components are made by chemical vapour deposition from chlorosilane or silane-ammonia-hydrogen mixtures. CVD and physical vapour deposition processes, and the resulting crystal structures, have been much studied and used to deposit silicon nitride films and grow epitaxial layers as electrical insulators and as masks for the deposition of other materials in electronic integrated circuitry.

Silicon Oxynitride. Si_2N_2O, orthorhombic crystals made by reaction-bonding or by crystallization from OXYNITRIDE GLASSES (q.v.). Density is 1.9 to 2.3 kg/l and flexural strength 30 to 90 mPa. Silicon oxynitride is used as a lining for cells for the electrolysis of $AlCl_3$ to Al and $ZnCl_2$ to Zn, and as crucibles and tubes for molten salts. It occurs as a constituent of nitride bonds (see SILICON NITRIDE, SILICON CARBIDE) and has been used to strengthen silica optical fibres by surface coating.

Silicone. Organo-silicon polymers some of which can be used up to a temperature of 300°C or even higher. Silicones have been used in the ceramic industry as mould-release agents, for the coating of glass-ware to increase its strength and chemical durability; a specification for silicone for spraying on brick-work to give it water-repellancy is provided in B.S. 6477 *Water Repellants for Masonry Surfaces.*

Silicosis. A lung disease caused by inhalation, over a long period, of siliceous dusts, particularly those containing a high proportion of free silica of a size between l and 2µm. See DUST; HEALTH & SAFETY.

Silit. Trade-mark: a heating element consisting principally of SiC; to maintain a constant resistance for a long period, Si is included in the batch and the shaped rods are fired in a controlled atmosphere to cause some nitridation and/or carbonation. (Siemens-Planiawerke A.G., Germany.)

Silk. See GLASS FIBRE.

Silk-screen Process. See SCREEN PRINTING.

Sill. A horizontal course of brickwork, or a precast slab of concrete, under an opening e.g. a window in a house or a door or port in a furnace wall. Sizes, materials (stone, concrete or clayware), workmanship and jointing and water-proofing of sills are specified in B.S. 5641 Pt 1.

Sillimanite. A mineral having the same composition (Al_2SiO_5) as kyanite and andalusite but with different physical properties. The chief sources are S. Africa and India. Sillimanite changes into a mixture of mullite and cristobalite when fired at a high temperature (1550°C); this change occurs without any significant alteration in volume (cf. KYANITE). The mineral is used as a refractory.

Sillimanite Refractory. A refractory material made from any of the SILLIMANITE (q.v.) group of minerals. Such refractories generally contain about 60% Al_2O_3; they have a high RuL (q.v.) and good spalling resistance. Sillimanite refractories are much used in glass-tank and frit-melting furnaces, electric steel-furnace roofs, rotary kilns and in kiln furniture for use in the firing of pottery-ware.

Silver Lustre. Because silver itself tends to tarnish, 'silver lustre' is, in fact, made from platinum, with or without the addition of gold.

Silver-marking of Glazes. Silver cutlery, or other relatively soft metal, will leave a very thin smear of metal on pottery-ware if the glaze is minutely pitted. A glaze may have this defective surface as it leaves the glost kiln, or it may subsequently develop such a surface as a result of inadequate chemical durability. The fault is also known as CUTLERY MARKING.

Silvering. The formation of a thin film of silver on glass; such a film can be produced by treatment of the glass surface with an ammoniacal solution of $AgNO_3$: together with a reducing agent. A trace of copper is also often included.

Simax A glass, of very high thermal endurance and good chemical resistance made by the Kavalier Glassworks, Sazava, Czechoslovakia .

Simplex Kiln. A type of annular kiln in which two barrel-arch galleries are each divided by transverse walls into eight or nine chambers, each of which has grates at the corners for the handfiring of solid fuel. This kiln can be used for the firing of facing bricks, roofing tiles, firebricks or (because the flue system permits each chamber to be isolated) blue engineering bricks.

SiMONs. A more general term for SIALONS (q.v.) M is 'metal' substituted in the silicon nitride structure. The term is not widely used.

Sindanyo. An asbestos-free heat resistant board (Tenmat Ltd, Manchester).
Singer's Test. A rough test for glaze-fit proposed by F. Singer *(Sprechsaal,* **50**, 779, 1917); the glaze is placed in a dish of the biscuit ware and fired to its normal maturing temperature; when cold, the glazed dish is examined for faults.
Single-bucket Excavator. There are two principal types of this machine: the MECHANICAL SHOVEL (q.v.) and the DRAG-LINE (q.v.).
Single Crystal. See CRYSTAL STRUCTURE.
Single Edge Notched Beam. See FRACTURE TOUGHNESS TESTS.
Single Edge Precracked Beam. See FRACTURE TOUGHNESS TESTS.
Single-fired Ware. See ONCE-FIRED WARE.
Single-lap Roofing Tile. This term, as defined in B.S. 1424 *'Clay Single-lap Roofing Tiles and Fittings'* includes pantiles, double Roman tiles, flat interlocking tiles and Pooles tiles.
Single Layer Setting. Ware is placed in the kiln in a single layer, to facilite rapid firing. Such setting is also conducive to mechanical handling or to automated setting and dehacking.
Single Roman Tile. See under ROMAN TILE.
Single-screened Ground Refractory Material. A US term defined (ASTM-C71) as: A refractory material that contains its original gradation of particle sizes resulting from crushing, grinding, or both, and from which particles coarser than a specified size have been removed by screening (cf. DOUBLE-SCREENED).
Single-taper Construction. Making brick arches or rings, using only one size of taper brick.
Single-toggle Jaw Crusher. A jaw crusher with one jaw fixed, the other jaw oscillating through an eccentric mounted near its top. This type of jaw crusher has a relatively high output and the product is of fairly uniform size.
Sinter. As a verb, see SINTERING. As a noun, a porous compact formed by heating a powder so that the contact points of the grains fuse together at a temperature below the melting point of the material.
Sinter-active Powders. See ULTRAFINE POWDERS.
Sinter-Canning. An ENCAPSULATED-HIP (q.v.) technique. SOL-GEL (q.v.) technology is used to coat (e.g.) silicon nitride with alumina or zirconia gel, which is dried and sintered, to form a dense layer of oxide, encapsulating the preform before HIPing.
Sinter-forging. See HOT FORGING.
Sinter-HIP. Also known as CLADLESS HOT ISOSTATIC PRESSING. A form of HIPing in which gas pressures of 70 to 200 MPa are applied direct to the surface of the compact, which has previously been sintered to that density (92–96% theoretical) at which all the remaining pores are closed pores. The gas must be chosen to inhibit dissociation of the sintered material. The technique is useful for large parts.
Sintered Filter. A filter made from sintered glass, sintered silica or unglazed ceramic. B.S. 1752 refers to sintered disk filters for laboratory use varying in maximum pore size from Grade 00 (250–500μm) to Grade 5 (<2μm). (cf. CERAMIC FILTER.)
Sintered Glass. Glass-ware of controlled porosity used for filtration, aeration, etc.; it is made by carefully heating powdered glass so that the surfaces of the particles begin to melt and adhere to one another.
Sintering. The general term for the DENSIFICATION by heat of a powder compact to produce a polycrystalline

body. Densification may be by the formation of liquid phases which fill pores; by solid state diffusion or other mechanisms which also reduce surface energy. The term was originally restricted to what is now known as SOLID-STATE SINTERING (q.v.) See also VITRIFICATION; LIQUID PHASE SINTERING; OSTWALD RIPENING.

Sintering Aid. A material, typically an oxide, which reacts with the material to be sintered to form small quantities of a liquid phase. See LIQUID PHASE SINTERING.

Siporex. Trade-name: A lightweight (0.5 kg/l) pre-cast concrete made from portland cement, fine sand, Al powder and water. The set aerated blocks are autoclaved. (Costain Concrete Co. Ltd., London.)

SIPRE. (*Si*derurgie et *PRE*) see PRE.

Sitall. Russian term for GLASS CERAMIC (q.v.).

Sitter-up. An assistant TEASER (q.v.) in a glass-works.

Size. A solution used to treat the surface of pottery ware or of plaster moulds. Mould-makers' size is commonly a solution of soft soap. Decorators' size is traditionally based on boiled linseed oil; after its application, and after it has become tacky, the lithograph is applied and brushed down.

SK Porosity Test. A method for the determination of the porosity of aggregates; the principle is to fill the voids in turn with mercury, air and water. The method, primarily developed for the testing of iron ores, was proposed by H. L. Saunders and H. J. Tress at South Kensington, hence the name 'SK'; (*J. Iron Steel Inst.,* **152,** No. 2, 291P, 1945).

Skewback. A refractory block having an inclined face, or a course of such blocks forming the top of a wall, from which an arch or furnace roof may be sprung; also known as SPRINGER.

Skids. Refractory skids for use in pusher-type reheating furnaces in the steel industry have been made from sintered or fused alumina; they have a long service life and permit more uniform heating of the ingots or billets than do water-cooled metal skids.

Skimmer. A single-bucket excavator in which the bucket travels along the boom, which is kept almost horizontal during operation. This machine is sometimes used for removing OVERBURDEN (q.v.).

Skimmer Block, Skimmer-dam. A special refractory block that is partly immersed in the molten glass in a tank furnace to prevent impurities from entering the feeder channel.

Skimming Pocket. One of the small recesses, on the sides of a tank furnace for the production of flat glass, by means of which impurities on the surface of the molten glass can be removed. The skimming pockets are usually located a short distance before the FLOATERS (q.v.).

Skintling. The setting of bricks in a kiln so that courses of bricks lie obliquely to the courses above and below.

Skittle Pot. A small POT (q.v.) for glass melting.

Skiving. US term for a finishing process, on a lathe, of partially dried ceramic ware such as H.T. insulators, ribbed formers or sparking-plug insulators.

Skull. Metal remaining in a steel-casting ladle at the end of teeming; its removal often causes damage to the refractory lining of the ladle. The same term is used for the glass left in a ladle after most of the molten glass has been poured in glass-making; in the glass industry the usual spelling is SCULL.

Skull Melting. The refractory material to be melted is contained in a water-cooled crucible. The melt forms inside a sintered shell (or *skull*) of the same composition, eliminating reaction with or contamination by the crucible material. (Intermat Corp., Cambridge, Mass. *Brick Clay Rec.* **168** (3) 1976 p24.

Sky Firing. Burning wood slivers in the top of an updraft biscuit kiln, to increase the draught at the finish of the firing. An American craft pottery term.

Skylight. A plate glass of low quality.

Slab Glass. A block of optical glass resulting from preliminary shaping of CHUNK GLASS (q.v.).

Slabbing. (1) The fixing of ceramic tiles to fireplace surrounds, etc., to produce a prefabricated unit.
(2) A form of failure of refractories, also known as SHELLING. See SHELL.

Slag. (1) The non-metallic fusion that floats on a metal during its extraction or refining.
(2) The product of reaction between fluxing materials and a refractory furnace lining.

Slag Attack Tests. Refractories are attacked by the corrosive slags formed during metallurgical processes. Various simulative tests have been devised to assess the resistance of refractories to slag attack.
In the ***Crucible Test*** a well, drilled or moulded in a block of the refractory to be tested is filled with the selected slag and the block heated to the required temperature, for a specified time. The block is then removed, cooled, sectioned and the slag/refractory interface examined. The test is simple, but is static, and does not allow a temperature gradient to be studied. The reaction layer built up at the interface may produce erroneous results. The ***Pill Test*** is similar – a pellet of slag can be measured. In the ***Drip Test*** a stream of slag pellets falls on to a refractory block maintained at high temperature in a furnace. The test is dynamic, maintaining slag/refractory interaction, but heating is uniform and the interface for examination is small. (ASTM C768 specifies a drip test). The ***Rotary Drum*** tests are dynamic and a temperature gradient can be maintained through the test-piece. Test bars or a test panel of refractories are mounted in a rotating furnace into which slag pellets are introduced, to pour down the test panel. B.S. 1902 Pt. 5.13 and ASTM C-874 specify tests of this type. In the ***Finger Dip*** test a bar of refractory material is immersed in molten slag, or several bars mounted on a rotating disc to provide a dynamic test. Slag chemistry is difficult to maintain and there is little or no temperature gradient across the test pieces, but the effects of slag and molten metal simultaneously can be assessed. ***Induction Melting*** provides erosive wear and dynamic slag conditions by the induction stirring effect, maintains slag chemistry and can provide a controlled temperature gradient across a test piece. (Cf. ASTM C621 and C622 which specify tests for the resistance of refractories to molten glass; C987 which specifies a test for vapour attack on refractories for furnace superstructures; C288 which specifies a test for carbon monoxide attack on refractories; C454 which describes the disintegration of carbon refractories by alkali. B.S. 1902 Pts 3.10 and 7.6 specify tests for the resistance of refractory materials and unshaped refractories to carbon monoxide.

Slag Cement. A mixture of granulated blast-furnace slag and lime, together with an ACCELERATOR (q.v.); very little is now made (cf. PORTLAND BLAST-FURNACE CEMENT).

Slag Line. The normal level of the slag/ metal interface in the working chamber of a metallurgical furnace. The refractory lining of the furnace is liable to be severely eroded at this level owing to the improbability of chemical equilibrium between slag metal and refractory.

Slag Notch. The hole in the refractory brickwork of the wall of the hearth of a blast furnace permitting molten slag to flow from the furnace as and when necessary. It is also sometimes known as the CINDER NOTCH.

Slag Pocket. A refractory-lined chamber at the bottom of the downtake of an open-hearth steel furnace, or of a glass-tank furnace, designed to trap slag and dust from the waste gases before they enter the regenerator.

Slagsitalls. Glass ceramics made in Eastern Europe from glasses from blast-furnace and other slags. They are used to make durable building products, such as tiles.

SLAM. Scanning Laser Acoustic Microscopy (q.v.)

Slant Mill. This type of BALL MILL consists of an inclined cylinder which rotates around a central horizontal axis, causing the charge of balls and feed to roll from one end of the mill to the other. The charge does not lift, the mill has no critical speeds, and less power is required for its operation compared to a normal ball mill. (H. E. Vivian *Industr. Ceram.* **727,** 1979 p.265).

Sled(ge) Kiln. See PUSHED BAT KILN.

Sleek. A fine, scratchlike mark, with smooth boundaries, made on glass by a foreign particle during polishing.

Sleeper Block. See THROAT.

Sleeper Wall. In structural brickwork, a low wall (generally built with openings checkerwise) built to carry floor joists.

Sleeve. See ROD COVER. *Sleeve Bricks* are refractory tubes used to line slag vents.

Slenderness Ratio. The ratio between height and thickness of a wall.

Slick. (1) To move a smooth metal blade, or piece of wood, over the surface of plastic clay; the combined action of pressure and movement tends to bring water, soluble salts and fine clay particles to the surface. If slicking occurs as a result of movement of the knives or screw in a pug, it causes lamination as the affected clay surfaces do not readily knit together again.
(2) In USA a ceramic raw material is said to be 'slick' if it is greasy to the touch but is not plastic; such a material is talc or pyrophyllite .

Slide-gate Valve. A device for controlling the flow of molten metal in casting. Two refractory plates slide laterally or rotate against one another, so that the solid plate opens or closes the hole in the other plate.

Slide-off Decal. The cover-coat and pattern layer are attached to the backing layer with a water soluble adhesive, such as dextrin. Also known as a *water-slide transfer.*

Slide-off Transfer. A type of transfer for the decoration of pottery. The pattern is SILK SCREEN (q.v.) printed on litho papcr and then covered with a suitable plastic medium. Prior to use. the transfer is soaked in water so that the pattern, still firmly fixed to the plastic, can be slid off the paper and applied to the ware. During the subsequent decorating fire, the plastic coating burns away.

Sliding. The faulty draining of wet-process enamel by the slipping downwards of patches of enamel; one cause is overflocculation of the enamel slip.

Sliding-bat Kiln. See PUSHED-BAT KILN.

Sliding-gate Valve. See SLIDE-GATE VALVE.

SLIM. Single Leaf Insulated Masonry. A thin brick external wall is restrained by brick partition walls and the roof. A narrow cavity separates the brick wall from 50mm thick expanded polystyrene insulation, lined internally with 10mm-thick plasterboard. The system and guidelines for architectural design details for its use, were developed by British Ceramic Research Ltd., Stoke-on-Trent.

Slinger Process. A method used in USA for the moulding of insulating refractories; the wet batch is thrown on to pallets by rotary machine -a 'slinger' – to form a column on a belt conveyor; the column is then cut to shape, dried and fired.

Slip. (1) The mechanism by which shear stress causes plastic deformation, by driving lines of dislocation across certain crystal planes, the *slip or glide planes.* See CRYSTAL STRUCTURE; FRACTURE. (2) A suspension in water of clay and/or other ceramic materials; normally a DEFLOCCULANT (q.v.) is added to disperse the particles and to prevent their settling out. In the whiteware industry, a slip is made either as a means of mixing the constituents of a body (in which case it is subsequently dewatered, e.g. by filter-pressing) or preparatory to CASTING (q.v). In vitreous enamelling, a slip is used for application of the enamel to the ware by spraying or dipping. (Cf. BRICK SLIP).

Slip Casting. See CASTING.

Slip Clay. A fusible clay (because of fluxing impurities) with a fine-grained structure and low firing shrinkage. It fires at a low temperature, producing a natural glaze.

Slip Decoration. Application of a contrasting layer of slip (of a body composition, rather than a glaze) to decorate ware. It may be applied through a RESIST (q.v.) See SLIPWARE.

Slip Glaze. US term for a glaze made essentially from a fusible clay.

Sliphouse. Defined by COSHH in the *'Production of Pottery, Approved Code of Practice'* 1990 to 'include any place where machinery is used for mixing clay and/or other materials to form slip.'

Slip Kiln. A heat resistant trough placed over a heated flue; water can be evaporated from slip placed on the trough. This method of dewatering was replaced by filter-pressing.

Slip Plane. See SLIP (1).

Slip Resistance. An important property of floors and tiled surrounds to swimming pools. It is measured by instruments such as the TORTUS (q.v.) and the TRRL PENDULUM (q.v.) which take account of (deliberate) macroscopic irregularities in tiles and tiled surfaces, as well as measuring the (microscopically determined) friction between them and typical footwear materials.

Slip Trailing. See TRAILING.

Slipware. An early type of pottery (revived by studio potters) usually having a red body and a lead glaze, decorated with white or coloured slip by dipping, trailing or sgraffito. English slipware was made in the 17th and early 18th century, the chief centres being Staffordshire, Kent, Sussex and the West Country.

Slit Kiln. A fast-fire kiln, usually a roller-hearth kiln, whose width is much greater than its height. The ware, usually tiles, is fired one-high.

Slop Glaze. A suspension of glaze-forming materials prepared for application to ceramic ware, usually by dipping or spraying. The materials are kept in suspension by the presence of dispersed clay and by the high concentration of solids. A small amount (about 0.02%) of calcium chloride is also added to prevent the glaze from setting in the glaze tub and to act as a thickening agent.

Slop Moulding. The hand-moulding of building bricks by a process in which the clay is first prepared at a water-content varying from 20 to 30% depending on the clay. The wet clay is thrown into a wooden mould, pressed into the corners, and the top surface is finally struck smooth with a wet wooden stick. The filled mould is set on a drying floor until the clay has dried sufficiently to maintain its shape; the mould is then removed and the drying process is completed.

Slop Peck. The volume of slip that contains 20 lb of dry material. See also STANDARD SLOP PECK.

Slop Weight. The weight (oz) per pint of a suspension of clay flint, etc., in water. See PINT WEIGHT.

Slotting Wheel. A thin abrasive wheel used for cutting slots or grooves; such wheels are usually made with an organic bond.

Slow. The bricks with smaller radial taper in a two-taper system of arch or ring construction. The opposite of SHARP.

Slow Crack Growth. CRACK PROPAGATION (q.v.) can occur in ceramics at values of STRESS INTENSITY FACTOR (q.v.) much below the critical value K_{IC} needed to initiate BRITTLE FRACTURE (q.v.). The susceptibility to cracking of glasses and many oxide ceramics is greatly increased by chemical reaction between stressed cracks, especially small surface cracks, and environmental agents, especially water. Under these conditions there are three regions of crack growth. The third is governed by the intrinsic properties of the material. The second (a plateau) occurs when the crack growth rate is limited by non-availability of reactant. The first is most important. In it, stresses are usually well below the critical stress intensity, and the crack growth is controlled by the reaction rate. Such slow crack growth is also termed *stress corrosion*, and may be more rapid at higher temperatures. The crack velocity υ is related to the stress intensity K_I by the power law $\upsilon = AK_I^n$ where A is a constant and *n* is the *slow-crack growth exponent.* A low value of *n* indicates that the material is susceptible to slow crack growth over a broad range of stresses. A high value of *n* means susceptibility over a narrow stress range. Constant load stress-rupture tests (static FATIGUE (q.v.) tests) plot the stress against time to failure on logarithimic axes and have a slope of $-1/n$. Strength plots against loading rate, again on logarithmic axes, have a slope of $1/(1 + n)$. (Dynamic FATIGUE tests.). Grain boundary sliding, diffusion and other intrinsic deformation processes lead to CRACK NUCLEATION (q.v.), CRACK PROPAGATION (q.v.) and coalescence. At moderate, rather than high, temperatures, these processes may only be effective in causing crack growth in the already highly-stressed region near the tip of an existing crack. This is also known as *slow crack growth.* At high temperatures, these processes cause CREEP (q.v.) fracture, and are those same atomic mobility processes active in SINTERING (q.v.).

Slug. (1) A rough piece of prepared clay body sufficient for making one piece of ware, by throwing or by jiggering for example.

(2) A fault in a glass-fibre product resulting from the presence of non-fibrous glass; also called SHOT.

Slugged Bottom. A fault in a glass container characterized by the bottom being thick on one side and thin on the other; also called a WEDGED BOTTOM and (in USA) HEEL TAP.

Slum. A US term for fireclay containing a high concentration of fine coal particles.
Slumping. Serious deformation of a brick or column setting of bricks under its own weight.
Slump Test. (1) A rough test for the consistency of freshly mixed concrete in terms of the subsidence of a truncated cone of concrete when upturned from a bucket; ASTM – C143 and B.S. 1881.
(2) A works' test for the consistency of vitreous enamel slip; for details see IRWIN SLUMP TEST.
Slurry. A suspension with high solids content.
Slurry infiltrated fibrous castable (SIFCA). A refractory castable containing up to 16% steel fibres. The fibres are filled into the mould first, followed by a fine-grained castable, this approach overcoming previous difficulties of mixing and placing which limited the fibre content to *c.* 2 vol% (US Pat. 4266255).
Slurry process The production of refractory fibres by extruding a suspension as a filament, drying and calcining.
Slushing. A method for the application of vitreous-enamel slip to ware, particularly to small awkwardly-shaped items. The slip is applied by dashing it on the ware to cover all its parts, excess then being removed by shaking the ware. A slip of 'thicker' consistency than normal is required for this process.
Smalt or Powder Blue. A fused mixture of cobalt oxide, sand and a flux, e.g. potash. It is sometimes used as a blue pigment for the decoration of pottery or for the colouring of glass or vitreous enamel (cf. ZAFFRE).
Smear. In the glass industry the word has the special meaning of a surface crack in the neck of a glass bottle (cf. CHECK).
Smectite. Obsolete name for the clay mineral HALLOYSITE (q.v.).
Smock, Potter's Smock. OVERALLS (q.v.) worn by (male) operatives in the pottery industry, which are designed without lapels, pockets and other stitched features; which have short sleeves; and which are made from the smooth artificial fibre, terylene. They are thus free of dust traps. See also DUST, RESPIRABLE DUST, HEALTH AND SAFETY.
Smoked. Glass or glaze discoloured by a reducing flame.
Snagging. Rough grinding with an abrasive wheel to remove large surface defects; wheels with an organic bond are generally used.
Snaking. (1) Progressive longitudinal cracking in continuous flat glass production.
(2) Variation in the width of the sheet in glass drawing.
Snakeskin Glaze. A decorative effect obtainable on potteryware by the application of a glaze of high surface tension, e.g. 320 dynes/cm. A quoted glaze formula for the production of a snakeskin glaze maturing at 1140°C is: (0.3 PbO; 0.3 MgO; 0.2 CaO; 0.2 ZnO). 1.5 SiO_2. 0.2 Al_2O_3.
Snap. A device for gripping formed glass during polishing.
Snap-header. See HALF-BAT.
Soaking. As applied to the firing of ceramic ware this term signifies the maintenance of the maximum temperature for a period to effect a desired degree of vitrification, chemical reaction and/or recrystallization
Soaking Area. The part of a cross-fired glass-tank furnace between the GABLE WALL (q.v.) and the first pair of ports; also known as the FRITTING ZONE.
Soaking Pit. In the steel industry, a refractory-lined furnace for the reheating

of ingots. In the glass industry, a furnace for bringing pots of glass to a uniform temperature.
Soap. See PUP.
Soapstone. A popular name for STEATITE (q.v.).
Socket. (of sewer pipe) see BELL.
Soda Ash. Anhydrous sodium carbonate, Na_2CO_3. A major constituent of most glass batches. Together with sodium silicate it is used to deflocculate clay slips.
Sodium Aluminate. An electrolyte used as a mill addition in the preparation of acid-resisting vitreous enamel slips; 0.10–0.25% is generally sufficient.
Sodium Antimonate. $Na_2O.Sb_2O_5.½ H_2O$ An opacifier for dry-process enamel frits for cast-iron enamel-ware and for some acid-resistant frits for sheet steel enamels. It is also a source of antimony for yellow ceramic colours.
Sodium Carboxymethylcellulose. See CARBOXYMETHYLCELLULOSE.
Sodium Cyanide. NaCN; used as a neutralizer in vitreous enamelling.
Sodium Fluorsilicate. See SODIUM SILICOFLUORIDE.
Sodium-line Reversal Method. A technique for the measurement of flame temperatures. If a black-body is viewed, by means of a spectroscope, through a flame that has been coloured by sodium, there will be some temperature of the black-body at which its brightness in the spectral region of the Na_D line will equal the brightness of the light transmitted in this region through the flame, plus the brightness of the Na_D lines from the flame itself. At this temperature the spectrum of the black-body as seen in the spectroscope is continuous; there is a reversal in contrast above or below this temperature. The method was first described by C. Fery *(Compt. Rend.*, **137**, 909, 1903).
Sodium Niobate. $NaNbO_3$; a compound believed to be ferroelectric and having potential use as a special electroceramic. The Curie temperature is 360°C.
Sodium Nitrate. $NaNO_3$; used as an oxidizing agent in some glass batches and enamel frits. The old name was Saltpetre.
Sodium Nitrite. $NaNO_2$; added in small quantities to vitreous enamels to to prevent rust-spotting and tearing; 0.1–0.25% is usually sufficient.
Sodium Phosphate. The hexa-metaphosphate (generally under the proprietary name CALGON) is used to control the viscosity of slips in the pottery and vitreous enamel industries.
Sodium Silicate. Name given to fused mixtures in which the $Na_2O:SiO_2$ ratio normally varies from about 1:2 to 1:3.5; care should be taken in selecting the grade best suited to the purpose in view, e.g. as a deflocculant or as an air-setting bond for refractory cements, etc. More alkaline silicates are used for cleaning metal, prior to enamelling for example. A classification of sodium silicates is given in B.S. 3984.
Sodium Silicofluoride. Na_2SiF_6; used to some extent as a flux in vitreous enamels and as a constituent of acid-resisting silicate cements.
Sodium Sulphate. See SALT CAKE.
Sodium Sulphate Test. A test claimed to indicate the resistance of a clay building material to frost action. The test-piece is soaked in a saturated solution of sodium sulphate and is then drained and dried; the cycle is repeated and the test-piece is examined for cracks after each drying. The principle underlying the test is that the stresses caused by the expansion of sodium sulphate as it crystallizes are, to some extent, similar to the stresses caused by water as it freezes. The test

has also been used to reveal laminations present in bricks.

Sodium Tannate. Sometimes used as a deflocculant for clay slips; the effect is marked, only a small proportion being required. The material used for this purpose is generally prepared from NaOH and tannic acid; the former should be in excess and the pH should be about 8–9.

Sodium Tantalate. $NaTaO_3$; a ferroelectric compound having the ilmenite structure at room temperature; the Curie temperature is approx. 475°C. Of potential interest as a special electroceramic.

Sodium Uranate. $Na_2O.2UO_3.nH_2O$; has been used as a source of uranium for coloured glazes; the colour can vary from ivory through yellow and orange to deep red depending on the glaze composition.

Sodium Vanadate. $NaVO_3$; a ferroelectric compound having potential use as a special electroceramic. The Curie temperature is approx. 330°C.

Sofim-Fichter Kiln. A gas-fired open-flame tunnel kiln; its novel feature, when introduced in about 1935, was the design of the pre-mix burners. The name derives from the initial letters of Société des Fours Industriels et Métallurgiques, the designers of the original kiln, and Fichter (of Sarreguemines) who designed the burners.

Soft. As applied to a glass or glaze, this word means that the softening temperature is low; such a glass or glaze, when cold, is also likely to be relatively soft, i.e. of lower than average hardness, in the normal sense.

Soft Ferrite. FERRITE (q.v.) materials whose magnetization is easily lost in the absence of an applied field.

Soft Fire. A cool flame, deficient in air.

Soft-mud Process. A process for the shaping of building bricks from clay at a water content of about 35%. The prepared wet clay is fed into sanded moulds which are then shaken or jolted until the clay fills the mould; because of its THIXOTROPY (q.v.), after the jolting ceases the clay stiffens sufficiently for the bricks to maintain their shape. The process can form the basis of hand-making or, more commonly, it can be mechanized as in the BERRY MACHINE (Berry & Son, Southend-on-Sea, England) or in the ABERSON MACHINE (Aberson, Olst. Holland).

Soft-paste or Fritted Porcelain. A type of porcelain made from a soft body containing a glassy frit and fired at a comparatively low temperature (1100°C). The most famous soft-paste ware was that produced in the 18th century at the Sèvres factory in France, and at Chelsea, Derby, Bow, Worcester and Longton Hall in England. The COMBINED NOMENCLATURE (q.v.) states that soft porcelain contains less alumina but more silica and fluxes than hard porcelain. See PORCELAIN.

Softening Point. Generally, the indefinite temperature at which a ceramic material begins to melt. The term has a definite meaning however, when referring to glass of density near 2.5 namely the temperature at which the viscosity is $10^{7.6}$ poises; this viscosity corresponds to the temperature at which tubes, for example, can be conveniently bent and was first proposed as a basis for definition by J. T. Littleton (*J. Soc. Glass Tech.* **24.** 176, 1940). Also known as the 7.6 TEMPERATURE or the LITTLETON SOFTENING POINT.

Solar Furnace. A particular type of IMAGE FURNACE (q.v.) in which the sun's rays are focussed to provide the heat.

Solarization. An effect of strong sunlight (or artificial ultraviolet radiation) on some glasses, causing a change in their transparency. Glasses free from arsenic and of low soda and potash contents are less prone to this defect.
Solder Glass. The glass powder used in making GLASS-TO-METAL SEALS (q.v.) is essentially lead borate.
Soldier Blocks. Refractory blocks set on end. In the glass industry the term is particularly applied to blocks that extend more than the depth of the glass in a furnace.
Sole. The refractory brickwork forming the floor of a coke oven; as the charge of coal-which is subsequently transformed into coke-rests on this brickwork, and as the coke is pushed out of the oven by means of a mechanical ram, the sole is subjected to severe abrasion.
Sole-flue Port Brick. See RIDER BRICKS.
Sol-gel Process. Ceramic mixed oxides are prepared by co-precipitating the component metal oxides or precursors as gels from an aqueous mixture of components. The gelled particles are separated and sintered. Close control of chemical composition is possible (UKAEA, Br. Pat. 1253807, 1971 and other later patents).
Solid Brick or Block. Defined in B.S. 3921 as a brick or block 'in which holes passing through, or nearly through, the brick (or block) do not exceed 25% of its volume, and frogs do not exceed 20% of its volume' (cf. SOLID MASONRY UNIT).
Solid Casting. See under CASTING.
Solid Electrolyte. A ceramic capable of conducting electricity which acts as the electrolyte in batteries, fuel cells or oxygen sensors. Electricity is conducted by the movement of electrons, holes or charged ions.
Solid Masonry Unit. Defined in ASTM – C43 as: 'A unit whose net cross-sectional area in every plane parallel to the bearing surface is 75% or more of its gross cross-sectional area measured in the same plane' (cf. SOLID BRICK).
Solid Solution. A crystalline phase, the composition of which can, within limits, be varied without the appearance of an additional phase. In Fig. 5 (p230), a solid solution of mullite is shown to exist ranging in composition from that of pure mullite (60 mol% Al_2O_3) to 63 mol% Al_2O_3. A complete series of solid solutions exists between forsterite (Mg_2SiO_4) and fayalite (Fe_2SiO_4); these solid solutions occur naturally as the mineral olivine and are important in the technology of forsterite refratories.
Solid-state Sintering. Densification of powder compacts by heat, below their melting points. Initially the compact shrinks because of material transport by viscous or plastic flow, and volume diffusion. Evaporation-condensation and surface diffusion change the shapes of the pores, so that the original particulate structure is replaced by a polycrystalline body containing a network of pores. Grain growth then occurs by grain boundary migration, producing closed pores which gradually shrink as vacancies diffuse to the surface. All processes take place in the solid state.
Solidus. In a binary equilibrium diagram without solid solution (see left-hand side of Fig. 5, p230, representing the binary between silica and mullite), the solidus is the line showing the temperature below which a given composition, when in chemical equilibrium, is completely solid. At temperatures between the solidus and the LIQUIDUS (q.v.) a composition will contain both solid and liquid material. In a ternary system without solid solutions, the solidus is a plane.
Soluble Salts. Salts, particularly sulphates or carbonates of Ca, Mg and

Na, present in some clays used in ceramic manufacture or subsequently entering the fired product from other sources. If present in the clay, these salts tend to migrate to the surface during drying. In the pottery industry this can cause trouble during glazing; in the wet process of body preparation however, these salts are mostly eliminated during filter-pressing. If present in fired products (particularly bricks or glazed tiles) or if entering these products from mortar, cement or other sources, soluble salts may cause EFFLORESCENCE (q.v.), SCUMMING (q.v.), or SALT EXPANSION (q.v.). B.S. 3921 specifies a method for determining soluble salts in bricks. Bricks with low (L) salt content classification have Ca 0.3%; Mg, K, Na 0.03% each; SO_4 0.5%. ASTM C867 specifies a photometric method for determining soluble sulphates in whitewares.

Solution Ceramics. A type of ceramic coating introduced by Armour Research Foundation, USA. In the original process (Brit. Pat., 776 443, 5/6/57) a solution containing a decomposable metal salt is sprayed on the hot surface that is to be coated. A subsequent development (Brit. Pat., 807 302, 14/1/59) refers to the application of a coating of vitreous enamel or thermoplastic resin to the surface that has been flame-sprayed with the solution; this is claimed to result in a vitreous-enamelled surface having improved resistance to thermal shock.

Sorel Cement. A cement, used particularly in monolithic flooring compositions, first made by Stanislaus Sorel *(Comp. Rend.,* **65,** 102, 1867). Calcined magnesia is mixed with a 20% solution of $MgCl_2$, interaction forming a strong bond of Mg oxychloride, $Mg_3(OH)_2Cl.4H_2O$. In flooring compositions, various fillers are used, e.g. sawdust, silica, asbestos, talc; colouring agents are also added. The relevant British specification is B.S. 776, there are ASTM specifications for sampling and testing the constituent materials.

SORSI (Self Organised Sieving System). An icosahedral sieving chamber can be rotated on two great circular supports. Ten triangular surface elements between these can carry several sieving surfaces. The system operates in continuous or in batch mode, the latter being preferable for powder characterization. The residue on the sieve after a prolonged period can be calculated.

Sorting. The removal, from pottery taken from the glost kiln, of adhering bedding material and/or particles that have become detached from the kiln furniture. Sorting is usually done with a small pneumatic tool.

Soundness. As applied to portland cement, this term refers to its volume stability after it has set (cf. UNSOUNDNESS).

Souring. (1) An alternative term for AGEING (q.v.).
(2) The storage for a short while of the moistened batch for making basic refractories; some magnesium hydroxide is formed and this acts as a temporary bond after the bricks have been shaped and dried. If souring is allowed to proceed too far, cracking of the bricks is likely during drying and the initial stages of firing. The high pressure exerted by modern brick presses generally gives sufficient dry-strength without the bricks being soured, and this process is therefore now generally omitted.

Space Group. The 3-dimensional symmetry of a CRYSTAL STRUCTURE (q.v.) which is developed by applying symmetry operations (translation, rotation, reflection) to the 14 BRAVAIS

LATTICES (q.v.). It is the group of 230 such operations which leave the infinitely extended regularly repeating pattern of a crystal unchanged.
Space Lattice. See CRYSTAL STRUCTURE; also RECIPROCAL LATTICE.
Spacer. The tapered section of a pug joining the barrel to the die; in this section beyond the shaft carrying the screw or blades the clay is compressed before it issues through the die.
Spacer Lug. One of the projections on the edge of a glazed ceramic tile or paving brick ensuring correct width of joint; for sizes see BS 6431 and ASTM – C7, respectively.
Spalling. (1) The cracking of a refractory product or, in severe cases, the breaking away of corners or faces. The principal causes are: thermal shock, crystalline inversion, a steep temperature gradient, slag absorption at the working face with a consequent change in properties, or pinching due to inadequate expansion allowance (see also HOT-PLATE SPALLING TEST; PANEL SPALLING TEST). (2) The chipping of vitreous enamelware in consequence of internal stress.
Spaltplatten. See SPLIT TILES.
Spandrel Glass. ARCHITECTURAL GLASS used in external cladding and curtain walls.
Spanish Tile. A fired clay roofing tile that, in section, is a segment of a circle; the tile tapers along its length so that the lower end of one tile will fit over the upper end of the tile below (cf. ITALIAN TILES).
Spark Plug; Sparking Plug (UK, obsolete). The older type of ceramic core for spark plugs was made of ELECTRICAL PORCELAIN (q.v.); the modern type is sintered alumina. The cores, formerly shaped by pressing, rotary tamping or injection moulding, are made by automated dry-bay isostatic pressing: they are fired at 1600–1650C. The standard size is 14 mm.
Spark Test. A method for the detection of pin-holes in a vitreous enamel surface by the discharge of an electric spark.
Sparrow-pecked. See STIPPLED.
Spathic Iron. Ferrous carbonate $FeCO_3$, used as the basis of many ceramic pigments.
Spatterware. A type of mottled ceramic artware; a white or coloured glaze is spattered irregularly over the dipped glaze before the glost firing.
Special Ceramics. A generic term for non-clay ceramics for electronic and engineering applications, more used when such materials were novel.
Special Oxides. See OXIDE CERAMICS.
Specific Gravity. See APPARENT SPECIFIC GRAVITY; BULK SPECIFIC GRAVITY; TRUE SPECIFIC GRAVITY. When the term is used without qualification it may be assumed that true specific gravity is meant.
Specific Heat. The mean specific heat of most traditional ceramics lies between 0.22 and 0.26 (20–500°C); 0.23–0.28 (20–1000°C). The principal exception is zircon: 0.17 (20-500°C) and 0.18 (20–1000°C).
Specific Surface. The total surface area per unit weight of a powder or porous solid; in the latter case the area of the internal surfaces of the pores is included. The usual units are m^2/g. Methods of determination include GAS ADSORPTION (q.v.) and PERMEABILITY (q.v.). See also SURFACE FACTOR.
Speckled Ware. Vitreous enamelware having small particles of one colour uniformly scattered in an enamel background of a different colour.
Spectroscopy. A great variety of spectroscopic techniques has been applied to elucidating the structure and properties of ceramics. Details must be

sought in general texts on physical and chemical techniques. Some of the more common are listed by their abbreviations.

Specular Gloss (45°). An operational definition of this property, relevant in the surface evaluation of glazed or enamelled surfaces is as follows (ASTM-C346): 'The ratio of reflected to incident light, times 1000, for specified apertures of illumination and reception when the axis of reception coincides with the mirror image of the axis of illumination'. A test for 60° Specular Gloss is given in ASTM – C584.

'Speedy' Moisture Meter. The apparatus measures the acetylene evolved when the moisture in the test sample is reacted with calcium carbide.

Speedycal. A process and equipment for the rapid application of decorative transfers, developed by Matthey Printed Products and F. Malkin & Co.

Speeton Clay. A clay occurring in thc Lower Cretaceous strata; the name was first applied to a blue clay outcrops at Speeton, between Bridlington and Filey (Yorkshire, England).

Spengler Press. Trade-name: (Spengler-Maschinenbau G.m.b.H., Berlin.) A cam-operated automatic rotary table dry press for bricks and tiles.

Spew. See DUST.

Sphalerite. Zinc Blende, (Zn,Fe)S, a mineral with a cubic structure. BN can also assume this structure.

Sphene (or Titanite). $CaO.SiO_2.TiO_2$; m.p. 1386°C. This mineral has been found in the ceramic colour known as CHROME-TIN PINK (q.v.).

Spherical. See MORPHOLOGY.

Spheroidizing. The production of spherical particles, e.g. by calcining gelled particles produced from droplets of pre-determined size, comprising polymer and refractory precursor.

Spider. (1) A metal unit with two or more radial arms; for example, a spider is used in the mouthpiece of a pug to hold a core or to break up laminations.
(2) The term is sometimes used as an alternative to CENTRE BRICK (q.v.).
(3) A star-shaped fracture in a vitreous enamelled surface.
(4) A multi-legged refractory brick arch capped with refractory castable, used in some vertical shaft preheaters in cement manufacture.

Spigot. The part of a sewer pipe joint which is overlapped by the socket of the next pipe.

Spike Ramming, Spiking. The installation of a refractory lining, particularly in an induction furnace, by compacting successive layers of dry aggregate with a spiked steel shaft, ensuring that each layer is bonded with the previous one.

Spiked Bottom. One form of bottom for a CONVERTER (q.v.); it is of monolithic refractory material, air passages being formed by spikes which pierce the bottom while the refractory material is being rammed into place (cf. TUYERE BLOCK BOTTOM and TUBE BOTTOM).

Spinel. A group of isometric minerals having the same general structure and formula – $R^{2+}O.R_2{}^{3+}O_3$, where R^{2+} is a divalent metal (Mg, Fe^{2+}, Zn, Co, Ni, etc.) and R^{3+} is a trivalent metal (Fe^{2+}, Cr, Al). In a normal spinel all the divalent cations occupy tetrahedral sites while the trivalent cations occupy ocathedral sites; in an inverse spinel the divalent cations occupy octahedral sites while the trivalent cations occupy both tetra- and octahedral sites. The type mineral, to which the term 'spinel' refers if not qualified, is magnesium aluminate, $MgO.Al_2O_3$; this mineral is sometimes synthesized, by firing a mixture of the constituent oxides at a high temperature, for use as a refractory material. Another important group of spinels, known collectively as FERRITES (q.v.), are used as FERROMAGNETICS (q.v.). For other spinels see CHROMITE, HERCYNITE, PICROCHROMITE, MAGNESIOFERRITE.

Spin-Flash Drying. This process is in essence fluidized-bed drying. Filter cake is introduced into the chamber by a screw feeder and drops to the bottom. A stirrer breaks up the particles until they are small enough to be carried up by hot air to a cyclone via a central constriction of variable diameter which acts as a classifier. The dried powder does not have the good flow properties required for automatic pressing.

Spinodal Decomposition. A diffusion limited process of phase separation which takes place in thermodynamically unstable multicomponent (solid) solutions. A *spinode* on an equilibrium diagram is a cusp, a point at which there are two real tangents.

Spiral Bricking. A method of bricking a LADLE (q.v.) in which the first course of the sidewall comprises special shapes forming an incline. The spiral is completed with standard (semi-universal) bricks. Closure bricks are not needed. The brickwork is a continuous spiral, instead of separate rings. B.S. 3056 Pt. 8 gives further details.

Spiralling. Circumferential movement of brick rings relative to the casing of a cylindrical kiln and to each other. Some axial movement may also take place. The effect may take years to become apparent, or may swiftly affect refractory stability and wear.

Spiratron. In this fettling machine biscuit tableware traverses a spiral path through a chamber, in which it is smoothed by tumbling hardwood blocks. (Central Ceramic Services Ltd. *Ceramics* **20**, (242) 13, 1969).

Spissograph. An automatic machine for checking the time of set of cement or mortar. *(Bull ASTM,* No. 170, 79, 1950.)

Spit-out. Pin-holes or craters sometimes occurring in glazed non-vitreous ceramics while they are in the decorating kiln. The cause of this defect is the evolution of water vapour, adsorbed by the porous body, during the period between the glost firing and the decorating firing, via minute cracks in the glaze.

Split. (1) A brick of the shape of a STANDARD SQUARE (q.v.) split down its length, see Fig. 1, p39; commonly a 9 × 4½ in. (228 × 114 mm) brick with a thickness of 1¼, 1½ or 2 in. (31.5, 37 or 51 mm) (Also known as a SCONE.)
(2) A glass bottle containing 6½ fl. oz (200 ml) (for mineral water).
(3) A crack that penetrates glass-ware, as distinct from a CHECK or VENT (q.v.).

Split Rock. US term for the grade of FIRESTONE (q.v.) that is most suitable for use as a refractory.

Split Tiles. These tiles are formed double and separated after firing, to obtain single tiles. They have characteristic ridges on their backs, and may be glazed or not. The German word is *Spaltplatten,* which has gained some currency as an import.

Splitter. A refractory restriction in a duct, to change the gas flow pattern.

Spluttering. Glaze fragments from the ware fusing to the setters and shelves.

Spodumene. A lithium alumino-silicate, $Li_2O.Al_2O_3.4SiO_2$; it occurs in Rhodesia, Canada, USA and Brazil, and the commercial product usually contains about 6.5% Li_2O. The α form changes irreversibly at approx. 900°C to the β form with a considerable expansion; the β form is a constituent of some special ceramic bodies of zero thermal expansion.

Sponge, Sponging. The smoothing out, with a moistened sponge, of slight surface blemishes on pottery-ware before it is dried.
Spongy Enamel. Vitreous enamelware that is faulty because of local high concentrations of bubbles.
Spoon Proof. A specimen of molten glass for analysis. These are taken at various stages of melting.
Spout. A refractory block shaped to carry molten glass, usually to a forming machine; see also FEEDER SPOUT and POT SPOUT.
Sprag. See TWIG.
Spray Drying. The process of dewatering a suspension, e.g. clay slip, by spraying the suspension into a heated chamber, the dried powder being removed from the bottom of the chamber. Spray drying is particularly used to prepare granulate for pressing tiles and tableware. The granulate comprises free-flowing round or toroidal particles.
Spray Frost. A ceramic or organic material, sprayed onto a glass surface and fired, to simulate acid-etching.
Spray – ICP. See INDUCTIVELY COUPLED PLASMA.
Spray Sagging. The appearance of wavy lines on the vertical surface of vitreous enamelware after it has been sprayed and before the slip coating is dried. The fault is caused by incorrect adjustment of the fluidity and thixotropy of the slip.
Spray Welding. A process for the localized repair of cracks in the refractory brickwork of a gas retort or coke oven; a refractory powder is carried by a stream of oxygen into an oxy/coal-gas flame so that the powder fuses on the brickwork to seal the selected damaged area on which it is projected. This technique was first used by T. F. E. Rhead, City of Birmingham Gas Dept. (*Trans. Inst. Gas Engrs.,* 81, 403, 1931) (cf. AIR-BORNE SEALING).
Spreader Block. See DEFLECTING BLOCK.
Sprigging. The decoration of pottery vases, etc., by affixing clay figures (frequently classical) or other motifs to form a bas-relief. The figures are pressed in a biscuit pottery mould separately from the ware and are made to adhere to the ware by means of clay slip (cf. EMBOSSING).
Springback. The expansion of a compact after dry-pressing, when the pressure is removed and the compact is ejected from the die. Springback is greater at higher compaction pressures. It may be prevented by applying axial pressure (*hold-down*) to the compact during ejection.
Springer. See SKEWBACK.
Springing. The radially outward movement or thrust of brick arches or rings when being constructed or removed, a problem if axial restraint is lacking.
Sprue. In vitreous enamelling this term is sometimes applied to the frit that builds up at the tap-hole of a frit-kiln. From time to time pieces fall into the quenched frit and cause trouble in milling because of their hardness.
Sprung Arch; Sprung Roof. An arch, or furnace roof, the brickwork of which is supported solely by SKEWBACKS (q.v.); (cf. SUSPENDED ARCH or ROOF).
Spur. An item of KILN FURNITURE (q.v.) in the form of a tetrahedron with concave faces and sharp points on which the ware is set; see Fig. 4, p177.
Sputtering. A DC or RF voltage is applied to ionize argon gas. The accelerated Ar ions hit a solid target, so that the material of the target is 'sputtered' on to the article to be coated. Targets of complex composition can be

used, and higher plating productivity is achieved at lower temperatures than with other methods.

Square. (1) See STANDARD SQUARE.
(2) A SQUARE of roofing tiles is 100 ft^2 (*c*.9m^2) as laid.

Square Pitch. See under PITCH.

Squatting. SLUMPING (q.v.), though slumping is used to refer to the most severe cases.

Squeegee Process. See SCREEN PRINTING. The term SQUEEGEE OIL is used, particularly in the US industry, for the mixture of oils used to suspend a ceramic or enamel colour for silk-screen printing. Similarly, the term SQUEEGEE PASTE is used for the mixture of oils, colours and flux used in this process of decoration.

Squeeze Casting. A hybrid metal-forming process, combines casting and forging, applying a high pressure (70 N/m^2) as the liquid metal solidifies. The process has been applied to produce metal-matrix composites.

Squint. A building brick with one end chamfered on both edges so that the brick can be used at an oblique quoin.

SRBSN. Sintered Reaction-Bonded Silicon Nitride.

SR Fusion-cast Refractory. A French fusion-cast refractory, e.g. glass-tank block, made by a process that largely eliminates the cavities that occur as a result of shrinkage while the block cools from the molten state. The block is made by a so-called high-filling process (SR = *Sur-Remplis*) and is then heat-treated. (cf. DCL, RO and RT).

SRA Load Test. Automatically controlled RuL (q.v.) equipment designed by the Special Refractories Association, USA *(Bull. Amer. Ceram. Soc.,* **42**, (12), 741, 1963).

SSN. Sintered Silicon Nitride.

Stabilized Zirconia. See ZIRCONIA.

Stack or Shaft. The upper part of a blast furnace, widening from the THROAT (q.v.) at the top to the LINTEL (q.v.) at the bottom. This part of the furnace is lined with refractories selected for their resistance to abrasion and to the disintegrating action of the carbon monoxide present in the stack atmosphere.

Stacked Fibre, Stacking. Ceramic fibre blanket aligned with the plane orientation normal to the lining.

Stack Kiln. An old type of up-draught intermittent kiln for the firing of bricks. It was usually round, the feature of the kiln being the brick cone built on top of the kiln; waste gases passed into this cone, and thence to the atmosphere, through openings in the kiln crown.

Staffordshire Blues. See under ENGINEERING BRICKS.

Staffordshire Cones. PYROMETRIC CONES (q.v.) made by Harrison Mayer Ltd., Stoke-on-Trent, England. For squatting temperatures see Appendix 2.

Staffordshire Kiln. A particular design of transverse-arch kiln: it is fired from the top into combustion spaces in the setting. Such kilns were used for firing building bricks. The original design was patented by Dean, Hetherington & Co. in 1904.

Stahlton System. Prefabricated building elements made from hollow clay blocks, a row of which is united by prestressed wires and cement filling; each reinforcing wire is pre-stressed individually and anchored singly. The elements are used in floors and ceilings. The name derives from the German words *Stahl* (steel) and *ton* (clay).

Stain. (1) A coloured imperfection on the surface of ceramic ware.
(2) An inorganic colouring material used in the preparation of under-glaze and

on-glaze pottery colours, for colouring pottery bodies and glazes, or for decorating the surface of glassware.
(3) To produce yellow, amber or red colours in glass by applying to its surface compounds of Cu and/or Ag as a powder, the latter usually being 'diluted' with an inert material, e.g. sillimanite.
(4) An organic colorant (which burns out during firing) used to identify batches of different slips.

Stamping. A process for the application, by hand or by machine, of decoration to pottery-ware; a rubber stamp with a sponge backing is used. Stamping is particularly suitable for the application of BACK STAMPS (q.v.) and for some forms of gold decoration.

Standard Black. A ceramic colour; a quoted composition is 30 parts Co_2O_3 56 parts Fe_2O_3, 48 parts Cr_2O_3, 8 parts NiO and 31 parts Al_2O_3.

Standard Slop Peck. The volume of a pottery slip that, at 32 oz/pint, contains 20lb of dry material of sp. gr. 2.5. See also SLOP PECK.

Standard Square. In the UK this term refers to bricks of the following sizes:
(1) Common Clay Building Bricks (B.S. 3921): 8 5/8ths × 4 1/8th × 2 5/8ths in. (219 × 105 × 67mm)
(2) Refractory Bricks (B.S. 3056): 9 × 4½ × 3 in. and 9 × 4½ × 2½ in. (229 × 114 × 76 or 64 mm); the ISO recommended size is 230 × 114 × 64 mm.

Standard Surface Factor. See SURFACE FACTOR.

Stanford Joint. A joint, for sewer-pipes, consisting of tar, sulphur and ground brick or sand; the proportions vary according to the nature of the tar, which should contain sufficient pitch to ensure that the joint will set (W. H. C. Stanford. *Brit. Pat* 1873).

Stannic Oxide. See TIN OXIDE.

Staple Fibre. See GLASS FIBRE.

Star Dresser. A tool for the DRESSING (q.v.) of abrasive wheels; it consists of a series of star-shaped metal cutters separated by washers, freely mounted on the spindle of a holder. Also known as a HUNTINGTON DRESSER.

Star Mark. A defect (star-shaped fracture) liable to occur in vitreous-enamel cover-coats if the ware is roughly handled as it is being placed on the pointed supports used in the firing process; the fault may also occur if these supports are either too cold or too hot. The marks result from fracture of the enamel coating.

Starred Glaze. Term sometimes applied to a glaze that has partially devitrified, star-shaped crystals appearing on the surface; the cause may be SULPHURING (q.v.).

Starved. (1) The poor quality of the glazed surface of ceramic ware if insufficient glaze has been applied; also sometimes called SHORT GLAZED.
(2) A blue brick is said to be 'starved' if there is an orange colour or dappled grey-brown colour over the blue.

Static Leach Test. See MCC-1P STATIC LEACH TEST.

STD Method. See SUMITOMO-TSURUMI DRY FORMING METHOD.

Steam Blowing Process. The process used to make most GLASS FIBRE (q.v.) for glass wool. Glass is poured through slots in a platinum rhodium bushing, and fiberized by a steam blower just below.

Steam Curing. The rapid CURING (q.v.) of pre-cast concrete units; this can be done at high pressure in an AUTOCLAVE (q.v.) or at atmospheric pressure in chambers or tunnels.

Steam Pressing. US term for REPRESSING (q.v.).

Steam Shovel. See MECHANICAL SHOVEL.

Steam Tempering. The treatment of clay (particularly brick clay) with steam to develop its plasticity for the shaping process: with hollow blocks an additional advantage is to reduce the time required to dry the shaped blocks. The clay generally reaches a temperature of 70–80°C. (Also known as HOT PREPARATION; cf. HOT EXTRUSION).

Steam Test. See under AUTOCLAVE.

Stearates. Organic compounds used as a lubricant in the dry pressing of ceramics; the stearates used are those of Al, Ca, Mg and Zn.

Steatite. Massive form of the mineral talc, $Mg_3Si_4O_{10}(OH)_2$. The chief sources are in USA, France, Italy, India, Austria and Norway. The natural rock can be machined and the shaped parts fired for use as electroceramics. A far greater proportion is ground and shaped in the usual way with a clay bond to produce low-loss electroceramics; thermal expansion (20–500°C) $7–9 \times 10^{-6}$; volume resistivity (300°C) $10^8 - 10^{11}$ ohm-cm.; dielectric constant (1 MHz) 5.5–6.5; power factor (1MHz) $5 - 25 \times 10^{-4}$. Steatite is also used as a raw material for cordierite electroceramics and for refractories containing cordierite.

Steelite. A high strength hotelware body, presumed to contain alumina, first introduced by Ridgeway Potteries, 1967. Now the name of a hotelware company, Steelite International Ltd, Stoke-on-Trent.

Steger's Crazing Test. A method for the assessment of the 'fit' between a ceramic body and a glaze by measuring any deformation, on cooling, of a thin bar glazed on one side only. (W. Steger, *Ber. Deut. Keram. Ges.*, **9**, 203, 1928; more precise details were given in a later paper-*ibid.*, **11**, 124, 1930.)

Steinbuhl Yellow. Barium Chromate.

Stemware. A general term for wine glasses, etc., having stems. The stem may be pulled or drawn from the bowl (PULLED STEM or DRAWN STEM), or it may be made separately (STUCK SHANK).

Stent. Cornish term for the mixture of sand and stone occurring with china clay in the china-clay rock.

Step Fracture. See STRIATION.

Stepping. The undesirable projection on one or other contact face of adjacent bricks in a ring or arch.

Steric Hindrance. See SURFACTANT.

Stern Plane. See ZETA POTENTIAL.

Sticking Up. The process of joining together, by means of SLIP (q.v.), the various parts of items of pottery-ware that cannot readily be made in one piece, e.g. putting the knob on a tureen cover, or spouts on teapots.

Stiff-mud Process. The equivalent US term for the WIRE-CUT PROCESS (q.v.).

Stiff-plastic Process. A process of brickmaking by mechanical presses; the clay is prepared to a moisture content of about 12% and the shaped bricks can be set direct in the kiln without preliminary drying. This is particularly common in brickworks sited on the clays and shales of the Coal Measures.

Stikons. Trade-name for surface thermocouple sensors, made by RDF Corp., New Hampshire, USA.

Stillage. A small platform on which shaped clayware may be placed to facilitate its handling within the factory to and from a dryer, for example (cf. PALLET).

Stilt or Wedge-stilt. An item of KILN FURNITURE (q.v.) in the form of a three-pointed star; the three arms are SADDLES (q.v.) with points. See Fig. 4, p177.

Stippled. (1) A surface texture on a clay facing brick produced by the rapid impingement on the green brick of a head, carried on a reciprocating arm, fitted with steel spikes; this texture is also called SPARROW-PECKED.

(2) A mottled decoration on pottery-ware or on vitreous enamelware, produced by applying colour with a sponge, brush, steel wool or specially prepared pad.

Stishovite. A high-pressure modification of SiO_2: sp. gr. 4.28; a = 0.417nm; c = 0.268nm (S. M. Stischow and S. V. Popova, *Geokhimiya*, **10,** 837, 1961).

Stockbarger Method. A technique for growing single crystals in a temperature-gradient furnace (D. C. Stockbarger, *Rev. Sci. Instr.,* **7**, 133, 1936; **10**, 205, 1939).

Stock Brick. Originally a term localized to S.E. England and meaning a clay building brick made by hand on a 'stock', i.e. the block of wood that defined the position of the mould on the moulding table. Stock bricks are now machine-made. The London Stock Brick is a yellow brick of rough texture. The term 'stock' in the sense of 'usual', is sometimes applied to bricks of other areas, however, e.g. a Lincoln Stock is a semi-dry-pressed brick from the Lias. A Belfast Stock is a pink wire-cut brick from the Keuper Marl.

Stockholm Black. An underglaze ceramic colour. A stated composition is 37.7 g. $CuSO_4.5H_2O$, 39.6 g. $MnSO_4.4H_2O$, 22.7 g. $CoSO_4$, $7H_2O$; these are dissolved in boiling water and the solution is made slightly alkaline by adding Na_2CO_3. The solution is boiled and filtered; the residue is then washed to remove sulphates and calcined at about 600°C.

Stoichiometric. A chemical compound, or a batch for synthesis, is said to be stoichiometric when the ratio of its constituents is exactly that demanded by the chemical formula. Of some interest in the field of special ceramics are NON-STOICHIOMETRIC (q.v.) compounds.

Stoker. A mechanical device for feeding solid fuel into a kiln firebox or through the roof of a Hoffmann-type kiln.

Stokes. A unit of kinematic viscosity: 1 Stokes = 1 poise ÷ (density of the fluid).

Stokes's Law. The force required to move a sphere of radius r at uniform velocity υ through a medium of viscosity η is equal to $6\pi\eta\upsilon r$. This law is applied in the determination of particle size by sedimentation, elutriation or centrifugation. For a particle of density d_1 settling in a medium of density d_2, g being the acceleration due to gravity, when equilibrium is reached the following equation applies:

$r^2 = [9\upsilon\eta/2(d_1-d_2)g]$

Stone. A crystalline inclusion present as a fault in glass; stones may result from incomplete reaction of particles of batch or from the pick-up of small fragments of the refractory lining of the pot or furnace in which the glass is melted. The most common constituents of stones are carnegieite, corundum, cristobalite, mullite, nephelite, tridymite and zirconia. See also CHINA STONE.

Stone China. A US term for IRONSTONE WARE (q.v.).

Stoner. A workman in a glass-works whose job is to smooth the rims of the ware; also called a FLATTER.

Stoneware. The COMBINED NOMENCLATURE (q.v.) is: 'Stoneware, which, though dense, impermeable and hard enough to resist scratching by a steel point, differs from porcelain because it is more opaque, and normally only partially vitrified. It may be vitreous or semi-vitreous. It is usually coloured grey or brownish because of impurities in the clay used for its manufacture, and is normally glazed'. A TRANSLUCENCY (q.v.) test is specified. Water absorption is to be 3 wt %. The body consists either of a naturally vitrifying clay (a stoneware clay) or of a mixture of suitable ball clays, filler and flux. Stoneware may be once-fired or it may be biscuit fired at about 900°C followed by glost-firing at 1200–1250°C. At the present day stoneware is produced on a commercial scale chiefly as kitchen-ware and tableware, for which purposes its high strength and freedom from crazing are valuable; on a smaller scale, stoneware is much favoured by studio potters. See also CHEMICAL STONEWARE.

Stoning. The removal, by means of a RUBBING STONE (q.v.), of excrescences from enamelware during the stages of manufacture. The need for stoning may arise as a result of splashes of enamel slip from the spray-gun, or of non-uniform draining of the slip.

Stopper. (1) The round-ended refractory (fireclay or graphite) shape that terminates the stopper-rod assembly in a steel-casting ladle and controls the rate of flow of metal through the nozzle.
(2) A movable refractory shape for the control of the flow of glass in the channel leading from a glass-tank furnace to a revolving pot of a suction machine.
(3) A refractory closure for the mouth of a covered glass-pot or for the working-hole outside an open pot.
An alternative term with meanings (2) and (3) is GATE.

Stopping. The filling up of any cracks in biscuit pottery-ware prior to the glost firing.

Storage Heater. A type of electric space heater in which a refractory block is heated at night and transmits its heat to the room as it cools during the day. Storage heater refractories need high density and high specific heat.

Storey-height Panel. A hollow clay block, whose cross section is usually the size of a normal brick, of sufficient length (c. 2.5m) to form the whole of a wall in a dwelling when used vertically. Such blocks are usually extruded, and are difficult to dry and fire without distortion.

Stormer Viscometer. A rotating cylinder viscometer of a type that has found considerable use in USA for the determination of the viscosity and thixotropy of clay slips. *(Industr. Engng. Chem.,* **1**, 317, 1909.)

Stove. See HOT-BLAST STOVE.

Stove Clay. A little used American term for FIRECLAY.

Stove Fillings. The special fireclay refractory shapes used as checker bricks in a HOT-BLAST STOVE (q.v.). Normally, these stoves operate at a max. temperature (at the top of the stove) of about 1200°C; the top 5m or so of fillings are therefore built with 40–42% Al_2O_3 refractories, with 35–37% Al_2O_3 refractories below. At higher hot-blast temperatures 50–65% Al_2O_3 stove fillings for the upper courses have been introduced.

St.P. Abbreviation for STRAIN POINT (q.v.).

Straightening Kilns. Electric intermittent kilns have been used specifically for straightening distorted ware, especially earthenware.

Straightening Ring. A SETTER (q.v.) designed to correct the shape of distorted flatware.

Strain Disk. A glass disk internally stressed to give a calibrated amount of birefringence; the disk is used as a comparative measure of the degree of annealing of glass-ware. The discs are air-quenched to produce a strain of 22×8 nm. each. The discs are piled up on top of the glass whose degree of strain is to be measured, until the total disc strain pattern matches the strain pattern of the glass.

Strain Hardening. Under concentrated strains, such as at crack tips, stress redistribution occurs so that localized fracture is delayed. A zone of microcracking extends around the crack tip, over a diffuse region much greater than the normal volume of microcracks around a crack tip. This expanded zone can be created by stress transfer by reinforcing fibres. It produces considerable energy absorption away from the plane of the crack. This off-plane inelastic absorption of energy gives rise to a ductile type fracture behaviour, and has developed strain capacities over 100 times those of the matrix in cementitious composites reinforced with polyethylene fibres.
Strain-line. Alternative term for the vitreous-enamel fault more generally known as HAIR-LINE (q.v.).
Strain Point. The temperature corresponding to the lower end of the annealing range and defined in USA (ASTM- C336) as: the temperature corresponding to a rate of elongation of 0.00043 cm/min when measured by the prescribed method. At this temperature the viscosity of a glass is approx. $10^{14.5}$ poises. Abbreviated to St.P.
Strainer Core. A porous refractory shape for use in foundries to control the flow of metal and to keep slag and sand inclusions out of the casting. High thermal-shock resistance is required.
Strand Count. The thickness of a strand of glass fibres expressed as the number of 100 yd lengths per pound weight.
Strapping. (1) the methods of fastening (loadbearing) brickwork to floors and roof.
(2) fastening conveniently-sized loads of bricks on pallets or into cubes, for transport, using steel or nylon bands.
Strass. A glass of high lead content and great brilliancy used for artificial jewelry; named from its 18th century German inventor, Josef Strasser.
Straw Stem. The slender hollow stem of a wine-glass.
Stress Corrosion. The susceptibility of glasses and many oxide ceramics to SLOW CRACK GROWTH (q.v.) due to interaction between stressed cracks and environmental agents.
Stress intensity factor, K_I. For a crack of length C, and an applied stress σ, the stress intensity K_I is given by $K_I = \gamma\sigma C^{1/2}$ where γ is a dimensionless constant depending on the specimen geometry and the crack shape.
The *critical stress intensity* K_{IC} is the value of K_I at the critical stress σ_c at which the crack propagates destructively. See also FRACTURE, CRACK PROPAGATION.
Stretched-membrane Theory. See MEMBRANE THEORY OF PLASTICITY.
Stretcher. A brick with its length parallel to that of the wall in which it has been laid (cf. HEADER).
Stretcher Bond. A brick wall in which all the courses are stretchers, the vertical joints in alternate courses being in line.
Striae. A fault, in glass, in the form of fine CORDS (q.v.); also known as VEIN.
Striation. The separation of the advancing crack front into separate fracture planes, giving a series of step fractures.
Striker Pad. An area of increased lining thickness or higher quality refractory which accepts the initial impact of hot metal into a ladle.
Striking. (1) The development of opacity or colour in glassware by a heat-treatment process, e.g. by the formation of a colloidal dispersion, within the glass, of a small amount of Cu, Ag or Au.
(2) The smoothing of the wet clay surface in a mould by means of a wooden or metal rod, as in the hand-moulding of special clay building bricks.

(3) A fault sometimes encountered with enamel colours on pottery-ware, the colour becoming detached from the ware during firing; a common cause is lack of control during HARDENING-ON (q.v.). (4) Setting out the first mould and the profile tool for the shaping of ware on a JIGGER (q.v.).

String. A fault in glass with the appearance of a straight or curled line; the usual cause is slow solution of a large sand grain or other coarse material.

String Course. A distinctive, usually projecting, course in a brick wall; its purpose is aesthetic.

String Dryer. A tunnel-type dryer, particularly for building bricks, that is operated intermittently; in the early stages of drying, the exits and exhaust ducts are closed so that a high humidity is built up. This type of dryer has been used particularly in Texas, USA.

Stripper. A person employed in the pottery industry to remove the dried ware from the plaster moulds.

Stronbal Glass. Name for *Stron*tium *B*oro-*Al*uminate glass proposed by M. Monneraye, J. Serindat and C. Jouwersma *(Glass Techn.,* **6**, (4), 132, 1965).

Strontium Boride. SrB_6; m.p. 2235°C; sp. gr., 3.42.

Strontium Oxide. SrO; m.p. 2420°C; sp. gr. 4.7; thermal expansion (20–1200°C) 13.9×10^{-6}.

Strontium Stannate. $SrSnO_3$; sometimes used as an additive to titanate bodies, one result being a decrease in the Curie temperature.

Strontium Titanate. $SrTiO_3$; used alone or in combination with $BaTiO_3$ as a ceramic dielectric. The power factor at low frequency is high. The Curie temperature is approx. 260°C. The dielectric constant is 230–250.

Strontium Zirconate. $SrZrO_3$; m.p. 2700°C; sp. gr. 5.48. Sometimes used in small amounts (3–5%) in ceramic dielectric bodies, one effect being to lower the Curie temperature.

Structon. A unit of atomic structure in an amorphous solid, such as glass, proposed by M. L. Huggins (*J. Phys. Chem.,* **58**, 1141, 1954) and defined as 'A single atom (or ion or molecule) surrounded in a specified manner by others'; examples for a sodium silicate glass would be Si(40), O(2Si,Na), etc. The concept aimed to reconcile the randomness of the long-range structure of a glass with the relative regularity in the short-range structure. The term has not achieved popularity (see VITRON).

Structural Ceramics. In Britain, ceramics for civil engineering structures. In the USA the term is so used, but also used for ceramics for mechanical engineering applications such as automotive engine parts.

Structural Clay Tile. In USA this term is defined (ASTM – C43) as a hollow burned-clay masonry building unit with parallel cells or cores or both; the equivalent term in the UK is 'Hollow Clay Block'.

Structural Spalling. A form of SPALLING (q.v.) defined in USA as: The spalling of a refractory unit caused by stresses resulting from differential changes in the structure of the unit.

Structure. As applied to abrasive wheels, this term refers to the proportion and distribution of the abrasive grains in relation to the bond. In USA the term GRAIN SPACING is also used.

Stub. An abrasive wheel that has been used until its diameter has been so much reduced by wear that it is no longer serviceable.

Stuck Shank. See STEMWARE.

Stuck Ware. Pottery-ware that has stuck to the kiln furniture during the glost firing and is therefore waste. The fault may be caused by careless placing, by the presence on the ware of too much glaze, or by firing at a temperature that is too high for the glaze being used, which therefore becomes too fluid.

Stud. A round piece of steel, welded inside a steel enclosure to retain a refractory lining.

Stupid. An extrusion machine in which clay is forced through the die by of a piston; such machines are now rare.

Submerged Entry Nozzle. A refractory component acting as both tundish nozzle and SUBMERGED ENTRY SHROUD (q.v.).

Submerged Entry Shroud. A LADLE SHROUD (q.v.) which may incorporate specially designed exit ports and facilities for gas injection.

Submerged Wall. A wall of refractory material below the level of the molten glass in a tank furnace and separating the melting zone from the refining zone.

Sub-porcelain. A class of WHITEWARE defined (Tariff Schedules of the United States, 1976) to be: fine grained ceramic ware (other than stoneware) whether or not glazed or decorated, having a fired body which is white (unless artificially coloured) and will absorb more than 0.5% but not more than 3% of its weight of water.

Subsolidus. That region of the phase diagram in which the only stable phases are solids.

Substance. In the glass industry this word means the thickness of glass sheets in oz/ft^2.

Substrate. The base on which a thin layer of different composition is deposited; an electroceramic, for example, can serve as the substrate of a thin-film resistor or of a printed circuit. ASTM F865 specifies the camber or curvature of substrates; F394 their strength (modulus of rupture).

Substitutional Defect. See CRYSTAL STRUCTURE.

Suck-and-Blow-Process. A method of shaping glass-ware, the PARISON (q.v.) is made by sucking molten glass into a mould, the final shape being subsequently produced in a BLOW-MOULD (q.v). The process is known in USA as VACUUM-AND-BLOW.

Sucking. Loss of volatile oxides, particularly lead oxide, from a glaze by volatilization during glost firing in unglazed saggars or adjacent to non-vitreous kiln furniture. The vapour being 'sucked' into the porous refractory. The fault is prevented by washing the insides of saggars with glaze or, in saggarless firing, by the use of kiln furniture of low porosity.

Suction Pyrometer or High Velocity Thermocouple. An instrument for the determination of the temperature of moving gases when it differs considerably from that of their surroundings: the hot gases are drawn rapidly past the junction of a fine-wire noble-metal thermocouple. Such an instrument is used, for example, in the determination of the temperature of hot kiln gases passing through a setting of relatively cool bricks.

Suffolk Brick. A white or pale yellow building brick made from the limy clays of Suffolk, England.

Suffolk Kiln. An early type of rectangular up-draught intermittent kiln. The fireboxes were below the kiln floor which was perforated for the upward passage of the hot gases: the waste gases escaped through openings in the kiln crown.

S.u.G Process. A German method for the shaping of highly grogged fireclay refractories; the bond-clay is added as a slip and shaping is by a compressed-air rammer. The name derives from the originators: Scheidhauer und Giessing A.G. (Brit. Pat., 262 383, 24/6/26).

Sugar. A fault on lead crystal glass resulting from inadequate control during acid polishing and revealed as crystallites on the surface of the glass.

Sugar Test. A quality test for cement; see MERRIMAN TEST.

Sugary Cut. Undue roughness of the edge of flat glass resulting from faulty cutting.

Sullivan Process. A process for the treatment of sheet steel prior to vitreous enamelling; it is based on the spraying of a suspension of NiO on the metal to give a deposit of about 0.01 – 0.1 g/m^2. (J. D. Sullivan, US Pat., 2 940 865. 14/6/60).

Sulmag Process. Registered tradename of Sulzer Bros. Ltd. Precalcined magnesite is dissolved by reaction with $CaCl_2$ and CO_2 to form $MgCl_2$ and $CaCO_3$. The solution is made alkaline with ammonia, and further CO_2 precipitates $MgCO_3.3H_2O$. The precipitate has excellent filtration and washing characteristics, and decomposes at low temperatures to reactive MgO with density 3.35–3.40 g/ml and over 99% purity.

Sulphate-resisting Cement. A portland cement specially made to resist exposure to water containing dissolved sulphates. The chief difference between this cement and ordinary portland cement is that the former has a low tricalcium aluminate content; B.S. 4027 stipulates 3.5 % (See also SUPERSULPHATED CEMENT).

Sulphate Scab. See SCAB.

Sulphides. Compounds of metals with the element sulphur. Some melt above 2000°C, though some decompose at much lower temperatures. Applications are as electroceramics, in solar cells and other energy conversion applications. Cadmium sulphide is a pigment, thorium sulphide a crucible material for cerium.

Sulphide Stain. Discolouration of enamel decoration by exposure to sulphide compounds, commonly by too long storage in corrugated packaging.

Sulphite Lye. A by-product of the paper industry used as a cheap temporary bond, e.g. for silica refractories; it has also been used as a plasticizer in making building bricks. It generally contains 50–70% of the ligno-sulphonate of Na, Ca or NH_4, together with 15–30% of a mixture of sugars; the ash content of the NH_4 type is very low but that of the Ca or Na types may amount to 30%.

Sulpho-aluminate Cement. A hydraulic cement made by grinding a mixture of high-alumina cement and gypsum (or anhydrite) (Lafarge, Brit. Pat., 317 783, 29/7/29).

Sulphuring. A surface bloom or dulling of glazed ceramic ware resulting from attack by SO_3 in the kiln, and more particularly by sulphuric acid condensed on the ware in cooler parts of the kiln in the early stages of firing. The sulphur compounds may originate in the fuel or in traces of sulphur compounds (e.g. pyrite) present in the ware itself. The fault is most common in glazes containing calcium or barium; glazes containing lead are usually more fluid and, although they may absorb more sulphur as sulphate, this may not result in a visible fault.

Sumitomo-Tsurumi Dry Forming Method. Blast furnace troughs are lined with a dry material (Al_2O_3-SiC with water content 0.5%), densely packed using pressure and vibration. Uniform packing is achieved, though unlimited repair by patching is possible, with short curing times.

Sump Throat. See THROAT.

Sunburner. A faulty piece of hand-blown glass-ware characterized by excessive local thickness.

Sunderland Splatter. A characteristic effect found on some of the old lustre pottery made in Sunderland, England; after the lustre had been applied it was 'splattered' with a second medium which caused the formation of irregular patches on the base lustre.

Superceram. Tableware containing added alumina, made by Hall China, USA.

Superconductor. A material with virtually no resistance to the passage of electric current.

Superconductivity is a quantum mechanical effect observed at low temperature. Roughly speaking, due to interactions with the crystal lattice, the conduction electrons in superconductors tend to form weakly bound pairs. Thermal vibration disrupts these pairs at other than very low temperatures, but at such temperatures the 'electron pairs' act as though they were single particles, subject to 'Bose statistics' – that is, they all tend to congregate in the lowest energy state In normal conduction, the electrons have a range of energies and it is relatively easy for successive electrons to be 'knocked out' of the regular flow by interaction with the crystal lattice. In the superconducting state, the electrons 'stick together' in the lowest energy state and maintain the current.

Superconductors also show the *Meissner effect,* expelling a magnetic field from within themselves if cooled below T_C, the critical temperature below which they exhibit superconductivity.

Superconductive behaviour is lost if the material is heated above T_C and also if the current density in the material is above a critical value J_C. Most metals show superconducting behaviour, but only at temperatures of a few degrees K. Some metal alloys have T_C s of some 30K, but ceramic superconductors exist with T_C's as high as 90–110K. The objectives of superconductor development are to have T_C and J_C as high as possible, with initially T_C above the temperature of liquid nitrogen.

Ceramic superconductors are usually oxygen-deficient materials with distorted PEROVSKITE (q.v.) crystal lattices. Two systems have been particularly studied: the 1:2:3 or $YBa_2Cu_3O_{7-x}$ system (in which other rare earths may be substituted for Y) and the lead/bismuth – rare earth-copper oxide systems.

Ceramic superconductors are made by the powder forming technique now usual for technical ceramics, but require extremely careful control of microstructure and chemical purity. In practical devices, ceramic-metal joining techniques are also of great importance.

Supercritical Solvents. Supercritical fluids are fluids above their critical points i.e. in the P-T region where the vapour cannot be condensed by pressure alone, but must also be cooled. They have high dissolving power, good diffusional properties, high density and low viscosity. Expansion of solutions based on supercritical fluids leads to rapid nucleation and precipitation of solid solute particles. Submicron mixed oxide and spinel powders can be made by decomposition of organometallic compounds in supercritical fluids such as ethanol. See RESS PROCESS.

Super-duty Fireclay Refractory. Defined in ASTM – C27 as a fireclay brick having a P.C.E. 33 and an after-contraction 1% when tested at 1600°C. There are three types: Regular, Spall Resistant, Slag Resistant. Limits are set for panel spalling loss, modulus of rupture and bulk density. ASTM C-673 specifies a super-duty plastic refractory with P.C.E. 32½.

Super-duty Silica Refractory. In the UK the National Silica Brickmakers'

Association has defined this type of refractory as SILICA REFRACTORY (q.v.) containing not more than 0.5% Al_2O_3 and with a total of Al_2O_3 plus alkalis not exceeding 0.7%.

Superheater. A refractory-lined chamber in a water-gas plant. It is filled with checkers and ensures completion of the decomposition of the oil vapours begun in the CARBURETTOR (q.v.).

Superlattice. The regular, ordered structure of the atoms of two elements in the crystal lattice of a SOLID SOLUTION.

Superplasticity. The ability of some ceramics to demonstrate large strains or deformation without cracking. Such ceramics usually have ultrafine (nanometre) grain size and have high strength across grain boundaries, so that cavities do not form. The phenomenon is related to diffusional creep, in which strain rate is inversely related to the cube of grain size, and has been reviewed by Y. Maehara and T.G. Langdon (*J Mater. Sci* **25,** 1990 p 2275–2286).

Super Staffordshire Kiln. A STAFFORDSHIRE KILN (q.v.) modified in design so that the fuel is charged on to a grate instead of within the setting of bricks. Kilns of this type can be used for temperatures up to about 1200°C.

Supersulphated Cement. A hydraulic cement made by finely grinding a mixture of 80–85% granulated blast furnace slag 10–15% calcined gypsum and 5% portland cement (or lime). Its principal feature is resistance to attack by water containing dissolved sulphates. (See also SULPHATE-RESISTING CEMENT).

Surface Combustion. That fuel gases burn more readily when brought into contact with a hot surface was first demonstrated by W. A. Bone and R. V. Wheeler *(Phil. Trans, A,* **206**, 1, 1906). The principle has been applied to furnace design, combustion taking place on the incandescent surface of refractory material; in one surface-combustion device, the fuel gas is passed through a porous refractory burner, 'flameless combustion' occurring over the surface of the refractory.

Surface Factor. A factor used in the ceramic industry to indicate the fineness of a powder. It is calculated from the equation:

$$S = \frac{6}{G}\left\{\frac{W_1}{d_1} + \frac{W_2}{d_2} + \frac{W_3}{d_3} + \dots\right\}$$

where G is the specific gravity of the powder; W_1, W_2 etc., are the fractional weights of material whose average diameters are d_1, d_2. etc.; $W_1+W_2+\dots = 1$.

Surface Tension. The effect of internal molecular attraction acting at the surface of a liquid to cause the surface to behave as though it were a tense skin. Some surface tension values are (dynes/cm): paraffin (kerosene), about 20 at 20°C; water, 74.2 at 10°C; glasses in their working temperature range, 200–300. See SESSILE DROP TEST.

Surfactant. A surface active agent; a chemical which reduces the surface tension of a liquid, and so improves the wetting of a surface. They are useful in dispersing powders, when surfactants adsorbed on particle surfaces produce repulsive forces between the particles when the adsorbed layers overlap (steric hindrance.)

Surkhi. Indian term for clay that has been lightly fired to give it some of the properties of POZZOLANA (q.v.).

Suspended Arch, Roof or Wall. An arch, furnace roof, or wall, in which some or all of the bricks are suspended by metal hangers from a steel framework. In an arch or roof, the object is to relieve the bricks from the mechanical stress resulting from the thrust of the arch; in a suspended furnace wall, the usual purpose is to permit refractory brickwork to be inclined inwards without danger of it falling into the furnace.

Suspending Agent. A material, such as clay, used to keep a vitreous enamel or glaze in suspension so that it can be conveniently used for application to the ware by dipping or spraying.

Svedala Kiln. A chamber kiln with a dryer built above it; designed in Scandinavia for use in the building brick industry.

Swab Test. An electrical test, carried out with low voltage, for exploration of such defects in vitreous enamelware as pin-holes or other discontinuities.

Sweep-fire System. In this system for firing stoneware pipes in periodic kilns, burners are arranged in opposed pairs, and the power to opposite burners is varied rhythmically. The system was developed by Paul Schneider & Co, and licensed to Bickley Co., in USA and W.Germany (see *Z.I. Int.* **40**, (12) 629 1987).

Sweep-gas System. As binder is burned out, a carrier gas is passed over the parts to entrain the binder vapour. This can be carried out at low pressures (a few Torr) instead of at atmospheric pressure, if binder is removed from the body under vacuum. (*Ind. Heating* **53** (4) 1986, p 28)

Sweet. A term applied to glass that is easy to shape.

Swept Valley. A form of roof tiling designed to cover a re-entrant corner of a roof without any sharp valley; careful cutting of the roofing tiles is necessary to ensure a symmetrical finish.

Swindell-Dressler Kiln. See DRESSLER KILN.

Swinger. The Pereco SW-850 rocking furnace for drying and calcining powder. (*Interceram* **24** (4) 261 1975).

Swing Press. A hand-operated screw press sometimes used, for example, in the shaping of a small quantity of special-shaped wall or floor tiles.

Swirlamiser. A design of oil and gas burner which replaces the fuel/air missing tube with a short choke venturi mixing port. The port is a refractory quarl to sustain high temperature and promote rapid and stable combustion. (Autocombustions, Telford).

Sworl. The marks made on the bottom of a thrown pot when it is cut from the wheel.

Syalon. Trade-name, for SIALON (q.v.) materials manufactured by Lucas, Cookson Syalon, UK. They have particular application in cutting tools (in which yttria is incorporated, but this is perhaps fortuitous).

Syndite. Trade-name. An abrasive-sintered diamond with a cobalt bond (Cf hard metals). De Beers Co., South Africa.

Syneresis. The phenomenon of the separation of liquid from a gel after the corresponding sol has set; the liquid expressed is termed the 'Syneretic liquid' (J. Ferguson and M. P. Appleby, *Trans. Farad. Soc.,* **26**, 642, 1930).

Synroc Process. High level radioactive waste from reprocessed nuclear fuel is immobilised in a synthetic mineral formed from a mixture of oxides, principally TiO_2, BaO, ZrO_2, Al_2O_3 and CaO. SYNROC-B contains the three phases barium hollandite ($BaAl_2Ti_6O_{16}$); perovskite ($CaTiO_3$) and zirconolite ($CaZrTi_2O_7$) chosen for their particularly great geochemical stability and for their ability to take into solid solution, as titanates, all the important elements in radioactive wastes. (*Australian Ceram. Soc.* **16,** (1) 10, 1980)

Synthetic Vitreous Silica. See under VITREOUS SILICA.

Syphon. A special type of refractory block for the bath of a glass-tank furnace.

'T' Glass. A glass composition for making glass fibre. 'T_1' glass, with 10% Na_2O, replaced the 'A' glass initially used, giving increased chemical durability and better fibre-forming properties. 'T_2'-type glasses (15% Na_2O) replaced 'T_1'-type with the introduction of modern rotary-spinner production technology.

T_a and T_b. See under REFRACTORINESS-UNDER-LOAD.

T_C – Critical Temperature, below which a superconducting material is a SUPERCONDUCTOR.

T_e. (1) The temperature at which the electrical resistivity is 1 megohm.cm. The T_e value of some ceramics is: porcelain, 350°C, cordierite, 780°C; steatite, 840°C; zircon, 870°C, forsterite, 1040°C; alumina, 1070°C.
(2) A temperature associated with REFRACTORINESS-UNDER LOAD (q.v.)

TA-Luft. The Federal Geraman Clean Air Guide of 1986, which sets emission limits for fluorides and the oxides of sulphur and nitrogen.

Taber Abraser. A device for assessing the abrasion resistance of a surface; the principle is contact with loaded abrasive wheels which are rotated against the surface to be tested. It is used for testing floor-tiles, see ASTM – C501. (Taber Instrument Corp. 111 Goundry St. North Tonawanda, NY, USA).

Tabular Alumina. See ALUMINA (TABULAR).

Tailings. The oversize material retained on a screen; normally the tailings are returned to the grinding plant for further grinding but in some cases the tailings contain a concentration of impurities and are discarded.

Talc. See STEATITE.

Talwalker-Parmelee Plasticity Index. This index, based on the results of tests on a clay in shear with a specially designed apparatus, is the ratio of the total deformation at fracture to the average stress beyond the proportional limit. (T. W. Talwalker and C. W. Parmelee, *J. Amer. Ceram. Soc.*, **10**, 670, 1927.)

Tammann Furnace. An electric high-temperature furnace having a carbon tube as resistor; it is used without a protective atmosphere but is sealed to prevent ingress of air (G. Tammann; see paper by E. Lowenstein. Z. *Anorg. Chem.*, **154**, 173, 1926.).

Tammann Temperature. The temperature at which reaction between two solids (or sintering within powder composed of a single solid) becomes appreciable. As originally defined by G. Tammann (*Z. angew. Chem.*, **39**, 869, 1926) this temperature lies between the two values obtained by multiplying the m.p. (K) of each reactant by 0.57. The temperature is only approximate and different multiplying factors have been proposed.

Tamping. The shaping of a semi-dry powder, e.g. of refractory material, in a mould by repeated blows delivered mechanically on the top mould-plate (cf. JOLT MOULDING and RAMMING).

Tank Block. A refractory block used in the lower part of a glass-tank furnace. These blocks are normally made of sillimanite, mullite or corundum; they are frequently made by electrofusion of the refractory, which is then cast in a mould to form a highly crystalline, virtually non-porous, block. See T. S. Busby, *Tank Blocks for Glass Furnaces* (Sheffield: Soc. Glass Tech., 1966).

Tank Furnace. A furnace for the continuous process of glass melting and refining; it is usually gas-fired and regenerative. The glass is melted in the bath of the furnace, which is lined with refractory blocks.

Tannin. A complex organic compound of C, H and O produced by metabolism in trees and plants. Sodium tannate is used to some extent as a deflocculant for clay slips.

Tantalum Borides. Several borides are known, including: TaB_2; m.p. 3000°C; sp. gr. 12.5; thermal expansion, 5.5×10^{-6}. TaB; m.p. 2400°C; sp. gr. 14.3. Ta_3B_4 : melts incongruently at 2650°C; sp. gr. 13.6.

Tantalum Carbides. Two carbides exist; TaC, m.p. 3800C; Ta_2C, m.p.3400°C. The monocarbide has the following properties: sp. gr. (theoretical), 14.5 g/ml; hardness, 850 (K100); modulus of rupture, 200 MNm^{-2} at 25°C and 120MNm^{-2} 2000°C; thermal expansion (25–2000°C) 8.4×10^{-6}.

Tantalum Nitrides. Two nitrides are known: TaN, m.p. 3000°C; Ta_2N, which loses nitrogen at 1900°C.

Tantalum Oxide. Ta_2O_5; m.p. 1870°C; sp. gr. 8.7. Transition temp. from low form (b) to the high (a) form is approx. 1340°C. The sintered oxide can be used as a special dielectric material; α-Ta_2O_5 ceramics exhibit large migrational losses, similar to those found in glasses, caused by migration of Ta^{5+}.

Tap Density. The density of a powder which has been tapped in a container under specified conditions, e.g. until the volume of powder does not decrease further (ISO 3953, 1977)

Tapered Cantilever Beam. See FRACTURE TOUGHNESS TESTS.

Tape Casting. A ceramic slip, containing suitable plasticizers if required, is filtered through a slot on to a moving continuous band. A gate or blade controls the thickness of the resultant cast tape, or sheet, which may be cut into appropriate thin shapes. The band is of plastics or other non-porous material, and removal of the liquid carrier medium, which is usually organic, is by evaporation. Good dispersion of the powder in the carrier liquid is essential before adding binders and plasticizers. Fatty acids, their esters, or synthetic polymer dispersants and deflocculants have been used.

Tapered Wheel. An abrasive wheel that is thicker at the hub than at the circumference.

Tapestry Brick. A US term for a brick with a rough textured surface.

Taphole. A small passage through the refractory lining of the hearth-wall of a metallurgical furnace, e.g. a blast furnace or open-hearth steel furnace.

Taphole Clay. The plastic refractory material used to plug the tap-hole of a blast furnace generally consists of a mixture of fireclay and grog together with carbon in the form of coal or coke (in USA pitch is also sometimes used). The range of chemical composition of a number of British taphole clays is (per cent): SiO_2, 40–60; Al_2O_3, 22–27; loss on ignition, 8–18. The material is adjusted to have a moisture content of 15–18% and is forced into the taphole by means of a 'gun'.

Taphole Gun. A gun-shaped device used to force TAPHOLE CLAY (q.v.) into the TAPHOLE (q.v.) of a blast furnace after it has been tapped.

Tapped Wirecut. This term has been used in the Ibstock (England) area for a wire-cut facing brick; these bricks were made on a side-cutting table and as the pallets were lifted from the table they were given a gentle tap to fettle the edges where the wires had emerged from the clay column.

Tapping Cone. A protective refractory fibre cone for the end of a tapping bar, used to seal a furnace TAPHOLE.

Tappit Hen. A 3-quart wine bottle.

Tar-bonded Refractory. Tar or pitch is added (4 to 6%) as a binder to a graded dolomite or magnesite refractory grain in a heated mixer. The mixed batch is stored at about 100°C and shaped by hydraulic or vibratory pressing. The resultant blocks may be 'tempered' at 250–350°C to improve hydration resistance and carbon retention. The finished blocks are sprayed with tar and wrapped in plastics to prevent moisture absorption. The blocks are used to line furnaces, particularly basic oxygen steelmaking furnaces, being heated in-situ, during which process the tar or pitch forms carbon and the refractory is fired, bonding the grains. (For relevant properties of tars and pitches, see W.D. Betts, *Trans. J. Br. Ceram. Soc.* **74** (3), 85, 1975).

Tar-impregnated Refractory. A fired refractory, particularly one subject to hydration, such as dolomite or magnesite, whose pores have been filled with tar by autoclaving at c. 200°C, evacuating and pouring in tar. Tar impregnated refractories are used to line basic oxygen furnaces.

Tarras Cement. TRASS (q.v.).

TCB. Tapered Cantilever Beam. See FRACTURE TOUGHNESS TESTS.

TD Diffusion Coating. Carbides are coated on to iron and steel by dipping in a molten salt bath for 1–10 hours. The bath is mainly borax added to low cost carbide precursors. Coatings 5 to 15mm thick are deposited, and cooled in air or by quenching in oil or water to obtain optimum mechanical properties. (Toyota R & D Laboratories, Japan. *Engrs. Dig.* **36,** (7), 1975, p27).

Tea-dust Glaze. An opaque stoneware glaze, greenish in colour from reduced iron compounds, sometimes used by studio potters.

Teapot Ladle. A ladle, lined with refractory material and used in foundries for the transfer of molten iron or steel; its distinctive feature is the refractory dam below which the metal passes to reach the ladle spout; the refractory dam prevents the outflow of slag.

Tear. (1) An open crack in glass-ware (cf. CHECK). (2) An excess spot of enamel on the edge of a vitreous-enamel sheet caused by too thick a coating or by firing at too high a temperature.

Tearing. Cracks (usually partially healed) in the cover-coat of vitreous enamelware. The cause is high drying shrinkage resulting, most commonly, from too-fine grinding, too wet a slip, or too heavy an application of slip. The amount of electrolyte in the slip is important; the presence of sodium nitrite is beneficial.

Teaser. A glass-worker who controls the temperature of a pot or tank-furnace to that required for feeding the batch; also called FOUNDER.

Tecalemit System. Trade-name: an impulse system for the oil-firing (from the top or side) of ceramic kilns; the impulses are controlled by a compressed-air device (Tecalemit Engineering Ltd., London).

Technate. A less common name for DIDYMIUM (q.v.).

Technical Ceramics. As well as modern ENGINEERING CERAMICS, this term also includes more traditional electrical porcelains, steatites etc.

Teem. To pour molten metal (or glass) from a ladle (or pot).

Tegula. See ITALIAN TILES.

Telefunken Process. See under METALLIZING.

Tellus System. A system for bricklayers working inside steel converters. A horizontal working platform is hydraulically raised or lowered by a central column, around which is provided a spiral automatic feed system to deliver bricks to the bricklayers.

TEM. Transmission Electron Microscopy.
Temmoku. See TENMOKU.
Temper. The residual stress in annealed glass-ware as measured by comparison with STRAIN DISKS (q.v.). See also TEMPERING.
Temperature. For conversion from FAHRENHEIT to CENTIGRADE, and vice versa, see these entries.
7.6 Temperature. See SOFTENING POINT.
13.0 Temperature. See ANNEALING POINT.
Temperature Gradient Furnace. An electric-resistance furnace in which the heating element is wound round the furnace tube in such a manner that there is a steady temperature gradient along the axis of the furnace. Such a furnace is useful in the ceramic laboratory in that it will expose a long test-piece placed within the furnace to different temperatures at different points, so that the effect of various firing temperatures can be studied in a single operation.
Temperature Indicating Crayons or Paints. The marks made by such paints or crayons change colour at a range of predetermined temperatures, up to several hundred °C.
Tempered Dolomite Block. A tarred dolomite block that has been preheated at about 400–500°C to eliminate volatile material from the tar; this prevents the blocks from slumping while they are being heated up to the operating temperature of the furnace or steel converter in which they are being used. The process is also claimed to improve the hydration resistance.
Tempered Glass. See TOUGHENED GLASS.
Tempering. The mechanical treatment of moistened clay, or body, to disperse the water more uniformly and so to improve the plasticity; tempering is usually done in a solid-bottom edge-runner mill with the mullers raised slightly above the pan bottom.
Tempering Tub. A combined pan and vertical pug mill for the preparation of clay for brickmaking. The mixing pan is about 2m dia., a central vertical shaft carrying the mixing blades; the shaft continues downward as the shaft of the pug mill. In some districts of England this equipment is known as a SLUDGE PAN.
Tenmoku. A glaze containing iron compounds sometimes used by studio potters; it is lustrous black except where it is thinner and has oxidised to a red colour. A quoted composition is (%): feldspar, 50; quartz, 23; whiting, 10.9; kaolin, 5.9; magnesite, 2.3; iron oxide, 7.6.
Tensile Strength. The force required to break a specimen, when applied to pull apart the material. Direct tensile strength tests for ceramics have notorious difficulties in aligning the test pieces and gripping them properly without introducing bending forces. *Indirect tensile tests* such as the BRAZILIAN TEST (q.v.) and others involving splitting rings or cylinders have been adopted. ASTM C1006 is a splitting test for the tensile strength of masonry.
Terminology. See Appendix A, B.
Terra Alba. Pure white uncalcined gypsum.
Terra Cotta. Unglazed fired clay building blocks and moulded ornamental building components. In England, the use of architectural terra cotta reached its peak in the 19th century (cf. FAIENCE).
Terra Rosa. Haematite used as a red glaze pigment.
Terra Sigillata. A type of pottery having a fine-textured red body and a glossy surface, usually with a raised decoration. The Latin name derives from the pastilles

made from certain clay deposits in Lemnos and Samos (hence the alternative name *Samian ware*) and impressed with the sacred symbol of Diana; these pastilles were used medicinally.

Tertiary Air. Preheated air introduced into the waste-gas flue of a kiln firing under reducing conditions, e.g. a blue-brick kiln; its purpose is to burn the combustible matter in the gases leaving the kiln chamber, thus helping to minimize smoke emission from the stack (cf. PRIMARY AIR and SECONDARY AIR).

Tessara(e). The small rectangular tile(s) used in mosaics.

Tessha. A version of the TENMOKU (q.v.) glaze; it is described by Bernard Leach (*A Potter's Book*, (Faber) London, 1945) as 'more metallic and broken than Tenmoku'.

Test Cone. A test-piece cut or moulded from a sample of refractory material that is to be tested for refractoriness or P.C.E. (q.v.). The shape is that of a pyramid on a triangular base; the dimensions vary according to which national specification is being followed.

Tetracalcium Aluminoferrite. See BROWNMILLERITE.

Tetrahedron. See SILICATE STRUCTURES.

Textile Ceramics. The achievement of artistic effects with combinations of textiles and ceramics, in particular by using viscose-silica fibre, a cellulose fibre containing up to 33% silica, used for fire-resistant clothing and upholstery. Bowls, wall hangings and other ornaments of delicate texture can be produced.

Texture. (1) The physical property of a ceramic product determined by the shapes and sizes of the pores and the grading of the solid constituents. The texture can to some extent be evaluated in terms of porosity and permeability; additional information is provided by pore size measurement and the total internal surface area (as measured by gas adsorption, for example).
(2) The general surface appearance of a clay facing brick.

Tg Point. See TRANSFORMATION POINTS.

Thénard's Blue. A dark blue colour for use under-glaze in pottery decoration; it consists of approx. four parts cobalt oxide to five parts alumina.

Theoretical Density. The density of a material, assuming 100% purity and zero porosity.

Therm. A unit of heat equivalent to 100 000 Btu. (1 Btu = 1.055 kJ)

Thermal Analysis. A group of techniques in which a physical property of a substance is measured as a function of temperature, whilst the substance is subjected to a controlled temperature programme. The Nomenclature Committee of the International Committee on Thermal Analysis (ICTA) defines a wide range of techniques and their abbreviations, these being incorporated in ASTM E473–85. Those common in ceramics are (q.v.) DIFFERENTIAL THERMAL ANALYSIS (DTA); DIFFERENTIAL THERMO-GRAVIMETRY (DTG); DIFFERENTIAL SCANNING CALORIMETRY (DSC).

Thermal Barrier. The zone where the temperature is highest in the melting end of a glass-tank furnace.

Thermal Checking. Thermal checking of a glass container occurs when large amounts of heat are transferred from its surface very rapidly, and the resulting stresses exceed the strength of the glass.

Thermal Conductivity. The quantity of heat transmitted through a material in unit time, per unit temperature gradient along the direction of flow, and per unit of cross-sectional area. The thermal conductivity of crystalline ceramics of low porosity decreases with increasing temperature; that of porous ceramics having a glassy bond generally increases with temperature. Some typical room-temperature values are: Silicon carbide, 14; magnesite, 4.5; chrome-magnesite, 2.2; fireclay refractory, 1.2; insulating refractory, 0.4 W/m.K.

Thermal Conductivity Tests. There are two main classes: *Direct Tests.* A constant, measured thermal gradient is established in a test-piece. The heat input required to maintain it is measured. The test-pieces are usually rectangular prisms or cylinders, with 'guards' to ensure that a uniform temperature is maintained up to the edges. Tests of this type are B.S. 1902: Pt 5.5 Method 1902–505; B.S. 1902 Pt. 6, for ceramic fibre products up to 1300°C, and ASTM C201 and C202, the latter being a modification of the former for lower conductivity materials (up to 28.8 W/m.K). These ASTM tests for refractories have special procedures for unfired monolithic refractories (C417); insulating firebrick (C182) and carbon refractories (C767).

In *Comparative Tests* the same constant heat flow is established through the specimen to be measured, and a standard sample of known conductivity. The different temperature gradients across the two samples are compared. (B.S. 1902 Pt. 5.8 specifies a test of this type for refractories, Pt.6 Section 6 a guarded hot-plate test up to 1200°C for ceramic fibre products; ASTM C408 one for ceramic whitewares, using a copper standard from 40 to 150°C.

For a more rapid method see HOT-WIRE METHOD; for rapid, transient heat flow methods see THERMAL DIFFUSIVITY.

Thermal Cycling Test. Refractories subject to repeated changes in temperature (thermal cycling) lose strength. ASTM C1171 evaluates this effect by using sonic techniques to measure modulus of elasticity, or by measuring cold modulus of rupture, before and after specified cyclic thermal treatment.

Thermal Diffusivity. A measure of the rate of change of temperature at a point in a material when heat is applied or removed at another point, e.g. the rate of rise of temperature at the centre of a furnace wall when the furnace is heated. The thermal diffusivity is the ratio of the thermal conductivity of the material to the product of the bulk density and the specific heat. Two methods of measurement used for ceramics are the *Angström Method,* in which a sinusoidal variation in temperature is applied to one face of a specimen, and the changes in amplitude and phase of the temperature wave are measured on the opposite (parallel) face; and the *Flash Method,* in which a short duration heat pulse (from a flash tube or laser) is applied to one face, and the temperature rise on the opposite face is recorded electronically as a function of time. (ASTM E1461–92). Both methods are rapid and suitable for relatively small samples, which may be heated in a small furnace to measure the variation of thermal diffusivity with temperature.

Thermal Endurance. The ability of a piece of glass-ware to resist thermal shock. An attempt to assess this in terms of the physical properties of the glass is offered by the WINKELMANN AND SCHOTT EQUATION (q.v.).

Thermal Expansion. The reversible increase in dimensions of a material when it is heated. Normally, the linear expansion is quoted, either as a percentage or as a coefficient, in either case over a stated temperature range; for example, the thermal expansion of a silica refractory may be quoted as 1·2%

or as 12×10^{-6} (per °C, implied), between 0 and 1000°C. B.S. 1902 describes a horizontal linear method (Pt. 5.3) for use up to 1100°C, and a comparative method for large test pieces, suitable up to 1600°C to provide data for furnace design calculations (B.S. 1902 Pt. 5.4) Pt. 5.14 describes horizontal and vertical dilatometer methods suitable up to 1500°C. ASTM has two tests, interferometric (C539) and a dilatometer method (C372) for vitreous enamel, glaze frits and fired whitewares. B.S. 7030 specifies a method for glass.

Thermal Expansion of Composites. Kerner derived an equation incorporating shear effect, to predict the thermal expansion of a two phase composite. Without shear, a simpler expression due to Turner can be used. (see W.D. Kingery, H.K. Bowen and D.R. Uhlman, *Introduction to Ceramics*, Wiley 1976, p 307).

Thermal Expansion Factors for Glass. Factors that have been proposed from time to time for the calculation of the coefficient of linear thermal expansion of a glass on the assumption that this is an additive property. The factors are inserted in the equation $\lambda (= 10^8 \Delta l / l \Delta T) = a_1 p_1 + a_2 p_2 + ... a_n p_n$ where p_1, etc., are the percentages of the oxides and a_1, etc., are the factors in the following table:

	Winkel-mann & Schott (1)	English & Turner (2)	Gilard & Dubrul (3)	Hall (4)
SiO_2	2.67	0.50	0.4	–
B_2O_3	0.33	–6.53	–4+0.1p	2.0
Na_2O	33.33	41.6	51–0.33p	38.0
K_2O	28.33	39.0	42–0.33p	30.0
MgO	0.33	4.5	0	2.0
CaO	16.67	16.3	7.5+0.35p	15.0
ZnO	6.0	7.0	7.75–0.25p	10.0
BaO	10.0	14.0	9.1+0.14p	12.0
PbO	10.0	10.6	11.5–0.05p	7.5
Al_2O_3	16.67	1.4	2	5.0

1. A. Winkelmann and O. Schott, *Ann. Physik.*, **51**, 735, 1894.
2. S. English and W. E. S. Turner, *J. Amer. Ceram. Soc.* **10**, 551, 1927; ibid., **12**, 760, 1929.
3. P. Gilard and L. Dubrul, *Verres Silicates Ind.* **5**, 122, 141, 1934.
4. F. P. Hall, *J. Amer. Ceram. Soc.* **13**, 182, 1930.

Thermal Shock. Sudden heating or cooling: the stresses set up by the differential expansion or contraction between the outside and inside of a thermally-shocked ceramic may cause it to crack.

Thermal Shock Tests. *Panel Test.* (ASTM C38–89 and B.S. 1902 Pt.5.11) Panels of refractory bricks are heated to the test temperature, cooled by water spray and the weight loss and appearance after a predetermined number of such cycles are observed. See PANEL SPALLING TEST.

Cylinder Test (Water Quench). (PRE/R5 Part 1) the number of cycles required to rupture a hollow cylinder 50mm diameter 50mm high by cycling between a furnace at 950°C and running water at 10/20°C is recorded. The test is severe; for types of refractory unsuited to it, there is:

Prism Test (Air Quench) (PRE/R5 Part 2), test pieces $114 \times 64 \times 64$ mm are cycled between 950°C and a cold air blast for 5 min. followed by a bend test under a load of 0–3MN/m^2. The cycle is repeated to rupture, or for 30 cycles.

Small Prism Test (B.S. 1902 S.5.11 1986 Method 1902–511. The test pieces are $75 \times 50 \times 50$ mm and subjected to 10 min. cycles between cold and 1000–1200°C, and subjected to an equal stress in each cycle. Testing is to rupture.

Water-Quench Test (for unshaped refractories) B.S. 1902 Pt.7.6. The test is a water quench cycle from 1200°C to cold running water, for 20 cycles or until 10% of the specimen weight has been lost. The test conditions are specifically designed for samples cast from monolithic insulating refractories.

Ribbon Test (B.S. 1902 Pt.5.12) The lower faces of a series of bricks are heated by a linear burner 1.52m long at 980°C for 15min. The bricks are then cooled to 200°C. The reduction in modulus of elasticity (using a non-destructive ultrasonic method) after 5–10 cycles provides a measure of themal shock resistance. ASTM C1100 also specifies a ribbon test.

ASTM C484 is a test for ceramic tiles, which are removed from an oven at 145°C and placed on a sheet of aluminium at room temperature. C149 is a test for glass containers involving transfer from hot to cold water. C600 is a water-quench test for glass pipes.

Thermal Spalling. SPALLING (q.v.) caused by stresses resulting from non-uniform dimensional changes of a brick or block produced by a difference in temperature.

Thermal Transmittance. (U-value). The heat flux density through a given structure divided by the difference in environmental temperatures on either side of the structure under steady-state conditions. The units are W/m^2K. BS 874 Pt3 describes a calculated hot-box method for its determination for large structural elements.

Thermal Wheel. See Munter Wheel.

Thermalite Ytong. Trade-name: A lightweight concrete made from portland cement, sand, and pulverized fuel ash; these are well mixed with water and a small proportion of aluminium powder is then added. This causes gas bubbles to form. Blocks of the cellulated material are then autoclaved. The material is a development of the Swedish material YTONG (q.v.). (Thermalite Ytong Ltd, Lea Marston, Sutton Coldfield, England.)

Thermistor. A material having a high temperature coefficient of electrical resistance. Ceramic thermistors having a negative temperature coefficient (the usual type) are made from mixtures of the oxides of Mn, Fe, Ni, Co, Cu and U; the batch is shaped and fired to low porosity. The electrical resistance of $BaTiO_3$ to which rare earths, Bi or Th have been added has a high positive temperature coefficient. The properties of thermistors make them useful in various instruments and controllers.

Thermocouple. A device for the measurement of temperature on the basis of the electric current generated when the junction between two dissimilar conductors is heated. The common types of thermocouple, with their maximum temperature of use (when protected) are: Copper/Constantan, 400°C; Iron-Constantan, 1000°C; Chromel/Alumel, 1200°C; Platinum Platinum-Rhodium, 1700°C. Thermocouples with Tungsten/Iridium, or with Silicon-Carbide/Carbon, elements have been used for special purposes at 2000°C. B.S. 1041 Pt. 4 gives guidance on the choice and use of an appropriate thermocouple.

Thermocromaticity. The phenomena of colour value readings changing as the temperature of the object being measured varies (R.L. Weintraub, *Ceram. Eng. Sci Proc.* **13**, (1/2), 1992, p 385.

Thermodyne Test. A test to determine the durability of optical glass in contact with moist air. Freshly broken or optically polished surfaces, sealed off in a flask together with a quantity of water, are subjected to a series of temperature cycles, each of 2 h duration, from 15 to 60°C in air saturated with water vapour for a period of 12 days. (W. M. Hampton, *Proc. Physical Soc.,* **54**, 391, 1942.)

Thermography. The observation of the temperature distribution across buildings or objects by infra-red photography. Photographic prints, or computer-processed images, show regions in various colours, corresponding to their temperatures.

Thermoluminescence. Emission of light by heated ceramics. The energy resulting from radioactive decay of small concentrations of radioactive materials in ceramic raw materials is stored in the crystal lattices by electron trapping in crystal flaws such as lattice vacancies and impurity sites. Some of these bonds are stable at room temperature, but disturbed by temperatures above c.300°C, when the energy is emitted as light. Firing would have released any prior trapped electrons, so the degree of thermoluminescence of fired pottery when it is re-heated gives some indication of the time since it was fired, (i.e. an estimate of the age of ancient pottery), when compared to the thermoluminescence produced by a known dose of radiation in the laboratory. Aitken Tite & Reid, *Nature* **202**, (4936), 1032, 1964.

Thermoplastic Decoration. Colours are dispersed in a thermoplastic medium (one which softens when heated and hardens when cooled, without changing other properties). They are then applied through a hot screen (see SCREEN PRINTING) on to cold ware, which hardens the design.

Tbermoscope. See PYROSCOPE.

Thick-bed. Mortar or adhesive layers for TILE-FIXING (q.v.) greater than 6mm thick.

Thick-film Electronics. Electronic circuitry produced by printing circuit components on to a substrate.

Thimble. (1) An item of kiln furniture, Fig. 4, p177; it is conical, hollow, and with a projection at the base to support the pottery-ware being fired; thimbles are inserted into one another so that a bung of ware can be built up.
(2) A refractory shape, usually resembling the letter L, used to stir optical glass in a pot.

Thimble-Bat. A refractory BAT (q.v.) of a type used in the firing of pottery; it is perforated to hold the ends of THIMBLES (q.v.).

Thin Bed. Mortar or adhesive layers for TILE-FIXING (q.v.) up to about 3mm thick.

Thin-film Electronics. Electronic circuitry produced by depositing circuit components of microscopic dimensions by CVD, vacuum and sputtering techniques.

Thin Section. A section of material that has been prepared, by grinding to extreme thinness (about 30μm), for examination by transmitted light under a polarizing microscope. For details of the technique see such books as *Thin Section Mineralogy of Ceramic Materials* by G. R. Rigby, 2nd Ed. 1953, and *Microscopy of Ceramics and Cements* by H. Insley and D. Frechette, 1955.

Thirsty Glass. POROUS GLASS (q.v.), so-called because of its affinity for moisture.

Thirting. Cutting ball clay into blocks about 0.25m cube.

Thivier Earth. A siliceous hydrated iron oxide from Thivier, 19 miles N.E. of Perigueux, France. A quoted composition is: 83% SiO_2, 10% Fe_2O_3, 2% Al_2O_3, 1% CaO, 1% alkalis, 3% loss on ignition. It has been used as a red colour for pottery decoration.

Thixotropy. The property, exhibited by many clay slips for example, of becoming more viscous when left undisturbed but more fluid when stirred. The name was first proposed by T. Peterfi (*Arch. Entwicklungsmech.* **112,** 689, 1927).

Thoria. See THORIUM OXIDE.

Thorium Borides. At least two borides are known: ThB_4 (grey) and ThB_6 (deep red). The more attention has been paid to the tetraboride, the properties of which are: m.p., 2200°C (but oxidizes slowly above 1000°C); thermal expansion, 5.9×10^{-6} (20–1000°C); sp. gr. 8.45 g/ml; modulus of rupture (20°C), 140 MNm^{-2}. Some properties of ThB_6 are: m.p. 2200°C; sp. gr. 6.1.

Thorium Carbides. Two carbides are known: ThC, m.p. 2625°C; ThC_2, m.p. 2700°C. Monoclinic ThC_2 inverts to the tetragonal form at 1410 $\pm$ 20°C and from tetragonal to cubic at 1500° $\pm$ 20°C. These special carbides are of potential interest in nuclear engineering.

Thorium Nitride. Three nitrides have been reported: ThN, m.p. 2650°C; Th_2N_3, decomposes at 1750°C; Th_3N_4.

Thorium Oxide or Thoria. ThO_2; m.p. approx. 3220°C; sp. gr. 9.8. Thermal expansion (20–1000°C) 9.3×10^{-6}; (20–2500°C)10.9×10^{-6}.

Thoria has been used for crucibles for the melting of refractory metals but it has poor thermal-shock resistance.

Thorium Sulphides. Three sulphides have been reported: Th_4S_7, Th_2S_3 and ThS. Crucibles made of these sulphides have been used as containers for molten Ce.

Thorpe's Ratio. A formula suggested by Prof. T. E. Thorpe in 1901 for assessing the probable solubility of a lead frit: the sum of the bases expressed as PbO divided by the sum of the acid oxides expressed as SiO_2 should not exceed 2.

Thread Guide. Porcelain thread guides are satisfactory for use with cotton, wool, or silk; man-made fibres, e.g. rayon and nylon, are more abrasive and sintered alumina or synthetic sapphire thread guides are used. Thread guides have been made from or coated with titanium dioxide heated in hydrogen. The white TiO_2 insulator turns black, and becomes conducting (see MAGNELI PHASE) which prevents the frictional build-up of static electricity on the guides.

Three-edge Bearing Method. In measuring the load-carrying capacity of concrete pipes, the load is applied at the centre of the pipe, resting on two outside edges.

Theshold Limit Value (TLV). That concentration of (possiblity harmful) substance to which it is believed nearly all workers may be exposed day after day without adverse effect. The values are time-weighted average concentrations for a 7 or 8 hour day and a 40 hour week.

Throat. (1) The part of a blast furnace at the top of the stack.

(2) The zone of decreased cross-section found between the port area and the furnace chamber in some designs of open-hearth steel furnace.

(3) The submerged passage connecting the melting end to the working end of a glass-tank furnace; the refractory blocks forming the sides of the throat are known as THROAT CHEEKS, SLEEPER BLOCKS or DICE BLOCKS, the refractories for the top are the THROAT COVER. A throat below the bottom of the melter is known variously as a *submerged, submarine or sump* throat.

Throwing. The method of shaping pottery hollow-ware in which a ball of the prepared body is thrown on a revolving potter's wheel, where it is centred and then worked into shape with the hands. The process is now chiefly used by studio potters, although a small amount of high-class commercial pottery is still made in this way.

Thwacking. The process by which clay pantiles are given their final curved shape. When partially dry, the tiles are placed on a wooden block of the correct curvature and beaten to that contour by means of a bevelled block of wood (hence the alternative name BLOCKING). Any distortion of the tile caused by unequal shrinkage during the preliminary drying is thus corrected.

Tickell Roundness Number. An index of the shape of a particle in terms of the ratio of the actual area of the projection of the grain to the area of the smallest circumscribing circle; (F. G. Tickell, *The Examination of Fragmental Rocks* 1931).

Tiering. See TORCHING.

Tiger Eye. A particular type of AVENTURINE (q.v.) glaze.

Tiger Skin. A decorative SALT GLAZE (q.v.), in which CRAWLING gives a striped or mottled appearance.

Tile. See FLOOR TILE, ROOFING TILE, WALL TILE. In Europe, the chief distinction is between glazed and unglazed tiles, rather than between floor tiles and wall tiles; a given range of tiles may be supplied for either application. B.S. 6431 Pt.1 classifies tiles according to method of manufacture (extruded, dust-pressed or cast) and their water absorption (group I low 3%; medium 3–10, subdivided IIa 3–6, IIb 6–10; high10%, Group III). Other parts of the standard specify a wide range of requirements and test methods. Standards for glazed wall and floor tiles in the USA deal separately with individual properties and test methods. See ASTM C482, C483, C484, C485, C499, C501, C502, C609, C627, C648, C650, C895, C1026, C1027, C1028. In the USA the word 'tile' also denotes hollow clay building blocks, in such self-evident combinations as 'load bearing tile,' 'partition tile;' DRAIN TILES and AGRICULTURAL TILES are FIELD DRAIN PIPES. The following ASTM standards relate to structural clay tile: C212, facing tile; C34 load-bearing wall tile; C56 non-loadbearing screen tile.

Tile Bridging. Shrinkage of the base on which tiles are laid leads to compression of the tiling, and the formation of a gap behind the tile(s), which arch clear of the base. (New concrete floors can undergo substantial shrinkage.)

Tilebacker. Extruded polystyrene foam, reinforced with glass-fibre mat, and plastic modified mortar used as an underlay for tiles.

Tile Fillet. Roofing tiles cut and fitted to form a fillet as an alternative to flashings. See FLASH (2).

Tile Hanging. The process of fixing roofing tiles (as distinct from wall tiles) on a vertical outside wall.

Tile Listing. Roofing tiles used to form a splayed fillet at abutments.

Tilemaster. An electronic system, based on an array of 512 × 512 light-emitting diodes, used to measure length, width and rectangularity of tiles, and to detect damaged edges and corners.

Tiling. The fixing of tiles to walls and floors is governed by several British Standards, B.S. 5383 is a code of practice for the design and installation of internal tiles and mosaics for walls and floors. B.S. 8000 specifies workmanship on building sites. Pt.11 specifying a code of practice for ceramic, terrazzo, mosaic and natural stone tiles. Pt.9 specifies floor screeds; Pt.6 slating and tiling of roofs. B.S. 5980 is the specification for wall tile adhesives. Proper specification of appropriate adhesives and fixing methods, and a high standard of workmanship are essential to the success of tiling.

Till. Older name for BOULDER CLAY (q.v.).

TI-LOC Process. Trade-name: a process for the treatment of steel prior to enamelling, it is claimed to improve adherence and to eliminate the need for a ground-coat. (Strong Mfg. Co., Sebring, Ohio, USA.)

Timbrell–Coulter Shearicon. Semi-automatic image-shearing equipment to size particles or features observed by optical microscopy, or of images from optical or electron photomicrographs, or a television picture. (*J.Powder Bulk Solids Technol.* **3** (1), 26, 1979).

Tin Ashes. Mixed oxides of tin and lead formerly used as a constituent of some glazes; the metals were mixed in a ratio that would produce SnO_2 and PbO in the required proportions when the alloy was oxidized in a preliminary calcination.

Tin Oil. A mixture of $SnCl_2$, $SnCl_4$, S and oils; it is used in the preparation of some types of LUSTRE (q.v.).

Tin Oxide. Two oxides exist but the more common is stannic oxide, SnO_2, m.p. approx. 1630°C; sp. gr. 7.0. It is used as an opacifier and in the preparation of colours, e.g. CHROME–TIN PINK (q.v.). Properties making this oxide of some interest as a special ceramic are its high thermal conductivity and low thermal expansion (20–1000°C) 4×10^{-6}; it is sensitive to reducing atmospheres. Tin oxide has been used for making refractory feeders for high-lead glasses.

Tin-Vanadium Yellow. See VANADIUM YELLOW.

Tinsel. Very thin glass that has been crushed and silvered for use as a decorative material (cf. GLASS FROST).

Tirschenreuther Pegmatite. A by-product of china-clay washing, comprising quartz and feldspar, used in porcelain bodies and in spray-dried prepared bodies. (Hutschenreuther AG, Germany).

Tit. A small imperfection on glassware.

Titanate Ceramics. A group of electro-ceramic materials generally based on the compound BARIUM TITANATE (q.v.) but often with the addition of other titanates, zirconates, stannates or niobates. These ceramics are notable for their high dielectric constant (up to, and even exceeding, 10 000 compared with a value of 5–10 for the more common ceramic materials); because of this property they find use in capacitors. Titanate ceramics are also used where piezoelectric properties are needed, i.e. in transducers.

Titania. See TITANIUM OXIDE.

Titanising. Trade-name (United Glass Ltd., England). A process for the strengthening of glassware by applying liquid titanium compounds to the glass surface while it is still hot; the glass thereby acquires a small amount of metal oxide in the surface layer, and this enables the article to retain a greater proportion of its strength when in use.

Titanite. See SPHENE.

Titanium Borides. The most important is TiB_2; m.p. 2920°C with excellent oxidation-resistance up to 1000°C; sp. gr. (theoretical). 4–5, Knoop hardness (K100), 2710; thermal expansion, 9.7×10^{-6} (200–1800°C). Modulus of elasticity (GNm^{-2}) of material of density 4.21: 43.5 (20°C), 42 (1000°C), 30 (2000°C). Transverse strength MNm^{-2} at 25°C is greatly dependent on crystal size: 20mm, 300; 35mm, 200; 100mm, 105. Thermal conductivity 0.10 c.g.s. (1000–1400°C). Electrical resistivity (20°C) 20 mΩ-cm increasing to 80 at 1200°C. Inert to molten Al and Zn but not to ferrous alloys. It has been used in rocket nozzles, as a constituent of cermets, and for boats for the evaporation of Al (for thin-film coatings). Other compounds that have been reported include TiB, Ti_2B and Ti_2B_5, but these are stable only in the absence of carbon.

Titanium Carbide. TiC; m.p. 3150°C; sp. gr. 4.9, thermal expansion (25–1000°C)

8×10^{-6} (25-2000°C) 9×10^{-6}. A hard refractory compound; it has found use as the ceramic component of some cermets.

Titanium Dioxide. See TITANIUM OXIDES.

Titanium Nitride. TiN; a special refractory material; m.p. approx. 3000°C; thermal expansion (25–1400°C) 9×10^{-6}. It can readily be produced from $TiCl_4$ and NH_3. TiN coatings on metal cutting tools act as a diffusion barrier to impurities. They have higher electrical conductivity and better adhesion than alumina, and are used as an undercoat for Al_2O_3 coatings on tungsten carbide. Thin TiN films show iridescent interference effects, and are used as decorative hard surfacing for watches and jewelry.

Titanium Oxides. The common oxide is TiO_2; m.p. approx. 1850°C. Used as an opacifier, particularly in vitreous enamels, and as a constituent of some ceramic colours. Titania and titanate electroceramics, for use in the radio-frequency field, are based on this oxide and its compounds. Titania occurs in three crystalline forms: ANATASE, BROOKITE and RUTILE (see under each mineral name). Ti_2O, TiO, Ti_2O_3 and Ti_3O_5 also exist.

Titanium Silicides. Several compounds exist. Ti_5Si_3; m.p. 2120°C; sp. gr. 4.3; has good resistance to high-temperature oxidation but not to thermal shock.

Titanium Suboxides. See MAGNELI PHASE.

Titanizing. See TITANISING.

Titzicon. $Na_{0.8+x}\ Zr_{1.55}Si_xP_{3-x}O_{11}$ see NASICON, of which titzicon is a modification usually with enhanced amount (30%) of glassy phase.

TLV. Threshold Limit Value.

Tobermorite. A hydrated calcium silicate approximating in composition to $5CaO.6SiO_2.5H_2O$. Tobermorite gel is the principal cementing compound in hardened portland cement.

Toggle Press. See MECHANICAL PRESS.

Toki. Japanese hard-fired eathenware. A quoted body recipe is: 70% clay, 10% pegmatite, 20% ground pitchers.

Tomography. Investigation of the internal structure of an article based on the detection of emissions from radioactive elements distributed within the item. A 3-dimensional representation of the internal structure is built up by computerised reconstruction from a succession of 2-dimensional cross-sections. Resolutions of 10μm can be achieved.

Tong Outcrop Clay. A fireclay associated with the Better Bed coal, Yorkshire, England. The raw clay contains about 65% SiO_2, 22% Al_2O_3, 1.5% Fe_2O_3 and 1.6% alkalis.

Tongue. See MIDFEATHER.

Tonstein. A hard kaolinitic deposit sometimes found as thin bands in the Coal Measures. The kaolinite crystals are large but are not orientated, having apparently been formed *in situ* from a gel. Tonstein differs from FLINT CLAY (q.v.) in that it is not a LATERITE (q.v.). Tonsteins were first observed in Germany; the name means 'Clay Stone'.

Tool (Ceramic). Ceramic tools for the machining of metals retain strength, hardness and wear-resistance at high temperatures, giving long life and clean cutting at high cutting speeds. Materials used are Sialons, boron carbide, cubic boron nitride, hard metals especially Co-bonded tungsten carbide, cemented oxide (alumina bonded with a transition metal) and pure alumina (to which a small amount of MgO is added to inhibit crystal growth to 10μm).

Tooling. The rubber or plastic bag used in ISOSTATIC PRESSING (q.v.); the term may also include the rigid core or former needed for the pressing of ware that is hollow or recessed.

Toothing. In structural brickwork, bricks left projecting at the temporary termination of a wall so that future extensions can be bonded in.

Top-Blown Converter. See CONVERTER.

Top-fired Kiln. A kiln with the burners in the roof, rather than in the walls, or fired by feeding coal or oil through apertures in the roof. The typical kiln of this type is the Hoffmann annular kiln, but the Monnier kiln provides an example of a topfired car-tunnel kiln.

Top-hat Kiln. An intermittent kiln of a type sometimes used in the firing of pottery. The ware is set on a refractory hearth, or plinth, over which a box-shaped cover is then lowered.

Top-Jet Firing System. For more complete utilisation of the firing space in a tunnel kiln, the flue-gas passage under the setting is omitted. High velocity burners, staggered longitudinally, are installed vertically in the sidewalls above the usable firing space. The burners are directed downwards and the combustion gases directly led into the car loading above the setting level. The system was introduced by Riedhammer (Germany) *(Ceram Forum Int/DKG* (6), 466, 1981).

Topping Material. Refractory above the metal level in an induction furnace.

Top Pouring. See DIRECT TEEMING.

Topaz. $Al_2SiO_4(OH,F)_2$. Occurs in economic quantities in Australia, Brazil, Nigeria and USA. After it has been calcined, the material has a composition similar to that of mullite and it has been used to a small extent, either alone or mixed with calcined kyanite, for making high-alumina refractories.

Torbed Process. A toroidal, rotating, fluidized bed of particulate feed some 20–50mm deep is established by a fan which directs a flow of air through a series of rotating apertures in a steel or ceramic ring. High specific throughputs are achievable from a compact plant, easily dust-sealed. The system can be used for continuous calcining, expansion, exfoliation, flash drying, combustion or liquid coating of particles, up to 1360°C.

Torching. The pointing, with cement or mortar, of the underside of a tiled roof; also sometimes known as TIERING (cf. SARKING) .

Torkret Process. German process for spraying refractory patching material on the walls of steel-furnace ladles.

Tornebohm's Minerals. See ALITE; BELITE; CELITE; and FELITE.

Torsion Viscometer. An instrument much used for works' control of the viscosity and thixotropy of clay slips. It consists of two concentric cylinders, the slip occupying the space between them; the inner cylinder is suspended from a torsion wire and is released from a position equivalent to a 360° twist in the wire. The *overswing* (past zero) gives a measure of viscosity; comparison of this degree of overswing with that after a specified lapse of time provides a measure of the thixotropy of the slip.

Tortoiseshell. A decorative effect produced on a lead glaze by dusting metal oxides (MnO_2, CoO or CuO) over the surface and firing.

Tortus. A floor friction test machine. A tracked trolley, about 30 cm long, drags a spring-loaded probe about 1cm diameter across the floor to be tested. Various materials (e.g. leather, rubber) are used to tip the probe. A continuous reading of the frictional force between the probe and floor is obtained. The method was devised by British Ceramic Research Ltd, and the machine made by Severn Science (Instruments) Ltd, Bristol.

Toseki. Japanese china stone (q.v.).

Total Energy System. The use of e.g. diesel generators to produce mechanical power, heat and electricity on-site.
Total Transfer Printing. A development from the MURRAY-CURVEX MACHINE (q.v.) in which the whole of the design is transferred from design plate to ware. This enables multicoloured printing by successive operations of the same machine. The design is prepared on a heated metal screen, and squeegeed on to a warmed silicone rubber substrate, from which it is picked up by a silicone pad, and transferred to cold ceramic ware. The key to the total transfer is that as the medium of the colour cools, it becomes more tacky, and careful development of the inks and pads, with proper combination of surface properties, allows complete transfer of the design. (British Ceramic Research Ltd, UK Pat 2118900 A, 1983 incorporated into a commercial machine by Service Engineers plc).
Totanin. Trade-name: an ammonium-based SYLPHITE LYE (q.v.) (Lambeth and Co. Ltd., Liverpool).
Toughened Glass. Glass that has been rapidly cooled so that the surface layers are in compression; the thermal and mechanical endurance are increased and, if the glass does break, it will shatter into small, granular, fragments rather than into large and dangerously jagged pieces.
Toughness. The ability of a material to resist FRACTURE (q.v.). Various mechanisms exist which absorb the energy required to form new fracture surfaces. These increase the toughness of the ceramic and may act concurrently, to enhance the toughening effect. See also FRACTURE TOUGHNESS; STRESS INTENSITY FACTOR; CRACK BRIDGING; QUASI-BRITTLE FRACTURE; STRAIN HARDENING; CRACK BRANCHING; CRACK DEFLECTION; TRANSFORMATION TOUGHENING; WHISKER REINFORCEMENT; FIBRE REINFORCEMENT; MULTIPLE CRACKING; DISPERSION STRENGTHENING.
Tourmaline. A complex alumino-borosilicate that will often contain small amounts of iron and alkalis. When heated, tourmaline loses water at 150–750°C; B_2O_3 is evolved at 950–970°C. Tourmaline is an accessory mineral in the granite from which the Cornish china clays are derived.
Tower Packing. Increased surface area for the reaction, and longer durations, are provided for chemical reactions if the liquid reagents are allowed to trickle down a tower filled with inert shapes of high surface area. Chemical porcelain or other ceramics are frequently used. Such tower packings may also act as CATALYST CARRIERS (q.v.). The ASTM Standard is C515.
Towing. The smoothing, generally on a powered wheel, of the outer edge of dried pottery flatware before it is fired. Tow is commonly used for the purpose, but sandpaper or a profile scraper are also sometimes used.
Trace Flue or Trace Hole. A small horizontal passage or flue left in a setting of bricks in a kiln to facilitate the movement of hot gases.
Trailing. A method of slip decoration sometimes used by the studio potter: a pattern is formed on the ware by means of a viscous slip squeezed through a fine orifice, e.g. a quill.
Tramp Iron. Accidental metallic contamination during manufacture.
Transducer. A device for the direct transformation of electrical energy into mechanical energy. PIEZOELECTRIC (q.v.) ceramics are used for this purpose.
Transfer. See DECAL.
Transfer Glass. US term for optical glass that has been cooled in the pot in which it was melted (cf. ROLLED GLASS).

Transfer Ladle or Hot-metal Ladle. A large ladle lined with refractory material (usually fireclay bricks) for the transport of molten pig-iron from a blast furnace to a hot-metal mixer or to a steelmaking furnace.

Transfer Printing. An intaglio process of decoration, particularly applicable to pottery-ware; a single-coloured pattern is transferred directly from a printing plate or roller by means of thin paper. The colour used is generally dispersed in linseed oil; soft-soap is the traditional size for the transfer paper but various synthetics have also been used (cf. LITHOGRAPHY).

Transfer Ring. See HOLDING RING.

Transformation Points. Temperatures at which the coefficient of thermal expansion of a glass changes. For any one glass, there are normally two such points known respectively as the Tg point and the Mg point: the Tg point is the first temperature at which there is a sudden change in expansion when the glass is heated at 4°C/min; the Mg point is the temperature at which the thermal expansion curve reaches a maximum and is usually equal to the softening temperature.

Transformation Toughening. Increasing the TOUGHNESS (q.v.) of a ceramic or composite by using a change in CRYSTAL STRUCTURE to absorb energy and inhibit FRACTURE (q.v.). The method is particularly applicable to ZIRCONIA (q.v.). Tetragonal zirconia crystals present in bulk zirconias are stable, except in highly stressed areas near crack tips. There, they undergo a MARTENSITIC TRANSFORMATION (q.v.) to monoclinic zirconia, absorbing energy and inhibiting CRACK PROPAGATION (q.v.). This toughening mechanism is effective in fine-grained single-phase tetragonal zirconia; in two-phase zirconia ceramics containing precipitated tetragonal zirconia; in two-phase alumina or silicon nitride containing a dispersed phase of tetragonal zirconia.

Transit-mixed Concrete. Concrete wholly mixed in the mixer lorry on the way to the site. (cf. SHRINK-MIXED CONCRETE).

Transition Aluminas. Metastable crystalline aluminas (there are several series) which contain minor amounts of OH^- ions. They are usually produced by gradual heating of alumina hydrates.

Transition Scarp. A RIB MARK when a crack changes from one mode of growth to another.

Translucency. Bodies are translucent if they transmit light, though not TRANSPARENT. (q.v.). Translucency is an important attribute, enhancing the appearance of certain types of high quality tableware. It agrees reasonably well with LAMBERT'S LAW (q.v.). B.S. 5416 1990 'China Tableware' includes a simple test for translucency, measuring the percentage of white light transmitted through a 2mm thick sample. The COMBINED NOMENCLATURE of the EC (q.v.) also includes a (different) test for translucency, based on the outline of an object being visible through ware 2 to 4mm thick, illuminated by a standard source.

Translucent Vitreous Silica. See under VITREOUS SILICA.

Transmission Electron Microscopy. See ELECTRON MICROSCOPY.

Transmutation Glaze. A term that has been applied to a glaze such as ROUGE FLAMBÉ (q.v.) or SANG DE BOEUF (q.v.).

Transparent Ceramics. Bodies are transparent if they transmit light without scattering or diffusion, so that objects on their far side can be seen clearly. For ceramics to be transparent, their crystal structure must not be disrupted by impurities, and careful sintering is required to remove any residual closed pores which might act as light scattering centres.

Transparent Vitreous Silica. See under VITREOUS SILICA.

Transverse-arch Kiln. An ANNULAR KILN (q.v.) that is divided into a series of chambers by fixed walls (hence the alternative name CONTINUOUS CHAMBER KILN). The axis of the arched roof of each chamber is transverse to the length of the kiln. This type of kiln finds use in the heavy-clay industry.

Transverse Strength. See MODULUS OF RUPTURE.

Trass. A product of the partial decomposition of volcanic ash; it is often consolidated as a result of infiltration of calcareous or siliceous solutions. Trass is used in POZZOLANA (q.v.).

Travelling Thermocouple. A continuous record of the temperature applied to ware in a (tunnel) kiln is obtained from a thermocouple arranged to travel with the ware through the kiln.

Traversing Rule. A wooden straight-edge used for floating a plane surface, by moving it, with a sawing motion, with its ends resting on screed battens or guides at the correct thickness. (B.S. 6100 Pt.6).

Trébuchon-Kieffer Annealing Schedule. A procedure for annealing the glass components of electron tubes; it is based on annealing at the transformation temperature for 20 min followed by cooling at a rate dependent on the nature of the glass and its thickness (G. Trebuchon and J. Kieffer, *Verres Refract.*, **4**, 230, 1950; *Glass Industry*, **32**, 240, 1951.)

Trefoil. A refractory structure in the preheaters of some rotary kilns, dividing the kiln into three channels to improve contact between the hot gases and the burden.

Tremolite. $2CaO.5MgO.8SiO_2.H_2O$. Tremolite is sometimes present in steatite rocks, rendering them unsuitable for use as a ceramic raw material.

Triangle Bar. One type of metal support for vitreous enamelware during firing.

Triaxial Test. A method of testing clay in which the test-piece, in a plastic state, is enclosed in a rubber envelope and is then subjected to uniform hydrostatic pressure while it is also being loaded axially. A stress/deformation diagram is plotted.

Triaxial Whiteware Body. One based on a three-component system of clay, filler and flux. Most traditional pottery bodies fall into this class. The filler is usually quartz, but may be alumina. Fluxes are usually feldspathic materials. Higher clay content improves unfired workability; higher flux content increases fusibility; the filler content affects thermal expansion and so glaze-body fit. In the triaxial quartz-feldspar-clay system, only a relatively small range of compositions are suitable for pottery bodies. (See A Dinsdale. *Pottery Science*, Ellis Horwood Ltd. 1986 for a concise discussion).

Tribomet Process. Trade-name. An electrodeposition process for the production of wear-resistant or lubricating coatings. The electrodeposited particles are metal particles. The composite particles may incorporate hard borides, oxides, nitrides or carbides; or molybdenium disulphide or other solid lubricant. (Bristol Aerojet. Br.Pats 1220078 and 1220331, 1971).

Tricalcium Aluminate. $3CaO.Al_2O_3$; melts incongruently at 1539 ± 5°C; thermal expansion (0–1200°C) $10{\cdot}5 \times 10^{-6}$ This compound is present in portland cement.

Tricalcium Disilicate. $3CaO.2SiO_2$ (see RANKINITE).

Tricalcium Pentaluminate. A compound, $3CaO.5Al_2O_3$, formerly believed to be present in high-alumina hydraulic cement. It is now known that a melt of this composition consists of a mixture of $CaO.2Al_2O_3$ and $CaO.Al_2O_3$, the latter compound being responsible for the hydraulic properties.

Tricalcium Silicate. $3CaO.SiO_2$ undergoes the following transformations: Triclinic I $\xrightarrow{620\,^{\circ}C}$ Triclinic II $\xrightarrow{920\,^{\circ}C}$ Triclinic III $\xrightarrow{950\,^{\circ}C}$ Monoclinic $\xrightarrow{990\,^{\circ}C}$ Orthorhombic $\xrightarrow{10050\,^{\circ}C}$ Hexagonal. Dissociates at approx. 1900°C to form CaO and $2CaO.SiO_2$; thermal expansion (0–1000°C) 12.9×10^{-6}. This compound is the principal cementing constituent of portland cement, small quantities of MgO and Al_2O_3 usually being present in solid solution. Tricalcium silicate is also present in some stabilized dolomite refractories.

Trichromatic Printing. The system of printing used to print coloured pictures on paper. The image is made up of tiny dots of three primary colours, present in such numbers as to give the impression of the desired colour by mixing. A fourth black pigment is usually added, to ensure dense blacks.

Tridymite. SiO_2; sp. gr. 2.28. According to the classical research of C. N. Fenner *(Amer. J. Sci.*, **36**, 331, 1913) tridymite is the form of silica that is stable between 870 and 1470°C; he considered that there are three crystalline varieties changing reversibly, one into another, as follows: α-tridymite (117°C) β_1-tridymite (163°C) β_2-tridymite.
More recent research has indicated a further inversion at about 250°C and has suggested that tridymite can be produced only in the presence of foreign ions, which enter the crystal lattice and cause disorder in the structure.

Trief Process. A process for making concrete with PORTLAND BLAST-FURNACE CEMENT (q.v.) first proposed by a Belgian, V. Trief (Brit. Pat., 673 866, 11/6/52; 674913, 2/7/52). The slag is wet ground and fed as a slurry to a concrete mixer together with portland cement and aggregate.

Trimmers. See FITTINGS.

Triple Brick. A brick 5½ × 4 × 12 in (135 × 102 × 305mm)

Triple Round Edge. A type of wall tile (see Fig. 7, p350).

Tripoli. A sedimentary rock consisting essentially of silica and having a porous and friable texture. A principal use is as an abrasive.

Tri-stimulus Values. A set of three chosen primary colours. All colours that can be made by additive mixtures of these three primaries are known as the *gamut* of the chosen primaries. The objective for TRICHROMATIC PRINTING (q.v.) is to find three primaries whose gamut covers the whole range of colours. The CIE (Commission International d'Eclairage) specified three colour wavelengths which when mixed give an equal-energy White: Red 700nm; Green 546.1nm; Blue 435.8nm. See COLOUR.

Triton. (1) A rotary pulveriser and mixer used to prepare clay slip (A. Faure & Co., Limoges).
(2) Trade-name, Wheatley & Co, Staffs, for FLOOR QUARRIES.
(3) Trade-name, under which KAOWOOL was originally sold in UK.
The tradename has also been used for a wetting agent, and, as *Triton Milan*, for roofing tiles.

Trommel. See REVOLVING SCREEN.

Trompe l'Oeil. Decoration, mainly applied as transfers, which deceives the eye to simulate e.g. bas-reliefs, or lustres to simulate silver or marble.

Tropenas Converter. See CONVERTER.

TRRL Pendulum Test. A test for the SLIP RESISTANCE of floor tiles and pavers, (originally developed by the Transport and Road Research

Laboratory to measure the skid resistance of road surfaces. Road Note 27, 1960). A weighted pendulum is adjusted to carry a rubber slider along a 5 inch (127mm) length of the surface to be tested. The pendulum is released from a horizontal position, carrying with it a pointer which records the upswing on the other side of the vertical. The greater the upswing, the lower the skid resistance of the surface (though the scale is marked in reverse to give low readings for low skid resistance).

Truck Chamber Kiln. See BOGIE KILN.

True Density. A term used when considering the density of a porous solid, e.g. a silica refractory. It is defined as the ratio of the mass of the material to its TRUE VOLUME (q.v.). (Sometimes referred to as POWDER DENSITY, though if the grains contain imprenetrable pores, the true density will be greater than the powder density, which is assessed by grinding the material to powder form. B.S. 1902 Pt. 3.4 specifies a PYCNOMETER (q.v.) method for refractories and Pt. 3.5 the REES-HUGILL FLASK method for powder density.

True Porcelain. HARD PORCELAIN (q.v.).

True Porosity. See under POROSITY.

True Specific Gravity. The ratio of the mass of a material to the mass of a quantity of water that, at 4°C, has a volume equal to the TRUE SOLID VOLUME of the material at the temperature of measurement. ASTM C604 specifies a gas-comparison pycnometer test, and C135 a water immersion test for refractories.

True Volume. A term used in relation to the density and volume of a porous solid, e.g. a brick. It is defined as the volume of the solid material only, the volume of any pores being neglected.

Trumpet or Bell. A fireclay refractory funnel placed at the top of the assembly of GUIDE TUBES (q.v.) to receive molten metal from the nozzle of a ladle in the BOTTOM POURING (q.v.) of steel.

Trutile. A plastics grid to aid tile-fixing. The grid is fixed to the base to be tiled, using a little tile adhesive. Its voids are filled with tiles, which are grouted without removing the grid.

T-T-T Curve. Time-temperature-transformation curves describing mineralogical changes.

TTZ. Transformation Toughened Zirconia (see ZIRCONIA and TRANSFORMATION TOUGHENING).

Tube Bottom. One form of bottom for a CONVERTER (q.v.); it is made of monolithic refractory material, the air passages being lined with copper tubes (cf. SPIKED BOTTOM and TUYERE BLOCK BOTTOM).

Tube Furnace. A type of furnace, particularly for vitreous enamelling, heated by tubes of heat-resisting metal in which gas is burned.

Tube Mill. A ball mill having a cylinder longer than usual, this usually being sub-divided internally so that the material to be ground passes from one compartment to the next, the grinding media in successive compartments being appropriately smaller than in preceding compartments.

Tube Press. A high pressure tubular filter press. The filter cake forms on the outside of a hollow perforated steel tube, around which the filter cloths are wrapped. Hydraulic pressure forces the liquid to the inside of the tube. (English Clays Lovering Pochin: various British Patents.

Tubelining. A process of decoration, particularly for wall tiles requiring 'one-off' designs. Lines of coloured slip are added to the tile by squeezing slip from a rubber bag through a narrow tube. The tile is then fired and the pattern between the raised lines is filled-in with various colours and refired; alternatively, the colours can be applied to the unfired tile and the once-fired process used.

Tuckstone; Tuckwall. A shaped refractory block fitting above the tank blocks of a glass furnace. The general purpose of the tuckstones is to protect the top of the tank blocks from the furnace gases and, in some types of tank furnace, to act as a seal between the tank blocks and the side- and end-walls. The course of tuckstones is sometimes called the TUCKWALL.

Tumbler. A refractory (or heat-resistant steel) projection inside a rotary kiln, to improve heat transfer and stir the feed. Tumblers can be arranged in series in various axial rows, to form *tumbling ledges.*

Tumbling. BARRELLING (q.v.).

Tundish. A rectangular trough lined with fireclay refractories and with one or more refractory nozzles in its base. Tundishes are sometimes used between the ladle and the ingot moulds in the teeming of steel.

Tungsten Borides. Data have been reported on five compounds. W_2B; m.p. approx. 2700°C; sp. gr. 16.7. WB exists in two crystalline forms: α-WB, sp. gr. 16.0; β-WB, sp. gr. 15.7 (both forms melt at approx. 2900°C). W_2B_5 m.p. 2200°C; sp. gr.13.1. WB_2 m.p. 2900°C. WB_6, m.p. 2920°C.

Tungsten-bronze Structure. This crystal structure is typified by certain alkali metal compounds of tungstic oxide. It is that of several niobate electroceramics. The structure is that of MO_6 groups (M=W,Ta,Ti,Nb etc) with spaces between, forming tunnels in the structure. These A-site tunnels are perpendicular to the *c*-plane, in which their section is surrounded by 3(A3) 4(A1) or 5(A2) oxygen atoms. Lower valency cations are located in these tunnels above and below the *c*-plane oxygen atom network. The MO_6 groups provide two distinguishable B sites. The general formula is $[(A1)_2(A2)_4A3][(B1)_2(B2)_8]O_{30}$. In ferroelectric tungsten bronzes, the polar axes are usually parallel to the A-site tunnels.

Tungsten Carbides. WC; m.p. 2865°C; hardness, 1880 (Kl00); sp. gr., 15.7; modulus of rupture. 350–550 MNm^{-2} at 25°C, thermal expansion (25–1000°C) 5×10^{-6}. The principal use of this carbide is in cutting tools. There is also a ditungsten carbide: W_2C; m.p. 2855°C; sp. gr. 17.2; hardness 2150 (K100); thermal expansion (25–1000°C) 4×10^{-6}

Tungsten Nitride. W_2N; exists in at least two crystalline forms.

Tungsten Oxide. The most common oxide is WO_3: m.p. 1470°C; inversion temp. approx. 300°C. Thermal expansion 13.5×10^{-6} (20–300°C); 11×10^{-6} (300–700°C).

Tungsten Silicides. There are least at three compounds. WSi m.p. 2150°C; WSi_2 m.p. 2160°C; sp. gr. 9.3; hardness (K100), 1090 kg/mm^2; thermal expansion (20–1000°C) 8.25×10^{-6}; W_3Si_2 m.p. 2340°C; sp. gr. 12.2; hardness (K100) 770 kg/mm^2.

Tuning-fork Test (for Glaze-Fit). A test-piece is made by joining, with clay slip, two bars of the extruded pottery body to a short piece of the same material. The test-piece is biscuit fired and the outer faces are then glazed. The test-piece is placed in a furnace and fired so that the glaze matures; it is then allowed to cool, while still in the furnace, and any relative movement of the two ends of the 'tuning fork' is measured by a micrometer telescope. From this measurement the magnitude of any stress in the glaze can be calculated. The test was devised by A. M. Blakely (*J. Amer. Ceram. Soc.,* **21**, 243, 1938).

Tunnel Dryer. A continuous dryer through which shaped clayware can be transported on cars; it is controlled so that the humidity is high at the entrance and low at the exit.

Tunnel Kiln. A continuous kiln of the type in which ware passes through a stationary firing zone near the centre of the kiln. In the most common type of tunnel kiln the ware is placed on the refractory-lined deck of a car, a continuous series of loaded cars being slowly pushed through a long, straight, tunnel.

Turbidimeter. An instrument for determining the concentration of particles in a suspension in terms of the proportion of light absorbed from a transmitted beam. An instrument of this type designed for particle-size analysis is the WAGNER TURBIDIMETER (q.v.)

Turbine Blades. See GAS TURBINE.

Turbostratic. A layered structure in which successive layers are oriented at random with respect to each other. One form of boron nitride has this structure. (US Pat 3,241,919 22/03/1966; E.I du Pont de Nemours & Co).

Turret Chain Machine. See PUFF AND BLOW PROCESS.

Tuscan Red. An iron oxide glaze pigment.

Tuscarora Quartzite. An important source of raw material for silica refractories occurring in Pennsylvania, USA. A typical analysis is (per cent): SiO_2, 97.8; Al_2O_3, 0.9; Fe_2O_3, 0.7; alkalis, 0–4.

Tuyere. A tube or opening in a metallurgical furnace through which air is blown as part of the extraction or refining process. In a blast furnace the tuyeres are water-cooled metal tubes which pass through the refractory lining of the BOSH (q.v.) (French word meaning 'a tube'). *Tuyere bricks* are refractory bricks containing holes to act as tuyeres.

Tuyere Block Bottom. One form of bottom for a CONVERTER (q.v.). The passages for the air blast are separate pre-formed tuyeres each having several holes; these tuyere blocks are interspersed with solid refractory blocks, the whole bottom then being finished by ramming refractory material into any spaces.

Twaddell Degrees (° Tw). A system for denoting the specific gravity of a liquid: Degrees Twaddell = (sp. gr. – 1) × 200 The specific gravity of solutions of sodium silicate, for example, is often quoted in this form; each Twaddell degree corresponds to a sp. gr. interval of 0.005. This scale is named after William Twaddell who, in Glasgow in 1809, made a hydrometer with this scale to the design of Charles Macintosh (of rainwear fame).

Tweel Block. A type of refractory block used in the glass industry for such purposes as protection of a newly-set pot, the construction of a furnace door or damper, or control of the flow of molten glass. (From French *tuile* a tile.)

12.0 Temperature. See ANNEALING POINT.

Twin-plate Process.

Twig. A small metal anchor for unshaped linings, usually V-shaped.

Twinning. Intergrowth of two (or more) crystals of near-symmetry, with a definite relation between their crystal orientations.

Twin-plate Process. A process for the simultaneous grinding and polishing of both faces of a continuously-produced ribbon of glass; the complete flow-line is over 400m long. The process was introduced by Pilkington Bros. Ltd, England, in 1952.

Twist Hackle. A HACKLE MARK that separates portions of the crack surface, each of which has rotated from the original crack plane due to rotation of the axis of principal tension.

Tyler Sieve. See under SIEVE; for mesh sizes see Appendix E.

Tyndall Beam Lighting. A technique of indirect lighting used to reveal normally invisible dust particles by sideways scattering of the light, as dust motes are revealed by a shaft of sunlight.

Tyranno. Trade-name. Ceramic fibres in the Si-Ti-C-O system.

TZP. Tetragonal Zirconia Polycrystals. See ZIRCONIA.

U-type Furnace or Hair-pin Furnace. A furnace for the firing of vitreous enamelware, which is carried along a U-shaped path so that ware enters and leaves the furnace at adjacent points.

U-value. See THERMAL TRANSMITTANCE.

Udden Grade Scale. A scale of sieve sizes introduced by J. A. Udden (*Augustana Library Pub.* No. l, 1898). The basic opening is 1 mm, the scale above and below being a geometrical series with a ratio of 2 (above 1 mm) and ½ (below 1 mm).

Ulexite. A boron mineral approximating in composition to $Na_2O.2CaO.5B_2O_3.16H_2O$. Ulexite occurs in Chile and Argentina. Trials have been made with this mineral as a flux in ceramic glazes.

Ultimate Analysis. The chemical analysis reported in terms of constituent oxides, as distinct from the RATIONAL ANALYSIS (q.v.), which is in terms of the minerals actually present.

Ultimate Compressive Strength. The compressive strength of a test piece with load bearing surface ground flat and smooth, and tested without packing materials. B.S. 1902 Pt. 4.2 specifies the test for dense refractory materials.

Ultrafine Powders. Powders in the size range 10–100nm have high surface areas and so enhanced reactivity in sintering. Additives are required to maintain good flow properties in this size range.

Ultragres. A floor tile comprising two inseperable layers, the upper being hard and impermeable to moisture. (Mosa BV, Holland).

Ultrasonic Equipment. The word 'ultrasonic' signifies vibration at a *frequency* greater than the maximum audible frequency, and should not be confused with 'supersonic', which signifies a *velocity* greater than that of sound. Ultrasonic vibrations can be generated by piezoelectric ceramics, by magnetostrictive devices, or by 'whistles' in which there is a steel blade vibrated by a high-pressure jet of fluid. Ultrasonic equipment has been used in the ceramic industry for the dispersion of clay slips, for metal cleaning prior to vitreous enamelling, and for flaw detection, particularly in large electrical porcelain insulators.

Ultrasonic Machining. A specially designed tool is vibrated at low amplitude (*c.* 25µm) but at high frequency (20 kHz). This tool combines with abrasive slurry to microscopically grind the workpiece.

Ultra-violet Absorbing Glass. Glasses can be made to absorb U.V. light, while transmitting visible light, by the inclusion of CeO_2 in the batch. Other elements absorbing U.V. light include Cr, Co, Cu, Fe, Pb, Mn, Nd, Ni, Ti, U and V.

Ultra-violet Transmitting Glass. For high transmittance of U.V. light, a glass must be free from Fe, Ti and S. Phosphate glasses and some borosilicate glasses have good U.V. transmittance. Uses include special windows and germicidal lamps.

Umber. A naturally-occurring hydrated iron oxide occasionally used as a colouring agent for pottery decoration.

Unaker. An old American Indian (Cherokee) term for clay from which sand and mica had been removed by washing.
Underarching. Underadjustment of the circumferential joints from true radial alignment in brick arches or rings. It may be due to the use of too many slow taper bricks.
Underclay. See SEAT EARTH.
Undercloak. A layer, of plain clay tiles for example, between the laths and the roof tiling proper at the VERGE (q.v.) of a tiled roof.
Undercutting. Faulty cutting of flat glass resulting in an edge that is oblique to the surface of the glass.
Under-glaze Decoration. Decoration applied to pottery before it has been glazed. Because it is finally covered by the glaze, such decoration is completely durable, but because the subsequent glost firing is at a high temperature the range of available colours for under-glaze decoration is limited.
Underloading. Having insufficient charge in a ball mill for proper grinding; or, deliberately reducing the charge to speed grinding.
Under-ridge Tile. A roofing tile for use at the top of a tiled roof. Such tiles are shorter than standard roofing tiles and are used to complete the roof along the ridge beneath the RIDGE TILES (q.v.).
UNI. Prefix to specifications of the Italian Standards Association: Ente Nazionale Italiano di Unificazione, Piazza Diaz, Milano, Italy.
Uni Process. A vitreous enamelling process which requires no pickling or nickel deposition prior to enamelling. Described in *Vitreous Enameller* **28**, (3), 48, 1977 and **30**, (2), 39, 1979.
Uniflame Process. The TRANSFER (q.v.) design layer contains a solid resin which differs in polarity from the resin in the covercoat, which protects the design in storage but may be removed manually or by air pressure after the transfer is applied to the ware, but before firing. This eliminates the need for two separate firings of transfer and glaze for underglaze transfers (Commercial Decal Inc., USA).
Unirotor Mill. Trade-name. A hammer mill for clay preparation. The single rotor is reversible to even out wear. (Hazemag, Germany).
Unit Cell. See CRYSTAL STRUCTURE.
Univertrave. A machine for prefabricating clay beams for roofs and floors. Pairs of hollow blocks are fed along the steel reinforcing rods set in feeder channels where mortar is vibrated into the joints between the blocks. A walking beam carries the blocks forward until the desired length of beam has been built up. (Univertex, Milan).
Unshaped Refractory. A mixture of refractory materials, that may contain a proportion of metallic constituents, supplied in granular or pliable form suitable for the construction of monolithic structures which will be subjected to elevated temperature. It may be placed in-situ by one or more of the pour casting, pumping, vibration casting, tamping, ramming or pneumatic gunning installation techniques, but not by trowelling or brushing. B.S. 1982 Pt. 7 defines various classes of unshaped refractories, and specifies test methods for their properties, and the preparation of test samples. ASTM Standards C973, C974, C975 describe preparing test specimens from basic gunning refractories (by pressing), from basic castables (casting) and from basic ramming mixes (pressing). C862 describes casting refractory concrete specimens, C903 their preparation by cold gunning; C865 their firing. C1054

describes pressing and drying plastic refractory and ramming mix specimens.

Unsoundness. As applied to portland cement, this term refers to slow expansion after the cement has set. The principal causes of this fault are the presence of free CaO, excess MgO, or excess sulphates.

Up-cast. Local term in England for positive pressure in the atmosphere of some zones of an annular kiln; the term is applied to the hot products of combustion that escape through any open feed caps in these zones.

Up-draught Kiln. An intermittent kiln in which the combustion gases pass from the fireboxes through the setting and thence through one or more chimneys in the roof. Such kilns are inefficient and are now obsolete.

Up-draw Process. The continuous vertical drawing of glass rod or tubing from an orifice; to produce tubing, the rod is drawn around a refractory cone. (This process has also been called the SCHULLER PROCESS, or WOOD'S PROCESS.)

UPEC Classification System. In this French system, floor tiles are assigned to various classes according to tests relating to four criteria: U (*usure*) wear due to walking; P (*poinçonnement*) marking due to fixed or moving furniture; E (*eau*) behaviour in water; C (*chimique*) resistance to chemicals and domestic staining. (*Industr. Ceramique* (746), 1981, p68).

Uphill Teeming. See BOTTOM POURING.

Upright. See POST.

Uprisings. Scraps recovered for re-use.

Uptake. See DOWNTAKE.

Upward Drilling. Wear on horizontal faces in glass-tank refractories, below the glass level, due to trapped gas bubbles.

Uranium Borides. Three borides are known: UB_2, UB_4 and UB_{12}. The most attention has been paid to the tetraboride, the properties of which are: m.p. > 2100°C (but oxidises rapidly above 600°C); sp. gr. 9–38 g/ml; thermal expansion, 7.1×10^{-6} (20–1000°C); modulus of rupture (20°C), 400 MNm^{-2}; electrical resistivity, 3×10^{-5} ohm.cm.

Uranium Carbides. Three carbides exist. UC: m.p. approx. 2500°C; sp. gr. 13.6; thermal expansion (25–1000°C) 11×10^{-6}. UC_2: m.p. approx. 2400°C; sp. gr. 11.7; thermal expansion (25–1000°C) 9.5×10^{-6}. U_2C_3: m.p. approx. 2400°C; sp. gr. 12.9; thermal expansion (25–1000°C) 9×10^{-6}. The powdered carbides can be made, by ceramic processes, into nuclear fuel elements.

Uranium Nitrides. UN decomposes at 2800°C at a N_2 pressure of 1 atm.; under the same conditions U_2N_3 decomposes at 1345°C. UN_2 also exists.

Uranium Oxides. The important oxides of uranium are UO_2, U_3O_8 and UO_3. The dioxide (m.p. 2880°C; theoretical density 10.96) is used as a nuclear fuel element. Uranium oxide has been used to produce red and yellow glazes and ceramic colours.

Uranium Red. A ceramic stain for coloured glazes suitable for firing temperatures up to 1000°C. Increasing the uranium oxide content strengthens the colour from orange, through red to tomato red. The glaze should be basic, preferably 0.5 mol SiO_2, 0.1–0.2 mol Al_2O_3, 0.1 mol K_2O and the remaining bases chiefly PbO; B_2O_3 should not be present in significant quantity.

Uranium Silicides. Several compounds have been reported; the most interest has been shown in USi_2.

Uvaroite. $3CaO.Cr_2O_3.3SiO_2$; the colouring agent in VICTORIA GREEN (q.v.).

UV-Curing. Rapid curing of organic materials by exposure to ultra-violet radiation. The process has been applied to the covercoats and decorative designs of ceramic transfers.

Uviol Glass. Glass highly transparent to ultraviolet wavelengths.
V-bricks. A series of perforated clay building bricks designed by the Building Research Station, England, in 1959–60; the name derives from the fact that the perforations are Vertical.
V-draining. See DOUBLE-DRAINING.
V-Stud. A V-shaped STUD OF ANCHOR for a refractory lining.
Vacancy. See CRYSTAL STRUCTURE.
Vacuity. The expansion space left above the liquid in a closed glass container.
Vacupress. A continuous machine which increases the bulk density of powders by deaeration and pre-compaction.
Vacuum-and-Blow Process. See SUCK-AND-BLOW PROCESS.
Vacuum Casting. A SLIP CASTING (q.v.) process in which the casting rate is increased by applying a vacuum to the outside of a porous mould. Cf. PRESSURE CASTING.
Vacuum Degassing. A SECONDARY STEELMAKING PROCESS in which the bath of molten steel in the ladle, or the stream of molten steel from it, is exposed to a vacuum to accelerate the removal of hydrogen from the steel.
Vacuum Drying. Reducing the air pressure over ware being dried speeds up moisture removal.
Vacuum Firing. A process for the firing of special types of ceramic either to prevent oxidation of the ware or to reduce its porosity. Vacuum firing is used, for example, in the firing of dental porcelain to produce teeth of almost zero porosity.
Vacuum Forming. A sheet of material is placed over a porous mould, heated if necessary, and the air between the sheet and the mould is pumped out through the mould, forcing the sheet to conform to its shape. A variety of rigid shapes are made from ceramic fibre blankets with suitable binders, using this technique.
Vacuum Mixer. A machine for the simultaneous de-airing and moistening of dry, prepared clay as it is fed to a pug. In the original design (L. Walker, *Claycraft*, **25**, 76, 1951) the clay fell as a powder through a vertical de-airing chamber where water was added as a fine spray; from the bottom of the de-airing chamber the moist, de-aired, clay passed into a pug. There have been several developments of this principle.
Vacuum Pug. A PUG (q.v.) with a vacuum chamber in which the clay is de-aired before it passes into the extrusion chamber.
VAD (vapour-phase axial deposition). See OPTICAL FIBRES.
Valiela Process. A decorating process for tableware. A table has a recess to receive the article to be decorated, alongside a silk-screen stencil which can slide into a predetermined position above a rubber sheet. This is on a panel hinged to superimpose it on to the curved surface of the ware, where it is forced down by compressed air to transfer the decoration. (J. Valiela and Interpace Corp, USA, Br. Pat. 746,100 1956).
Vallendar Clay. A clay from Westerwald, Germany. The best known grade is that used in the vitreous enamel industry; a quoted analysis is (%): SiO_2, 54; Al_2O_3, 35; Fe_2O_3, 0.8; alkalis, 1.8; loss-on-ignition, 8.3.
Valley Tiles. Specially-shaped roofing tiles for use in the 'valley' where two roof slopes meet; these tiles are made to fit into the angle. They lap and course in with the normal tiling.
VAMAS Programme. The Versailles Project on Advanced Materials and Standards is a European Community inititative to develop standard test

methods for the evaluation of advanced ceramics.

Vanadium Borides. Several borides have been reported, including the following: VB_2; m.p. 2100°C; sp. gr. 5.0; the thermal expansion is highly anisotropic. VB; m.p. 2250°C; sp. gr. 5.3. V_3B_4; melts incongruently at 2300°C; sp. gr. 5.5.

Vanadium Carbide. VC, m.p. 2830°C; sp. gr. 5.8.

Vanadium Nitrides VN, m.p. approx. 2300°C; sp. gr. 6.1. V_3N is relatively unstable.

Vanadium Oxide. The common oxide is V_2O_5; m.p. 675°C.

Vanadium Yellow or Vanadium-Tin Yellow. A ceramic colour produced by the calcination, at about 1000°C, of a mixture of 10–20% V_2O_5 (as ammonium metavanadate) and 80–90% SnO_2. A stronger yellow results if a small amount of TiO_2 is added. These colours can be used in most glazes and either SnO_2 or zircon can be used as opacifier.

Vanadium–Zirconium Blue (or Turquoise). See ZIRCONIUM–VANADIUM BLUE.

Vanal. Trade-name: Hagenberger-Schwalb A.G., Hettenleidelheim/Pfalz, Germany. A coating for the protection of refractories against slag attack developed by A. Staerker *(Ber. Deu. Keram. Ges., 28, 390, 1951;* 29, 122, 1952). It contains vanadium and is claimed to prevent slags wetting the refractory.

Van der Waal's Forces. Electrostatic forces between polarized molecules or groups of atoms, which are considerably weaker than the normal interatomic bonding forces giving rise to solid crystal structures. They are important in producing aggregates of particles in dense suspensions, and spontaneous agglomeration in dry powders of very small (sub-micron) particle sizes.

Vane Feeder. A device for feeding dry ground clay from a hopper to a tempering machine or mixer, for example. Vanes fixed to a horizontal shaft at the base of the hopper rotate to discharge the material.

VAR. Vacuum Arc Refining. See SECONDARY STEELMAKING.

Varistor. A material having an electrical resistance that is sensitive to changes in applied voltage. A typical example is the varistor made from a batch consisting of granular silicon carbide, mixed with carbon, clay and water; the shaped components are fired at 1100–1250°C in H_2 or N_2.

Varved Clay. A clay deposited in layers, some coarse, others fine and silty.

VAS. Vacuum Arc Spraying. A thermally sprayed coating process.

VE. Abbreviation for VITREOUS ENAMEL (q.v.).

Vebe Apparatus. A device developed by the Swedish Cement Association for the measurement of the consistency of concrete. It is a slump test in which the consistency is expressed in degrees, the value being obtained by multiplying the ratio of the volume of the test-piece after vibration to that before vibration by the number of vibrations required to cause the test-piece to settle.

Vegard's Law. States that in a binary system forming a continuous series of solid solutions, the lattice parameters are linearly related to the atomic percentage of one of the components. This law has been applied, for example, in the study of mixed spinels of the type formed in chrome-magnesite refractories. (L. Vegard, *Z. Physik*, **5**, 17, 1921.)

Veiling. A US term for gold and organic colours sprayed on glass in a thread-like texture.

Vein. See STRIAE.

Vein Quartz. An irregular deposit of quartz, often of high purity, intruded into other rocks. In consequence of its mode of formation, vein quartz usually contains occluded gas bubbles and is unsuitable as a raw material for silica refractories.

Vello Process. A method for the production of glass tubing; molten glass flows vertically through an annular orifice; the central refractory pipe within the orifice is hollow and rotates. The process will produce up to 1 tonne/h of tubing, 1.5–75mm dia. cf. DANNER PROCESS.

Vellum Glaze. See SATIN GLAZE.

Velvet Finish. An embossed appearance formed on glass by acid treatment which obscures transmission of light.

Veneered Wall. A wall having a facing (of faience panels, for example) which is attached to the backing but not in a way to transmit a full share of any imposed load; the veneer and the backing do not exert a common action under load (cf. FACED WALL).

Venetian Red. A colour consisting chiefly of ferric oxide.

Vent. See CHECK.

Verge. The gable edge of a tiled roof. At the verge the roofing tiles are edge-bedded, preferably on a single or double undercloak of plain tiles. This form of undercloak gives a neat appearance to the verge and slightly inclines the verge tiles so that rainwater is turned back on to the main roof.

Vermiculite. A group name for certain biotite micas that have been altered hydrothermally. When rapidly heated to 800–950°C vermiculite exfoliates as the combined water is expelled; the volume increase is about 15%. Large deposits of vermiculite occur in USA, Transvaal, Uganda, Australia and the Ural Mountains of Russia. The exfoliated material is used as loose-fill insulation and in the manufacture of vermiculite insulating bricks. Standards for the loose-fill are laid down in US Govt. Fed. Spec. HH-I-00585. Vermiculite bricks are made by bonding the graded material with about 30% of clay, shaping, and firing at 1000°C. The bulk density of such bricks is 500–600kg.m^{-3}; thermal conductivity 0.2–0.25W/mK. They can be used up to 1000°C.

Verneuil Process. A method for the production of ceramic BOULES (q.v.) by feeding powdered material, e.g. Al_2O_3, into an oxy-hydrogen flame. (A. Verneuil, *Ann. Chim. Phys.,* **3**, 20, 1904.)

Versatran. A programmable machine for positioning, transfer and handling or objects, suitable for bricks. (Licensed from the American Machine & Foundry Co).

Vesicular. A solid is said to be vesicular if it contains a lot of small bubbles; a bloated clay may be vesicular, for example.

Verson System. A continuous moulding process for clayware, particularly sanitaryware. A thin mould liner is supported by a rigid female mould. A forming die gives preliminary compaction to the charge of plastic clay between it and the liner. The liner and charge can then be moved on to a second (and subsequent) female mould, the last of which imparts the finished shape. (Verson Manufacturing Co, USA Br. Pat. 1004 540, 1964).

Vibrating Ball Mill; A BALL MILL (q.v.) supported on springs so that an out-of-balance mechanism can impart vibration to the mill, usually in the vertical plane and typically at about 25 Hz. The feed should not be coarser than 100 mesh; the charge should be at least 85% of the mill volume. It is particularly suitable for grinding to less than 40μm. Advantages over the ordinary ball mill are increased rate of grinding (particularly with very hard materials), lower energy consumption per ton of product, and less wear.

Vibrating Screen. A screen, set at an angle of 25–35° to the horizontal, and vibrated by an eccentric, a cam or hammer, an out-of-balance pulley, or an electro-magnet. When operating with 10-mesh cloth on damp clay, such a screen should have an output of about 5 ton/h/m^2. The screening efficiency is generally 80–85%.

Vibrating Table. A table which oscillates rapidly to remove air and compact a refractory castable placed on it.

Vibro-casting. Pouring concrete into a mould, and densifying by vibration. Large concrete pipes can be vibrocast.

Vibro-compaction. Vibration and pressure are applied simultaneously, to densify powder mixtures. The usual techniques produce particle accelerations of some 50g, and require an appropriate particle size distribution to obtain dense compaction. To densify mixed powders whose components differ greatly in density, very strong vibrations (accelerations c. 20 000g) have been studied.

Vibro-energy Mill. Trade-name: a VIBRATING BALL MILL (q.v.) designed to oscillate both horizontally and vertically, the vertical motion being of small but sufficient amplitude to prevent the charge from becoming tightly packed. (W. Podmore Ltd., and W. Boulton Ltd., Stoke-on-Trent, England.)

Vicat Needle. An instrument for evaluating the consistency of cement in terms of the depth of penetration of a 'needle' of standard shape and under a standard load; it was designed by L. J. Vicat, a Frenchman, in the early 19th century. Details of dimensions and method of use are given in B.S. 12 and its application to refractory mortars in B.S. 1902 Pt. 11. (cf. GILLMORE NEEDLE).

Vickers Hardness. An indentation test in which a diamond pyramid is used. The diamond is loaded mechanically by a lever, the application of the load being hydraulically controlled to give the correct duration. The symbol is HV; it should be supplemented by a number indicating the load used. The Vickers Hardness of most pottery glazes lies between 570 and 630 kg/mm^2; that of vitreous enamels is approx. 500 kg/mm^2. See also FRACTURE TOUGHNESS TESTS.

Victoria Green. A bright green ceramic colour. A quoted batch composition is: 38% $K_2Cr_2O_7$, 20% $CaCO_3$, 22% $CaFz_2$, 20% SiO_2. This batch is calcined, washed free from soluble chromates, and ground. The colouring agent is stated to be uvarovite ($3CaO.Cr_2O_3.3SiO_2$).

Vignetting. The decoration of a glass surface by firing-on a metal or other suitable powder; the surface is first coated with sodium silicate solution and the powder is then dusted on and the article fired.

Vinsol Resin. Trade-name: a thermoplastic powder used as an AIR-ENTRAINING (q.v.) agent in the mixing of concrete. (Hercules Powder Co. Ltd., London.)

Visc 2B. This viscosity meter suited to determine the flow properties of enamel suspensions measures the dependence of the deformation speed gradient as a function of sliding tension measured by turning a roller in the enamel suspension. *Vit Enameller* **29** (1) 5 1978).

Viscometer. The commonest types of viscometer, as used in the ceramic industry, depend on measurement of the flow of the test liquid through an orifice or of the drag on one of a pair of

concentric cylinders when the other is rotated, the test-liquid occupying the space between the cylinders. B.S. 7034 specifies tests for viscosity and viscometric fixed points for glasses. See also FLOW CUP, TORSION VISCOMETER.
Viscone. Trade-name. Silicone rubber for pads and membranes for ceramic printing, supplied by Blythe Colours Ltd, Stoke on Trent.
Viscosity. The viscosity of a true ('Newtonian') liquid is the ratio of the shearing stress to the rate of shear, of which the viscosity itself is independent; this is not generally true of clay slips, vitreous-enamel slips, or glaze suspensions, all of which exhibit THIXOTROPY (q.v.).
Viscous Plastic Processing (VPP). A green forming technique for advanced ceramics which minimises the size and number of flaws by avoiding agglomeration of the starting powders through high-shear mixing and careful choice of binder. Articles are then shaped by plastic forming. The process was developed from MDF CEMENT (q.v.) technology by British Ceramic Research Ltd.
Visil. Trade-name. A viscose silica fibre – a cellulose fibre containing up to 33% silica. Kemira Oy, Finland.
Vitramic Ware. A vitrified red-brown body, oven and freezer proof, chip-resistant and dishwasher proof, was made by Hornsea Pottery (Lancaster) to a basically earthenware recipe. *Ceram Ind. J.* **84** (1991) 20 1975.
Vitreous. This term, meaning 'glassy', is applied to ceramic ware that, as a result of a high degree of vitrification (as distinct from sintering) has an extremely low porosity. In the USA the term is defined (ASTM – C242) as generally signifying that the ware has a water absorption below 0.5%, except for floor tiles, wall tiles and low-tension electrical porcelain which are considered to be vitreous provided that the water absorption does not exceed 3%.
Vitreous Carbon. See GLASS-LIKE CARBON.
Vitreous-china Sanitaryware. Defined in B.S. 3402 as: 'A strong high-grade ceramic ware used for sanitary appliances and made from a mixture of white-burning clays and finely ground minerals. After it has been fired at a high temperature the ware will not, even when unglazed, have a mean value of water absorption greater than 0.5% of the ware when dry. It is coated on all exposed surfaces with an impervious non-crazing vitreous glaze giving a white or coloured finish.' A typical batch for this type of body is: 20–30% ball clay, 20–30% china clay, 10–20% feldspar, 30–40% flint, 0–3% talc; sometimes nepheline syenite is used instead of feldspar. B. S. 3402 also includes tests for chemical resistance to various acids, alkalis and detergents, (the visual assessment of reflectivity to be unchanged by the chemical treatment); an autoclave test for crazing resistance, and definitions of various blemishes and faults affecting sanitaryware.
Vitreous Enamel. Defined in B.S. 1344 as: 'An inorganic glass which is fused on to a metal article in the form of a relatively thin coating and provides protection against corrosion. A glazed surface finish produced by the application of a powdered inorganic glass, dry or suspended in water, to metal parts, and its subsequent fusion.' Iron is the usual metal so treated, but aluminium, copper and other metals can be enamelled with glasses with appropriate properties. The equivalent US term is PORCELAIN ENAMEL (q.v.). B.S. 1344 also specifies various tests for

chemical resistance; fluidity behaviour, a low voltage electrical test for defects in coatings on metal, and a high voltage test for use when the enamelled articles will face severe conditions.

Vitreous Silica. The glassy (vitreous) form of silica. There are two major commercial forms: TRANSLUCENT VITREOUS SILICA (which has often been referred to as *Fused Silica)* and TRANSPARENT VITREOUS SILICA (often referred to as *Fused Quartz).*

Translucent Vitreous Silica is made by the fusion of quartz sand in a furnace open to the atmosphere; it is translucent, i.e. not fully transparent, because of the numerous small bubbles that it contains.

Transparent Vitreous Silica is made by the fusion (frequently under vacuum) of quartz crystal, or by the hydrolysis or oxidation of a volatile silicon compound and subsequent fusion of the silica so formed; the product of the latter process is known as *Synthetic Vitreous Silica.*

Vitreous Slip. A US term defined (ASTM – C242) as a slip coating matured on a ceramic body to produce a vitrified surface.

Vitriam. A type of sanitary earthenware, made by Société Amandinoise de Faìencerie et de Produits Refractaires et Sanitaires S.a.r.l.

Vitrification. The progressive partial fusion of a clay, or of a body, as a result of a firing process or, in the case of a refractory material, of the conditions of use in a furnace lining. As vitrification proceeds the proportion of glassy bond increases and the apparent porosity of the fired product becomes progressively lower. The body is said to be completely vitrified if the glass fills all the pores, and forms a bond between the grains at the grain boundaries. The VITRIFICATION RANGE is the temperature interval between the beginning of vitrification of a ceramic body and the temperature at which the body begins to become deformed.

Vitrified. See VITREOUS.

Vitrified Wheel. An abrasive wheel made from a batch consisting of abrasive grains and a ceramic bond formed by kiln firing at 1200–1300°C. Over half the abrasive wheels currently produced are of this type.

Vitrite. The black glass used in the caps of electric lamps.

Vitroceramic. One of several terms proposed for the type of ceramic product formed by the controlled devitrification of a glass; see GLASS CERAMIC.

Vitroderm. Sections of glazed ware prepared for microscopic examination, by grinding away the body from below the glaze. The term, which means 'glass skin' was coined by F. Smithson (*Trans. Br. Ceram. Soc.* **47,** 191, 1948).

Vitron. A unit of atomic structure, particularly in silica glass, proposed by L. W. Tilton (*J. Res. Nat. Bur. Stand.* **59**, 139, 1957). Its basis is a pentagonal ring of five SiO_4 tetrahedra; these rings can be built up into three-dimensional clusters but only to a limited extent because of increasing distortional stress; a cluster of the pentagonal SiO_4 rings is a VITRON. Its most important property, as a basis for the understanding of the properties of glass, is its fivefold symmetry which precludes the formation of crystals. (cf. STRUCTON).

Vogel's Red. A pure ferric oxide produced by precipitating ferrous oxalate which is then calcined. It has been used as a basis for some 'iron' colours on porcelain.

Void. (1) The spaces between grains, or other cavities in shaped, especially pressed, ware.

(2) A porous or hollowed-out region, representing the last area of solidification of a fusion cast refractory.

Volclay. A sodium bentonite from Wyoming, USA. (Tradename.)
Voluphant. A derivative of the ELEPHANT (q.v.) electrophoretic casting machine, to make large pieces of ware, such as sanitaryware. The anodes form the upper and lower dies of a press moulding. The cathode is shaped to fit between them and build up the cast before it is removed and the press die closed to shape the article. (*Industrie Ceram* (718) 423 1978)
Von Kobells' Scale. An empirical scale of fusibility for minerals, based on the flames required to fuse the solid. It is 1. Stibnite (candle flame); 2. Natrolite (luminous Bunsen); 3. Garnet (Blowpipe, readily); 4. Hornblende (Blowpipe, readily at edges); 5. Orthoclase (Blowpipe, edges fuse with difficulty); 6. Bronzite (Blowpipe, only thin particles fused); 7. Infusible.
VOR. Vacuum oxygen refining. See SECONDARY STEELMAKING.
VPB Kiln. A kiln for the firing of building bricks; it consists of two groups of chambers in which the fire travel follows a zig-zag course. The bricks are set on refractory bats and are put into, and subsequently drawn from, the kiln without the workmen having to enter the hot chambers. The name is from the initials of the inventor, V. P. Bodin (*French Pat.*, 1 156 918, 22/5/58).
VPS. Vacuum plasma spraying. See PLASMA SPRAYING.
Vycor. Trade-name. A POROUS GLASS (q.v.), made by a process patented by H.P. Hood and M.E Nordberg, Corning Glass (US Pat. 2106744, 1938). Vycor is almost pure vitreous silica. A borosilicate glass is first made and this solidifies in two phases, one of which is soluble in dilute acid and is thus removed, leaving a highly siliceous skeleton. The porous ware is then heated at about 1000°C; it shrinks considerably and non-porous high-silica glass (96% SiO_2) is produced. This trade-name is derived from Viking and Corning.
Wad. An extruded strip or rod of fireclay (with or without the addition of fine grog) used in the firing of pottery to seal the joints between saggars, or to level the supporting surfaces of saggars in a bung.
Wad Box. A simple hand-extrusion device for producing cylindrical fireclay WADS (q.v.) for use during the setting of bungs of saggars. A wad box was also used in the old method of pressing handles and similar shapes of pottery ware; used for this purpose, the device was sometimes called a DOD BOX.
Wadhurst Clay. A Cretaceous clay used for brickmaking in parts of Kent and Sussex, England.
Waechter's Gold Purple. A colour, of various shades, that has been used in the decoration of porcelain.
Waelz Furnace. A rotary furnace used particularly for the calcination of non-ferrous ores; chrome-magnesite linings have been used in Waelz furnaces producing ZnO.
Wafers. Small sheets of electroceramic material (e.g. $BaTiO_3$, 0.025–0.25 mm thick for use in electronic equipment, particularly in miniature capacitors, transistors, resistors and other circuit components.
Wagner Turbidimeter. Apparatus for the determination of the fineness of a powder by measurement of the turbidity, at a specified level and after the lapse of a specified time, of a suspension of particles that are settling by gravity according to Stokes' Law. The method was proposed by L. A. Wagner (*Proc. ASTM,* **33, Pt. 2, 553, 1933).**
Waist. See BELLY.
Wake Hackle. See GULL WINGS.

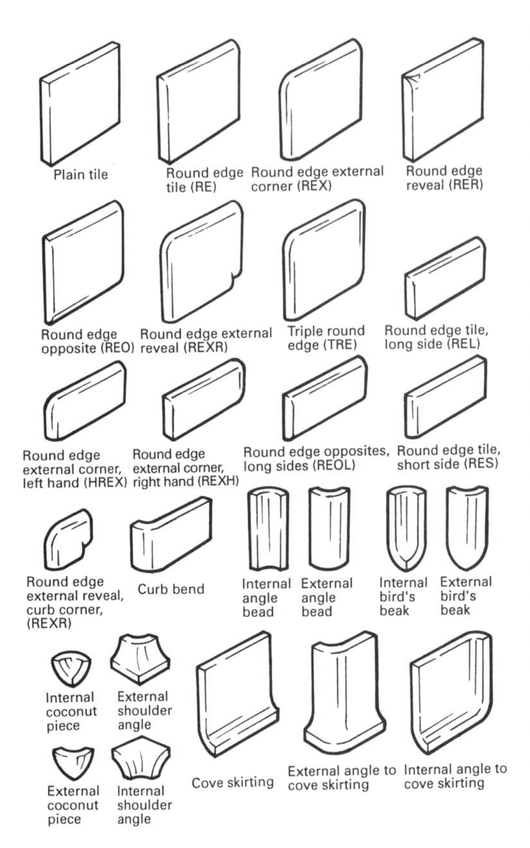

Fig. 7 Wall Tile Shapes.

Walker Vacuum Mixer. See VACUUM MIXER.

Walking-beam Kiln. A tunnel kiln of unusual type, the ware (set on bats) being moved through the kiln in steps by a mechanism that alternately lifts the bats and sets them down further along the kiln.

Wallner Line. Wavelike lines on the FRACTURE SURFACE (q.v.) caused by disturbance of the fracture front by sonic waves. The Wallner line is the locus of interception of the spreading elastic pulse with successive points along the propagating crack front.

Wall Tie. A metal wire, or sometimes plastics, shape used to link together the inner and outer leaves of a cavity wall.

Wall Tiles and Fittings. Glazed wall tiles (which are also used in fireplace surrounds) are made by a highly mechanized dust pressing process from white or buff bodies; to reduce the firing contraction and MOISTURE EXPANSION (q.v.) these bodies often contain lime compounds. Wall tiles are made in a wide variety of colours and glazes; a satin or matt glaze is often used. Some of the shapes of tiles and fittings are shown in Fig. 7; for standard sizes and other properties see B.S. 6431; for fixing procedures see B.S. 5385. See TILE and TILING.

Wall White. See EFFLORESCENCE.

Warm Block. A building block of good thermal and accoustic insulation and high heat capacity. It is perforated by a series of narrow slits arranged in close-spaced lines parallel to the faces of the block. Trade-name Wienerberger Baustoffindustrie AG, Austria.

Warning Limit. See ACTION LIMIT.

Warpage. A test for warpage (departure from flatness) of vitreous-enamelled flatware is given in ASTM-C314; a test for the warpage of refractories is included in B.S. 1902, Pt. lA. ASTM C485 is a test for warpage of ceramic tiles.

Wash. See REFRACTORY COATING.

Wash-back. See under WASH-MILL.

Washbanding. A form of pottery decoration, usually on-glaze in which a thin layer of colour is applied over a large surface of the ware by means of a brush.

Washboard. Unintentional waviness on the surface of glassware; also known as LADDERS.

Wash-gate. See under WASH-MILL.

Washing Off. Removing printing paper from pottery ware that has been decorated by the transfer process.

Wash-mill. A large tank fitted with stirrers (known as HARROWS or WASH-GATES) for the cleaning of the impure surface clays used in the manufacture of STOCK BRICKS (q.v.). From the wash-mill the clay slurry, together with a slurry of any lime or chalk that is to be added, is pumped into a settling tank known as a WASHBACK.

Waste-heat Dryer. A dryer for clayware that derives its heat from the cooling ware in the kilns. Such dryers are common in the brick industry and, even though the fuel consumption in the kilns may be slightly increased, this method of heat utilization results in an overall economy in fuel. Hot-floors, chamber dryers or tunnel dryers can be operated on this principle and the necessary heat can be derived from intermittent, annular or tunnel kilns.

Waster. A brick, structural or refractory, that is defective as drawn from the kiln; wasters in the refractories industry were crushed and re-used as GROG (q.v.).

Water Absorption. The weight of water absorbed by a porous ceramic material, under specified conditions, expressed as a percentage of the weight of the dry material. This property is much quoted when referring to structural clay products. B.S. 3921 specifies a test method and limits for ENGINEERING BRICKS and for bricks for DAMP-PROOF COURSES. B.S. 6431 classifies TILES (q.v.) on the basis of their water absorption. Limits for water absorption are also specified for vitreous china sanitaryware (B.S. 3402); vitrified hotelware (B.S. 4034); china and porcelain tableware (B.S. 5416). The US tests for refractories are ASTM C20 (a boiling water test) and ASTM C830 (A VACUUM PRESSURE TEST); the apparent porosity is more commonly quoted for refractories and whitewares. The two properties are related by the equation:

Apparent Porosity = Water Absorption × Bulk Density

(Note: In the USA, the term 'Absorption' is preferred to 'Water Absorption').

Water Gain. See BLEEDING.

Water Glass. Popular name for soluble grades of SODIUM SILICATE (q.v.).

Water Mark or Water Spot. (1) A shallow depressed spot sometimes appearing as a defect in vitreous enamelware.
(2) During transfer-printing on pottery, a water-mark may form if a drop of water dries on the ware, leaving a deposit of soluble salts.

Water Streak. A fault in vitreous enamelware arising from drops of water running down the ware, while it is being dried, and partially removing the enamel coating. The obvious cause is the use of a slip that is too wet, when water-streaks are liable to occur in any sharp angle of the ware; condensation of drops of water on parts of the ware during the drying process is another cause.

Water Test. See FLOC TEST.

Waterfall Process. A method for the application of glaze materials to a ceramic body by mechanically conveying the ware through a continuously flowing (recirculated) vertical stream of the glaze suspension. The process is used in the glazing of wall tiles.

Waterford Glass. Cut or gilded glass made in the Waterford district of Ireland and characterized by a slight blueness resulting from the presence of a trace of cobalt.

Waterjet Cutting. See ABRASIVE JET MACHINING.

Water-line. A defect in vitreous enamelware in the form of a line marking the limit of water penetration from wet beading enamel into the unfired enamel coating.

Waterproofing. (1) Concrete can be made more waterproof by surface treatment of the set concrete or by the addition of an integral waterproofer. For surface treatment a solution of sodium silicate or of a silicofluoride may be used; silicones, drying oils and mineral oils are also sometimes employed. Integral waterproofers include calcium chloride solution and/or various stearates.
(2) Water repellants for above-ground brickwork are recommended in B.S. 6477.

Water-slide Transfer = SLIDE OFF DECAL (q.v.).

Water-smoking Period. The stage in the firing of heavyclayware and of fireclay refractories when the mechanically-held water in the clay is being evolved, i.e. the temperature range 100–250°C. The temperature distribution in the setting should be such as to prevent condensation of any of this water vapour, and the temperature should not be raised further until all the water has been evolved.

Waterstruck Brick. A soft-mud brick made in a damp mould to prevent sticking.

Watkin Heat Recorders. Cylindrical pellets (9mm high, 6mm dia.) made of a blend of ceramic materials and fluxes so proportioned that, when heated under suitable conditions, they will fuse at stated temperatures. They are numbered from l(600°C) to 59 (2000°C) and were introduced by H. Watkin (Stoke-on-Trent, England) in 1899 and supplied by Harrison Mayer, Ltd, Stoke-on-Trent.

Wauk. A US term for a plastic clay body rolled and beaten to the rough shape of the mould, before hand moulding.

Wave. An optical defect in glass caused by uneven glass distribution or by STRIAE (q.v.).

Waviness. Long range departure from flatness, on a scale greater than a specified surface finish. See also KINK. (ASTM F109).

Weald Clay. A Cretaceous clay, often variable in composition even within the same clay-pit, used for brickmaking in parts of Surrey, Kent and Sussex, England.

Wear diagram. A 3-co-ordinate representation of the dependence of wear on the applied pressure and the sliding velocity.

Weathering. (1) The preparation, particularly of clay, by exposure to the weather for a long period. This helps to oxidize any pyrite present, rendering it soluble, so that this and other soluble impurities are to some extent leached out; the water content also becomes more uniform and agglomerates of clay are broken down with a consequent increase in plasticity.
(2) The attack of glass or other ceramics by the atmosphere.
(3) An inclined surface that facilitates the running-off of rainwater from brickwork.

Web. One of the clay partitions dividing a hollow building block into cells.

Webb Effect. The increase in volume of a pottery slip as deflocculation proceeds. (H. W. Webb, *Trans. Brit. Ceram. Soc.,* **33**, 129, 1934.) After a detailed experimental study, P. H. Dal and W. J. H. Berden (*Science of Ceramics,* G. H. Stewart (Ed.) **2**, 59, 1965) concluded that this effect is non-existent.

Wedge Brick. US term for KEY BRICK (q.v.).

Wedge Pyrometer. An instrument for the approximate measurement of high temperatures. It depends on a wedge of coloured glass, the position of which is adjusted until the source of heat is no longer visible when viewed through the glass; movement of the wedge operates a scale calibrated in temperatures.

Wedge-stilt. See STILT.

Wedged Bottom. See SLUGGED BOTTOM.

Wedging. (1) A procedure for preparing clay or a clay body by hand: the lump of clay is repeatedly thrown down on a workbench; between each operation the lump is turned and sometimes cut through and rejoined in a different orientation. The object is to disperse the water more uniformly, to remove lamination and to remove air.
(2) A fault in dust-pressed tiles if the powder is not charged to a uniform depth in the die. In ASTM C502, wedging refers to finished tiles in which tiles, not uniformly pressed and fired, are out-of-square. The standard specifies a procedure to measure wedging.

Wedgwood Pyrometer. A device for the determination of high temperatures on the basis of the approximate relationship between the contraction of clay test-pieces and the temperature to which they have been exposed; this pyrometer was introduced by Josiah Wedgwood (*Phil. Trans.*, **72**, 305, 1782) and now forms the basis of the BULLERS' RING (q.v.).

Weibull Distribution. This empirical statistical distribution function describes the scatter in strength values of brittle materials. It is used to assign mechanical properties to brittle materials in probabilistic terms, and to define design requirements in terms of strength and reliability. If $p\ (\sigma_c)$ is the probability that the material will fail at a measured critical stress σ_c then $\ln \ln [1 - p\ (\sigma_c)]^{-1} = M \ln (\sigma_c) + C$ where M is the Weibull Modulus. High values of M correspond to little scatter in the test data. Metals typically have values of M ~100, while traditional ceramics have values ~5 and engineering ceramics are in the range 10–25. The goodness-of-fit of Weibull statistics to experimental data can be assessed by the K–S TEST (q.v.) and others. See also CARES program.

Weibull's Theory. A statistical theory of the strength of materials proposed by W. Weibull (*Ing. Vetenskaps Akad. Hand.*, No. *151*, 1939); its basic postulate is that the probability of fracture of a solid body depends on the volume under stress and on the stress distribution. This theory has been applied in studies of the strength of ceramic materials and their resistance to the stresses induced by thermal shock.

Weissenberg Camera. A camera for X-ray diffraction analysis of crystal structures. The crystal is rotated, in the X-ray beam. The film is rotated and moved parallel to the axis of rotation.

Weissenberg Rheogoniometer. In essence a CONE AND PLATE VISCOMETER (q.v.), but with accessories and control systems to allow it to measure, as well as viscosity, elasticity and a wide range of flow properties such as dilatancy, thixotrophy, relaxation phenomena, etc in steady rotation or in torsional oscillation, over wide but controlled ranges of temperature and shear.

Weiss Field and Curie-Weiss Law. See CURIE LAW.

Weld Mark. A deep groove or fissure (in glass or whiteware) formed by incomplete union of two or more particles or streams of material flowing together. (ASTM F109).

Well. The term sometimes used for the lowest part of a blast furnace, i.e. the part in which molten iron collects.

Well-hole Pipe. One of the short fireclay pipes that were used to carry the flame upwards from the well-hole in the bottom of a BOTTLE OVEN (q.v.).

Wentworth Grade Scale. An extension of the UDDEN (q.v.) scale of sieve openings proposed by C. K. Wentworth (*J. Geol.,* **30**, 377, 1922).

Westerwald Clay. A refractory clay occurring in an area east of the Rhine between Coblenz and Marburg. These clays are of Oligocene origin and vary widely in composition from highly siliceous to clays that contain (raw) over 35% Al_2O_3.

Westlake Process. An automatic method for making glass-ware in paste moulds; the process closely imitates hand-making and was invented in 1916 by the Westlake European Machine Co., Toledo, USA.

Wet Bag Process. See ISOSTATIC PRESSING.

Wet Felt. Flexile ceramic fibre insulation, impregnated with organic liquid binder which dries to a strong rigid lightweight shape.

Wet Pan. An EDGE-RUNNER MILL (q.v.) used for grinding relatively wet material in the refractories and structural clay-wares industries. The bottom has slotted grids with a proportion of solid plates on which the mullers can grind.

Wet Pressing. Alternative term in USA for PLASTIC PRESSING (q.v.).

Wet Process. (1) The method of blending the constituents of a whiteware body in the form of SLIPS (q.v.).
(2) The process of portland cement manufacture in which the limestone (or chalk) and clay are fed to the kiln as a slurry.
(3) The method of applying vitreous enamel as a slip, usually by spraying.

Wet-rubbing Test. A test to determine the degree of attack of a vitreous-enamelled surface after an acid-resistance test; (see ASTM – C282).

Wetting Off. The severing of a hand-made glass bottle from the blow-pipe by means of a fine jet of water.

Whales. See DRAIN LINES.

Wheelabrator. A shot-blasting machine of a type used for cleaning castings prior to vitreous enamelling. This equipment has also been adapted to the testing of refractory bricks for abrasion resistance (see *Trans. Brit. Ceram. Soc.,* **50**, 145, 1951).

Whelp. A refractory brick of the same thickness and breadth as a standard square but of greater length, e.g. 12 × 4½ × 3 in. (305 × 114 × 76 mm) (see Fig. 1, p39).

Whirler. (1) A piece of tableware that has warped slightly during drying and/or firing; in consequence, such ware will 'whirl' on its foot if spun on a flat surface.
(2) A turntable used for checking the symmetry of a model in pottery making, or for the hand-making of a SAGGAR (q.v.).

Whirlering. The plaster moulds for bone-china hollow-ware are often revolved on a turntable while they are being filled with slip; this is known as 'whirlering'. The object is to prevent WREATHING (q.v.).

Whisker. A hair-like single crystal; the thickness can vary considerably but is typically 1–2μm; the length can attain several cm. The most important property of ceramic whiskers is high strength, e.g. alumina whiskers have been made with a tensile strength of 14000 MNm^{-2}. Ceramic whiskers have been dispersed in metals to increase their strength, particularly at high temperatures (cf. FIBRE (CERAMIC) and FILAMENT).

Whisker Reinforcement. The addition of ceramic WHISKERS (q.v.) to metals or ceramics increases their strength and TOUGHNESS (q.v.) by several possible

mechanisms. The whiskers are themselves strong, and may inhibit cracking by STRAIN HARDENING (q.v.) or CRACK BRIDGING (q.v.) mechanisms.
Whiskers = STRIATION (q.v.)
White Acid. An etchant for glass, being a mixture of HF and ammonium bifluoride.
White Alumina. A recrystallized alumina abrasive.
White Cement. Portland cement made from non-ferruginous raw materials, i.e. chalk (or low-iron limestone) and china clay. The Fe_2O_3 content is <1%.
White Dirt. A fault sometimes occurring during the manufacture of tableware as a result of the protrusion of white particles through the glaze; usually, if the glaze layer is adequately thick, any such 'dirt' is hidden. Sources include the bedding medium (alumina or sand), fragments of the body itself, and scale derived from hard water.
White Firing. A clay of low iron content which produces a white biscuit ware on firing.
White Flint. See FLINT GLASS.
White Glass. Glass that has a uniform low absorption coefflcient over the whole visible spectral range; for containers and for flat glass this is usually achieved by the addition to the glass batch of a DECOLORIZER (q.v.).
White Granite Ware. A US term for IRONSTONE WARE (q.v.).
White Ground-coat. Term sometimes used for a white vitreous enamel of high opacity used for one-coat application.
White Hard. Clay from which surface water has evaporated.
White Lead. Basic lead carbonate, $2PbCO_3.Pb(OH)_2$. Used to some extent in USA as a glaze constituent.
White Spot. A fault sometimes occurring in pottery colours, e.g. in chrome-tin pink and in manganese colours. It is caused by evolution of gas during firing, the glaze subsequently flowing over the crater left by the gas bubble without carrying with it sufficient colour to match the surrounding area.
White's Test. A method for the detection of free lime, for example in portland cement or dolomite refractories. A few mg of the powdered sample is placed on a glass microscope-slide and wetted with a solution of 5g phenol dissolved in 5 ml nitrobenzene with the addition of two drops of water. Micro-examination (× 80) will reveal the formation of long birefringent needles if free CaO is present. (A. H. White, *Industr. Engng. Chem.* **1**, 5, 1909.)
Whiteware. A general term for all those varieties of pottery that usually have a white body, e.g. tableware, sanitary ware and wall tiles. See also CERAMIC WHITEWARE, which has an ASTM definition.
Whitewash. (1) Local term for EFFLORESCENCE (q.v.) or SCUM (q. v.) on bricks.
(2) A fault in glass; see SCAB.
Whiting. Finely-ground Cretaceous chalk, $CaCO_3$. British whiting is 97–98% pure and practically all finer than 25 µm. It is used as a source of lime in pottery bodies and glazes, and to a small extent in glasses and vitreous enamels.
Wicket. A wall built of refractories to close an opening into a kiln or furnace; it is of a temporary nature, serving as a door, for example, in intermittent or annular kilns.
Wiegner Sedimentation Tube. Apparatus for particle-size analysis by sedimentation from a suspension in a tube (or cylinder) of relatively large diameter; the rate of sedimentation is indicated by the movement of the meniscus in a narrow side-tube, joined

to the large tube near the base of the latter and itself containing the dispersing liquid free from solid particles. It was designed by G. Wiegner (*Landw. Vers.-Stat.,* **91**, 41, 1918); a later development was the KELLY SEDIMENTATION TUBE (q.v.).

Wilkinite. A particular type of bentonitic clay; it has been used as a suspending agent for glazes.

Wilkinson Oven. A pottery BOTTLE OVEN (q.v.) designed so that the hot gases rise through the bag-walls and the central well-hole, and descend between the saggars to leave the kiln through flue openings in the floor mid-way along radii. (A. J. Wilkinson, Brit. Pat. 4356, 20/3/1890.)

Willemite. Zn_2SiO_4; formed in crystalline glazes that are loaded with ZnO.

Williams' Plastometer. A parallel-plate compression apparatus designed by I. Williams (*Industr. Eng. Chem.,* **16**, 362, 1924) for the testing of rubber; it has since been used quite considerably in the testing of clay.

Williamson Kiln. A tunnel kiln of the combined direct-flame and muffle type designed by J. Williamson (*Trans. Brit. Ceram. Soc.,* **27**, 290, 1928) and first used, in England, for the firing of wall-tiles and sewer-pipes. This kiln differed from earlier tunnel kilns in that the hot combustion gases passed across, rather than along, the kiln.

Willow Blue. Cobalt blue diluted with white ingredients such as ground silica; a quoted recipe is 40% cobalt oxide, 40% feldspar and 20% flint.

Willow Pattern. This well-known pseudo-Chinese scene was first engraved, in 1780, for the decoration of pottery ware, by Thomas Minton for Thomas Turner of the Caughley Pottery, Shropshire, England.

Wimet. Trade-name. Cemented tungsten carbide tool metal. Wickman Wimet Ltd, Coventry.

Winchester. A straight-sided glass bottle of the type used for transporting laboratory liquids; the two British Standard Winchesters have volumes of 80 and 90 fl. oz respectively, i.e. approx. 2¼ and 2½ litres. (B.S. 830.)

Winchester Cutting. A method of splay-cutting roofing tiles for the top course(s) in the vertical tiling of exterior walls, so that the final course of tiles is perpendicular to the overhanging verge.

Wind Loading. Lateral loads exerted on (brick) structures by the wind.

Wind Ridge. A type of RIDGE TILE (q.v.).

Wing Walls. Narrow refractory walls in a gasifier, built radially to form several vertical chambers.

Winkelmann and Schott Equation. An equation proposed by A. Winkelmann and O. Schott (*Ann. Phys. Chem.,* **51**, 697, 730, 735, 1894) for assessing the thermal endurance (F) of glass-ware on the basis of the tensile strength (P), modulus of elasticity (E), coefficient of linear thermal expansion (α), thermal conductivity (K), specific heat (C) and specific gravity (S): $F = (P\sqrt{K})/\alpha E\sqrt{(SC)}$ See also THERMAL EXPANSION FACTORS FOR GLASS.

Winning. The combined process of getting (i.e. excavating) and transporting a raw material such as clay to a brickworks or stockpile.

Winnofos. Aluminium phosphate, used as a binder for refractories, made by ICI, Runcorn until 1976.

Wire-cut Process (UK); **Stiff-mud Process** (USA). The shaping of bricks by extruding a column of clay through a die, the column being subsequently cut to the size of bricks by means of taut wires. The bricks are usually *end cut*, but *side cut* bricks are also made.

Wired Glass. Flat glass that has been reinforced by the incorporation of wire mesh. One use is as a 'fire-stop': whereas, in a fire, ordinary window panes crack, fall out and allow flames to spread, wired glass will crack but hold together.
WIST. A porous cold-setting resin for mouldmaking, which hardens by reaction with water. It is much lighter (70–80%) than plaster. The name is an acronym of the German for the formation of an intermediate structure with water (. . . *die mit Hilfe von* **W***asser eine* **I***ntermediär***st***ructur bildet*) (*Keram Z.* **37**, (12), 1985, p 683).
Witenite. Trade-name. Slip resistant floor tiles for wet situations. (Certels, UK).
Withe. A partition between adjacent flues in a domestic chimney; also known as a MIDFEATHER (q.v.).
Witherite. Natural BARIUM CARBONATE (q.v.).
Wohl Block. A hollow clay building block designed for the construction of walls of various thicknesses and for use in ceilings. (E. Wohl, *Ziegelindustrie,* **2**, 99, 1949.)
Wollanite. A material claimed to increase the hardness and strength of earthenware, and to comprise crystals of wollastonite and cristoballite sintered together. (Oreste Bitossi, Livorno, Italy.)
Wollastonite. Calcium metasilicate, $CaSiO_3$; m.p. 1544°C; There are two principal forms: the natural, β, form sp. gr. 2.92; thermal expansion (0–1000°C), 11×10^{-6}; the synthetic, α, form which is also known as pseudo-wollastonite sp. gr. 2.91; thermal expansion (0–1000°C), 6×10^{-6}. The α form is slowly produced when the β form is heated above 1200°C. There are economic deposits in New York State, Finland and USSR. Wollastonite has been used in some pottery bodies (particularly wall tiles), in glazes and in special low-loss electroceramics.
Wonderstone. A popular name (particularly in S. Africa) for PYROPHYLLITE (q.v.).
Wood's Glass. A special glass that transmits ultra-violet but is almost opaque to visible light; such a glass was first made by Prof. R. W. Wood, John Hopkins University, USA, and was used for invisible signalling during World War 1.
Woods Hole Sediment Analyser. A method of particle-size analysis based on measurement of pressure changes resulting from sedimentation in a suspension of the particles in water. It is applicable to coarse silts and fine gravels, and permits 150 tests to be made in a day. The instrument was designed by Woods Hole Oceanographic Institution (*J. Sed. Petrology,* **30**, 490, 1960).
Wood's Process. See UP-DRAW PROCESS.
Woodhall-Duckham Kiln. See ROTARY-HEARTH KILN.
Wool. See GLASS FIBRE; MINERAL WOOL.
Wool-drag. A fault in GROUND LAYING (q.v.) resulting from accidental smearing of the colour.
Worcester Shape. A tea- or coffee-cup having the general shape of a plain cylinder, rounded sharply near the bottom, which has a broad but shallow foot. For specification see B.S. 3542.
Work-board. A board, about 6 ft long and 9 in. wide, on which pottery-ware may be placed and carried from one process to the next.
Work Size. See CO-ORDINATING SIZE.
Work Speed. A term relating to the process of grinding with abrasive wheels. In surface grinding, the work speed is the rate of table traverse, usually expressed in m/s. In centreless, cylindrical and internal grinding, the

work speed is the rate at which the object being ground (the 'work') revolves; this may be expressed either in rev/min or in m/s.

Workability. A measure of the ease of deformation of dough-like pastes. A test for the workability of unshaped refractories, involving measurement of the percentage change in the height of a prepared cylinder subjected to specified blows, is described in BS 1902 Pt. 7.2. ASTM C181 specifies a similar test for the workability index of fireclay and high-alumina plastic refractories.

Working End. Part of a glass-tank furnace: (1) in such a furnace as used for container glass, the term signifies the compartment following the melting end and separated from it by the bridge; (2) the end from which the glass is withdrawn in a glasstank furnace that has no bridge.

Working Mould. See under MOULD.

Working Range. The temperature range within which glass is amenable to shaping; at higher temperatures the glass is too fluid (viscosity less than about 10^3 poises) and at lower temperatures too rigid (viscosity above 10^7 poises). For comparative purposes, the Working Range is equated to a viscosity range from 10^4 to $10^{7.6}$ poises (see SOFTENING POINT).

Wreathing. (1) A fault that sometimes occurs on the inside of cast whiteware as a slightly raised crescent or snake-like area; it is probably a result of orientation of the plate-shaped clay particles and can usually be prevented by increasing the viscosity and the thixotropy of the casting slip.
(2) Wreathing can also occur in a glaze: (a) as a result of the presence of soluble salts which can segregate as the glaze dries; (b) as a result of the glaze not being applied uniformly to the ware.

Wyko Trace. A surface profile contour trace.

Wythe. A continuous vertical section of masonry one unit thick. Also the thickness of masonry separating flues in a chimney.

Xerography. A method of copying designs. An image is projected on to a charged metallic plate coated with photoconductive material (usually selenium). A pigment powder is dispersed over the plate. It adheres more thickly in proportion to the charge on the plate, which is turn depends on the darkness of the image. The image is then transferred electrostatically to paper and fused by heat. It has been applied in the production of transfers for the decoration of glass and ceramic ware.

Xonolite. A hydrated calcium silicate with better thermal stability than other calcium silicates. Large panels can be fabricated for use up to 800°C.

X-Ray Crystallography. Because the wavelength of X-rays is comparable to the separations between atoms in their regular arrangement which makes up a CRYSTAL STRUCTURE (q.v.), the X-rays are diffracted. That is, a pattern of interference is produced between the original X-ray beam, and the parts of it scattered by the atoms in the crystal. The pattern produced is characteristic of the arrangement of atoms, which can be deduced from it. (See RECIPROCAL LATTICE). Strong diffracted beams are produced at an angle θ to the incident beam when $2d \sin \theta = nd$, where d is the distance between planes of atoms, λ is the X-ray wavelength and n is a whole number. (*Bragg's Law*). Various techniques have been used. The *Debye-Sherrer* or powder technique surrounds a sample of crystalline powder (randomly-oriented crystals) with a circular photographic film, and illuminates the powder with X-rays of a single wavelength.

XRD X-ray Diffraction. See X-RAY CRYSTALLOGRAPHY.

XRF – X-Ray Fluorescence Spectrometry. A now standard method for chemical analysis, dependent on the characteristic X-ray spectra emitted by materials subjected to a beam of X-rays.

YAG. Yttrium Aluminium Garnet (q.v.).

YBCO. Yttrium-barium-copper oxide SUPERCONDUCTORS.

Yellow Ware. US term for CANE WARE (q.v.).

YIG. Abbreviation for YTTRIUM-IRON GARNET (q.v.).

Young's Modulus. The MODULUS OF ELASTICITY (q.v.) in tension or compression; named from T. Young, an English physicist (1773–1829). Some typical values for this property are: earthenware, 40 kNm^{-2}.; electrical porcelain, 80kNm^{-2}; sintered alumina, 300 kNm^{-2}. B.S. 1902 Pt. 4.11 specifies the test for refractories. ASTM C885 specifies a sonic resonance test for refractories; C848 is a sonic resonance test for whitewares, and C1198 a sonic resonance test for advanced ceramics; these also measure SHEAR MODULUS AND POISSON'S RATIO. C623 is a similar test for glass and glass-ceramics.

Y-Stud. A piece of round or rectangular section metal, Y-shaped, used to anchor a refractory lining.

YSZ. Yttria stablised ZIRCONIA.

Ytong. A cellular (lightweight) concrete made in block form from shale and lime, and subsequently hardened by autoclave treatment. It was first produced in 1918, by A. Eriksson in Sweden. (cf. THERMALITE YTONG).

Yttralox. Transparent yttria.

Yttria. YTTRIUM OXIDE Y_2O_3 (q.v.).

Yttrium Aluminium Garnet. $Y_3Al_5O_{12}$ YAG occurs as a grain-boundary phase in β-sialon ceramics used as cutting tools. Doped with Nd^{3+} it is a laser material.

Yttrium Iron Garnet. A ferromagnetic. A well-vitrified special ceramic can be produced by firing the compacted oxide at 1800°C, up to which temperature the cubic form is stable.

Yttrium Oxide. Y_2O_3; m.p. approx. 2450°C; sp. gr. 4.84; thermal expansion (25–1500°C) 9×10^{-6}. An oxide extracted from MONAZITE (q.v.) though not itself, strictly, a rare earth. An infra-red transmitter in the wavelength range 1 to 8 μm.

ZAC Block. A zirconia alumina fusion-cast refractory, resistant to corrosion in glass tanks.

Zachariasen's Theory. See under GLASS.

Zaffre. A roasted mixture of cobalt ore and sand, formerly used as a blue colouring material for pottery and glass (cf. SMALT).

Zahn Cup. An orifice-type viscometer; it has been used for the determination of the viscosity of glaze suspensions. (E. A. Zahn, *Chem. Industries,* **51**, (2), 220, 1942.)

ZAN. A form of pressure-sintered silicon nitride using ZrN and AlN powders as sintering aids, (referred to as AZ powders).

ZAS Refractory. See AZS.

ZCG. Zirconia-lime-graphite refractories.

Zebra Roof. A type of roof for basic Open Hearth steel furnaces, the feature being alternate rings of chrome-magnesite and of silica refractories, hence the name from the dark and light stripes across the roof. The Zebra Roof was introduced in 1947 with a view to combining the merits of the two types of refractory. By 1952 there were 300 such roofs in service in USA alone, but the Zebra Roof was later displaced by the all-basic roof.

Zeissig Green. An underglaze colour that has been used for pottery decoration. It is made by calcining a mixture of 10 parts barium chromate, 8 parts whiting and 5 parts boric acid.

Zeolites. A group of aluminosilicates of alkali or alkaline earth metals with ring structures (see SILICATE STRUCTURES). Water or other substances can enter the cavities within the structure. Zeolites occur naturally, or can be made by sol-gel processes. They are used as molecular sieves, catalyst carriers and in ion exchange.

Zerodur. Trade-name. A transparent glass-ceramic with very low thermal expansion. (Jenaer Glaswerk Schott u. Gen, Mainz).

Zeta Potential. The electrical potential ζ set up by the double layer of charged ions in a colloidal solution (such as a clay slip) where it is close to a solid surface:

$\zeta = 4\pi ed/D$

where e = density of charge on the surface of the particles; d = thickness of the double layer; D = dielectric constant of the liquid medium. It is the potential at the limit of the rigidly-held layer of polarized water molecules adjacent to the particle surface – the liquid shear plane or *Stern plane*. If the zeta potential falls below a critical value, particles agglomerate and settle out. Control of the zeta potential is important for maintaining stable suspensions for casting slips.

Zettlitz Kaolin. A Czechoslovakian kaolin; a quoted analysis of the standard grade is (%):SiO_2 46.9; Al_2O_3 38.0; Fe_2O_3, 0.9; TiO_2, 0.25; CaO 0.65; MgO 0.15; alkalis 0.4; loss-on-ignition 12.6.

Zig-zag Kiln. A TRANSVERSE-ARCH KILN (q.v.) with staggered dividing walls, the fire-travel thus being forced to follow a zig-zag path. Such kilns find use in the firing of structural clay products.

Zinc Aluminate (Gahnite). $ZnAl_2O_4$; m.p. 1950°C. This spinel, when made from industrial-grade oxides, has a P.C.E.> 1900°C and RuL > 1700°C. Russian experiments indicate that it can be used as a refractory lining for electric furnaces melting Al, Zn, Pb or Sn. It is rapidly attacked by alkalis.

Zinc Flash. Yellow to green colours formed on the surface of red bricks by the deposition of zinc vapour from zinc compounds introduced into the kiln at the end of firing. (see FLASHING).

Zinc Oxide. ZnO; sp. gr. 5.65: m.pt. 1970°C; thermal expansion (20–1000°C) 6.7×10^{-6}. Used in glasses, glazes, enamels, and more recently in special ferromagnetic ceramics and varistors.

Zircon. Zirconium orthosilicate. $ZrO_2.SiO_2$; sp. gr. 4.56. The principal source is along the most easterly part of the coast of Australia, on both sides of the border between Queensland and New South Wales, where it occurs abundantly as beach sands. When heated, zircon dissociates at 1600–1800°C into SiO_2 and ZrO_2; it is nevertheless used as a refractory, in the lining of aluminium-melting furnaces and nozzles for the continuous casting of steel, for example. Two types of zircon refractory are defined in ASTM – C545, Type A with bulk density < 3.84 g/cm^3 and Type B ≥ 3.84 g/cm^3. Zircon is also used as an opacifier in vitreous enamels and glazes, and as a constituent of special electrical porcelains.

Zircon Colours. The $ZrSiO_4$ lattice is extremely stable, and lends itself to the production of ceramic colours, both LATTICE COLOURS (q.v.) and ENCAPSULATED COLOURS (q.v.). See also PIGMENT.

Zircon Porcelain. An electroceramic made from a batch consisting of 60–70% zircon, 20–30% flux and 10–20% clay; the flux may be a complex Ca-Mg-Ba-Zr silicate or other alkaline-earth composition. Zircon porcelain has high mechanical strength (Young's Modulus) and good thermal-shock resistance over a wide temperature range. Electrical properties are: volume resistivity (300°C) 10^{10}–10^{12} ohm-cm.; dielectric constant (1 MHz) 8.5–9.5: power factor (1 MHz) 10–15 × 10^{-4}.

Zirconia. Zirconium Oxide, ZrO_2; M.pt.approx. 2700°C; density 5.6 kg.m^{-3}. The high temperature crystal form is cubic with a fluorite structure. Below 2370°C this transforms to a tetragonal form, which undergoes a MARTENSITIC TRANSFORMATION (q.v.) to the low temperature monoclinic phase. There is a 3% volume change at this transformation, causing cracking in pure zirconia. Heating zirconia with small amounts of cubic oxides (CaO, MgO) produces solid solutions with the ZrO_2 which is *stabilized* in the cubic structure.
Fully-stabilized zirconia (with *c.* 15 mole % CaO) has the cubic fluorite structure at room temperature.
Partially-stabilized zirconia(with *c.* 8% MgO) has a cubic structure containing a fine dispersed precipitate of *c.* 0.1 µm tetragonal (and/or monoclinic) zirconia crystals.
Tetragonal zirconia polycrystals comprise *c.* 0.1µm crystals of wholly tetragonal crystal structure, made by the controlled sintering of very fine starting powders with *c.* 3% Y_2 O_3 sintering aid. In tetragonal zirconia, toughening is by the Martensitic transformation to monoclinic zirconia. This absorbs energy and sets up compressive stresses around the crack tip, inhibiting further crack growth. With sufficient microclinic zirconia present, the resulting microcracking throughout the structure causes CRACK BRANCHING, absorbing crack propagation energy, when the dispersed particles are large enough. Zirconia is used as a second phase, usually in its tetragonal form, in dispersed-phase ceramic-ceramic composites such as *zirconia-toughened alumina* and silicon nitride. About 20% ZrO_2 is usual. Zirconia is of particular interest as an engineering ceramic, as its thermal expansion coefficient approximates that of cast iron, simplifying ceramic-metal joining.
Fused zirconia is used as kiln furniture for titanates.

Zirconium Borides. ZrB, m.p. 2950°C; sp. gr. 6.7. ZrB_2, m.p. approx. 3050°C; sp. gr. 6.1. A hard, refractory, chemically resistant material. When shaped by ceramic processes it may find use in rocket nozzles, in the casting of alloys, and for thermocouple sheaths. It has a high electrical conductivity; the thermal expansion is approx. (25–1000°C) 6.0 × 10^{-6}. A less well-known compound is ZrB_{12}; m.p. 2680°C; sp. gr. 3.6.

Zirconium Carbide. ZrC; m.p. 3530°C; sp. gr. 6.7; thermal expansion (25–2000°C) 7.5 × 10^{-6}. The face-centred cubic form is stable at least up to 2650°C. This special refractory material has been used in the manufacture of a flame deflector operating at very high temperatures.

Zirconium-Iron Pink. A ceramic stain suitable for the colouring of a variety of glazes maturing at 1220–1280°C. This firing range is greater than that permissible with chrome-tin pink or with chrome-alumina pink.

Zirconium Nitride. ZrN; m.p. 2980 ± 50°C; sp. gr. 7.3; thermal expansion (25–1400°C) 8 × 10^{-6}. A brown compound Zr_3N_4, sp. gr. 5.9, has also been reported.

Zirconium Oxide. See ZIRCONIA.

Zirconium Phosphate. Normal zirconium phosphate, ZrP_2O_7, has a reversible inversion at 300°C and at 1550°C dissociates into zirconyl phosphate, $(ZrO)_2P_2O_7$, with loss of P_2O_5 as vapour. Zirconyl phosphate is stable up to about 1600°C and has a very low thermal expansion (20–1000°C) 1×10^{-6}.

Zirconium Silicate. See ZIRCON.

Zirconium Silicides. Eight compounds have been reported; the following appear to be the most important. ZrSi, m.p. >2000°C, sp. gr. 5.5, hardness (K100) 1030 kg/mm^2. $ZrSi_2$, decomposes >1500°C, sp. gr. 4.9, hardness (K100) 1030 kg/mm^2. Zr_5Si_3, m.p. 2250°C, sp. gr. 5.9.

Zirconium-Vanadium Blue (or Turquoise). A pigment, for use in ceramic glazes, introduced by Harshaw Chemical Co. (US Pat., 2 441 447, 11/5/48; Brit. Pat. 625 448, 28/6/49). The composition is (parts by wt.): ZrO_2, 60–70; SiO_2, 26–36; V_2O_5, 3–5. Alkali must also be present, e.g. 0.5–5% Na_2O. In the absence of alkali a green colour is produced.

Zirkite. A term used in the mineral trade for the natural, impure, zirconia ore (baddeleyite) that occurs in Brazil.

Zirlane. Trade-name. Ceramic fibres containing zirconia (Kerlane, France).

Zone Melting. A heat treatment controlled in such a way as to cause a short molten zone to travel slowly through a relatively long solid charge of crystalline material, gathering up impurities more soluble in the liquid phase as it goes. The process was introduced by W. G. Pfann (*Trans. A.I.M.E.*, **194**, 747, 1952).

ZTA. Zirconia Toughened Alumina. See ZIRCONIA.

ZTS. Zirconium Titanium Stannate has temperature-stable high-frequency permittivity and very low dieectric losses, and is used in microwave resonators and filters.

Zytan Block. Lightweight clay blocks of precise dimensions. See PERFLUENT SINTERING.

Zyttrite. Translucent yttria-stabilized zirconia.

Appendix A
The Definition of 'Ceramics'

The definition of ceramics has proved perennially difficult. Scientific or technological definitions have attempted to be accurate and exhaustive, whether based on the type of material or whether operational definitions based on the processes used to make ceramics.

Definitions for commercial purposes (usually the collection of tariffs, or restriction of imports) have concentrated on precision at the expense of scientific accuracy or generality.

Dictionary definitions must reflect the actual use of words, without straying too far from what is precisely and technically correct.

Historically, usage has broadened from 'largely crystalline products made of fired clay' to include non-clay materials available as articles, or as coatings or powders. This extension of meaning began in 1822 when silica refractories were first made by the normal ceramic process of shaping, drying and firing. It has gathered pace in the last half century to encompass many novel materials shaped by new processes. In 1920 the ***American Ceramic Society*** extended the scope of the word ceramics to include all the silicate industries, bringing in glass, vitreous enamel and hydraulic cement. The problem today is to provide a definition which is sufficiently broad, without including a whole range of naturally occurring materials such as rocks, ores and minerals. (One of the latest general compendiums of information on ceramics – the ***Engineered Materials Handbook Vol 4*** – see Appendix B – abandons this struggle and regards 'ceramics' as a synonym for all 'non-metallic, inorganic materials.')

The 'everyday' definition from the ***Concise Oxford Dictionary*** is: 'Of (the art of) pottery; of (substances produced by) process of strong heating of clay etc. materials. An article made of pottery; substance made by firing clay etc. minerals at high temperatures.'

The ***Illustrated Dictionary of Ceramics*** for collectors of antiques and connoisseurs, says that ceramics is 'the broad term for all objects made of fired clay' (further subdivided into earthenware, stoneware and porcelain on the basis of firing temperature).

In British Standards, definitions usually relate to the content of a particular standard. ***BS 5416*** defines ceramics as 'materials generally made from a mixture of clays and other materials, distinguished from glass and glass ceramics by the fact that they are first shaped and then rendered permanent by firing at a temperature generally well in excess of 1000°C.' (This definition, framed in the context of defining china tableware for mainly commercial purposes, is not notable for generality, accuracy or precision).

The American Society for Testing and Materials adopts an operational approach, in terms of a 'ceramic process.' (***ASTM C-242***)

Ceramics – a general term applied to the art or technique of producing articles by a ceramic process, or to the articles so produced.

Ceramic Process – the production of articles or coatings from essentially inorganic, nonmetallic materials, the article or coating being made permanent and suitable for utilitarian and decorative purposes by the action of heat at temperatures sufficient to cause sintering, solid-state reactions, bonding, or conversion partially or wholly to the glassy state.

Ceramic Article — an article having a glazed or unglazed body or crystalline or partly crystalline structure, or of glass, which body is produced from essentially inorganic, non-metallic substances and either is formed from a molten mass which solidifies on cooling, or is formed and simultaneously or subsequently matured by the action of the heat.

Definitions for commercial purposes include those of the EC Combined (Tariff) Nomenclature, and the Tariff Schedules of the United States (Annotated Schedule 5 – Non-metallic Minerals and Products, Pt 2, Ceramic Products, Headnotes, 1976). The EINECS definition, aiming at defining existing substances in contrast to 'new' ones, adopts an 'exhaustive listing' approach.

Combined Nomenclature

Ceramic Product – products obtained by agglomerating (by firing) earths or other materials with a high melting point generally mixed with binders, all of which materials have been previously reduced to powders or, from rock fired after shaping. (The explanatory notes describe body preparation, shaping, drying, firing and finishing). Firing, after shaping, is the essential distinction between ceramic products and mineral or stone articles, which are generally not fired, and glass articles in which the vitrifiable compound has undergone complete fusion.

US Tariff Schedules

Ceramic Article – a shaped article having a glazed or unglazed body of crystalline or substantially crystalline structure, which body is composed essentially of inorganic non-metallic substances and either is formed from a molten mass which solidifies on cooling, or is formed and subsequently hardened by such heat treatment that the body, if reheated to pyrometric cone 020, would not become more dense, harder, or less porous, but does not include any glass article.

EINECS

Ceramic Materials and Wares, Chemicals – this category encompasses the various chemical substances manufactured in the production of ceramics. For purposes of this category, a ceramic is defined as a crystalline or partially crystalline, inorganic, non-metallic, usually opaque substance consisting principally of combinations of inorganic oxides of aluminium, or zirconium which conventionally is formed first by fusion or sintering at very high temperatures, then by cooling, generally resulting in a rigid, brittle monophase or multiphase structure. (Those ceramics which are produced by heating inorganic glass, thereby changing its physical structure from amorphous to crystalline but not its chemical identity are not included n this definition.) This category consists of chemical substances other than by-products of various ceramics and concurrently incorporated into a ceramic mixture. Its composition may contain any one or a combination of these substances. Trace amounts of oxides and other substances may be present. The following representative elements are principally present as oxides but may also be present as borides, carbides, chlorides, fluorides, nitrides, silicides, or sulfides in multiple oxidation states, or in more complex compounds:

Al, B, Ba, Be, C, Ca, Cd, Ce, Co, Cr, Cs, Cu, Fe, Hf, K, Li, Mg, Mn, Na, P, Si, Sn, Th, Ti, U, Y, Zn, Zr.

Appendix B
Other Sources of Information

These documents are grouped under three headings.

B.1 Works which provide additional technical information, either over a wider range of topics (eg Chambers) or in greater depth and detail.

B.2 Standard glossaries, in which the definitions of terms may differ from those given in this dictionary (having been developed in a different context) and in which a wider range of related terms may also be defined.

B.3 Some specialised multilingual dictionaries and glossaries of ceramic terms.

B.1

Chambers Science and Technology Dictionary.
Ed. P.M.B. Walker, Chambers, Edinburgh, 1991.
Provides brief definitions over the whole range of science and technology. (c. 1000 pp).

Engineered Materials Handbook, Vol 4, Ceramics and Glasses.
Ed. S.J. Schneider, ASM International, Materials Information Society, 1991.
Provides short technical descriptions (up to a few pages) of ceramics materials and processes (1 Vol, c. 1200 pp A4).

Encyclopaedia of Materials Science and Engineering.
Ed. M.B. Bever, Pergamon, 1986 (8 Vols).
Similar brief technical descriptions, over a wider field of materials technology.

An Illustrated Dictionary of Ceramics.
G. Savage and H. Newman, Thames and Hudson, London, 1974.
Definitions from the viewpoint of the artist, collector and connoisseur.

Ceramics Handbook.
Cookson Ceramics Ltd, Stoke-on-Trent, 1990.
A brief compendium of useful factual data relating to ceramics.

Technological and Economic Trends in the Steel Industries.
B. Wilshire, D. Horner, N.L. Cooke, Pineridge Press, Swansea, 1983.
Useful background information on the principal uses of refractories. Appendix D contains an extensive list of abbreviations for steelmaking processes.

B.2

BS 3130 Glossary of packaging terms. Pt 3 Glass containers and closures.
BS 3447 Glossary of terms used in the glass industry.
ISO R836 1968 Vocabulary for the Refractory Industry (Eng. Fr. Russian).
BS 6711 Vocabulary relating to laboratory apparatus made essentially from glass, porcelain or vitreous silica. Pt. 1. Names for items of apparatus.
BS 6045 Ceramic and Glass Insulating Materials. Pt. 1. Definitions and classifications.
BS 3402 'Quality of Vitreous China Sanitaryware', contains definitions of various blemishes and faults.
BS 6100 – 1992 ff. Glossary of building and civil engineering terms.
UDC 1000 BS 1000 [666] 1984. Universal Decimal Classifications. Glass Industry, Ceramics, Cement and Concrete.
BS 4642 1984 Industrial Furnace Terms.
ASTM D2946, C1154 and 460 give terminology for asbestos, asbestos and

fibre cement and asbestos-cement respectively.

ASTM C-904 Terminology relating to chamical-resistant non-metallic materials.

ASTM C-1180 Terminology for mortars for unit masonry.

ASTM C-896 Terminology relating to clay products.

ASTM C-43 Terminology relating to structural clay products.

ASTM C-162 Terminology relating glass and glass products.

ASTM C-242 Terminology relating to ceramic whiteware and related products.

ASTM C-1209 Terminology of concrete masonry units.

Terminology – B.3

Glossaire PRE – Refractories. (E, F, G, It.)

Federation Europeenne des Fabricant de Produits Refractaires, Case Postale 3361, Löwenstrasse 31, CH 8023, Zurich, 1973.

CEC Ceramic Tile Dictionary. (E, F, G, Sp, It, Sw.)

Conseil Européenne de Ceramique, 1970.

Vilmy Montanari, *Collection of Ceramic Technical Words* (tiles). (E, It.)

Casella Postale 8, 42013 Casalgrande Alto, Italy, 1989.

Terminologia Alafar. Refractories. (E, Sp, Port, F.)

Asociacion Latinoamericana de Fabricants de Refractarios.

Glossaire Feugrès: Ceramic Sewer Pipes. (G, F, E, It.)

Europaische Vereinigung der Steinzeugröhrenindustrie, Zürich, 1962.

ICG Dictionary of Glassmaking. (E, F, G.)

Sub-Committee A-1, International Commission on Glass, Elsevier, 1983.

European Ceramics Glossary. (E, F, G, Sp, It, Sw)

Gruppo Editoriale Faenza Editrice, s.p.A., 1992 48018 Faenza, Italy.

Appendix C
Hardness of materials

Hardness tests are at best semi-empirical, and comparisons between different scales unreliable, as the test conditions differ. Some indications of comparative values for materials of ceramic interest can be gleaned from the following tables. The Mohs reference minerals are in bold type.

Substance	Mohs Hardness No. (*Original scale*)	Mohs Hardness No. (*Extended scale*)	Knoop Hardness
Talc	**1**	**1**	
Rock salt	**2**		
or **Gypsum**	**2**	**2**	32
Kaolinite	(2–2.5)		
Mica	(2.5–3)		
Calcite	**3**	**3**	135
Marble	(3–4)		
Dolomite	(3.5–4)		
Fluorite	**4**	**4**	163
Glass	(4.5–6)		
Asbestos	(5)		
Apatite	**5**	**5**	370
Soda-lime glass	(5.5)		
Pumice	(6)		
Feldspar	**6**		560
(orthoclase)		**6**	
Agate	(6–7)		
Steel	(5–8.5)		
Vitreous Silica		**7**	
Quartz	**7**	**8**	820
(or **Stellite**)		**8**	
Zirconia	(7)		1160
Beryllia	(7)		1250
Topaz	**8**	**9**	
Garnet	(8)	**10**	
Fused Zirconia	(8)	**11**	
Co-bonded WC	(8–9)		1400–1800
Zirconium Boride	(8)		1550
Corundum	**9**		
Titanium Nitride	(9)		1800
Tungsten Carbide	(9)		1880
Tantalum Carbide	(9)		2000
Zirconium Carbide	(9)		2100
Fused Aluminia		**12**	2100
Beryllium Carbide			2410
Titanium Carbide			2470
Silicon Carbide		**13**	2480
Aluminium Boride			2500
Boron Carbide		**14**	2750
Diamond	**10**	**15**	7000

APPROXIMATE CONVERSION OF HARDNESS VALUES
SINTERED CARBIDES

Vickers diamond hardness, 50 kg load	*Rockwell A scale 60 kg load: diamond cone*	*Rockwell C scale 150 kg load: diamond cone*
1750	92.4	80.5
1700	92.0	79.8
1650	91.7	79.2
1600	91.3	78.4
1550	90.9	77.7
1500	90.5	77.0
1450	90.1	76.2
1400	89.7	75.4
1350	89.3	74.6
1300	88.9	73.8
1250	88.5	73.0
1200	88.1	72.2
1150	87.6	71.3
1100	87.0	70.4
1050	86.4	69.4
1000	85.7	68.2
950	85.0	66.6
900	84.0	64.6
850	82.8	–

Reference: In *Butterworths Metal Handbook* Vol 3.

Appendix D
Nominal Temperature (°C) Equivalents of Pyrometric Cones

CONE No.	*BRITISH (Staffordshire)*	*GERMAN (Seger)*	*AMERICAN (Orton)*		
	4°C/min		*Large 1°C/min*	*Large 2½°C/min*	*Small 5°C/min*
022	600	600	585	600	630
022A	625	–	–	–	–
021	650	650	602	614	643
020	670	670	625	635	666
019	690	690	668	683	723
018	710	710	696	717	752
017	730	730	727	747	784
016	750	750	767	792	825
015	790	–	790	804	843
015A	–	790	–	–	–
014	815	–	834	838	870
014A	–	815	–	–	–
013	835	–	869	852	880
013A	–	835	–	–	–
012	855	–	866	884	900
012A	–	855	–	–	–
011	880	–	886	894	915
011A	–	880	–	–	–
010	900	–	887	894	919
010A	–	900	–	–	–
09	920	–	915	923	955
09A	–	920	–	–	–
08	940	–	945	955	983
08A	950	940	–	–	–
07	960	–	973	984	1008
07A	970	960	–	–	–
06	980	–	991	999	1023
06A	990	980	–	–	–
05	1000	–	1031	1046	1062
05A	1010	1000	–	–	–
04	1020	–	1050	1060	1098
04A	1030	1020	–	–	–
03	1040	–	1086	1101	1131
03A	1050	1040	–	–	–
02	1060	–	1101	1120	1148
02A	1070	1060	–	–	–
01	1080	–	1117	1137	1178
01A	1090	1080	–	–	–
1	1100	–	1136	1154	1179
1A	1110	1100	–	–	–
2	1120	–	1142	1162	1179
2A	1130	1120	–	–	–
3	1140	–	1152	1168	1196
3A	1150	1140	–	–	–
4	1160	–	1168	1186	1209
4A	1170	1160	–	–	–
5	1180	–	1177	1196	1221
5A	1190	1180	–	–	–
6	1200	–	1201	1222	1255
6A	1215	1200	–	–	–
7	1230	1230	1215	1240	1264
7A	1240	–	–	–	–
8	1250	1250	1236	1263	1300
8A	1260	–	–	–	–
8B	1270	–	–	–	–
9	1280	1280	1260	1280	1317
9A	1290	–	–	–	–
10	1300	1300	1285	1305	1330
10A	1310	–	–	–	–
11	1320	1320	1294	1315	1336

CONE No.	*BRITISH (Staffordshire)*	*GERMAN (Seger)*	*AMERICAN (Orton)*		
	4°C/min		*Large 1°C/min*	*Large 2½°C/min*	*P.C.E. Cones 2½°C/min*
12	1350	1350	1306	1326	1337
13	1380	1380	1321	1346	1349
14	1410	1410	1388	1366	1398
15	1435	1435	1424	1431	1430
16	1460	1460	1455	1473	1491
17	1480	1480	1477	1485	1512
18	1500	1500	1500	1506	1522
19	1520	1520	1520	1528	1541
20	1530	1530	1542	1549	1564
23	–	–	1586	1590	1605
26	1580	1580	1589	1605	1621
27	1610	1610	1614	1627	1640
28	1630	1630	1614	1633	1646
29	1650	1650	1624	1645	1659
30	1670	1670	1636	1654	1665
31	1690	1690	1661	1679	1683
31½	–	–	1685	1700	1699
32	1710	1710	1706	1717	1717
32½	–	–	1718	1730	1724
33	1730	1730	1732	1741	1743
34	1750	1750	1757	1759	1763
35	1770	1770	1784	1784	1785
36	1790	1790	1798	1796	1804
37	1825	1825	–	–	1820
38	1850	1850	–	–	1835*
39	1880	1880	–	–	1865*
40	1920	1920	–	–	1885*
41	1960	1960	–	–	1970*
42	2000	2000	–	–	2015*

* *10°C/min*

Appendix E
Comparison Table for Sieve Sizes

ISO/ (USA)	*ASTM*		*BRITISH*		*IMM*		*TYLER*		*FRENCH*		*GER-MAN*	*CANADIAN* metric		*imperial*
Size	*No*	*Size*	*No*	*Size*	*No*	*Size*	*No*	*Size*	*No*	*Size*	*Size*	*Size*	*No*	*Size*
125		5"												
106		4.24"												
(100)		4"												
90		3½"												
75		3"												
63		2½"												
53		2.12"												
(50)		2"												
45		1¾"												
37.5		1½"												
31.5		1¼"												
26.5		1.06"						1.05"				26.9		1.06"
(25.0)		1"									25.0			
22.4		⅞"						.883"				22.6		⅞"
19.0		¾"						.742"			20.0	19.0		¾"
											18.0			
16.0		⅝"						.624"			16.0	16.0		⅝"
13.2		.530"						.525"				13.5		.530"
(12.5)		½"									12.5			
11.2		7/16"						.441"				11.2		7/16"
											10.0			
9.5		⅜"						.371"				9.51		⅜"
8.0		5/16"					2½				8.0	8.00		5/16"
6.7		.265"					3					6.73		.265"
(6.3)		¼"									6.3			
5.6	3½						3½					5.66	3½	
									38	5.000	5.0			
4.75	4	4.76					4	4.7				4.76	4	
4.00	5						5	3.96	37	4.000	4.0	4.00	5	
3.35	6	3.36	5	3.35			6	3.33				3.36	6	
									36	3.150	3.15			
2.80	7	2.83	6	2.80			7	2.79				2.83	7	
2.36	8	2.38	7	2.40	5	2.54	8	2.36	35	2.500	2.5	2.38	8	
2.00	10		8	2.00			9	1.98	34	2.000	2.0	2.00	10	
1.70	12		10	1.68	8	1.57	10	1.65	33	1.600	1.6	1.68	12	
1.40	14	1.41	12	1.40			12					1.41	14	
1.18					10	1.27			32	1.250	1.25			
1.00	16	1.19	14	1.20			14	1.17				1.19	16	
	18		16	1.00	12	1.06	16	.99	31	1.000	1.0	1.00	18	
.850	20	0.84	18	.850			20	.83				.841	20	
					16	.792			30	.800	.800			
.710	25	.707	22	.710			24	.701				.707	25	
					20	.635			29	.630	.630			
.600	30	.595	25	.600			28	.589				.595	30	
.500	35		30	.500			32	.495	28	.500	.500	.500	35	
.425	40		36	.420	30	.422	35	.417				.420	40	
									27	.400	.400			
.355	45	.354	44	.355			42	.351				.354	45	
					40	.317			26	.315	.315			
.300	50	.297	52	.300			48	.295				.297	50	
.250	60		60	.250	50	.254	60	.246	25	.250	.250	.250	60	
.212	70		72	.210			65	.208				.210	70	
					60	.211			24	.200	.200			
.180	80	.177	85	.180	70		80	.175				.177	80	
					80	.157			23	.160	.160			

ISO/ (USA)	*ASTM*		*BRITISH*		*IMM*		*TYLER*		*FRENCH*		*GER-MAN*	*CANADIAN*		
												metric		*imperial*
Size	*No*	*Size*	*No*	*Size*	*No*	*Size*	*No*	*Size*	*No*	*Size*	*Size*	*Size*	*No*	*Size*
.150	100	.149	100	.150	90	.139	100	.147				.149	100	
.125	120		120	.125	100	.127	115	.124	22	.125	.125	.125	120	
.106	140		150	.105	120	.107	150	.104				.105	140	
									21	.100	.100			
.090	170	.088	170	.090	150	.084	170	.088			.090	.088	170	
									20	.080	.080			
.075	200	.074	200	.075			200	.074				.074	200	
											.071			
.063	230		240	.063	200	.063	250	.061	19	.063	.063	.063	230	
											.056			
.053	270		300	.053			270					.053	270	
									18	.050	.050			
.045	325		350	.045			325	.043			.045	.044	325	
									17	.040	.040			
.038	400						400	.038				.037	400	
.032														
(.025)														
(.020)														

ISO/(USA) ISO 3310/ASTM E-11 metric*

BRITISH	BS 410
IMM	Institution of Mining & Metallurgy, London
FRENCH	AFNOR X-11-501
GERMAN	DIN 4188
ASTM	ASTM E-11 (USA)
TYLER	Tyler Sieve Sizes (USA)
CANADIAN	Standard sieve series 8-GP-1b

NB ISO Standard Sieve Sizes are now in general use, and national standards conform to them. The figures in the left-hand column are the ISO sizes, plus a few additional sizes (bracketed) retained in the metric version of ASTM E11 on grounds of extensive use. The remaining columns give designations and actual aperture sizes (in mm unless inches (") are specifically noted) of older sieve series likely to be found in the literature.